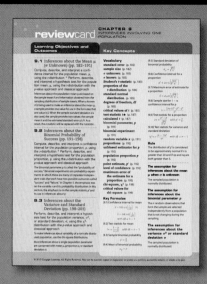
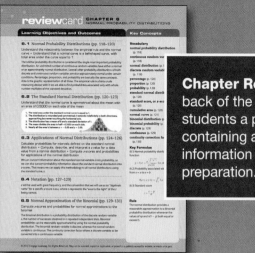
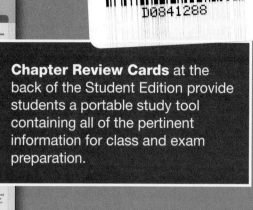

THE SOLUTION

STAT
Are you in?

ONLINE RESOURCES INCLUDED!

CourseMate Engaging. Trackable. Affordable.

CourseMate brings course concepts to life with interactive learning, study, and exam preparation tools that support STAT2.

FOR INSTRUCTORS:
- First Day of Class Instructions
- Instructor's manual
- PowerPoint® Slides
- Engagement Tracker
- Instructor Prep Cards

FOR STUDENTS:
- Interactive eBook
- Student Solutions Manual
- Data Sets
- Technology Manuals
- Applets
- Auto-Graded Quizzes
- Videos
- Tutorials
- Chapter Projects
- Flashcards
- Student Review Cards

Students sign in at
login.cengagebrain.com

BROOKS/COLE
CENGAGE Learning™

STAT, 2e
Robert Johnson
Patricia Kuby

Publisher: Richard Stratton

Senior Sponsoring Editor: Molly Taylor

Senior Development Editor: Renee Deljon

Associate Editor: Daniel Seibert

Senior Editorial Assistant: Shaylin Walsh

Media Editor: Andrew Coppola

Executive Brand Marketing Manager,
    4LTR Press: Robin Lucas

Project Manager, 4LTR Press: Kelli Strieby

Marketing Coordinator: Erica O'Connell

Marketing Communications Manager:
    Mary Anne Payumo

Content Project Manager: Susan Miscio

Senior Art Director: Linda Helcher

Senior Print Buyer: Diane Gibbons

Rights Acquisition Specialist, Image:
    Don Schlotman

Rights Acquisition Specialist, Text:
    Tim Sisler

Production Service: B-books, Ltd.

Text Designer: Stratton Design

Cover Designer: Ke Design

Cover Image: © Veer

For product information and technology assistance, contact us at
Cengage Learning Customer & Sales Support, 1-800-354-9706

For permission to use material from this text or product,
submit all requests online at **www.cengage.com/permissions**
Further permissions questions can be emailed to
permissionrequest@cengage.com

Library of Congress Control Number: 2010934153

ISBN-13: 978-0-538-73841-5
ISBN-10: 0-538-73841-3

Brooks/Cole Cengage Learning
20 Davis Drive
Belmont, CA 94002-3098
USA

Cengage Learning products are represented in Canada by Nelson Education, Ltd.

For your course and learning solutions, visit **www.cengage.com**.
Purchase any of our products at your local college store
or at our preferred online store **www.cengagebrain.com**.

Printed in the United States of America
3 4 5 16 15 14

Brief Contents

STUDY YOUR WAY

At no additional cost, students have access to online learning resources that include **tutorial videos, printable flashcards, data sets, technology manuals, applets**, and **auto-graded quizzes**!

Watch videos that offer step-by-step conceptual explanation and guidance for each chapter in the text.

With the online printable flashcards, they also have two additional ways to check their comprehension of key statistical concepts.

Students can find the videos and flashcards at **www.cengagebrain.com**.

Contents

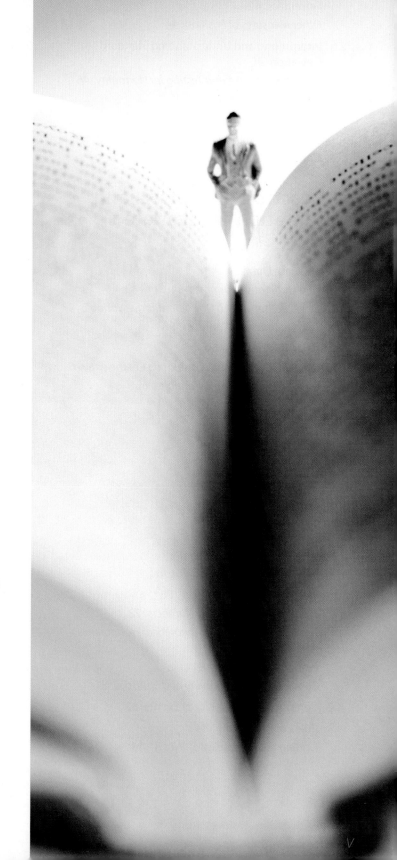

Contents

Contents

Contents

Contents

Contents

Part 4
More Inferential
Statistics 236

11 Applications of Chi-Square 236

12 Analysis of Variance 254

13 Linear Correlation and Regression Analysis 272

Contents

LEARN YOUR WAY

With **STAT2**, students have a multitude of study aids at their fingertips.

The **Students Solutions Manual** contains the worked-out solutions to every odd-numbered exercise, further reinforcing students understanding of statistical concepts in **STAT2**.

In addition to the Student Solutions Manual, **CourseMate** offers exercises and questions that correspond to every section and chapter in the text for students to practice the concepts they learned.

Students can sign in at **www.cengagebrain.com**.

aplia™

ONLINE
HOMEWORK

Cengage Learning's **Aplia** for General Introductory Statistics is an online, interactive learning solution that improves comprehension and outcomes by increasing student engagement.

Aplia for Statistics supplies students with:

- Engaging questions based on real scenarios to help student understand how statistics applies to everyday life

- Interactive applets allow students to explore fundamental statistical concepts

- Automatic grading and detailed explanations for chapter topics to help students learn from every question

Cengage Learning's **Aplia** is available for professor's consideration. To learn more, please visit **www.aplia.com/statistics**.

Chapter 1

Statistics

The U.S. Census Bureau annually publishes the *Statistical Abstract of the United States*, a 1,000+-page book that provides us with a statistical insight into many of the most obscure and unusual facets of our lives. This is only one of thousands of sources for all kinds of things you have always wanted to know about but never thought to ask. Are you interested in how many hours we work and play? How much we spend on snack foods? How the price of Red Delicious apples has gone up? All this and more—much more—can be found in the *Statistical Abstract* (http://www.census.gov/statab).

The statistical tidbits that follow come from a variety of sources and represent only a tiny sampling of what can be learned about Americans statistically. These are only some of thousands of measures used to describe life in the United States. Take a look.

objectives

1.1 What Is Statistics?

1.2 Measurability and Variability

1.3 Data Collection

1.4 Statistics and Technology

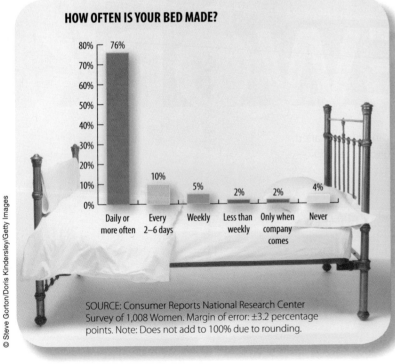

HOW OFTEN IS YOUR BED MADE?

- Daily or more often: 76%
- Every 2–6 days: 10%
- Weekly: 5%
- Less than weekly: 2%
- Only when company comes: 2%
- Never: 4%

SOURCE: Consumer Reports National Research Center Survey of 1,008 Women. Margin of error: ±3.2 percentage points. Note: Does not add to 100% due to rounding.

© Steve Gorton/Doris Kindersley/Getty Images

ARE YOU FRETTING OVER MESSAGES?

How Wi-Fi users responded when asked how long they go before they get "antsy" about checking e-mail, instant messaging, and social networking sites:

- 47% One hour or less
- 7% One week
- 46% One day

SOURCE: Impulse Research for Qwest Communications online survey of 1,063 adult Wi-Fi users in April 2009.

© Zeffss1/iStockphoto.com

1.1 What Is Statistics?

AS WE EMBARK ON OUR JOURNEY INTO THE STUDY OF STATISTICS, WE MUST BEGIN WITH THE DEFINITION OF *STATISTICS* AND EXPAND ON THE DETAILS INVOLVED.

Statistics has become the universal language of the sciences. As potential users of statistics, we need to master both the "science" and the "art" of using statistical methodology correctly. Careful use of statistical methods will enable us to obtain accurate information from data. These methods include (1) carefully defining the situation, (2) gathering data, (3) accurately

summarizing the data, and (4) deriving and communicating meaningful conclusions.

Statistics involves information, numbers, and visual graphics to summarize this information, as well as interpretation of the numbers and graphics. The word *statistics* has different meanings to people of varied backgrounds and interests. To some people it is a field of "hocus-pocus" in which a person attempts to overwhelm others with incorrect information and conclusions. To others it is a way of collecting and displaying information. And to still another group it is a way of making decisions in the face of uncertainty. In the proper perspective, each of these points of view is correct.

The field of statistics can be roughly subdivided into two areas: descriptive statistics and inferential statistics. *Descriptive statistics* is what most people think of when they hear the word *statistics*. It includes the collection, presentation, and description of sample data. The term *inferential statistics* refers to the technique of interpreting the values resulting from the descriptive techniques and making decisions and drawing conclusions about the population.

Statistics is more than just numbers: it is data, what is done to data, what is learned from the data, and the resulting conclusions. So, **statistics** is the science of collecting, describing, and interpreting data.

> **Statistics** The science of collecting, describing, and interpreting data.

Examples of Statistics in Everyday Life

From sports to politics to business, statistics are a near-daily presence in our lives. Let's look at a few illustrations of how and when statistics can be applied. For example, when you went to kindergarten, your first concerns were most likely having a good time and making friends. But what about your teacher?

Consider the information included in the following graphic. It describes the skills that kindergarten teachers consider essential or very important. Eight hundred kindergarten teachers (only a fraction of all kindergarten teachers) were surveyed, resulting in the skills and percentages reported. Leading the list are "Paying attention" and "Not being disruptive." Of the 800 teachers surveyed, 86% considered these skills essential or very important. Looking at all the percentages, it is clear that

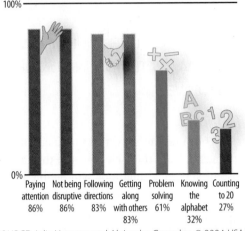

EVEN IN KINDERGARTEN, SOCIAL SKILLS TRUMP

Percentage of 800 kindergarten teachers surveyed who say these skills are essential or very important:

Paying attention	86%
Not being disruptive	86%
Following directions	83%
Getting along with others	83%
Problem solving	61%
Knowing the alphabet	32%
Counting to 20	27%

SOURCE: Julia Neyman and Alejandro Gonzalez, © 2004 *USA Today.*

they add up to more than 100%. Apparently, the teachers surveyed were allowed to give more than one skill as an answer.

In addition to describing our concerns, statistics can describe transitions in our behavior. For example, teenagers' classroom and driving behavior have changed with the prevalence of cell phone and MP3 technology. As the graphics about teens and young drivers show on the next page, a vast majority of teens have cell phones and use them all the time, including in the classroom and on the road. Consider the information collected to formulate these graphics: first and foremost, cell phone status; number of text messages per week; number of text messages during class per week; and types of activities while driving. How would the organizations responsible for the surveys use that collected information to come up with the 84% and 83% shown in the graphs? The percentages in the Busy Behind the Wheel graph add up to well over 100%,

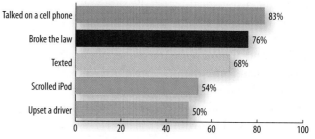

BUSY BEHIND THE WHEEL

Activities done by 16–20-year-olds while driving:

Activity	Percent
Talked on a cell phone	83%
Broke the law	76%
Texted	68%
Scrolled iPod	54%
Upset a driver	50%

SOURCE: National Organization for Teen Safety, Allstate Foundation online survey of 605 drivers age 16–20, June 16, 2009.

TEENS USING CELL PHONES IN CLASS

Teens without cell phones 16%

Teens with cell phones 84%

<u>Average Texts</u>
per week: 440
sent during school hours: 110
per class period: 3

SOURCE: Common Sense Media Survey of 1,013 teens, May/June 2009.

WHAT EMPLOYERS ARE LOOKING FOR

What do employers look for in a seasonal employee?

Previous experience 20%

Positive attitude and eagerness 39%

Ability to work the required schedule 28%

Commitment for the entire season 13%

SOURCE: SnagAJob.com survey of 1,043 hiring managers. Margin of error: ±3 percentage points.

suggesting that respondents were allowed to select more than one answer.

Always take note of the source for published statistics (and any other provided detail); it will tell you much about the information being presented. In these cases, both are national organizations. Common Sense Media is a respected leader on children and media issues, and Allstate is a funding partner for the National Youth Health and Safety Coalition. These source details can give you information about the quality of the information. Keep in mind that information may be biased or incomplete, but knowing about the source and data collection method helps you recognize biases.

One of the primary vehicles for delivering statistics is the media. Newspapers in particular publish graphs and charts telling us how various organizations or people think as a whole. Do you ever wonder how much of what we think is directly influenced by the information we read in these articles?

The graphic "What Employers Are Looking For" reports that 39% of employers look for a positive attitude and eagerness in seasonal employees. Where did this information come from? Note the source, SnagAJob.com. How did the researchers collect the information? They conducted a survey of 1,043 hiring managers. A margin of error is given at ±3 percentage points. (Remember to check the small print, usually at the bottom of a statistical graph or chart.) Based on this additional detail, the 39% on the graph becomes "between 36% and 42% of employers look for a positive attitude and eagerness in their seasonal employees."

Even though reputable media outlets regularly publish statistics, that doesn't mean you should blindly believe the statistics. Remember to consider the source when reading a statistical report. Be sure you are looking at the complete picture. Having said that, also remember that "One ounce of statistics technique requires one pound of common sense for proper application."

Consider the International Shark Attack File (ISAF). The ISAF, administered by the American Elasmobranch Society and the Florida Museum of Natural History, is a compilation of all known shark attacks. It is shown in the graph and chart below.

Using common sense while reviewing the shark attack graph, one would certainly stay away from the United States if he or she enjoys the ocean. Nearly two-fifths of the world's shark attacks occur in the United States. U.S. waters must be full of sharks, and the sharks must be mad! Common sense—remember? Is the graph a bit misleading? What else could be influencing the statistics shown here? First, one must consider how much of a country's or continent's border comes in contact with an ocean. Second, who is tracking these attacks? In this case, it is stated at the top of the chart—The Florida Museum of Natural History, a museum in the United States. Apparently, the United States is trying to keep track of unprovoked shark attacks. What else is different about the United States compared with the other areas? Is the ocean a recreational area in the other places? What is the economy of these other areas, and/or who is keeping track of their shark attacks? Remember to

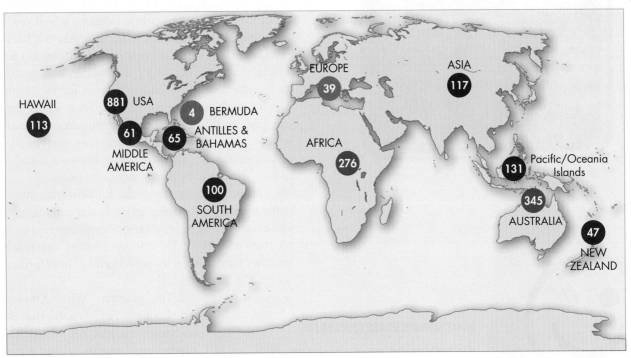

Territory	Total Attacks	Fatal Attacks	Last Fatality
● USA (w/out Hawaii)	881	38	2005
● Australia	345	135	2006
● Africa	276	70	2004
● Asia	117	55	2000
● Pacific/Oceania Islands (w/out Hawaii)	131	50	2007
● Hawaii	113	15	2004
● South America	100	23	2006

Territory	Total Attacks	Fatal Attacks	Last Fatality
● Antilles & Bahamas	65	19	1972
● Middle America	61	31	1997
● New Zealand	47	9	1968
● Europe	39	19	1984
● Bermuda	4	0	
● Unspecified	20	6	1965
WORLD	2,199	470	2007

SOURCE: http://www.flmnh.ufl.edu/fish/sharks/statistics/GAttack/World.htm

consider the source when reading a statistical report. Be sure you are looking at the complete picture.

Where Statistics Are Used

The uses of statistics are unlimited. It is much harder to name a field in which statistics is not used than it is to name one in which statistics plays an integral part. The following are a few examples of how and where statistics are used:

- In education, descriptive statistics are frequently used to describe test results.
- In science, the data resulting from experiments must be collected and analyzed.
- In government, many kinds of statistical data are collected all the time. In fact, the U.S. government is probably the world's greatest collector of statistical data.

A very important part of the statistical process is that of studying the statistical results and formulating appropriate conclusions. These conclusions must then be communicated accurately—nothing is gained from research unless the findings are shared with others. Statistics are being reported everywhere: newspapers, magazines, radio, and television. We read and hear about all kinds of new research results, especially in the health-related fields.

The Language of Statistics

To further continue our study of statistics, we need to "talk the talk." Statistics has its own jargon, terms beyond *descriptive statistics* and *inferential statistics,* that needs to be defined and illustrated. The concept of a population is the most fundamental idea in statistics.

The **population** is the complete collection of individuals or objects that are of interest to the sample collector. The population of concern must be carefully defined and is considered fully defined only when its membership list of elements is specified. The set of "all students who have ever attended a U.S. college" is an example of a well-defined population.

Typically, we think of a population as a collection of people. However, in statistics the population could be a collection of animals, manufactured objects, or whatever. For example, the set of all redwood trees in California could be a population.

There are two kinds of populations: finite and infinite. When the membership of a population can be (or could be) physically listed, the population is said to be **finite.** When the membership is unlimited, the population is **infinite.** The books in your college library form a finite population; the OPAC (Online Public Access Catalog, the computerized card catalog) lists the exact membership. All the registered voters in the United States form a very large finite population; if necessary, a composite of all voter lists from all voting precincts across the United States could be compiled. On the other hand, the population of all people who might use aspirin and the population of all 40-watt light bulbs to be produced by General Electric are infinite. Large populations are difficult to study; therefore, it is customary to select a *sample,* or a subset of a population, and study the data in that sample. A **sample** consists of the individuals, objects, or measurements selected from the population by the sample collector.

Statisticians are interested in particular **variables** of a sample or a population. That is, they examine one or more characteristics of interest about each individual element of a population or sample. Things such as age, hair color, height, and weight are variables. Each variable associated with one element of a population or sample has a value. That value, called the **data value,** may be a number, word, or symbol. For example, when Bill Jones entered college at age "23," his hair was "brown," he was "71 inches" tall, and he weighed "183 pounds." These four data values are the values for the four variables as applied to Bill Jones.

The set of values collected from the variable from each of the elements that belong to the sample is called **data.** Once the data is collected, it is common practice to refer to the set of data as the sample. The set of 25 heights (or weights, ages, and hair colors) collected from 25 students is an example of a set of data. To collect a set of data, a statistician would conduct an **experiment,** which is a planned activity whose results yield a set of data. An experiment includes the activities for both selecting the elements and obtaining the data values.

The "average" age at time of admission for all students who have ever attended our college and the "proportion" of students who were older than 21 years of age when they entered college are examples of two population parameters. A **parameter** is a numerical value that summarizes the entire population. Often a Greek letter is used to symbolize the name of a parameter. These symbols will be assigned as we study specific parameters.

For every parameter there is a *corresponding sample statistic*. The **statistic** is a numerical value summarizing the sample data and describing the sample the same way the parameter describes the population.

The "average" height, found by using the set of 25 heights, is an example of a sample statistic. A statistic is a value that describes a sample. Most sample statistics are found with the aid of formulas and are typically assigned symbolic names that are letters of the English alphabet (for example, $\bar{x}$, s, and r).

FYI: *Parameters describe the population; a statistic describes the sample.*

Kinds of Variables

There are basically two kinds of variables: (1) **qualitative variables** result in information that describes or categorizes an element of a population, and (2) **quantitative variables** result in information that quantifies an element of a population.

A sample of four hair-salon customers was surveyed for their "hair color," "hometown," and "level of satisfaction" with the results of their salon treatment. All three variables are examples of qualitative (attribute) variables because they describe some characteristic of the person, and all people with the same attribute belong to the same category. The data collected were {blonde, brown, black, brown}, {Brighton, Columbus, Albany, Jacksonville}, and {very satisfied, satisfied, somewhat satisfied, dissatisfied}.

By contrast, the "total cost" of textbooks purchased by each student for this semester's classes is an example of a quantitative (numerical) variable. A sample resulted in the following data: $238.87, $94.57, $139.24. [To find the "average cost," simply add the three numbers and divide by 3: $(238.87 + 94.57 + 139.24) \div 3 = \$157.56$.] As you can see, arithmetic operations, such as addition and averaging, are meaningful for data that result from a quantitative variable (and would be useless in examining qualitative variables).

Each of these types of variables (qualitative and quantitative) can be further subdivided as illustrated in the following diagram.

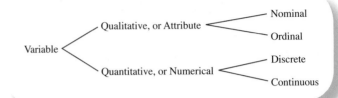

Qualitative variables may be characterized as nominal or ordinal. A **nominal variable** is a qualitative variable that characterizes (or describes, or names) an element of a population. Not only are arithmetic operations not meaningful for data that result from a nominal variable, but an order cannot be assigned to the categories.

In the survey of four hair-salon customers, two of the variables, "hair color" and "hometown," are examples of nominal variables because both name some characteristic of the person, and it would be meaningless to find the sample average by adding and dividing by 4. For example, (blonde + brown + black + brown) $\div$ 4 is undefined. Furthermore, color of hair and hometown do not have an order to their categories.

An **ordinal variable** is a qualitative variable that incorporates an ordered position, or ranking. In the survey of four hair-salon customers, the variable "level of satisfaction" is an example of an ordinal variable because it does incorporate an ordered ranking: "Very satisfied" ranks ahead of "satisfied," which ranks ahead of "somewhat satisfied." Another illustration of an ordinal variable is the ranking of five landscape pictures according to someone's preference: first choice, second choice, and so on.

Quantitative or numerical variables can also be subdivided into two classifications: *discrete* variables and *continuous* variables. A **discrete variable** is a quantitative variable that can assume a countable number of values. Intuitively, the discrete variable can assume any values corresponding to isolated points along a line interval. That is, there is a gap between any two values. By contrast, a **continuous variable** is a quantitative variable that can assume an uncountable number of values. Intuitively, the continuous variable can assume any value along a line interval, including every possible value between any two values.

Discrete variable A quantitative variable that can assume a countable number of values.

Continuous variable A quantitative variable that can assume an uncountable number of values.

BASIC TERMS

POPULATION: A collection, or set, of individuals, objects, or events whose properties are to be analyzed.

FINITE POPULATION: A population whose membership can or could be physically listed.

INFINITE POPULATION: A population whose membership is unlimited.

SAMPLE: A subset of a population that will be used to produce data.

VARIABLE: A characteristic of interest about each individual element of a population or sample.

DATA VALUE: The value of the variable associated with one element of a population or sample. This value may be a number, a word, or a symbol.

DATA: The set of values collected from the variable from each of the elements that belong to the sample.

EXPERIMENT: A planned activity whose results yield a set of data.

PARAMETER: A numerical value summarizing all the data of an entire population.

STATISTIC: A numerical value summarizing the sample data.

QUALITATIVE (OR ATTRIBUTE OR CATEGORICAL) VARIABLE: A variable that describes or categorizes an element of a population.

QUANTITATIVE (OR NUMERICAL) VARIABLE: A variable that quantifies an element of a population.

NOMINAL VARIABLE: A qualitative variable that characterizes (or describes, or names) an element of a population.

ORDINAL VARIABLE: A qualitative variable that incorporates an ordered position, or ranking.

EX 1-1 Applying the Basic Terms

A statistics student is interested in finding out something about the average dollar value of cars owned by the faculty members of your college. Eight of the terms just described can be identified in this situation.

1. The *population* is the collection of all cars owned by all faculty members at your college.
2. A *sample* is any subset of that population. For example, the cars owned by members of the mathematics department is a sample.
3. The *variable* is the "dollar value" of each individual car.
4. One *data value* is the dollar value of a particular car. Mr. Jones's car, for example, is valued at $9,400.
5. The *data* are the set of values that correspond to the sample obtained (9,400; 8,700; 15,950; . . .).
6. The *experiment* consists of the methods used to select the cars that form the sample and to determine the value of each car in the sample. It could

be carried out by questioning each member of the mathematics department, or in other ways.

7. The *parameter* about which we are seeking information is the average value of all cars in the population.
8. The *statistic* that will be found is the average value of the cars in the sample.

FYI: *Parameters are fixed in value, whereas statistics vary in value.*

Note: If a second sample were to be taken, it would result in a different set of people being selected—say, the English department—and therefore a different value would be anticipated for the statistic "average value." The average value for "all faculty-owned cars" would not change, however.

thinking stat

© PhotoDreams/Alamy

In many cases, the two types of variables can be distinguished by deciding whether the variables are related to a count or a measurement. The variable "number of courses for which you are currently registered" is an example of a discrete variable; the values of the variable may be found by counting the courses. (When we count, fractional values cannot occur; thus, there are gaps between the values.) The variable "weight of books and supplies you are carrying as you attend class today" is an example of a continuous random variable; the values of the variable may be found by measuring the weight. (When we measure, any fractional value can occur; thus, every value along the number line is possible.)

When trying to determine whether a variable is discrete or continuous, remember to look at the variable and think about the values that might occur. Do not look at only data values that have been recorded; they can be very misleading.

Consider the variable "judge's score" at a figure-skating competition. If we look at some scores that have previously occurred (for example, 9.9, 7.4, 8.8, 10.0), and we see the presence of decimals, we might think that all fractions are possible and conclude that the variable is continuous. This is not true, however. A score of 9.134 is impossible; thus, there are gaps between the possible values and the variable is discrete.

Remember to inspect the individual variable and one individual data value, and you should have little trouble distinguishing among the various types of variables.

1.2 Measurability and Variability

WITHIN A SET OF MEASUREMENT DATA, WE ALWAYS EXPECT VARIATION. IF LITTLE OR NO VARIATION IS FOUND, WE WOULD GUESS THAT THE MEASURING DEVICE IS NOT CALIBRATED WITH A SMALL ENOUGH UNIT.

NOTE: Don't let the appearance of the data fool you in regard to their type. Qualitative variables are not always easy to recognize; sometimes they appear as numbers. The sample of hair colors could be coded: 1 = black, 2 = blonde, 3 = brown. The sample data would then appear as {2, 3, 1, 3}, but they are still nominal data. Calculating the "average hair color" [(2 + 3 + 1 + 3)/4 = 9/4 = 2.25] is still meaningless. The hometowns could be identified using ZIP codes. The average of the ZIP codes doesn't make sense either; therefore, ZIP code numbers are nominal, too.

For example, we take a carton of a favorite candy bar and weigh each bar individually. We observe that each of the 24 candy bars weighs $\frac{7}{8}$ ounce, to the nearest $\frac{1}{8}$ ounce. Does this mean that the bars are all identical in weight? Not really! Suppose we were to weigh them on an analytical balance that weighs to the nearest ten-thousandth of an ounce. Now the 24 weights will most likely show **variability**.

It does not matter what the response variable is; there will most likely be variability in the data if the tool of measurement is precise enough. One of the primary objectives of statistical analysis is measuring variability. For example, in the study of quality control, measuring variability is absolutely essential. Controlling (or reducing) the variability in a manufacturing process is a field all its own—namely, statistical process control.

> **Variability** The extent to which data values for a particular variable differ from each other.

1.3 Data Collection

BECAUSE IT IS GENERALLY IMPOSSIBLE TO STUDY AN ENTIRE POPULATION (EVERY INDIVIDUAL IN A COUNTRY, ALL COLLEGE STUDENTS, EVERY MEDICAL PATIENT, ETC.), RESEARCHERS TYPICALLY RELY ON *SAMPLING* TO ACQUIRE THE INFORMATION, OR *DATA*, NEEDED.

It is important to obtain "good data" because the inferences ultimately made will be based on the statistics obtained from these data. These inferences are only as good as the data.

Although it is relatively easy to define "good data" as data that accurately represent the population from which they were taken, it is not easy to guarantee that a particular **sampling method** will produce "good data." As statisticians, we need to be on guard against **biased sampling methods** that produce data that systematically differ from the sampled population. We need to use sampling (data collection) methods that will produce data that are representative of the population and are **unbiased,** that is, not biased.

> **Sampling method** The process of selecting items or events that will become the sample.
>
> **Biased sampling method** A sampling method that produces data that systematically differ from the sampled population. Repeated sampling will not correct the bias.
>
> **Unbiased sampling method** A sampling method that is not biased and produces data that are representative of the sampled population.

Two commonly used sampling methods that often result in biased samples are the *convenience* and *volunteer samples*. A convenience sample, sometimes called a *grab* sample, occurs when items are chosen arbitrarily and in an unstructured manner from a population, whereas a volunteer sample consists of results collected from those elements of the population that chose to contribute the needed information on their own initiative.

Did you ever buy a basket of fruit at the market based on the "good appearance" of the fruit on top, only to later discover that the rest of the fruit was not as fresh? It was too inconvenient to inspect the bottom fruit, so you trusted a convenience sample. Has your teacher used your class as a sample from which to gather data? As a group, the class is quite convenient, but is it truly representative of the school's population? (Consider the differences among day, evening, and/or weekend students; type of course; etc.)

Have you ever mailed back your responses to a magazine survey? Under what conditions did (would) you take the time to complete such a questionnaire? Most people's immediate attitude is to ignore the survey. Those with strong feelings will make the effort to respond; therefore, representative samples should not be expected when volunteer samples are collected.

The Data Collection Process

The collection of data for statistical analysis is an involved process and includes the following steps:

1. Define the objectives of the survey or study. Examples: compare the effectiveness of a new drug to the effectiveness of the standard drug; estimate the average household income in the United States.

2. Define the variable and the population of interest. Examples: length of recovery time for patients suffering from a particular disease; total income for households in the United States.

3. Define the data collection and data measuring schemes. This includes sampling frame, sampling procedures, sample size, and the data measuring device (questionnaire, telephone, and so on).

4. Collect your sample. Select the subjects to be sampled and collect the data.

5. Review the sampling process upon completion of collection. Often an analyst is stuck with data already collected, possibly even data collected for other purposes, which makes it impossible to determine whether the data are "good." Using approved techniques to collect your own data is much preferred. Although this text is concerned chiefly with various data analysis techniques, you should be aware of the concerns of data collection.

Two methods commonly used to collect data are *experiments* and *observational studies*. In an experiment, the investigator controls or modifies the environment and observes the effect on the variable under study. Sometimes laboratory results are obtained by using white rats that are given medicine or other substances to test side effects. The experimental treatments were designed specifically to obtain the data needed to study the effect on the variable. In an observational study, the investigator does not modify the environment and does not control the process being observed. The data are obtained by sampling some of the population of interest. *Surveys* are observational studies of people.

If every element in the population can be listed, or enumerated, and observed, then a *census* is compiled. However, censuses are seldom used because they are often difficult and time-consuming to compile, and therefore very expensive. Imagine the task of compiling a census of every person who is a potential client at a brokerage firm. In situations similar to this, a *sample survey* is usually conducted.

Experiment or Observational Study?

SURGICAL INFECTION IS A MATTER OF TIME

Many surgical patients fail to get timely doses of the right medications, raising the risk of infection, researchers report in the *Archives of Surgery*. Of 30 million operations performed each year in the United States, about 2% are complicated by an on-site infection, the report says. The study of 34,000 surgical patients at nearly 3,000 hospitals in 2001 found that only 56% got prophylactic medications within an hour of surgery, when they can be effective.

SOURCE: *USA Today*, February 22, 2005.

This study is an example of an observational study. The researchers did not modify or try to control the environment. They observed what was happening and wrote up their findings.

Sampling Frame and Elements

When selecting a sample for a survey, it is necessary to construct a **sampling frame**, or a list, or set, of the elements belonging to the population from which the sample will be drawn. Ideally, the sampling frame should be identical to the population, with every element of the population included once and only once. In this case, a census would become the sampling frame. In other situations, a census may not be so easy to obtain, because a complete list is not available. Lists of registered voters or the telephone directory are sometimes used as sampling frames of the general public. Depending on the nature of the information being sought, the list of registered voters or the telephone directory may or may

= Population

Ideally, the sampling frame should be identical to the population.

not serve as an unbiased sampling frame. Because only the elements in the frame have a chance to be selected as part of the sample, it is important that the sampling frame be **representative** of the population.

Once a representative sampling frame has been established, we proceed with selecting the sample elements from the sampling frame. This selection process is called the sample design. There are many different types of sample designs; however, they all fit into two categories: *judgment samples* and *probability samples*. **Judgment samples** are samples that are selected on the basis of being judged "typical."

When a judgment sample is collected, the person selecting the sample chooses items that he or she thinks are representative of the population. The validity of the results from a judgment sample reflects the soundness of the collector's judgment. This is not an acceptable statistical procedure.

Probability samples are samples in which the elements to be selected are drawn on the basis of probability. Each element in a population has a certain probability of being selected as part of the sample. The inferences that will be studied later in this textbook are based on the assumption that our sample data are obtained using a probability sample. There are many ways to design probability samples. We will look at two of them, single-stage methods and multistage methods, and learn about a few of the many specific designs that are possible.

Judgment samples Samples that are selected on the basis of being judged "typical."

Probability samples Samples in which the elements to be selected are drawn on the basis of probability. Each element in a population has a certain probability of being selected as part of the sample.

Single-Stage Methods

Single-stage sampling is a sample design in which the elements of the sampling frame are treated equally and there is no subdividing or partitioning of the frame. Two single-stage designs that statisticians use are the simple random sample and the systematic sample.

Simple Random Sample

One of the most common single-stage probability sampling methods used to collect data is the **simple random sample**, or a sample selected in such a way that every

Single-stage sampling A sample design in which the elements of the sampling frame are treated equally and there is no subdividing or partitioning of the frame.

Simple random sample A sample selected in such a way that every element in the population or sampling frame has an equal probability of being chosen. Equivalently, all samples of size *n* have an equal chance of being selected.

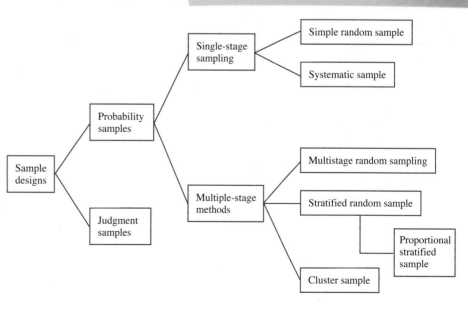

element in the population or sampling frame has an equal probability of being chosen. Equivalently, all samples of size *n* have an equal chance of being selected.

NOTE: Random samples are obtained either by sampling with replacement from a finite population or by sampling without replacement from an infinite population.

Inherent in the concept of randomness is the idea that the next result (or occurrence) is not predictable. When a random sample is drawn, every effort must be made to ensure that each element has an equal probability of being selected and that the next result does not become predictable. The proper procedure for selecting a simple random sample requires the use of random numbers. Mistakes are commonly made because the term *random* (equal chance) is confused with *haphazard* (without pattern).

To select a simple random sample, first assign an identifying number to each element in the sampling frame. This is usually done sequentially using the same number of digits for each element. Then using random numbers with the same number of digits, select as many numbers as are needed for the sample size desired. Each numbered element in the sampling frame that corresponds to a selected random number is chosen for the sample.

For example, imagine that the admissions office at your college wishes to estimate the current "average" cost of textbooks per semester, per student. The population of interest is the "currently enrolled student body," and the variable is the "total amount spent for textbooks" by each student this semester. Because a random sample is desired, the dean of admissions has obtained a computer list of this semester's full-time enrollment. Say there were 4,265 student names on the list and the dean numbered the students 0001, 0002, 0003, and so on, up to 4265; then, using four-digit random numbers, the dean identified a sample: 1288, 2177, 1952, 2463, 1644, 1004, and so on. (Go online to <u>cengagebrain.com</u> for a discussion of the use of random numbers.)

Why create a random sample? Because a simple random sample is the first step toward an unbiased sample. Random samples are required for most of the statistical procedures presented in this book. Without a random design, the conclusions we draw from the statistical procedures may not be reliable.

SYSTEMATIC SAMPLES

In concept, the simple random sample is the simplest of the probability sampling techniques, but it is sel-

Process for Collecting Data

Consider the What Employers Are Looking For graphic on page 5 and the five steps of the data-collection process.

1. *Define the objectives of the survey or study.* Determine the opinion of employers regarding what qualities they look for when hiring seasonal employees.

2. *Define the variable and the population of interest.* The variable is the opinion or response to a ques-

dom used in practice because it often is an inefficient technique. One of the easiest methods to use for approximating a simple random sample is the **systematic sampling method,** which involves selecting every *k*th item of the sampling frame, starting from a first element, which is randomly selected from the first *k* elements.

To select an *x* percent (%) systematic sample, we will need to randomly select 1 element from every $\frac{100}{x}$ elements. After the first element is randomly located within the first $\frac{100}{x}$ elements, we proceed to select every $\frac{100}{x}$th item thereafter until we have the desired number of data values for our sample.

Systematic sampling method A sample in which every *k*th item of the sampling frame is selected, starting from a first element, which is randomly selected from the first *k* elements.

3. *Define the data collection and data measuring schemes.* Based on the graphic itself, it can be seen that the source for the percentages presented was SnagAJob.com. Upon further investigation, IPSOS Public Affairs, a third-party research firm, conducted the survey on behalf of the "hourly job Web site" SnagAJob.com between February 20 and 25, 2009. It was an online survey of 1,043 hiring managers with responsibility for hiring summer and seasonal hourly employees.

4. *Collect your sample.* The information collected from each hiring manager was what he or she thought was the single "most" essential quality/characteristic that a seasonal employee should possess.

5. *Review the sampling process upon completion of collection.* Since the sampling process was an online survey, were only hiring managers who conduct their business online aware of this survey? Were various areas of the country and types of businesses represented? Perhaps you can think of additional concerns.

For example, if we desire a 3% systematic sample, we would locate the first item by randomly selecting an integer between 1 and 33 ($\frac{100}{x} = \frac{100}{3} = 33.33$, which becomes 33 when rounded). Suppose 23 was randomly selected. This means that our first data value is obtained from the subject in the 23rd position in the sampling frame. The second data value will come from the subject in the 56th ($23 + 33 = 56$) position; the third, from the 89th ($56 + 33$); and so on, until our sample is complete.

The systematic technique is easy to describe and execute; however, it has some inherent dangers when the sampling frame is repetitive or cyclical in nature. For example, a systematic sample of every kth house along a long street might result in a sample disproportional with regard to houses on corner lots. The resulting information would likely be biased if the purpose for sampling is to learn about support for a proposed sidewalk tax. In these situations the results may not approximate a simple random sample.

Multistage Methods

When sampling very large populations, sometimes it is necessary to use a *multistage sampling* design to approximate random sampling. **Multistage random sampling** is a sample design in which the elements of the sampling frame are subdivided and the sample is chosen in more than one stage.

Multistage sampling designs often start by dividing a very large population into subpopulations on the basis of some characteristic. These subpopulations are called *strata*. These smaller, easier-to-work-with strata can then be sampled separately. One such sample design is the stratified random sampling method. This method produces a sample by stratifying the population, or sampling frame, and then selecting a number of items from each of the strata by means of a simple random sampling technique.

A **stratified random sample** results when the population, or sampling frame, is subdivided into various strata, usually some already occurring natural subdivision, and then a subsample is drawn from each of these strata. These subsamples may be drawn from the various strata by using random or systematic methods. The subsamples are summarized separately first and then combined to draw conclusions about the entire population.

When a population with several strata is sampled, we often require that the number of items collected from each stratum be proportional to the size of the strata; this method is called a **proportional stratified sampling**. After stratifying the population or sampling frame, the researcher then selects a number of items in proportion to the size of the strata from each strata by means of a simple random sampling technique.

Multistage random sampling A sample design in which the elements of the sampling frame are subdivided and the sample is chosen in more than one stage.

Stratified random sample A sample obtained by stratifying the population, or sampling frame, and then selecting a number of items from each of the strata by means of a simple random sampling technique.

Proportional stratified sampling A sample obtained by stratifying the population, or sampling frame, and then selecting a number of items in proportion to the size of the strata from each strata by means of a simple random sampling technique.

A convenient way to express the idea of proportional sampling is to establish a quota. For example, the quota, "1 for every 150" directs you to select 1 data value for each 150 elements in each strata. That way, the size of the strata determines the size of the subsample from that strata. The subsamples are summarized separately and then combined to draw conclusions about the entire population.

A cluster sample is another multistage design. A **cluster sample** is obtained by stratifying the population, or sampling frame, and then selecting some or all of the items from some, but not all, of the strata. The cluster sample uses either random or systematic methods to select the strata (clusters) to be sampled (first stage) and then uses either random or systematic methods to select elements from each identified cluster (second stage). The cluster sampling method also allows the possibility of selecting all of the elements from each identified cluster. Either way, the subsamples are summarized separately, and then the information is combined.

> **Cluster sample** A sample obtained by stratifying the population, or sampling frame, and then selecting some or all of the items from some, but not all, of the strata.

To illustrate a possible multistage random sampling process, consider that a sample is needed from a large country. In the first stage, the country is divided into smaller regions, such as states, and a random sample of these states is selected. In the second stage, a random sample of smaller areas within the selected states (counties) is then chosen. In the third stage, a random sample of even smaller areas (townships) is taken within each county. Finally in the fourth stage, if these townships are sufficiently small for the purposes of the study, the researcher might continue by collecting simple random samples from each of the identified townships. This would mean that the entire sample was made up of several "local" subsamples identified as a result of the several stages.

Sample design is not a simple matter; many colleges and universities offer separate courses in sample surveying and experimental design. The topic of survey sampling is a complete textbook in itself. The preceding information should provide you with an overview of sampling and put its role in perspective.

1.4 Statistics and Technology

IN RECENT YEARS, ELECTRONIC TECHNOLOGY HAS HAD A TREMENDOUS IMPACT ON ALMOST EVERY ASPECT OF LIFE.

The field of statistics is no exception. As you will see, statistics uses many techniques that are repetitive in nature: calculations of numerical statistics, procedures for constructing graphic displays of data, and procedures that are followed to formulate statistical inferences. Computers and calculators are very good at performing these sometimes long and tedious operations. If your computer has one of the standard statistical packages or if you have a statistical calculator, then it will make the analysis easy to perform.

The Tech Cards at the end of the book correspond to the chapters and contain information about using various types of technology to complete statistical calculations.

Garbage In, Garbage Out!

There is a great temptation to use the computer or calculator to analyze any and all sets of data and then treat the results as though the statistics are correct. Remember the adage: "Garbage in, garbage out!" Responsible use of statistical methodology is very important. The burden is on the user to ensure that the appropriate methods are correctly applied and that accurate conclusions are drawn and communicated to others.

problems

Objective 1.1

1.1 Postyour.info is a worldwide service where Internet users from around the world can take part in questionnaires. [http://postyour.info/] Below is a graph depicting the combined summary of how users answered one of the posted questions. Results are given in percent (count).

HOW OFTEN DO YOU EAT FRUIT? (irrespective of the reasons why)

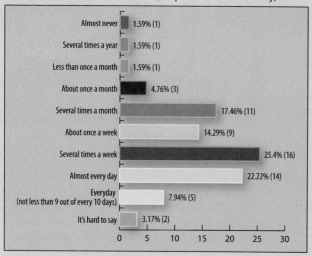

SOURCE: http://postyour.info/

a. What question was asked and answered in order to gather the information presented on this graph?

b. Who was asked the question?

c. How many people answered the question?

d. Verify the percentages, 1.59%, 17.46%, and 25.4%.

e. Are the percentages reported in this graph likely to be representative of all people? Explain why or why not.

1.2 Do you work hard for your money? Java professionals think they do, reporting long working hours at their jobs. Java developers from around the world were surveyed about the number of hours they work weekly. Listed here are the average number of hours worked weekly in various regions of the United States and the world.

Region	Hours Worked	Region	Hours Worked
U.S	48	California	50
Northeast	47	Pacific NW	47
Mid-Atlantic	49	Canada	43
South	47	Europe	48
Midwest	47	Asia	47
Central Mt	51	South America and Africa	49

SOURCE: Jupitermedia Corporation

a. How many hours do you work per week (or anticipate working after you graduate)?

b. What happened to the 40-hour workweek? Does it appear to exist for the Java professional?

c. Does the information in this chart make a career of being a Java professional seem attractive?

1.3 a. Both of the statistical graphics presented on page 2 seem to suggest that information is about what population? Is that the case? Justify your answer.

b. Describe the information that was collected and used to determine the statistics reported in Are You Fretting over Messages?

c. "47%—One hour or less" was one specific statistic reported in Are You Fretting over Messages? Describe what that statistic tells you.

1.4 a. Consider the graphic How Often Is Your Bed Made? If you had been asked, how would you have responded? What does the percentage associated with your answer mean? Explain.

b. How do you interpret the "5%—Weekly" reported in How Often Is Your Bed Made?

1.5 a. Write a 50-word paragraph describing what the word "statistics" means to you right now.

b. Write a 50-word paragraph describing what the word "random" means to you right now.

c. Write a 50-word paragraph describing what the word "sample" means to you right now.

1.6 *Statistics* is described on page 4 as "the science of collecting, describing, and interpreting data." Using your own words, write a sentence describing each of the three statistical activities.

1.7 Determine which of the following statements is descriptive in nature and which is inferential. Refer to the data below in How Old Is My Fish?

How Old Is My Fish?							
Average age by length of largemouth bass in New York State							
Length	8	9	10	11	12	13	14
Age (yrs)	2	3	3	4	4	5	5

SOURCE: *NYS DEC Freshwater Fishing Guide*

a. All 9-inch largemouth bass in New York State are an average of 3 years old.

b. Of the largemouth bass used in the sample to make up the *NYS DEC Freshwater Fishing Guide,* the average age of 9-inch largemouth bass was 3 years.

1.8 Determine which of the following statements is descriptive in nature and which is inferential. Refer to the Teens Using Cell Phones in Class graphic on page 5.

a. 84% of the teens surveyed in May and June of 2009 had cell phones.

b. In May and June 2009, 16% of all teens did not have a cell phone.

1.9

Setting a Date for Date Night

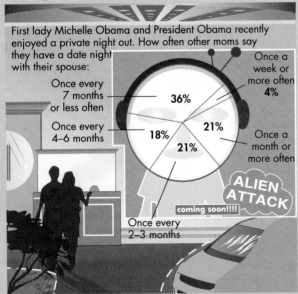

First lady Michelle Obama and President Obama recently enjoyed a private night out. How often other moms say they have a date night with their spouse:

- Once every 7 months or less often: **36%**
- Once a week or more often: **4%**
- Once a month or more often: **21%**
- Once every 2–3 months: **21%**
- Once every 4–6 months: **18%**

SOURCE: Frigidaire Motherload Index survey of 1,170 married women, ages 25–50 who have two or more children.

a. What group of people were polled?
b. How many people were polled?
c. What information was obtained from each person?
d. Explain the meaning of "18% say once every 4–6 months."
e. How many people answered "Once every 4–6 months"?

1.10 International Communications Research (ICR) conducted the 2008 Spring Cleaning Survey for The Soap and Detergent Association. ICR questioned 777 American adults who spring clean about which spring-cleaning chore they would like to hire someone to do for them. The "chore" results were: 47% washing windows, 23% cleaning the bathroom, 12% cleaning the kitchen, 8% dusting, 7% mopping, 3% other. The survey has a margin of error of $\pm 3.52\%$.

a. What is the population?
b. How many people were polled?
c. What information was obtained from each person?
d. Using the information given, estimate the number of surveyed adults who would gladly hire someone to wash the windows, if they could.
e. What do you think the "margin of error of $\pm 3.52\%$" means?
f. How would you use the margin of error in estimating the percentage of all adults who would like to hire out the spring cleaning chore of "Cleaning the kitchen"?

1.11 Opinion Research Corporation conducted the 2008 Lemelson-MIT Invention Index survey of 501 teens, ages 12–17. The teens were asked what everyday invention they think would be obsolete in five years.

In One Day, Out the Next

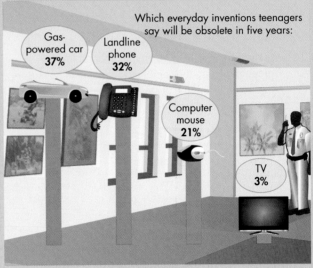

Which everyday inventions teenagers say will be obsolete in five years:

- Gas-powered car **37%**
- Landline phone **32%**
- Computer mouse **21%**
- TV **3%**

SOURCE: 2009 Lemelson-MIT Invention Index survey of 501 teens, ages 12–17, by Opinion Research Corp. Margin of error: ±4.3 percentage points.

a. What is the population?
b. How many people were polled?
c. What information was obtained from each person?
d. Estimate the number of surveyed teens who think the computer mouse will be obsolete in five years.
e. What do you think the "margin of error ±4.3 percentage points" means?
f. How would you use the margin of error in estimating the percentage of all teens who think the computer mouse will be obsolete in five years?

1.12

What Do You Plan to Spend Your Tax Refund on?

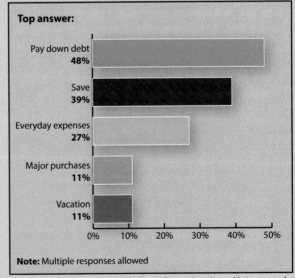

Top answer:

- Pay down debt **48%**
- Save **39%**
- Everyday expenses **27%**
- Major purchases **11%**
- Vacation **11%**

Note: Multiple responses allowed

SOURCE: National Retail Federation 2009 Tax Returns Consumer Intentions and Actions survey of 8,426 consumers. Margin of error: ±1 percentage point.

a. Describe the population of interest.
b. Describe the sample most likely used to collect this information.

c. Identify the variable used to collect this information.

d. What are the majority of people going to do with their tax refund? How does this majority show on the graph?

1.13 During a radio broadcast a few years ago, David Essel reported the following three statistics: (1) the U.S. divorce rate is 55%, (2) when married adults were asked if they would re-marry their spouse, 75% of the women said yes, and 3) 65% of the men said yes.

a. What is the "stay married" rate?

b. There seems to be a contradiction in this information. How is it possible for all three of these statements to be correct? Explain.

1.14 A working knowledge of statistics is very helpful when you want to understand the statistics reported in the news. The news media and our government often make a statement that says something like, "Crime rate jumps 50% in your city."

a. Does an increase in rate from 4 to 6 represent an increase of 50%? Explain.

b. Why would anybody report an increase from 4 to 6 as a "50% rate jump"?

1.15 Of the adult U.S. population, 36% has an allergy. In a sample of 1,200 randomly selected adults, 33.2% had an allergy.

a. Describe the population.

b. What is the sample?

c. Describe the variable.

d. Identify the statistic and give its value.

e. Identify the parameter and give its value.

1.16 In your own words, explain why the parameter is fixed and the statistic varies.

1.17 Is "football jersey number" a quantitative or a categorical variable? Support your answer with a detailed explanation.

1.18 a. Name two customer-related attribute variables that a newly opened department store might find informative to study.

b. Name two customer-related numerical variables that a newly opened department store might find informative to study.

1.19 a. Name two customer-related nominal variables that a newly opened department store might find informative to study.

b. Name two customer-related ordinal variables that a newly opened department store might find informative to study.

1.20 a. Explain why the variable "score" for the home team at a basketball game is discrete.

b. Explain why the variable "number of minutes to commute to work" is continuous.

1.21 The severity of side effects experienced by patients while being treated with a particular medicine is under study. The severity is measured as "none," "mild," "moderate," "severe," or "very severe."

a. Name the variable of interest.

b. Identify the type of variable.

1.22 A nationwide poll of adults on cell phone use and driving was conducted during May, 2009 by The Harris Poll. Their responses to "How dangerous is it for a driver to use a cell phone while driving?" were categorized as "very dangerous," "dangerous," "somewhat dangerous," "slightly dangerous," or "not dangerous at all."

a. Name the variable of interest.

b. Identify the type of variable.

1.23 Students are being surveyed about the weight of books and supplies they are carrying as they attend class.

a. Identify the variable of interest.

b. Identify the type of variable.

c. List a few values that might occur in a sample.

1.24 Below is a small sample of the one hundred sixty-five 2009 pickup trucks listed on MPGoMatic.com and available to the buying public.

Manufac-turer	Model	Drive	Engine Size (no. cylin-ders)	Engine Size, Dis-placement (liters)	Trans-mission	City MPG	Hwy MPG
Chevrolet	Colorado	2WD	4	2.9	Manual	18	24
GMC	Canyon	2WD	5	3.7	Auto	17	23
Hummer	H3T	4WD	8	5.3	Auto	13	16
Mitsubishi	Raider	4WD	8	4.7	Auto	9	12
Suzuki	Equator	2WD	4	2.5	Auto	17	22
Toyota	Tacoma	4WD	6	4.0	Manual	14	19

SOURCE: http://www.mpgomatic.com/

a. What is the population from which this sample was taken?

b. How many individuals are in the population? How many are in the sample?

c. How many variables are there?

d. Name the qualitative/categorical variables.

e. Which of the qualitative variables are nominal? Which are ordinal?

f. Name the quantitative variables.

g. Which of the quantitative variables are discrete? Which are continuous?

1.25 Identify each of the following as examples of (1) nominal, (2) ordinal, (3) discrete, or (4) continuous variables:

a. A poll of registered voters as to which candidate they support

b. The length of time required for a wound to heal when a new medicine is being used

c. The number of televisions within a household

d. The distance first-year college women can kick a football

e. The number of pages per job coming off a computer printer

f. The kind of tree used as a Christmas tree

1.26 Suppose a 12-year-old asked you to explain the difference between a sample and a population.

    a. What information should your answer include?

    b. What reasons would you give him or her for why one would take a sample instead of surveying every member of the population?

Objective 1.2

1.27 Suppose we measure the weights (in pounds) of the individuals in each of the following groups:

> Group 1: cheerleaders for National Football League teams
>
> Group 2: players for National Football League teams

For which group would you expect the data to have more variability? Explain why.

1.28 Suppose you were trying to decide which of two machines to purchase. Furthermore, suppose the length to which the machines cut a particular product part was important. If both machines produced parts that had the same length on the average, what other consideration regarding the lengths would be important? Why?

1.29 Consumer activist groups for years have encouraged retailers to use unit pricing of products. They argue that food prices, for example, should always be labeled in $/ounce, $/pound, $/gram, $/liter, and so on, in addition to $/package, $/can, $/box, $/bottle. Explain why.

1.30 A coin-operated coffee vending machine dispenses, on the average, 6 oz of coffee per cup. Can this statement be true of a vending machine that occasionally dispenses only enough to fill the cup half full (say, 4 oz)? Explain.

1.31 Teachers use examinations to measure students' knowledge about their subject. Explain how "a lack of variability in the students' scores might indicate that the exam was not a very effective measuring device." Thoughts to consider: What would it mean if all students attained a score of 100% on an exam? What would it mean if all attained a 0%? What would it mean if the grades ranged from 40% to 95%?

Objective 1.3

1.32 *USA Today* regularly asks readers, "Have a complaint about airline baggage handling, refunds, advertising, customer service? Write: … ."

    a. What kind of sampling method is this?

    b. Are the results likely to be biased? Explain.

1.33 *USA Today* conducted a survey asking readers, "What is the most hilarious thing that has ever happened to you en route to or during a business trip?"

    a. What kind of sampling method is this?

    b. Are the results likely to be biased? Explain.

1.34 In a survey about families, Ann Landers—a well known advice columnist—asked parents if they would have kids again; 70% responded "No." An independent random survey asking the same question yielded a 90% "Yes" response. Give at least one explanation as to why the resulting percent from the Landers's survey is so much different than the resulting percent from the random sample.

1.35 We all know that exercise is good for us. But can exercise prevent or delay the symptoms of Parkinson's disease? A recent study by the Harvard School of Public Health studied 48,000 men and 77,000 women who were relatively healthy and middle-aged or older. During the course of the study, 387 people developed the disease. The study found that men who had participated in some vigorous activity at least twice a week in high school, college, and up to age 40, had 60% reduced risk of getting Parkinson's. The study found no such reduction for women. What type of sampling does this represent?

SOURCE: "Exercise May Prevent Parkinson's," *USA Today*, February 22, 2005, p. 7D

1.36 A wholesale food distributor in a large metropolitan area would like to test the demand for a new food product. He distributes food through five large supermarket chains. The food distributor selects a sample of stores located in areas where he believes the shoppers are receptive to trying new products. What type of sampling does this represent?

1.37 In December 2008, NBC posted the question below on their Web site to survey the public.

> Live Vote March 16, 2009 with 12,810,699 responses tallied
>
> Should the motto "In God We Trust" be removed from U.S. currency?
>
> Yes. It's a violation of the principle of separation of church and state.
> 14%
>
> No. The motto has historical and patriotic significance and does nothing to establish a state religion.
> 86%

At the same time, the e-mail below was being circulated to help "get out the vote."

> Here's your chance to let the media know where the people stand on our faith in God, as a nation. NBC is taking a poll on "In God We Trust" to stay on our American currency.
>
> Please send this to every Christian you know so they can vote on this important subject. Please do it right away, before NBC takes this off the Web page.
>
> This is not sent for discussion; if you agree, forward it; if you don't, delete it. By my forwarding it, you know how I feel. I'll bet this was a surprise to NBC.

Describe two reasons why the results of the "In God We Trust" survey should not be expected to be representative of the population.

1.38 Consider a simple population consisting of only the numbers 1, 2, and 3 (an unlimited number of each). There are nine different samples of size two that could be drawn from this population: (1, 1), (1, 2), (1, 3), (2, 1), (2, 2,), (2, 3), (3, 1), (3, 2), (3, 3).

a. If the population consists of the numbers 1, 2, 3, and 4, list all the samples of size two that could possibly be selected.

b. If the population consists of the numbers 1, 2, and 3, list all the samples of size three that could possibly be selected.

1.39 a. What is a sampling frame?

b. In the example on page 14, what did the admissions office use for a sampling frame?

c. In the same example, where did the number 1,288 come from, and how was it used?

1.40 A random sample could be very difficult to obtain. Why?

1.41 Why is the random sample so important in statistics?

1.42 Sheila Jones works for an established marketing research company in Cincinnati, Ohio. Her supervisor just handed her a list of 500, 4-digit random numbers extracted from a statistical table of random digits. He told Sheila to conduct a survey by calling 500 Cincinnati residents on the telephone, provided the last 4 digits of their phone number matched one of the numbers on the list. If Sheila follows her supervisor's instructions, is he assured of obtaining a random sample of respondents? Explain.

1.43 Describe in detail how you would select a 4% systematic sample of the adults in a nearby large city in order to complete a survey about a political issue.

1.44 a. What body of the federal government illustrates a stratified sampling of the people? (A random selection process is not used.)

b. What body of the federal government illustrates a proportional sampling of the people? (A random selection process is not used.)

1.45 Suppose that you have been hired by a group of all-sports radio stations to determine the age distribution of their listeners. Describe in detail how you would select a random sample of 2,500 people from the 35 listening areas involved.

1.46 Explain why the polls that are so frequently quoted during early returns on Election Day TV coverage are an example of cluster sampling.

1.47 The telephone book might not be a representative sampling frame. Explain why.

1.48 The election board's voter registration list is not a census of the adult population. Explain why.

1.49 "Replace incandescent light bulbs with compact fluorescent bulbs that use up to 75% less energy and last up to 10 times as long."

SOURCE: "Simple Ways to Save Energy," *NYSEG Energy Lines*, February 2009

a. What two claims are made in the above statement by the New York State Electric and Gas (NYSEG) Company? State them in terms of a statistical parameter.

b. Do you feel the two statements by NYSEG are reasonable and likely to be true? Explain.

c. If you feel a claim is reasonable and likely true, would you be driven to find evidence to verify its truth? Explain.

d. If you feel a claim is not reasonable and likely not true, would you be driven to find evidence to verify it to be incorrect? Explain.

e. Would you be more likely to research the situation described in (c) or (d)? Explain.

f. How would you proceed to attempt to verify, "up to 75% less energy"?

g. How would you proceed to attempt to verify, "last up to 10 times as long"?

Objective 1.4

1.50 How have computers increased the usefulness of statistics to professionals such as researchers, government workers who analyze data, statistical consultants, and others?

1.51 How might computers help you in statistics?

1.52 What is meant by the saying "Garbage in, garbage out!" and how have computers increased the probability that studies may be victimized by the adage?

*ONLINE DATA

Throughout STAT, problems for which there are online data sets are marked with an asterisk (*). We didn't start working with data in Chapter 1, but remember the asterisk for later chapters.

Descriptive Analysis
and Presentation of Single-Variable Data

Ever wonder if your typical day measures up to that of other college students? If you look at the graph below, think about what you do during the day and how much time you spend doing those activities. Does your day break up into the categories shown below? Or do you have an extra category or two? On the average, how does the amount of time you spend compare? Perhaps you have different categories. You may wish that you too could average 8.3 hours of sleep!

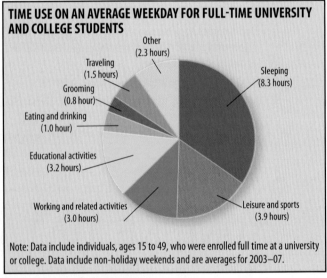

TIME USE ON AN AVERAGE WEEKDAY FOR FULL-TIME UNIVERSITY AND COLLEGE STUDENTS

Other (2.3 hours)
Traveling (1.5 hours)
Grooming (0.8 hour)
Eating and drinking (1.0 hour)
Educational activities (3.2 hours)
Working and related activities (3.0 hours)
Sleeping (8.3 hours)
Leisure and sports (3.9 hours)

Note: Data include individuals, ages 15 to 49, who were enrolled full time at a university or college. Data include non-holiday weekends and are averages for 2003–07.

SOURCE: Bureau of Labor Statistics

Can you imagine all this information written out in sentences? Graphical displays can truly be worth a thousand words. This one pie chart summarizes "Time Use" information from a 2003–2007 American Time Use Survey (ATUS) of over 50,000 Americans. ATUS is a federally administered, continuous survey on time use in the United States sponsored by the Bureau of Labor Statistics and conducted by the U.S. Census Bureau. Since it is a cross-sectional survey, this graph included only the full-time college students who participated.

Now that you know the source and see the overall sample size, you may feel that these data portray a relatively accurate picture of a college student's day. Or do some of the numbers seem off to you? The 0.8 hour per day grooming may have a gender difference

objectives

2.1 Graphs, Pareto Diagrams, and Stem-and-Leaf Displays

2.2 Frequency Distributions and Histograms

2.3 Measures of Central Tendency

2.4 Measures of Dispersion

2.5 Measures of Position

2.6 Interpreting and Understanding Standard Deviation

2.7 The Art of Statistical Deception

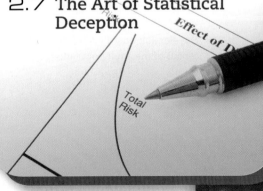

and strike you as inaccurate. As you work through Chapter 2, you will begin to learn how to organize and summarize data into graphical displays and numerical statistics in order to accurately and appropriately describe data.

2.1 Graphs, Pareto Diagrams, and Stem-and-Leaf Displays

ONCE THE SAMPLE DATA HAVE BEEN COLLECTED, WE MUST "GET ACQUAINTED" WITH THEM.

One of the most helpful ways to become acquainted with the data is to use an initial exploratory data analysis technique that will result in a pictorial representation of the data. The display will visually reveal patterns of behavior of the variable being studied. There are several graphic (pictorial) ways to describe data. The type of data and the idea to be presented determine which method is used.

NOTE: There is no single correct answer when constructing a graphic display. The analyst's judgment and the circumstances surrounding the problem play a major role in the development of the graph.

Qualitative Data

Graphs can be used to summarize qualitative, or attribute, or categorical, data. **Pie charts (circle graphs)** show the amount of data that belong to each category as a proportional part of a circle. **Bar graphs** show the amount of data that belong to each category as a proportionally sized rectangular area. Any graphic representation used, regardless of type, needs to be completely self-explanatory. That includes a descriptive, meaningful title and proper identification of the quantities and variables involved. To appreciate the differences between these two types of graphical representations, let's compare them by using the same data set to create one of each.

To get a better sense of what's involved in graphing qualitative data, let's consider an example about surgeries at a hospital. Table 2.1 lists the number of cases of each type of operation performed at General Hospital last year.

*Table 2.1 **Operations Performed at General Hospital Last Year**

Type of Operation	Number of Cases
Thoracic	20
Bones and joints	45
Eye, ear, nose, and throat	58
General	98
Abdominal	115
Urologic	74
Proctologic	65
Neurosurgery	23
Total	498

* Tables marked with an asterisk have data sets online at cengagebrain.com.

The data in Table 2.1 are displayed on a pie chart in Figure 2.1, with each type of operation represented by a relative proportion of a circle, found by dividing the number of cases by the total sample size—namely, 498. The proportions are then reported as percentages (for example, 25% is $\frac{1}{4}$ of the circle). Figure 2.2 displays the same "type of operation" data but in the form of a bar graph. Bar graphs of attribute data should be drawn with a space between bars of equal width.

When the bar graph is presented in the form of a *Pareto diagram*, it presents additional and very helpful information. That's because in a **Pareto diagram** the bars are arranged from the most numerous category to the least

Pie charts (circle graphs) Graphs that show the amount of data belonging to each category as a proportional part of a circle.

Bar graphs Graphs that show the amount of data belonging to each category as a proportionally sized rectangular area.

Pareto diagram A bar graph with the bars arranged from the most numerous category to the least numerous category. It includes a line graph displaying the cumulative percentages and counts for the bars.

numerous category. A Pareto diagram also includes a line graph displaying the cumulative percentages and counts for the bars. The Pareto diagram is popular in quality-control applications. A Pareto diagram of types of defects will show the ones that have the greatest effect on the defective rate in order of effect. It is then easy to see which defects should be targeted in order to most effectively lower the defective rate.

Pareto diagrams can also be useful in evaluating crime statistics. The FBI reported the number of hate crimes by category for 2003 (www.fbi.gov). The Pareto diagram in Figure 2.3 shows the 8,715 categorized hate crimes, their percentages, and cumulative percentages.

Quantitative Data

One major reason for constructing a graph of quantitative data is to display its **distribution,** or the pattern of variability displayed by the data of a variable. The distribution displays the frequency of each value of the variable. Two popular methods for displaying distribution of quantitative data are the **dotplot** and the **stem-and-leaf** display.

Distribution The pattern of variability displayed by the data of a variable. The distribution displays the frequency of each value of the variable.

Figure 2.1 **Pie Chart**

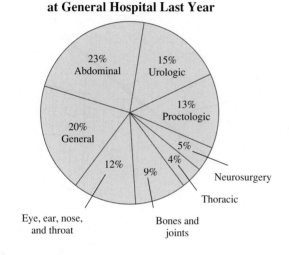

Operations Performed at General Hospital Last Year

Figure 2.2 **Bar Graph**

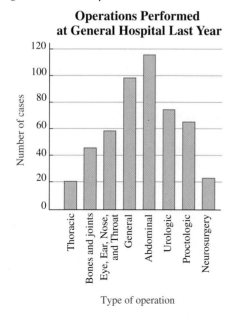

Figure 2.3 **Pareto Diagram**

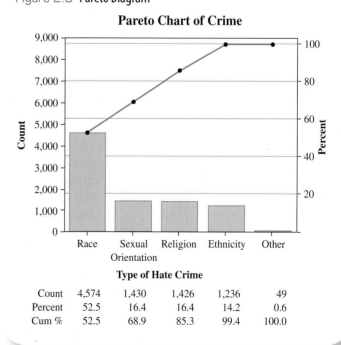

	Race	Sexual Orientation	Religion	Ethnicity	Other
Count	4,574	1,430	1,426	1,236	49
Percent	52.5	16.4	16.4	14.2	0.6
Cum %	52.5	68.9	85.3	99.4	100.0

DOTPLOT

One of the simplest graphs used to display a distribution is the **dotplot display.** The dotplot displays the data of a sample by representing each data value with a dot positioned along a scale. This scale can be either horizontal or vertical. The frequency of the values is represented along the other scale. The dotplot display is a convenient technique to use as you first begin to analyze the data. It results in a picture of the data and it sorts the data into numerical order. (To *sort* data is to list the data in rank order according to numerical value.) Table 2.2 provides a sample of 19 exam grades randomly selected from a large class. Notice how the data in Figure 2.4 are bunched near the center and more spread out near the extremes.

*Table 2.2 Sample of 19 Exam Grades

76	74	82	96	66	76	78	72	52	68
86	84	62	76	78	92	82	74	88	

Figure 2.4 **Dotplot**

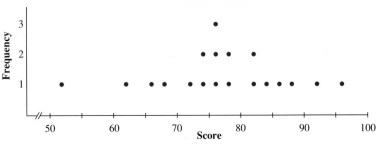

STEM-AND-LEAF DISPLAY

In recent years a technique known as the **stem-and-leaf display** has become very popular for summarizing numerical data. It is a combination of a graphic technique and a sorting technique. These displays are simple to create and use, and they are well suited to computer applications. A stem-and-leaf display presents the data of a sample using the actual digits that make up the data values. Each numerical value is divided into two parts: The leading digit(s) becomes the stem, and the trailing digit(s) becomes the leaf. The stems are located along the main axis, and a leaf for each data value is located so as to display the distribution of the data.

Dotplot display Displays the data of a sample by representing each data with a dot positioned along a scale. This scale can be either horizontal or vertical. The frequency of the values is represented along the other scale.

Stem-and-leaf display A display of the data of a sample using the actual digits that make up the data values. Each numerical value is divided into two parts: The leading digit(s) becomes the stem, and the trailing digit(s) becomes the leaf. The stems are located along the main axis, and a leaf for each data value is located so as to display the distribution of the data.

Let's construct a stem-and-leaf display for the 19 exam scores from Table 2.2. At a quick glance we see that there are scores in the 50s, 60s, 70s, 80s, and 90s. Let's use the first digit of each score as the stem and the second digit as the leaf. Typically, the display is constructed vertically. We draw a vertical line and place the stems, in order, to the left of the line.

$$
\begin{array}{c|}
5 \\
6 \\
7 \\
8 \\
9 \\
\end{array}
$$

Next we place each leaf on its stem. This is done by placing the trailing digit on the right side of the vertical line opposite its corresponding leading digit. Our first data value is 76; 7 is the stem and 6 is the leaf. Thus, we place a 6 opposite the 7 stem:

$$7\,|\,6$$

The next data value is 74, so a leaf of 4 is placed on the 7 stem next to the 6.

$$7\,|\,6\ 4$$

The next data value is 82, so a leaf of 2 is placed on the 8 stem.

$$
\begin{array}{c|l}
7 & 6\ 4 \\
8 & 2 \\
\end{array}
$$

We continue until each of the other 16 leaves is placed on the display. Figure 2.5A shows the resulting stem-and-leaf display; Figure 2.5B shows the completed stem-and-leaf display after the leaves have been ordered.

From Figure 2.5B, we see that the grades are centered around the 70s. In this case, all scores with the same tens digit were placed on the same branch, but this may not always be desired. Suppose we reconstruct the display; this time instead of grouping ten possible values on each stem, let's group the values so that only five possible values could fall on each stem. Do you notice a difference in the appearance of Figure 2.6? The general shape is approximately symmetrical about the high 70s. Our information is a little more refined, but basically we see the same distribution.

It is fairly typical of many variables to display a distribution that is concentrated (mounded) about a central value and then in some manner dispersed in one or both directions. Often a graphic display reveals something that the analyst may or may not have anticipated. The example that follows demonstrates what generally occurs when two populations are sampled together.

Figure 2.5A Unfinished Stem-and-Leaf Display

19 Exam Scores

5	2
6	6 8 2
7	6 4 6 8 2 6 8 4
8	2 6 4 2 8
9	6 2

Figure 2.5B Final Stem-and-Leaf Display

19 Exam Scores

5	2
6	2 6 8
7	2 4 4 6 6 6 8 8
8	2 2 4 6 8
9	2 6

Figure 2.6 Stem-and-Leaf Display

19 Exam Scores

(50–54)	5	2
(55–59)	5	
(60–64)	6	2
(65–69)	6	6 8
(70–74)	7	2 4 4
(75–79)	7	6 6 6 8 8
(80–84)	8	2 2 4
(85–89)	8	6 8
(90–94)	9	2
(95–99)	9	6

OVERLAPPING DISTRIBUTIONS

Let's examine overlapping distributions by considering a random sample of 50 college students. Their weights were obtained from their medical records. The resulting data are listed in Table 2.3. Notice that the weights range from 98 to 215 pounds. Let's group the weights on stems of ten units using the hundreds and the tens digits as stems and the units digit as the leaf (see Figure 2.7). The leaves have been arranged in numerical order.

*Table 2.3 Weights of 50 College Students

Student	1	2	3	4	5	6	7	8	9	10
Male/Female	F	M	F	M	M	F	F	M	M	F
Weight	98	150	108	158	162	112	118	167	170	120
Student	11	12	13	14	15	16	17	18	19	20
Male/Female	M	M	M	F	F	M	F	M	M	F
Weight	177	186	191	128	135	195	137	205	190	120
Student	21	22	23	24	25	26	27	28	29	30
Male/Female	M	M	F	M	F	F	M	M	M	M
Weight	188	176	118	168	115	115	162	157	154	148
Student	31	32	33	34	35	36	37	38	39	40
Male/Female	F	M	M	F	M	F	M	F	M	M
Weight	101	143	145	108	155	110	154	116	161	165
Student	41	42	43	44	45	46	47	48	49	50
Male/Female	F	M	F	M	M	F	F	M	M	M
Weight	142	184	120	170	195	132	129	215	176	183

Figure 2.7 Stem-and-Leaf Display

**Weights of
50 College Students (lb)
Stem-and-Leaf of WEIGHT
$N = 50$ Leaf Unit = 1.0**

9	8
10	1 8 8
11	0 2 5 5 6 8 8
12	0 0 0 8 9
13	2 5 7
14	2 3 5 8
15	0 4 4 5 7 8
16	1 2 2 5 7 8
17	0 0 6 6 7
18	3 4 6 8
19	0 1 5 5
20	5
21	5

Figure 2.8 "Back-to-Back" Stem-and-Leaf Display

Weights of 50 College Students (lb)

Female		Male
8	09	
1 8 8	10	
0 2 5 5 6 8 8	11	
0 0 0 8 9	12	
2 5 7	13	
2	14	3 5 8
	15	0 4 4 5 7 8
	16	1 2 2 5 7 8
	17	0 0 6 6 7
	18	3 4 6 8
	19	0 1 5 5
	20	5
	21	5

Close inspection of Figure 2.7 suggests that two overlapping distributions may be involved. That is exactly what we have: a distribution of female weights and a dis-tribution of male weights. Figure 2.8 shows a "back-to-back" stem-and-leaf display of this set of data and makes it obvious that two distinct distributions are involved.

Figure 2.9, a "side-by-side" dotplot (same scale) of the same 50 weight data, shows the same distinction between the two subsets.

Based on the information shown in Figures 2.8 and 2.9, and on what we know about people's weight, it seems reasonable to conclude that female college students weigh less than male college students. Situations involving more than one set of data are discussed further in Chapter 3.

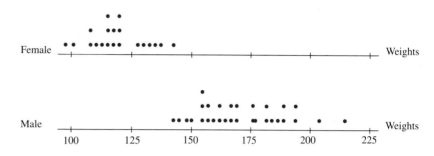

Figure 2.9 Dotplots with Common Scale

Weights of 50 College Students

2.2 Frequency Distributions and Histograms

LISTS OF LARGE SETS OF DATA DO NOT PRESENT MUCH OF A PICTURE.

Sometimes we want to condense the data into a more manageable form. This can be accomplished by creating a **frequency distribution** that pairs the values of a variable with their frequency. Frequency distributions are often expressed in chart form.

To demonstrate the concept of a frequency distribution, let's use this set of data:

| 3 | 2 | 2 | 3 | 2 | 4 | 4 | 1 | 2 | 2 |
| 4 | 3 | 2 | 0 | 2 | 2 | 1 | 3 | 3 | 1 |

If we let x represent the variable, then we can use a frequency distribution to represent this set of data by listing the x values with their frequencies. For example, the value 1 occurs in the sample three times; therefore, the

Frequency distribution A listing, often expressed in chart form, that pairs values of a variable with their frequency.

Frequency The number of times the value x occurs in the sample.

frequency for $x = 1$ is 3. The complete set of data is shown in the frequency distribution in Table 2.4.

The **frequency** f is the number of times the value x occurs in the sample. Table 2.4 is an **ungrouped frequency distribution**—"ungrouped" because each value of x in the distribution stands alone. When a large set of data has many different x values instead of a few repeated values, as in the previous example, we can group the values into a set of classes and construct a **grouped frequency distribution**. The stem-and-leaf display in Figure 2.5B (page 27) shows, in picture form, a grouped frequency distribution. Each stem represents a class. The number of leaves on each stem is the same as the frequency for that same *class* (sometimes called a *bin*). The data represented in Figure 2.5B are listed as a grouped frequency distribution in Table 2.5.

Table 2.4 Ungrouped Frequency Distribution

x	f
0	1
1	3
2	8
3	5
4	3

Table 2.5 Grouped Frequency Distribution

		Class	Frequency
50 or more to less than 60	⟶	$50 \le x < 60$	1
60 or more to less than 70	⟶	$60 \le x < 70$	3
70 or more to less than 80	⟶	$70 \le x < 80$	8
80 or more to less than 90	⟶	$80 \le x < 90$	5
90 or more to less than 100	⟶	$90 \le x < 100$	2
			19

The stem-and-leaf process can be used to construct a frequency distribution; however, the stem representation is not compatible with all *class widths*. For example, class widths of 3, 4, and 7 are awkward to use. Thus, sometimes it is advantageous to have a separate procedure for constructing a grouped frequency distribution.

Constructing Grouped Frequency Distribution

To illustrate this grouping (or classifying) procedure, let's use a sample of 50 final exam scores taken from last semester's elementary statistics class. Table 2.6 lists the 50 scores.

*Table 2.6 **Statistics Exam Scores**

60	47	82	95	88	72	67	66	68	98
90	77	86	58	64	95	74	72	88	74
77	39	90	63	68	97	70	64	70	70
58	78	89	44	55	85	82	83	72	77
72	86	50	94	92	80	91	75	76	78

PROCEDURE

1. Identify the high score ($H = 98$) and the low score ($L = 39$), and find the range:

   $$\text{range} = H - L = 98 - 39 = 59$$

2. Select a number of classes ($m = 7$) and a class width ($c = 10$) so that the product ($mc = 70$) is a bit larger than the range (range $= 59$).

3. Pick a starting point. This starting point should be a little smaller than the lowest score L. Suppose we start at 35; counting from there by tens (the class width), we get 35, 45, 55, 65, ... , 95, 105. These are called the **class boundaries**. The classes for the data in Table 2.6 are:

35 or more to less than 45	$\longrightarrow$	$35 \leq x < 45$
45 or more to less than 55	$\longrightarrow$	$45 \leq x < 55$
55 or more to less than 65	$\longrightarrow$	$55 \leq x < 65$
65 or more to less than 75	$\longrightarrow$	$65 \leq x < 75$
	$\vdots$	$75 \leq x < 85$
		$85 \leq x < 95$
95 or more to and including 105	$\longrightarrow$	$95 \leq x \leq 105$

NOTES:

1. At a glance you can check the number pattern to determine whether the arithmetic used to form the classes was correct (35, 45, 55, ... , 105).

2. For the interval $35 \leq x < 45$, 35 is the lower class boundary and 45 is the upper class boundary. Observations that fall on the lower class boundary stay in that interval; observations that fall on the upper class boundary go into the next higher interval.

3. The class width is the difference between the upper and lower class boundaries.

4. Many combinations of class widths, numbers of classes, and starting points are possible when classifying data. There is no one best choice. Try a few different combinations, and use good judgment to decide on the one to use.

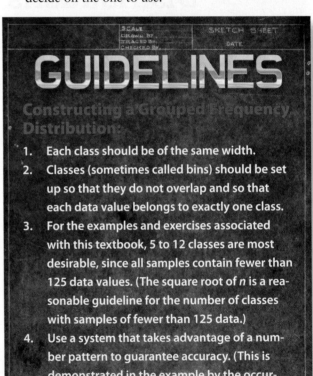

GUIDELINES

Constructing a Grouped Frequency Distribution:

1. Each class should be of the same width.
2. Classes (sometimes called bins) should be set up so that they do not overlap and so that each data value belongs to exactly one class.
3. For the examples and exercises associated with this textbook, 5 to 12 classes are most desirable, since all samples contain fewer than 125 data values. (The square root of n is a reasonable guideline for the number of classes with samples of fewer than 125 data.)
4. Use a system that takes advantage of a number pattern to guarantee accuracy. (This is demonstrated in the example by the occurrence of the 5s in every class boundary.)
5. When it is convenient, an even-numbered class width is often advantageous.

Once the classes are set up, we need to sort the data into those classes. The method used to sort will depend on the current format of the data: If the data are ranked, the frequencies can be counted; if the data are not ranked, we will **tally** the data to find the frequency

Table 2.7 Standard Chart for Frequency Distribution

Class Number	Class Tallies	Boundaries	Frequency
1	||	$35 \leq x < 45$	2
2	||	$45 \leq x < 55$	2
3	|||| ||	$55 \leq x < 65$	7
4	|||| |||| |||	$65 \leq x < 75$	13
5	|||| |||| |	$75 \leq x < 85$	11
6	|||| |||| |	$85 \leq x < 95$	11
7	||||	$95 \leq x \leq 105$	4
			50

Table 2.8 Frequency Distribution with Class Midpoints

Class Number	Class Boundaries	Frequency f	Class Midpoints x
1	$35 \leq x < 45$	2	40
2	$45 \leq x < 55$	2	50
3	$55 \leq x < 65$	7	60
4	$65 \leq x < 75$	13	70
5	$75 \leq x < 85$	11	80
6	$85 \leq x < 95$	11	90
7	$95 \leq x \leq 105$	4	100
		50	

Notes: ✗✗✗✗✗ ✗✗✗✗✗

1. If the data have been ranked (list form, dot-plot, or stem-and-leaf), tallying is unnecessary; just count the data that belong to each class.

2. If the data are not ranked, be careful as you tally.

3. The frequency f for each class is the number of pieces of data that belong in that class.

4. The sum of the frequencies should equal the number of pieces of data n ($n = \Sigma f$). This summation serves as a good check.

numbers. When classifying data, it helps to use a standard chart (see Table 2.7).

Now you can see why it is helpful to have an even class width. An odd class width would have resulted in a class midpoint with an extra digit. (For example, the class 45–54 is 9 wide and the class midpoint is 49.5.)

Each class needs a single numerical value to represent all the data values that fall into that class. The **class midpoint** (sometimes called the *class mark*) is the numerical value that is exactly in the middle of each class. It is found by adding the class boundaries and dividing by 2. Table 2.8 shows an additional column for the class midpoint, x. As a check of your arithmetic, successive class midpoints should be a class width apart, which is 10 in this example (40, 50, 60, ... , 100 is a recognizable pattern).

When we classify data into classes, we lose some information. Only when we have all the raw data do we know the exact values that were actually observed for each class. For example, we put a 47 and a 50 into class 2, with class boundaries of 45 and 55. Once they are placed in the class, their values are lost to us and we use the class midpoint, 50, as their representative value.

Histograms

One way statisticians visually represent frequency counts of a quantitative variable is to use a bar graph called a **histogram**. A histogram is made up of three components:

1. A title, which identifies the population or sample of concern.

2. A vertical scale, which identifies the frequencies in the various classes.

3. A horizontal scale, which identifies the variable x. Values for the class boundaries or class midpoints may be labeled along the x-axis. Use whichever method of labeling the axis best presents the variable.

Class midpoint (class mark) The numerical value that is exactly in the middle of each class.

Histogram A bar graph that represents a frequency distribution of a quantitative variable.

The frequency distribution from Table 2.8 appears in histogram form in Figure 2.10.

Sometimes the **relative frequency** of a value is important. The relative frequency is a proportional measure of the frequency for an occurrence. It is found by dividing the class frequency by the total number of observations. Relative frequency can be expressed as a common fraction, in decimal form, or as a percentage. In our example about the exam scores, the frequency associated with the third class (55–65) is 7. The relative frequency for the third class is $\frac{7}{50}$, or 0.14, or 14%. Relative frequencies are often useful in a presentation because nearly everybody understands fractional parts when they are expressed as percentages. Relative fre-

quencies are particularly useful when comparing the frequency distributions of two different size sets of data. Figure 2.11 is a **relative frequency histogram** of the sample of the 50 final exam scores from Table 2.8.

A stem-and-leaf display contains all the information needed to create a histogram, for example, Figure 2.5B (page 27). In Figure 2.12A the stem-and-leaf has been rotated 90° and labels have been added to show its relationship to a histogram. Figure 2.12B shows the same set of data as a completed histogram.

Histograms are valuable tools. For example, the histogram of a sample should have a distribution shape very similar to that of the population from which the sample was drawn. If the reader of a histogram is at all

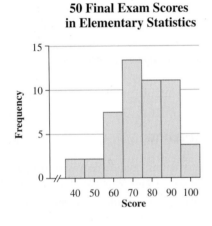

Figure 2.10 Frequency Histogram

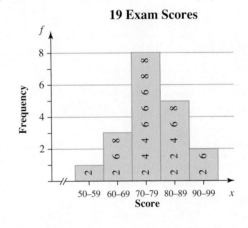

Figure 2.12A Modified Stem-and-Leaf Display

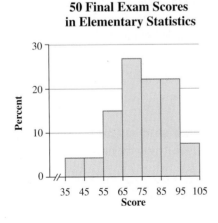

Figure 2.11 Relative Frequency Histogram

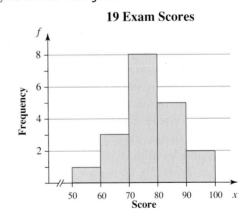

Figure 2.12B Histogram

Figure 2.13 **Shapes of Histograms**

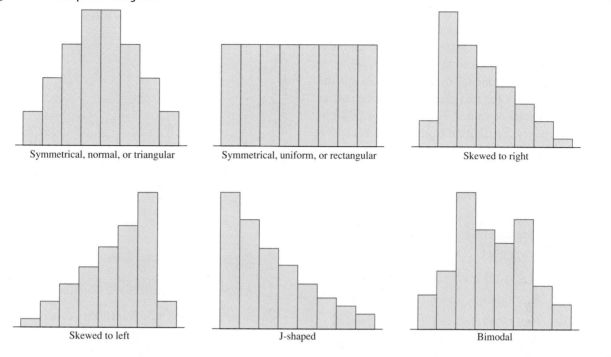

Symmetrical, normal, or triangular

Symmetrical, uniform, or rectangular

Skewed to right

Skewed to left

J-shaped

Bimodal

familiar with the variable involved, he or she will usually be able to interpret several important facts. Figure 2.13 presents histograms with descriptive labels resulting from their geometric shape.

Briefly, the terms used to describe histograms are as follows:

Symmetrical: Both sides of this distribution are identical (halves are mirror images).

Normal (triangular): A symmetrical distribution is mounded up about the mean and becomes sparse at the extremes. (Additional properties are discussed later.)

Uniform (rectangular): Every value appears with equal frequency.

Skewed: One tail is stretched out longer than the other. The direction of skewness is on the side of the longer tail.

J-shaped: There is no tail on the side of the class with the highest frequency.

Bimodal: The two most populous classes are separated by one or more classes. This situation often implies that two populations are being sampled. (See Figure 2.7, page 28.)

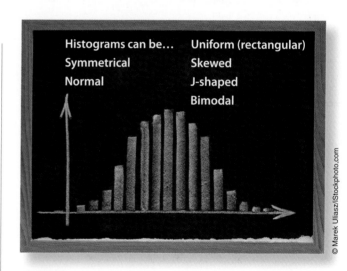

Histograms can be... Uniform (rectangular)
Symmetrical   Skewed
Normal   J-shaped
                   Bimodal

NOTES:

1. The mode is the value of the data that occurs with the greatest frequency. (Mode will be discussed in Objective 2.3)

2. The modal class is the class with the highest frequency.

3. A bimodal distribution has two high-frequency classes separated by classes with lower frequencies. It is not necessary for the two high frequencies to be the same.

Cumulative Frequency Distribution and Ogives

Another way to express a frequency distribution is to use a **cumulative frequency distribution** to pair cumulative frequencies with values of the variable.

The cumulative frequency for any given class is the sum of the frequency for that class and the frequencies of all classes of smaller values. Table 2.9 shows the cumulative frequency distribution from Table 2.8.

Table 2.9 **Using Frequency Distribution to Form a Cumulative Frequency Distribution**

Class Number	Class Boundaries	Frequency f	Cumulative Frequency
1	$35 \leq x < 45$	2	2 (2)
2	$45 \leq x < 55$	2	4 (2 + 2)
3	$55 \leq x < 65$	7	11 (7 + 4)
4	$65 \leq x < 75$	13	24 (13 + 11)
5	$75 \leq x < 85$	11	35 (11 + 24)
6	$85 \leq x < 95$	11	46 (11 + 35)
7	$95 \leq x \leq 105$	4	50 (4 + 46)

The same information can be presented by using a *cumulative relative frequency distribution* (see Table 2.10). This combines the cumulative frequency and the relative frequency ideas.

Cumulative distributions can be displayed graphically using an ogive. Whereas a histogram is a bar graph, an **ogive** is a line graph of a cumulative frequency or cumulative relative frequency distribution. An ogive has the following three components:

Figure 2.14 **Ogive**

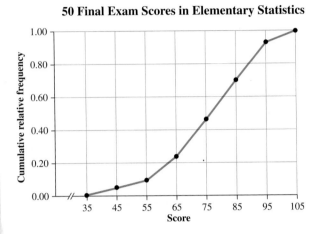

1. A title, which identifies the population or sample.

2. A vertical scale, which identifies either the cumulative frequencies or the cumulative relative frequencies. (Figure 2.14 shows an ogive with cumulative relative frequencies.)

3. A horizontal scale, which identifies the upper class boundaries. Until the upper boundary of a class has been reached, you cannot be sure you have accumulated all the data in that class. Therefore, the horizontal scale for an ogive is always based on the upper class boundaries.

The ogive can be used to make percentage statements about numerical data much like a Pareto diagram does for attribute data. For example, suppose we want to know what percent of the final exam scores were not passing if scores of 65 or greater are considered passing.

Table 2.10 **Cumulative Relative Frequency Distribution**

Class Number	Class Boundaries	Cumulative Relative Frequency	Cumulative frequencies are for the interval 35 up to the upper boundary of that class.
1	$35 \leq x < 45$	2/50, or 0.04	⟵ from 35 up to less than 45
2	$45 \leq x < 55$	4/50, or 0.08	⟵ from 35 up to less than 55
3	$55 \leq x < 65$	11/50, or 0.22	⟵ from 35 up to less than 65
4	$65 \leq x < 75$	24/50, or 0.48	
5	$75 \leq x < 85$	35/50, or 0.70	
6	$85 \leq x < 95$	46/50, or 0.92	
7	$95 \leq x < 105$	50/50, or 1.00	⟵ from 35 up to less than 105

Cumulative frequency distribution A frequency distribution that pairs cumulative frequencies with values of the variable.

Ogive (pronounced ō´jĭv) A line graph of a cumulative frequency or cumulative relative frequency distribution.

Following vertically from 65 on the horizontal scale to the ogive line and reading from the vertical scale, approximately 22% of the final exam scores were not passing grades.

NOTE: Every ogive starts on the left with a relative frequency of zero at the lower class boundary of the first class and ends on the right with a cumulative relative frequency of 100% at the upper class boundary of the last class.

2.3 Measures of Central Tendency

MEASURES OF CENTRAL TENDENCY
ARE NUMERICAL VALUES THAT LOCATE, IN SOME SENSE, THE CENTER OF A SET OF DATA.

The term *average* is often associated with all measures of central tendency, including the mean, median, mode, and midrange.

Finding the Mean

The **mean,** also called the **arithmetic mean,** is the average with which you are probably most familiar. The sample mean is represented by $\bar{x}$ (read "*x*-bar" or "sample mean"). The mean is found by adding all the values of the variable *x* (this sum of *x* values is symbolized Σx) and dividing the sum by the number of these values, *n* (the "sample size"). We express this in formula form as

$$\text{sample mean:} \quad \text{x-bar} = \frac{sum\ of\ all\ x}{number\ of\ x}$$

$$\bar{x} = \frac{\Sigma x}{n} \tag{2.1}$$

NOTE: The population mean, μ (lowercase mu, Greek alphabet), is the mean of all *x* values for the entire population.

> **FYI:** *The mean is the middle point by weight.*

Let's work on finding the mean using a set of data consisting of the five values 6, 3, 8, 6, and 4. To find the mean, we'll first use formula (2.1). Doing that, we find

$$\bar{x} = \frac{\Sigma x}{n} = \frac{6 + 3 + 8 + 6 + 4}{5} = \frac{27}{5} = 5.4$$

A physical representation of the mean can be constructed by thinking of a number line balanced on a fulcrum. A weight is placed on a number on the line corresponding to each number in the sample of our example above. In Figure 2.15 there is one weight each on the 3, 8, and 4 and two weights on the 6, since there are two 6s in the sample. The mean is the value that balances the weights on the number line—in this case, 5.4.

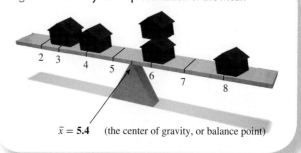

Figure 2.15 **Physical Representation of the Mean**

$\bar{x} = \textbf{5.4}$ (the center of gravity, or balance point)

© Onur Dongel/iStockphoto.com

Finding the Median

The value of the data that occupies the middle position when the data are ranked in order according to size is called the **median.** The sample median is represented by $\tilde{x}$ (read "*x*-tilde" or "sample median"). The population median, *M* (uppercase mu in the Greek alphabet), is the data value in the middle position of the entire ranked population.

Finding the median involves three basic steps. First, you need to rank the data. Then you determine the depth of the median. The **depth** (number of positions from either end), or position, of the median is determined by the formula

$$depth\ of\ median = \frac{sample\ size + 1}{2}$$

$$d(\tilde{x}) = \frac{n + 1}{2} \tag{2.2}$$

The median's depth (or position) is found by adding the position numbers of the smallest data (1) and the largest data (*n*) and dividing the sum by 2 (*n* is the number of pieces of data). Finally, you must determine the value of the median. To do this, you count the ranked data, locating the data in the $d(\tilde{x})$th position. The median will be the same regardless of which end of the ranked data (high or low) you count from. In fact, counting from both ends will serve as an excellent check.

The following two examples demonstrate this procedure as it applies to both odd-numbered and even-numbered sets of data.

> **FYI:** *The value of* $d(\tilde{x})$ *is the depth of the median, NOT the value of the median* $\tilde{x}$.

MEDIAN FOR ODD *n*

Let's practice finding the median by first working with an odd number *n*. We'll find the median for the set of data {6, 3, 8, 5, 3}. First, we rank the data. In this case, the data, ranked in order of size, are 3, 3, 5, 6, and 8. Next, we'll find the depth of the median: $d(\tilde{x}) = \frac{n+1}{2} = \frac{5+1}{2} = 3$ (the "3rd" position). We can now identify the median. The median is the third number from either end in the ranked data, or $\tilde{x} = 5$.

Notice that the median essentially separates the ranked set of data into two subsets of equal size (see Figure 2.16).

As in the above example, when *n* is odd, the depth of the median, $d(\tilde{x})$, will always be an integer. When *n* is even, however, the depth of the median, $d(\tilde{x})$, will always be a half-number, as shown next.

The median is the middle point by count.

MEDIAN FOR EVEN *n*

We can now compare the process we just completed with one in which we have an even number of points in our data set. Let's find the median of the sample 9, 6, 7, 9, 10, 8.

As before, we'll first rank the data by size. In this case, we have 6, 7, 8, 9, 9, and 10.

The depth of the median now is: $d(\tilde{x}) = \frac{n+1}{2} = \frac{6+1}{2} = 3.5$ (the "3.5th" position).

Finally, we can identify the median. The median is halfway between the third and fourth data values. To find the number halfway between any two values, add the two values together and divide the sum by 2. In this case, add the third value (8) and the fourth value (9) and then divide the sum (17) by 2. The median is $\tilde{x} = \frac{8+9}{2} = 8.5$, a number halfway between the "middle" two numbers (see Figure 2.17). Notice that the median again separates the ranked set of data into two subsets of equal size.

Figure 2.16 **Median of {3, 3, 5, 6, 8}**

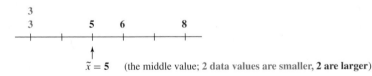

$\tilde{x} = 5$ (the middle value; **2 data values are smaller, 2 are larger**)

Figure 2.17 **Median of {6, 7, 8, 9, 9, 10}**

$\tilde{x} = 8.5$ (value in middle; **3 data values are smaller, 3 are larger**)

Finding the Mode

The **mode** is the value of x that occurs most frequently. In the set of data we used to find the median for odd n, {3, 3, 5, 6, 8}, the mode is 3 (see Figure 2.18).

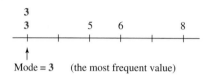

Figure 2.18 **Mode of {3, 3, 5, 6, 8}**

Mode = 3   (the most frequent value)

In the sample 6, 7, 8, 9, 9, 10, the mode is 9. In this sample, only the 9 occurs more than once; in our earlier data set {6, 3, 8, 5, 3}, only the 3 occurs more than once. If two or more values in a sample are tied for the highest frequency (number of occurrences), we say there is *no mode*. For example, in the sample 3, 3, 4, 5, 5, 7, the 3 and the 5 appear an equal number of times. There is no one value that appears most often; thus, this sample has no mode.

Finding the Midrange

The number exactly midway between a lowest data value L and a highest data value H is called the **midrange**. To find the midrange, average the low and the high values:

$$midrange = \frac{low\ value + high\ value}{2}$$

$$midrange = \frac{L + H}{2} \tag{2.3}$$

For the set of data {3, 3, 5, 6, 8}, $L = 3$ and $H = 8$ (see Figure 2.19).

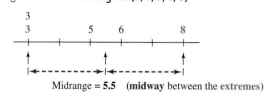

Figure 2.19 **Midrange of {3, 3, 5, 6, 8}**

Midrange = 5.5   (**midway** between the **extremes**)

Therefore, the midrange is

$$midrange = \frac{L + H}{2}$$

$$= \frac{3 + 8}{2} = 5.5$$

The four measures of central tendency represent four different methods of describing the middle. These four values may be the same, but more likely they will be different.

For the sample data set {6, 7, 8, 9, 9, 10}, the mean $\bar{x}$ is 8.2, the median $\tilde{x}$ is 8.5, the mode is 9, and the midrange is 8. Their relationship to one another and to the data is shown in Figure 2.20.

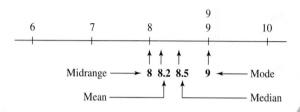

Figure 2.20 **Measures of Central Tendency for {6, 7, 8, 9, 9, 10}**

Midrange ⟶ 8 8.2 8.5   9 ⟵ Mode

Mean   Median

ROUND-OFF RULE

When rounding off an answer, let's agree to keep one more decimal place in our answer than was present in the original information. To avoid round-off buildup, round off only the final answer, not the intermediate steps. That is, avoid using a rounded value to do further calculations. In our previous examples, the data were composed of whole numbers; therefore, those answers that have decimal values should be rounded to the nearest tenth.

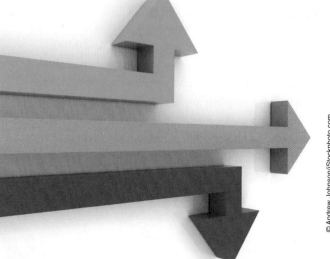

"Average" Means Different Things

When it comes to convenience, few things can match that wonderful mathematical device called *averaging*. With an **average** you can take a fistful of figures on any subject and compute one figure that will represent the whole fistful.

But there is one thing to remember. There are several kinds of measures ordinarily known as averages, and each gives a different picture of the figures it is called on to represent. Take an example: Table 2.11 contains the annual incomes of ten families.

*Table 2.11 Annual Income of 10 Families

| $54,000 | $39,000 | $37,500 | $36,750 | $35,250 |
| $31,500 | $31,500 | $31,500 | $31,500 | $25,500 |

What would this group's "typical" income be? Averaging would provide the answer, so let's compute the typical income by the simpler and more frequently used kinds of averaging.

- **The arithmetic mean.** It is the most common form of average, obtained by adding items in the series and then dividing by the number of items: $35,400. The mean is representative of the series in the sense that the sum of the amounts by which the higher figures exceed the mean is exactly the same as the sum of the amounts by which the lower figures fall short of the mean.

- **The median.** As you may have observed, six families earn less than the mean, four earn more. You might very well wish to represent this varied group by the income of the family that is right smack dab in the middle of the whole bunch. The median works out to $33,375.

- **The midrange.** Another number that might be used to represent the group is the midrange, computed by calculating the figure that lies halfway between the highest and lowest incomes: $39,750.

- **The mode.** So far we've seen three kinds of averages, and not one family actually has an income matching any of them. Say you want to represent the group by stating the income that occurs most frequently. That is called a mode. $31,500 would be the modal income.

Four different averages, each valid, correct, and informative in its way. But how they differ!

Arithmetic Mean	Median	Midrange	Mode
$35,400	$33,375	$39,750	$31,500

And they would differ still more if just one family in the group were a millionaire—or one were jobless!

So there are three lessons: First, when you see or hear an average, find out which average it is. Then you'll know what kind of picture you are being given. Second, think about the figures being averaged so you can judge whether the average used is appropriate. And third, don't assume that a literal mathematical quantification is intended every time somebody says "average." It isn't. All of us often say "the average person" with no thought of implying a mean, median, or mode. All we intend to convey is the idea of other people who are in many ways a great deal like the rest of us.

SOURCE: Reprinted by permission from *Changing Times* magazine (March 1980 issue). Copyright by The Kiplinger Washington Editors.

BASIC TERMS - TAKE TWO

Mean (arithmetic mean) The mean, also called the arithmetic mean, is the average with which you are probably most familiar. The sample mean is represented by $\bar{x}$ (read "x-bar" or "sample mean"). The mean is found by adding all the values of the variable x (this sum of x values is symbolized Σx) and dividing the sum by the number of these values, n (the "sample size").

Median The value of the data that occupies the middle position when the data are ranked in order according to size. The sample median is represented by $\tilde{x}$ (read "x-tilde" or "sample median").

Mode The mode is the value of x that occurs most frequently.

Midrange The number exactly midway between the lowest-valued data L and the highest-valued data H.

Range The difference in value between the highest data value (H) and the lowest data value (L).

Deviation from the mean A deviation from the mean, $x - \bar{x}$, is the difference between the value of x and the mean $\bar{x}$.

Sample variance The sample variance, s^2, is the mean of the squared deviations.

Sample standard deviation The standard deviation of a sample, s, is the positive square root of the variance.

2.4 Measures of Dispersion

HAVING LOCATED THE "MIDDLE" WITH THE MEASURES OF CENTRAL TENDENCY, OUR SEARCH FOR INFORMATION FROM DATA SETS NOW TURNS TO THE MEASURES OF DISPERSION (SPREAD).

The measures of dispersion include the range, variance, and standard deviation. These numerical values describe the amount of spread, or variability, that is found among the data: Closely grouped data have relatively small values, and more widely spread out data have larger values. The closest possible grouping occurs when the data have no dispersion (all data are the same value); in this situation, the measure of dispersion will be zero. There is no limit to how widely spread out the data can be; therefore, measures of dispersion can be very large. The simplest measure of dispersion is **range**, which is the difference in value between the highest data value (H) and the lowest data value (L):

$$range = high\,value - low\,value$$

$$range = H - L \tag{2.4}$$

The sample 3, 3, 5, 6, 8 has a range of $H - L = 8 - 3 = 5$. The range of 5 tells us that these data all fall within a 5-unit interval (see Figure 2.21).

The other measures of dispersion to be studied in this chapter are measures of dispersion about the mean.

Figure 2.21 **Range of {3, 3, 5, 6, 8}**

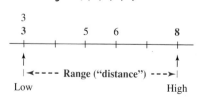

To develop a measure of dispersion about the mean, let's first answer the following question: How far is each x from the mean? The difference between the value of x and the mean $\bar{x}$, or $x - \bar{x}$, is called a **deviation from the mean.** Each individual value x deviates from the mean by an amount equal to $(x - \bar{x})$. This deviation $(x - \bar{x})$ is zero when x is equal to the mean $\bar{x}$. The deviation $(x - \bar{x})$ is positive when x is larger than $\bar{x}$ and negative when x is smaller than $\bar{x}$.

When you square the deviations and take an average of those, you get something called the **sample variance**, s^2. It is calculated using $n - 1$ as the divisor:

sample variance:

$$s\text{-squared} = \frac{sum\ of\ (deviations\ squared)}{number - 1}$$

$$s^2 = \frac{\Sigma(x - \bar{x})^2}{n - 1} \qquad (2.5)$$

where n is the sample size—that is, the number of data values in the sample.

The variance of the sample 6, 3, 8, 5, 3 is calculated in Table 2.12 using formula (2.5).

NOTES:

1. The sum of all the x values is used to find $\bar{x}$.

2. The sum of the deviations, $\Sigma(x - \bar{x})$, is always zero, provided the exact value of $\bar{x}$ is used. Use this fact as a check in your calculations, as was done in Table 2.12 (denoted by ✓).

3. If a rounded value of $\bar{x}$ is used, then $\Sigma(x - \bar{x})$ will not always be exactly zero. It will, however, be reasonably close to zero.

4. The sum of the squared deviations is found by squaring each deviation and then adding the squared values.

To graphically demonstrate what variances of data sets are telling us, consider a second set of data: {1, 3, 5, 6, 10}. Note that the data values are more dispersed than the data in Table 2.12. Accordingly, its calculated variance is larger at $s^2 = 11.5$. An illustrative side-by-side graphical comparison of these two samples and their variances is shown in Figure 2.22.

Sample Standard Deviation

Variance is instrumental in the calculation of the **standard deviation of a sample, s,** which is the positive square root of the variance:

sample standard deviation:
$$s = square\ root\ of\ sample\ variance$$

$$s = \sqrt{s^2} \qquad (2.6)$$

For the samples shown in Figure 2.22, the standard deviations are $\sqrt{4.5}$ or 2.1, and $\sqrt{11.5}$ or 3.4.

NOTE: The numerator for the sample variance, $\Sigma(x - \bar{x})^2$, is often called the *sum of squares for x* and

Table 2.12 **Calculating Variance Using Formula (2.5)**

Step 1. Find Σx	Step 2. Find $\bar{x}$	Step 3. Find each $x - \bar{x}$	Step 4. Find $\Sigma(x - \bar{x})^2$	Step 5. Find s^2
6	$\bar{x} = \dfrac{\Sigma x}{n}$	$6 - 5 = \ \ 1$	$(1)^2 = 1$	$s^2 = \dfrac{\Sigma(x - \bar{x})^2}{n - 1}$
3		$3 - 5 = -2$	$(-2)^2 = 4$	
8		$8 - 5 = \ \ 3$	$(3)^2 = 9$	
5	$\bar{x} = \dfrac{25}{5}$	$5 - 5 = \ \ 0$	$(0)^2 = 0$	$s^2 = \dfrac{18}{4}$
3		$3 - 5 = -2$	$(-2)^2 = 4$	
$\Sigma x = 25$	$\bar{x} = 5$	$\Sigma(x - \bar{x}) = \ \ 0$ ✓	$\Sigma(x - \bar{x})^2 = 18$	$s^2 = 4.5$

Figure 2.22 **Comparison of Data**

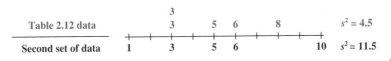

symbolized by SS(x). Thus, formula (2.5) can be expressed as

sample variance: $\quad s^2 = \dfrac{SS(x)}{n-1} \qquad$ (2.7)

The formulas for variance can be modified into other forms for easier use in various situations.

The arithmetic becomes more complicated when the mean contains nonzero digits to the right of the decimal point. However, the **sum of squares for x,** the numerator of formula (2.5), can be rewritten so that $\bar{x}$ is not included:

sum of squares: $\quad SS(x) = \Sigma x^2 - \dfrac{(\Sigma x)^2}{n} \qquad$ (2.8)

Combining formulas (2.7) and (2.8) yields the "shortcut formula" for sample variance:

$$s\text{-}squared = \frac{(sum\ of\ x^2) - \left[\dfrac{(sum\ of\ x)^2}{number}\right]}{number - 1}$$

sample variance: $\quad s^2 = \dfrac{\Sigma x^2 - \dfrac{(\Sigma x)^2}{n}}{n-1} \qquad$ (2.9)

Formulas (2.8) and (2.9) are called "shortcuts" because they bypass the calculation of $\bar{x}$. The computations for SS(x), s^2, and s using formulas (2.8), (2.9), and (2.6) are performed as shown in Table 2.13.

The unit of measure for the standard deviation is the same as the unit of measure for the data. For example, if our data are in pounds, then the standard deviation s will also be in pounds. The unit of measure for variance might then be thought of as *units squared*. In our example of pounds, this would be *pounds squared*. As you can see, the unit has very little meaning.

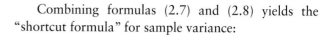

2.5 Measures of Position

MEASURES OF POSITION ARE USED TO DESCRIBE THE POSITION A SPECIFIC DATA VALUE POSSESSES IN RELATION TO THE REST OF THE DATA.

Table 2.13 **Calculating Standard Deviation Using the Shortcut Method**

Step 1. Find Σx	Step 2. Find Σx^2	Step 3. Find SS(x)	Step 4. Find s^2	Step 5. Find s
6	$6^2 = 36$	$SS(x) = \Sigma x^2 - \dfrac{(\Sigma x)^2}{n}$	$s^2 = \dfrac{\Sigma x^2 - \dfrac{(\Sigma x)^2}{n}}{n-1}$	$s = \sqrt{s^2}$
3	$3^2 = 9$			$s = \sqrt{5.7}$
8	$8^2 = 64$	$SS(x) = 138 - \dfrac{(24)^2}{5}$		$s = 2.4$
5	$5^2 = 25$		$s^2 = \dfrac{22.8}{4}$	
2	$2^2 = 4$	$SS(x) = 138 - 115.2$		
$\Sigma x = 24$	$\Sigma x^2 = 138$	$SS(x) = 22.8$	$s^2 = 5.7$	

Quartiles and percentiles are two of the most popular measures of position. Other measures of position include midquartiles, 5-number summaries, and standard scores, or z-scores.

Quartiles

Quartiles are values of the variable that divide the ranked data into quarters; each set of data has three quartiles. The *first quartile*, Q_1, is a number such that at most 25% of the data are smaller in value than Q_1 and at most 75% are larger. The *second quartile* is the median. The *third quartile*, Q_3, is a number such that at most 75% of the data are smaller in value than Q_3 and at most 25% are larger (see Figure 2.23).

The procedure for determining the values of the quartiles is the same as that for **percentiles,** which are the values of the variable that divide a set of ranked data into 100 equal subsets; each set of data has 99 percentiles (see Figure 2.24). The *k*th percentile, P_k, is a value such that at most *k*% of the data are smaller in value than P_k and at most $(100 - k)$% of the data are larger (see Figure 2.25).

NOTES:

1. The first quartile and the 25th percentile are the same; that is, $Q_1 = P_{25}$. Also, $Q_3 = P_{75}$.

2. The median, the second quartile, and the 50th percentile are all the same: $\tilde{x} = Q_2 = P_{50}$. Therefore, when asked to find P_{50} or Q_2, use the procedure for finding the median.

Percentiles

The procedure for determining the value of any *k*th percentile (or quartile) involves four basic steps as outlined in Figure 2.26.

Using the sample of 50 elementary statistics final exam scores listed in Table 2.14, find the first quartile Q_1, the 58th percentile P_{58}, and the third quartile Q_3.

> **Quartiles** Values of the variable that divide the ranked data into quarters; each set of data has three quartiles.
>
> **Percentiles** Values of the variable that divide a set of ranked data into 100 equal subsets; each set of data has 99 percentiles.

Figure 2.23 **Quartiles**

Ranked data, increasing order

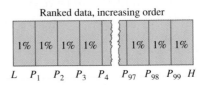

Figure 2.24 **Percentiles**

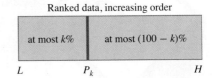

Figure 2.25 ***k*th Percentile**

Ranked data, increasing order

| at most *k*% | at most $(100 - k)$% |

L P_k H

Table 2.14 **Raw Scores for Elementary Statistics Exam**

60	47	82	95	88	72	67	66	68	98
90	77	86	58	64	95	74	72	88	74
77	39	90	63	68	97	70	64	70	70
58	78	89	44	55	85	82	83	72	77
72	86	50	94	92	80	91	75	76	78

Figure 2.26 **Finding P_k Procedure**

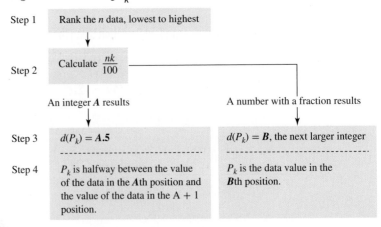

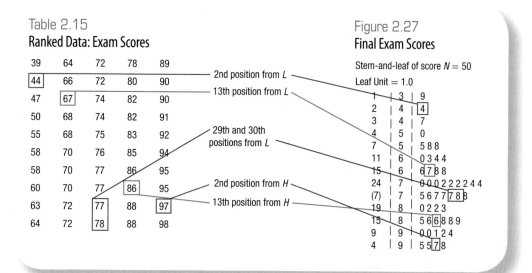

Table 2.15
Ranked Data: Exam Scores

39	64	72	78	89
44	66	72	80	90
47	67	74	82	90
50	68	74	82	91
55	68	75	83	92
58	70	76	85	94
58	70	77	86	95
60	70	77	86	95
63	72	77	88	97
64	72	78	88	98

2nd position from L
13th position from L
29th and 30th positions from L
2nd position from H
13th position from H

Figure 2.27
Final Exam Scores

Stem-and-leaf of score $N = 50$

Leaf Unit = 1.0

```
  1   | 3 | 9
  2   | 4 | 4
  3   | 4 | 7
  4   | 5 | 0
  7   | 5 | 5 8 8
 11   | 6 | 0 3 4 4
 15   | 6 | 6 7 8 8
 24   | 7 | 0 0 0 2 2 2 2 4 4
 (7)  | 7 | 5 6 7 7 7 8 8
 19   | 8 | 0 2 2 3
 15   | 8 | 5 6 6 8 8 9
  9   | 9 | 0 0 1 2 4
  4   | 9 | 5 5 7 8
```

PROCEDURE

Step 1 Rank the data: A ranked list may be formulated (see Table 2.15), or a graphic display showing the ranked data may be used. The dotplot and the stem-and-leaf are handy for this purpose. The stem-and-leaf is especially helpful since it gives depth numbers counted from both extremes when it is computer generated (see Figure 2.27). Step 1 is the same for all three statistics.

Find Q_1:

Step 2 Find $\frac{nk}{100}$: $\frac{nk}{100} = \frac{(50)(25)}{100} = 12.5$
($n = 50$ and $k = 25$, since $Q_1 = P_{25}$.)

Step 3 Find the depth of Q_1: $d(Q_1) = 13$
(Since 12.5 contains a fraction, B is the next larger integer, 13.)

Step 4 Find Q_1: Q_1 is the 13th value, counting from L (see Table 2.15 or Figure 2.27), $Q_1 = 67$

Find P_{58}:

Step 2 Find $\frac{nk}{100}$: $\frac{nk}{100} = \frac{(50)(58)}{100} = 29$
($n = 50$ and $k = 58$ for P_{58}.)

Step 3 Find the depth of P_{58}: $d(P_{58}) = 29.5$
(Since $A = 29$, an integer, add 0.5 and use 29.5.)

Step 4 Find P_{58}: P_{58} is the value halfway between the values of the 29th and the 30th pieces of data, counting from L (see Table 2.15 or Figure 2.27), so

$$P_{58} = \frac{77 + 78}{2} = 77.5$$

FYI: $d(P_k) = $ *depth or location of the* k^{th} *percentile.*

Optional technique: When k is greater than 50, subtract k from 100 and use $(100 - k)$ in place of k in Step 2. The depth is then counted from the largest data value H.

Find Q_3 using the optional technique:

Step 2 Find $\frac{nk}{100}$: $\frac{nk}{100} = \frac{(50)(25)}{100} = 12.5$
($n = 50$ and $k = 75$, since $Q_3 = P_{75}$, and $k > 50$; use $100 - k = 100 - 75 = 25$.)

Step 3 Find the depth of Q_3 from H: $d(Q_3) = 13$

Step 4 Find Q_3: Q_3 is the 13th value, counting from H (see Table 2.15 or Figure 2.27), $Q_3 = 86$

Therefore, it can be stated that "at most 75% of the exam grades are smaller in value than 86." This is also equivalent to stating that "at most 25% of the exam grades are larger in value than 86."

Therefore, it can be stated that "at most 58% of the exam grades are smaller in value than 77.5." This is also equivalent to stating that "at most 42% of the exam grades are larger in value than 77.5."

NOTE: An ogive of these grades would graphically determine these same percentiles, without the use of formulas.

Other Measures of Position

Let's now examine three other measures of position: midquartile, 5-number summary, and standard scores.

MIDQUARTILES

Using the fundamental calculations of quartiles, you can now calculate the measure of central tendency known as the **midquartile**, or the numerical value midway between the first quartile and the third quartile.

$$midquartile = \frac{Q_1 + Q_3}{2} \qquad (2.10)$$

So, to find the midquartile for the set of 50 exam scores given in our exam score example, you would simply add 67 to 86 and divide by 2.

$Q_1 = 67$ and $Q_3 = 86$, thus,

$$midquartile = \frac{Q_1 + Q_3}{2} = \frac{67 + 86}{2} = \mathbf{76.5}$$

The median, the midrange, and the midquartile are not necessarily the same value. Each is the middle value, but by different definitions of "middle." Figure 2.28 summarizes the relationship of these three statistics as applied to our set of 50 exam scores.

5-NUMBER SUMMARY

Another measure of position based on quartiles and percentiles is the **5-number summary**. Not only is the 5-number summary very effective in describing a set of data, it is easy information to obtain and is very informative to the reader.

The 5-number summary is composed of:

1. L, the smallest value in the data set,

2. Q_1, the first quartile (also called P_{25}, the 25th percentile),

3. $\tilde{x}$, the median,

4. Q_3, the third quartile (also called P_{75}, the 75th percentile), and

5. H, the largest value in the data set.

The 5-number summary for our set of 50 exam scores is

39	67	75.5	86	98
L	Q_1	$\tilde{x}$	Q_3	H

Notice that these five numerical values divide the set of data into four subsets, with one-quarter of the data in each subset. From the 5-number summary we can observe how much the data are spread out in each of the quarters. We can now define an additional measure of dispersion. The **interquartile range** is the difference between the first and third quartiles. It is the range of the middle 50% of the data. The 5-number summary makes it very easy to see the interquartile range.

The 5-number summary is even more informative when it is displayed on a diagram drawn to scale. A

Midquartile The numerical value midway between the first quartile and the third quartile.

5-number summary The presentation of 5 numbers that give a statistical summary of a data set: the smallest value in the data set, the first quartile, the median, the third quartile, and the largest value in the data set.

Interquartile range The difference between the first and third quartiles. It is the range of the middle 50% of the data.

Figure 2.28 **Final Exam Scores**

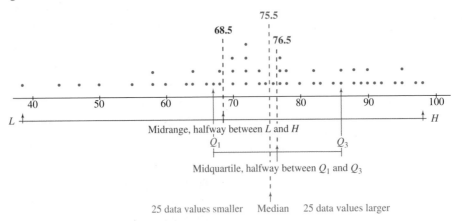

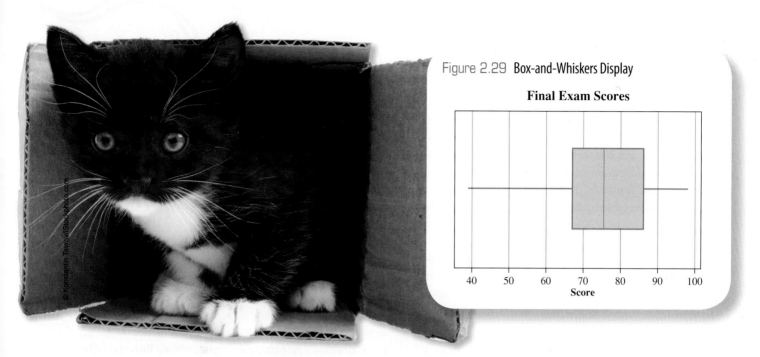

Figure 2.29 Box-and-Whiskers Display

Final Exam Scores

computer-generated graphic display that accomplishes this is known as the **box-and-whiskers display.** In this graphic representation of the 5-number summary, the five numerical values (smallest, first quartile, median, third quartile, and largest) are located on a scale, either vertical or horizontal. The box is used to depict the middle half of the data that lie between the two quartiles. The whiskers are line segments used to depict the other half of the data: One line segment represents the quarter of the data that are smaller in value than the first quartile, and a second line segment represents the quarter of the data that are larger in value than the third quartile.

Figure 2.29 is a box-and-whiskers display of the 50 exam scores.

STANDARD SCORES (Z-SCORES)

So far, we've examined *general* measures of position, but sometimes it is necessary to measure the position of a *specific* value in terms of the mean and standard deviation. In those cases, the *standard score,* commonly called the *z-score,* is used. The **standard score** (or *z*-score) is the position a particular value of *x* has

> **Box-and-whiskers display** A graphic representation of the 5-number summary.
>
> **Standard score or z-score** The position a particular value of *x* has relative to the mean, measured in standard deviations.

relative to the mean, measured in standard deviations. The *z*-score is found by the formula:

$$z = \frac{value - mean}{st.\ dev.} = \frac{x - \bar{x}}{s} \qquad (2.11)$$

Let's apply this formula to finding the standard scores for (a) 92 and (b) 72 with respect to a sample of exam grades that has a mean score of 74.92 and a standard deviation of 14.20.

SOLUTION

a. $x = 92, \bar{x} = 74.92, s = 14.20.$

Thus, $z = \frac{x - \bar{x}}{s} = \frac{92 - 74.92}{14.20} = \frac{17.08}{14.20} = \mathbf{1.20.}$

b. $x = 72, \bar{x} = 74.92, s = 14.20.$

Thus, $z = \frac{x - \bar{x}}{s} = \frac{72 - 74.92}{14.20} = \frac{-2.92}{14.20} = \mathbf{-0.21.}$

This means that the score 92 is approximately one and one-fifth standard deviations above the mean, while the score 72 is approximately one-fifth of a standard deviation below the mean.

NOTES:

1. Typically, the calculated value of *z* is rounded to the nearest hundredth.

2. *z*-scores typically range in value from approximately −3.00 to +3.00.

Because the *z*-score is a measure of relative position with respect to the mean, it can be used to help us compare two raw scores that come from separate populations. For example, suppose you want to compare a grade you received on a test with a friend's grade on a comparable exam in her course. You received a raw score of 45 points; she got 72 points. Is her grade better? We need more information before we can draw a conclusion. Suppose the mean on the exam you took was 38 and the mean on her exam was 65. Your grades are both 7 points above the mean, but we still can't draw a definite conclusion. The standard deviation on the exam you took was 7 points, and it was 14 points on your friend's exam. This means that your score is one (1) standard deviation above the mean ($z = 1.0$), whereas your friend's grade is only one-half of a standard deviation above the mean ($z = 0.5$). Since your score has the "better" relative position, you conclude that your score is slightly better than your friend's score. (Again, this is speaking from a relative point of view.)

Empirical rule If a variable is normally distributed, then: within one standard deviation of the mean there will be approximately 68% of the data; within two standard deviations of the mean there will be approximately 95% of the data; and within three standard deviations of the mean there will be approximately 99.7% of the data.

2.6 Interpreting and Understanding Standard Deviation

STANDARD DEVIATION IS A MEASURE OF VARIATION (DISPERSION) IN THE DATA.

It has been defined as a value calculated with the use of formulas. Even so, you may be wondering what it really is and how it relates to the data. It is a kind of yardstick by which we can compare the variability of one set of data with another. This particular "measure" can be understood further by examining two statements that tell us how the standard deviation relates to the data: the *empirical rule* and *Chebyshev's theorem*.

The Empirical Rule and Testing for Normality

The **empirical rule** states that if a variable is normally distributed, then: within one standard deviation of the mean there will be approximately 68% of the data; within two standard deviations of the mean there will be approximately 95% of the data; and within three standard deviations of the mean there will be approximately 99.7% of the data. This rule applies specifically to a *normal (bell-shaped) distribution,* but it is frequently applied as an interpretive guide to any mounded distribution.

Figure 2.30 shows the intervals of one, two, and three standard deviations about the mean of an approximately normal distribution. Usually these proportions do not occur exactly in a sample, but your observed values will be close when a large sample is drawn from a normally distributed population.

If a distribution is approximately normal, it will be nearly symmetrical and the mean will divide the distribution in half (the mean and the median are the same in a symmetrical distribution). This allows us to refine the empirical rule, as shown in Figure 2.31.

Figure 2.30 **Empirical Rule**

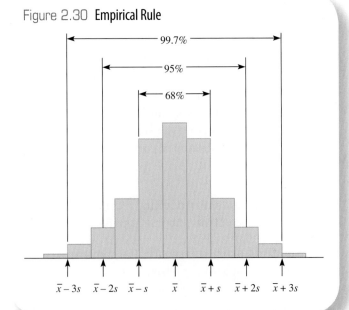

Figure 2.31 Refinement of Empirical Rule

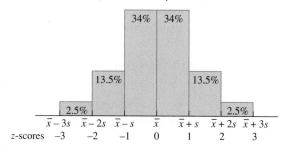

The empirical rule can be used to determine whether or not a set of data is approximately normally distributed. Let's demonstrate this application by working with the distribution of final exam scores that we have been using throughout this chapter. The mean, $\bar{x}$, was found to be 74.92, and the standard deviation, s, was 14.20. The interval from one standard deviation below the mean, $\bar{x} - s$, to one standard deviation above the mean, $\bar{x} + s$, is $74.92 - 14.20 = 60.72$ to $74.92 + 14.20 = 89.12$. This interval (60.72 to 89.12) includes 61, 62, 63, ... , 89. Upon inspection of the ranked data (Table 2.15, page 43), we see that 34 of the 50 data, or 68%, lie within one standard deviation of the mean. Furthermore, $\bar{x} - 2s = 74.92 - (2)(14.20) = 74.92 - 28.40 = 46.52$ to $\bar{x} + 2s = 74.92 + 28.40 = 103.32$ gives the interval from 46.52 to 103.32. Of the 50 data, 48, or 96%, lie within two standard deviations of the mean. All 50 data, or 100%, are included within three standard deviations of the mean (from 32.32 to 117.52). This information can be placed in a table for comparison with the values given by the empirical rule (see Table 2.16).

Table 2.16 Observed Percentages versus the Empirical Rule

Interval	Empirical Rule Percentage	Percentage Found
$\bar{x} - s$ to $\bar{x} + s$	≈ 68	68
$\bar{x} - 2s$ to $\bar{x} + 2s$	≈ 95	96
$\bar{x} - 3s$ to $\bar{x} + 3s$	≈ 99.7	100

The percentages found are reasonably close to those predicted by the empirical rule. By combining this evidence with the shape of the histogram, we can safely say that the final exam data are approximately normally distributed.

Another method for testing normality is to draw a probability plot using a computer or graphing calculator.

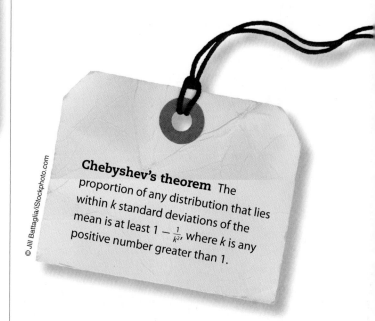

Chebyshev's theorem The proportion of any distribution that lies within k standard deviations of the mean is at least $1 - \frac{1}{k^2}$, where k is any positive number greater than 1.

Chebyshev's Theorem

In the event that the data do not display an approximately normal distribution, **Chebyshev's theorem** gives us information about how much of the data will fall within intervals centered at the mean for all distributions. It states that the proportion of any distribution that lies within k standard deviations of the mean is at least $1 - \frac{1}{k^2}$, where k is any positive number greater than 1. This theorem applies to all distributions of data.

This theorem says that within two standard deviations of the mean ($k = 2$), you will always find at least 75% (that is, 75% or more) of the data:

$$1 - \frac{1}{k^2} = 1 - \frac{1}{2^2} = 1 - \frac{1}{4} = \frac{3}{4} = 0.75, \textbf{ at least 75\%}$$

Figure 2.32 shows a mounded distribution that illustrates at least 75%.

Figure 2.32 Chebyshev's Theorem with $k = 2$

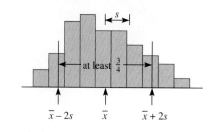

If we consider the interval enclosed by three standard deviations on either side of the mean ($k = 3$), the theorem says that we will always find at least 89% (that is, 89% or more) of the data:

$$1 - \frac{1}{k^2} = 1 - \frac{1}{3^2} = 1 - \frac{1}{9} = \frac{8}{9} = 0.89, \textbf{ at least 89\%}$$

Figure 2.33 shows a mounded distribution that illustrates at least 89%.

Figure 2.33 **Chebyshev's Theorem with $k = 3$**

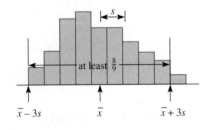

Imagine that all the third graders at Roth Elementary School were given a physical-fitness strength test. Their test results are listed below in rank order and are shown on the histogram (*data set).

1	2	2	3	3	3	4	4	4	5	5	5	5	6	6	6
8	9	9	9	9	9	9	10	10	11	12	12	12	13	14	14
14	15	15	15	15	16	16	16	17	17	17	17	18	18	18	18
19	19	19	19	20	20	20	21	21	21	22	22	22	23	24	24

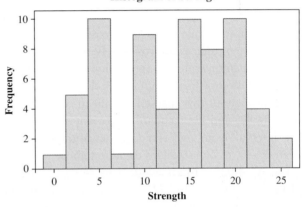

Histogram of Strength

Some questions of interest are: Does this distribution satisfy the empirical rule? Does Chebyshev's theorem hold true? Is this distribution approximately normal?

To answer the first two questions, we need to find the percentages of data in each of the three intervals about the mean. The mean is 13.0, and the standard deviation is 6.6.

Mean ± (k × Std. Dev.)	Interval	Percentage Found	Empirical	Chebyshev
13.0 ± (1 × 6.6)	6.4 to 19.6	36/64 = 56.3%	68%	—
13.0 ± (2 × 6.6)	−0.2 to 26.2	64/64 = 100%	95%	At least 75%
13.0 ± (3 × 6.6)	−6.8 to 32.8	64/64 = 100%	99.70%	At least 89%

It is left to you to verify the values of the mean, the standard deviation, the intervals, and the percentages.

The three percentages found (56.3, 100, and 100) do not approximate the 68, 95, and 99.7 percentages stated in the empirical rule. The two percentages found (100 and 100) do agree with Chebyshev's theorem in that they are greater than 75% and 89%. Remember, Chebyshev's theorem holds for all distributions. With the distribution seen on the histogram and the three percentages found, it is reasonable to conclude that these test results are not normally distributed.

2.7 The Art of Statistical Deception

"THERE ARE THREE KINDS OF LIES—LIES, DAMNED LIES, AND STATISTICS."

These remarkable words spoken by Benjamin Disraeli (19th-century British prime minister) represent the cynical view of statistics held by many people. Most people are on the consumer end of statistics and therefore have to "swallow" them.

Good Arithmetic, Bad Statistics

Let's explore an outright statistical lie. Suppose a small business employs eight people who earn between $300 and $350 per week. The owner of the business pays himself $1,250 per week. He reports to the general public that the average wage paid to the employees of his firm is $430 per week. That may be an example of good arithmetic, but it is bad statistics. It is a misrepresentation of the situation because only one employee, the owner, receives more than the mean salary. The public will think that most of the employees earn about $430 per week.

Graphic Deception

Graphic representations can be tricky and misleading. The frequency scale (which is usually the vertical axis) should start at zero in order to present a total picture. Usually, graphs that do not start at zero are used to save space. Nevertheless, this can be deceptive. Graphs in which the frequency scale starts at zero tend to emphasize the size of the numbers involved, whereas graphs that are chopped off may tend to emphasize the variation in the numbers without regard to the actual size of the numbers. The labeling of the horizontal scale can be misleading also. You need to inspect graphic presentations very carefully before you draw any conclusions from the "story being told."

SUPERIMPOSED MISREPRESENTATION

The "clever" graphic overlay from the *Ithaca Times* (December 7, 2000) has to be the worst graph ever to make

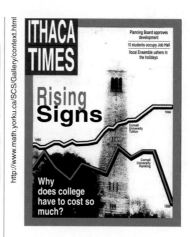

a front page. The cover story, "Why does college have to cost so much?" pictures two graphs superimposed on a Cornell University campus scene. The two broken lines represent "Cornell's Tuition" and "Cornell's Ranking," with the tuition steadily increasing and the ranking staggering and falling. A very clear image is created: Students get less and pay more!

Now view the two complete graphs separately. Notice: (1) The graphs cover two different time periods. (2) The vertical scales differ. (3) The "best" misrepresentation comes from the impression that a "drop in rank" represents a lower quality of education. Wouldn't a rank of 6 be better than a rank of 15?

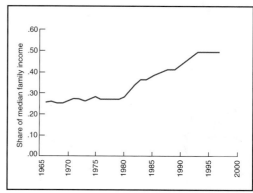

BY THE NUMBERS: OVER 35 YEARS, CORNELL'S TUITION HAS TAKEN AN INCREASINGLY LARGER SHARE OF ITS MEDIAN STUDENT FAMILY INCOME

SOURCE: http://www.math.yorku.ca/SCS/Gallery/context.html

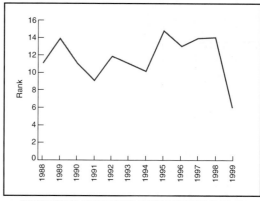

PECKING ORDER: OVER 12 YEARS, CORNELL'S RANKING IN *US NEWS & WORLD REPORT* HAS RISEN AND FALLEN ERRATICALLY.

What it all comes down to is that statistics, like all languages, can be and is abused. In the hands of the careless, the unknowledgeable, or the unscrupulous, statistical information can be as false as "damned lies."

problems

Objective 2.1

2.1 Results from Self.com poll on "What is your top cold-weather beauty concern?" were reported in the December 2008 issue of *Self* magazine: dry skin—57%, chapped lips—25%, dull hair—10%, rough feet—8%.
   a. Construct a pie chart showing the top cold-weather beauty concerns.
   b. Construct a bar graph showing the top cold-weather beauty concerns.
   c. In your opinion, does the pie chart in part (a) or the bar graph in part (b) result in a better representation of the information? Explain.

2.2 Some cleaning jobs are disliked more than others. According to the July 17, 2009 *USA Today* Snapshot on a survey of women by Consumer Reports National Research Center, the cleaning tasks women dislike the most are presented in the following Pareto diagram.

Cleaning Tasks That Women Dislike the Most

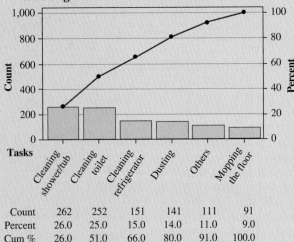

Tasks	Cleaning shower/tub	Cleaning toilet	Cleaning refrigerator	Dusting	Others	Mopping the floor
Count	262	252	151	141	111	91
Percent	26.0	25.0	15.0	14.0	11.0	9.0
Cum %	26.0	51.0	66.0	80.0	91.0	100.0

   a. How many total women were surveyed?
   b. Verify the 15% listed for "Cleaning refrigerator."
   c. Explain how the "cum % for dusting" value of 80% was obtained and what it means.
   d. What three tasks would make no more than 75% of the women surveyed happy if those tasks were eliminated?

2.3 A shirt inspector at a clothing factory categorized the last 500 defects as: 67—missing button, 153—bad seam, 258—improperly sized, 22—fabric flaw. Construct a Pareto diagram for this information.

*2.4 Shown below are the heights (in inches) of the basketball players who were the first round picks by the National Basketball Association professional teams for 2009.

82	86	76	77	75	72	75	81	78	74
77	77	81	81	82	80	76	72	74	74
73	82	80	84	74	81	80	77	74	78

SOURCE: http://www.mynbadraft.com/NBA-Draft-Results/

   a. Construct a dotplot of the heights of these players.
   b. Use the dotplot to uncover the shortest and the tallest players.
   c. What is the most common height and how many players share that height?
   d. What feature of the dotplot illustrates the most common height?

*2.5 Every year *Fortune* magazine ranks America's top 100 employers. The top 15 companies to work for and their job growth are listed below.

Company	Job Growth (%)	Company	Job Growth (%)
NetApp	12	Goldman Sachs	2
Edward Jones	9	Nugget Market	22
Boston Consulting Group	10	Adobe Systems	9
Google	40	Recreational Equipment (REI)	11
Wegmans Food Markets	6	Devon Energy	11
Cisco Systems	7	Robert W. Baird	4
Genentech	5	W. L. Gore & Associates	5
Methodist Hospital System	1		

SOURCE: http://money.cnn.com

   a. Construct a stem-and-leaf display of the data.
   b. Based on the stem-and-leaf display, describe the distribution of percentages of growth.

2.6 Given the following stem-and-leaf display:

Stem-and-Leaf of C1 N = 16
Leaf Unit = 0.010
1   59   7
4   60   148
(5) 61   02669
7   62   0247
3   63   58
1   64   3

   a. What is the meaning of Leaf Unit = 0.010?
   b. How many data values are shown on this stem-and-leaf display?
   c. List the first four data values.
   d. What is the column of numbers down the left-hand side of the figure?

Objective 2.2

*2.7 The U.S. Women's Olympic Soccer Team had a great year in 2008. One way to describe the players on that team is by their individual heights.

Height (inches)								
70	68	65	64	68	66	66	67	68
68	67	65	65	66	64	69	66	65

SOURCE: http://www.ussoccer.com

   a. Construct an ungrouped frequency distribution for the heights.
   b. Construct a frequency histogram of this distribution.
   c. Prepare a relative frequency distribution for this same data.

d. What percentage of the team is at least 5 ft 6 in tall?

*2.8 A survey of 100 resort club managers on their annual salaries resulted in the following frequency distribution:

Annual Salary ($1,000s)	15–25	25–35	35–45	45–55	55–65
No. of Managers	12	37	26	19	6

a. The data value "35" belongs to which class?

b. Explain the meaning of "35–45."

c. Explain what "class width" is, give its value, and describe three ways that it can be determined.

d. Draw a frequency histogram of the annual salaries for resort club managers. Label class boundaries.

*2.9 During the Spring 2009 semester, 200 students took a statistics test from a particular instructor. The resulting grades are given in the following table.

Test Grades	Number
50–60	13
60–70	44
70–80	74
80–90	59
90–100	9
100–110	1
Total	200

a. What is the class width?

b. Draw and completely label a frequency histogram of the statistics test grades.

c. Draw and completely label a relative frequency histogram of the statistics test grades.

d. Carefully examine the two histograms in parts (b) and (c), and explain why one of them might be more useful to a student and to the instructor.

*2.10 The speeds of 55 cars were measured by a radar device on a city street:

27	23	22	38	43	24	35	26	28	18	20
25	23	22	52	31	30	41	45	29	27	43
29	28	27	25	29	28	24	37	28	29	18
26	33	25	27	25	34	32	36	22	32	33
21	23	24	18	48	23	16	38	26	21	23

a. Classify these data into a grouped frequency distribution by using class boundaries 12–18, 18–24, . . . ,48–54.

b. Find the class width.

c. For the class 24–30, find the class midpoint, the lower class boundary, and the upper class boundary.

d. Construct a frequency histogram of these data.

*2.11 A survey of 100 resort club managers on their annual salaries resulted in the following frequency distribution.

Annual Salary ($1,000s)	15–25	25–35	35–45	45–55	55–65
No. of Managers	12	37	26	19	6

a. Prepare a cumulative frequency distribution for the annual salaries.

b. Prepare a cumulative relative frequency distribution for the annual salaries.

c. Construct an ogive for the cumulative relative frequency distribution found in part (b).

d. What value bounds the cumulative relative frequency of 0.75?

e. 75% of the annual salaries are below what value? Explain the relationship between (d) and (e).

Objective 2.3

*2.12 The cost for taking your pet aboard a flight with you in the continental United States varies according to airline. The prices charged by 14 of the major U.S. airlines in June 2009 were (in dollars):

69	100	100	100	125	150	100	60	100	125	75	100	125	100

Find the mean cost for flying your pet with you.

2.13 For those 7th graders with cell phones, the number of programmed numbers in their phones were:

100	37	12	20	53	10	20	50	35	30

a. Find the mean number of programmed numbers on a 7th grader's cell phone.

b. Find the median number of programmed numbers on a 7th grader's cell phone.

c. Explain the difference in values of the mean and median.

d. Remove the most extreme value and answer (a) through (c) again.

e. Did removing the extreme value have more of an effect on the mean or median? Explain why.

2.14 The number of cars owned per apartment in a sample of tenants in a large complex is 1, 2, 1, 2, 2, 2, 1, 2, 3, 2. What is the mode?

2.15 Each year around 160 colleges compete in the American Society of Civil Engineer's National Concrete Canoe Competition. Each team must design a seaworthy canoe from concrete, a substance not known for its capacity to float. The canoes must weigh between 100 and 350 pounds. When last year's entries weighed in, the weights ranged from 138 to 349 pounds.

a. Find the midrange.

b. The information given contains 4 weight values, explain why you did use two of them in (a) and did not use the other two.

2.16. Consider the sample 2, 4, 7, 8, 9. Find the following:

a. mean, $\bar{x}$ b. median, $\tilde{x}$

c. mode   d. midrange

Objective 2.4

2.17 a. The data value $x = 45$ has a deviation value of 12. Explain the meaning of this.

b. The data value $x = 84$ has a deviation value of -20. Explain the meaning of this.

2.18 All measures of variation are nonnegative in value for all sets of data.

a. What does it mean for a value to be "nonnegative"?

b. Describe the conditions necessary for a measure of variation to have the value zero.

c. Describe the conditions necessary for a measure of variation to have a positive value.

2.19 Consider the sample 2, 4, 7, 8, 9. Find the following:

a. Range

b. Variance s^2, using formula (2.5)

c. Standard deviation, s

2.20 Fifteen randomly selected college students were asked to state the number of hours they slept the previous night. The resulting data are 5, 6, 6, 8, 7, 7, 9, 5, 4, 8, 11, 6, 7, 8, 7. Find the following:

a. Variance s^2, using formula (2.5)

b. Variance s^2, using formula (2.9)

c. Standard deviation, s

2.21 Consider the following two sets of data:

Set 1	45	80	50	45	30
Set 2	30	80	35	30	75

Both sets have the same mean, which is 50. Compare these measures for both sets: $\Sigma(x - \bar{x})$, SS(x), and range. Comment on the meaning of these comparisons relative to the distribution.

2.22 Comment on the statement: "The mean loss for customers at First State Bank (which was not insured) was $150. The standard deviation of the losses was −$125."

Objective 2.5

2.23 Refer to the table of exam scores in Table 2.15 on page 43 for the following.

a. Using the concept of depth, describe the position of 91 in the set of 50 exam scores in two different ways.

b. Find P_{20} and P_{35} for the exam scores in Table 2.15.

c. Find P_{80} and P_{95} for the exam scores in Table 2.15.

*2.24 The U.S. Geological Survey collected atmospheric deposition data in the Rocky Mountains. Part of the sampling process was to determine the concentration of ammonium ions (in percentages). Here are the results from the 52 samples:

2.9	4.1	2.7	3.5	1.4	5.6	13.3	3.9	4.0
2.9	7.0	4.2	4.9	4.6	3.5	3.7	3.3	5.7
3.2	4.2	4.4	6.5	3.1	5.2	2.6	2.4	5.2
4.8	4.8	3.9	3.7	2.8	4.8	2.7	4.2	2.9
2.8	3.4	4.0	4.6	3.0	2.3	4.4	3.1	5.5
4.1	4.5	4.6	4.7	3.6	2.6	4.0		

a. Find Q_1

b. Find Q_2

c. Find Q_3

d. Find the midquartile

e. Find P_{30}

f. Find the 5-number summary

g. Draw the box-and-whiskers display

2.25 An exam produced grades with a mean score of 74.2 and a standard deviation of 11.5. Find the z-score for each test score x:

a. $x = 54$

b. $x = 68$

c. $x = 79$

d. $x = 93$

2.26 A sample has a mean of 120 and a standard deviation of 20.0. Find the value of x that corresponds to each of these standard scores:

a. $z = 0.0$

b. $z = 1.2$

c. $z = -1.4$

d. $z = 2.05$

2.27 The ACT Assessment® is designed to assess high school students' general educational development and their ability to complete college-level work. The table lists the mean and standard deviation of scores attained by the 3,908,557 high school students from the 2006 to 2008 graduating classes who took the ACT exams.

2006–2008	English	Mathematics	Reading	Science	Composite
Mean	20.6	21.0	21.4	20.9	21.1
Standard deviation	6.0	5.1	6.1	4.8	4.9

SOURCE: American College Testing

Convert the following ACT test scores to z-scores for both English and Math. Compare placement between the two tests.

a. $x = 30$

b. $x = 23$

c. $x = 12$

d. Explain why the relative positions in English and Math changed for the ACT scores of 30 and 12.

e. If Jessica had a 26 on one of the ACT exams, on which one of the exams would she have the best possible relative score? Explain why.

Objective 2.6

2.28 The empirical rule indicates that we can expect to find what proportion of the sample included between the following?

a. $\bar{x} - s$ and $\bar{x} + s$

b. $\bar{x} - 2s$ and $\bar{x} + 2s$

c. $\bar{x} - 3s$ and $\bar{x} - 3s$

2.29 The mean lifetime of a certain tire is 30,000 miles and the standard deviation is 2,500 miles.

a. If we assume the mileages are normally distributed, approximately what percentage of all such tires will last between 22,500 and 37,500 miles?

b. If we assume nothing about the shape of the distribution, approximately what percentage of all such tires will last between 22,500 and 37,500 miles?

2.30 Using the empirical rule, determine the approximate percentage of a normal distribution that is expected to fall within the interval described.

a. Less than the mean

b. Greater than 1 standard deviation above the mean

c. Less than 1 standard deviation above the mean

d. Between 1 standard deviation below the mean and 2 standard deviations above the mean

*2.31 Each year, NCAA college football fans like to learn about the up-and-coming freshman class of players. Following are the heights (in inches) of the nation's top 100 high school football players for 2009.

73	75	71	76	74	77	74	72	73	72
74	72	74	72	72	78	73	76	75	72
77	76	73	72	76	72	73	70	75	72
71	74	77	78	74	75	71	75	71	76
70	76	72	71	74	74	71	72	76	71
75	79	78	79	74	76	76	76	75	73
74	70	74	74	75	75	75	75	76	71
74	75	74	78	72	73	71	72	73	72
74	75	77	73	77	75	77	71	72	70
74	76	71	73	76	76	79	77	74	78

SOURCE: http://www.takkle.com/

a. Construct a histogram and one other graph of your choice that displays the distribution of heights.
b. Calculate the mean and standard deviation.
c. Sort the data into a ranked list.
d. Determine the values of $\bar{x} \pm s$, $\bar{x} \pm 2s$, and $\bar{x} \pm 3s$, and determine the percentage of data within one, two, and three standard deviations of the mean.
e. Do the percentages found in (d) agree with the empirical rule? What does this imply? Explain.
f. Do the percentages found in (d) agree with Chebyshev's theorem? What does that mean?
g. Does the graph show a distribution that agrees with your answers in part (e)? Explain.
h. Utilize one of the "testing for normality" technology instructions on your Chapter 2 Tech Card. Compare the results with your answer to part (e).

Objective 2.7

2.32

Clipping Coupons

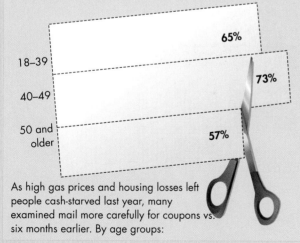

18–39   65%

40–49   73%

50 and older   57%

As high gas prices and housing losses left people cash-starved last year, many examined mail more carefully for coupons vs. six months earlier. By age groups:

SOURCE: DMNews for Pitney Bowes, survey conducted online among 1,003 adults, September 9–16, 2008.

a. Is the graph a bar graph or a histogram? Explain how you determined your answer.
b. The age grouping used in the Clipping Coupons graphic does not lead to a very informative graph. Describe how the age groups might have been formed and how your suggested grouping would give additional meaning to the graph.

2.33 What kinds of financial transactions do you do online? Are you worried about your security? According to Consumer Internet Barometer, the source of a March 25, 2009, *USA Today* Snapshot titled "Security of Online Accounts," the following transactions and percent of people concerned about their online security were reported.

What	Percent
Banking	72
Paying bills	70
Buying stocks, bonds	62
Filing taxes	62

SOURCE: *USA Today* and Consumer Internet Barometer

Prepare two bar graphs to depict the percentage data. Scale the vertical axis on the first graph from 50 to 80. Scale the second graph from 0 to 100. What is your conclusion concerning how the percentages of the four responses stack up based on the two bar graphs, and what would you recommend, if anything, to improve the presentations?

Descriptive Analysis
and Presentation of Bivariate Data

3.1 Bivariate Data

NOT ALL SAMPLE DATA CAN BE GRAPHICALLY DISPLAYED WITH ONE VARIABLE. TO GRAPHICALLY DISPLAY AND NUMERICALLY DESCRIBE SAMPLE DATA THAT INVOLVE TWO PAIRED VARIABLES, WE NEED TO USE **BIVARIATE DATA,** WHICH ARE THE VALUES OF TWO DIFFERENT VARIABLES THAT ARE OBTAINED FROM THE SAME POPULATION ELEMENT.

objectives

3.1 Bivariate Data

3.2 Linear Correlation

3.3 Linear Regression

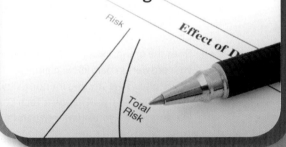

Bivariate data The values of two different variables that are obtained from the same population element.

Each of the two variables may be either *qualitative* or *quantitative*. As a result, three combinations of variable types can form bivariate data:

1. Both variables are qualitative (both attribute).

2. One variable is qualitative (attribute), and the other is quantitative (numerical).

3. Both variables are quantitative (both numerical).

Two Qualitative Variables

When bivariate data result from two qualitative (attribute or categorical) variables, the data are often arranged on a **cross-tabulation** or **contingency table.** To see how this works, let's use information on gender and college major.

CROSS-TABULATION

→ Thirty students from our college were randomly identified and classified according to two variables: gender (M/F) and major (liberal arts, business administration, technology), as shown in Table 3.1. These 30 bivariate data can be summarized on a 2 × 3 cross-tabulation table, where the two rows represent the two genders, male and female, and the three columns represent the three major categories of liberal arts (LA), business administration (BA),

*Table 3.1 Genders and Majors of 30 College Students

Name	Gender	Major	Name	Gender	Major	Name	Gender	Major
Adams	M	LA	Feeney	M	T	McGowan	M	BA
Argento	F	BA	Flanigan	M	LA	Mowers	F	BA
Baker	M	LA	Hodge	F	LA	Ornt	M	T
Bennett	F	LA	Holmes	M	T	Palmer	F	LA
Brand	M	T	Jopson	F	T	Pullen	M	T
Brock	M	BA	Kee	M	BA	Rattan	M	BA
Chun	F	LA	Kleeberg	M	LA	Sherman	F	LA
Crain	M	T	Light	M	BA	Small	F	T
Cross	F	BA	Linton	F	LA	Tate	M	BA
Ellis	F	BA	Lopez	M	T	Yamamoto	M	LA

Table 3.2 Cross-Tabulation of Gender and Major
(tallied)

	Major		
Gender	LA	BA	T
M	|||| (5)	||||| (6)	||||||| (7)
F	|||||| (6)	|||| (4)	|| (2)

Table 3.3 Cross-Tabulation of Gender and Major
(frequencies)

	Major			
Gender	LA	BA	T	Row Total
M	5	6	7	18
F	6	4	2	12
Col. Total	11	10	9	30

and technology (T). The entry in each cell is found by determining how many students fit into each category. Adams is male (M) and liberal arts (LA) and is classified in the cell in the first row, first column. See the red tally mark in Table 3.2. The other 29 students are classified (tallied, shown in black) in a similar fashion.

The resulting 2 × 3 cross-tabulation (contingency) table, Table 3.3, shows the frequency for each cross-category of the two variables along with the row and column totals, called *marginal totals* (or *marginals*).

The total of the marginal totals is the *grand total* and is equal to *n*, the *sample size*.

Contingency tables often show percentages (relative frequencies). These percentages can be based on the entire sample or on the subsample (row or column) classifications.

FYI:

m = n (rows)
n = n (cols)
for an m × n contingency table.

PERCENTAGES BASED ON THE GRAND TOTAL (ENTIRE SAMPLE)

The frequencies in the contingency table shown in Table 3.3 can easily be converted to percentages of the grand total by dividing each frequency by the grand total and multiplying the result by 100. For example, 6 becomes 20%:

$$\left(\frac{6}{30}\right) \times 100 = 20.$$

From the table of percentages of the grand total (see Table 3.4), we can easily see that 60% of the sample are male, 40% are female, 30% are technology majors, and so on. These same statistics (numerical values describing sample results) can be shown in a bar graph (see Figure 3.1).

Table 3.4 Cross-Tabulation of Gender and Major
(relative frequencies; % of grand total)

Gender	Major			
	LA	BA	T	Row Total
M	17%	20%	23%	60%
F	20%	13%	7%	40%
Col. Total	37%	33%	30%	100%

Figure 3.1 Bar Graph

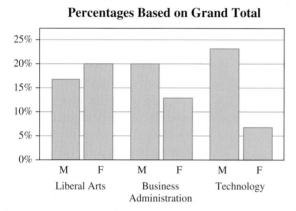

Percentages Based on Grand Total

Table 3.4 and Figure 3.1 show the distribution of male liberal arts students, female liberal arts students, male business administration students, and so on, relative to the entire sample.

PERCENTAGES BASED ON ROW TOTALS

The frequencies in the same contingency table, Table 3.3, can be expressed as percentages of the row totals (or gender) by dividing each row entry by that row's total and multiplying the results by 100. Table 3.5 is based on row totals. From Table 3.5 we see that 28% of the male students are majoring in liberal arts, whereas 50% of the female students are majoring in liberal arts.

Table 3.5 Cross-Tabulation of Gender and Major
(% of row totals)

Gender	Major			
	LA	BA	T	Row Total
M	28%	33%	39%	100%
F	50%	33%	17%	100%
Col. Total	37%	33%	30%	100%

PERCENTAGES BASED ON COLUMN TOTALS

The frequencies in the contingency table, Table 3.3, can be expressed as percentages of the column totals (or major) by dividing each column entry by that column's total and multiplying the result by 100. Table 3.6 is based on column totals. From Table 3.6 we see that 45% of the liberal arts students are male, whereas 55% of the liberal arts students are female.

Table 3.6 Cross-Tabulation of Gender and Major
(% of column totals)

Gender	Major			
	LA	BA	T	Row Total
M	45%	60%	78%	60%
F	55%	40%	22%	40%
Col. Total	100%	100%	100%	100%

One Qualitative and One Quantitative Variable

When bivariate data result from one qualitative and one quantitative variable, the quantitative values are viewed as separate samples, each set identified by levels of the qualitative variable. Each sample is described using the techniques from Chapter 2, and the results are displayed side by side for easy comparison.

To see how a side-by-side comparison works, let's use the example of stopping distance. The distance required to stop a 3,000-pound automobile on wet pavement was measured to compare the stopping capabilities of three tire tread designs (see Table 3.7). Tires of each design were tested repeatedly on the same automobile on a controlled patch of wet pavement.

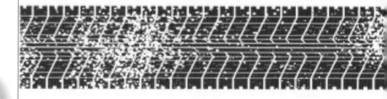

*Table 3.7 Stopping Distances (in Feet) for Three Tread Designs

Design A (n = 6)			Design B (n = 6)			Design C (n = 6)		
37	36	38	33	35	38	40	39	40
34	40	32	34	42	34	41	41	43

The design of the tread is a qualitative variable with three levels of response, and the stopping distance is a quantitative variable. The distribution of the stopping distances for tread design A is to be compared with the distribution of stopping distances for each of the other tread designs. This comparison may be made with both numerical and graphic techniques. Some of the available options are shown in Figure 3.2, Table 3.8, and Table 3.9.

Figure 3.2 **Dotplot and Box-and-Whiskers Display Using a Common Scale**

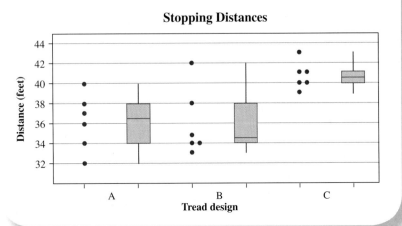

Table 3.8 **Five-Number Summary for Each Design**

	Design A	Design B	Design C
High	40	42	43
Q_3	38	38	41
Median	36.5	34.5	40.5
Q_1	34	34	40
Low	32	33	39

Table 3.9 **Mean and Standard Deviation for Each Design**

	Design A	Design B	Design C
Mean	36.2	36.0	40.7
Standard deviation	2.9	3.4	1.4

Two Quantitative Variables

When the bivariate data are the result of two quantitative variables, it is customary to express the data

mathematically as **ordered pairs** (x, y), where x is the **input variable** (sometimes called the **independent variable**) and y is the **output variable** (sometimes called the **dependent variable**). The data are said to be *ordered* because one value, x, is always written first. They are called *paired* because for each x value, there is a corresponding y value from the same source. For example, if x is height and y is weight, then a height and a corresponding weight are recorded for each person. The input variable x is measured or controlled in order to predict the output variable y. Suppose some research doctors are testing a new drug by prescribing different dosages and observing the lengths of the recovery times of their patients. The researcher can control the amount of drug prescribed, so the amount of drug is referred to as x. In the case of height and weight, either variable could be treated as input and the other as output, depending on the question being asked. However, different results will be obtained from the regression analysis, depending on the choice made.

CONSTRUCTING A SCATTER DIAGRAM

In problems that deal with two quantitative variables, we present the sample data pictorially on a **scatter diagram**, or a plot of all the ordered pairs of bivariate data on a coordinate axis system. On a scatter diagram, the input variable, x, is plotted on the horizontal axis, and the output variable, y, is plotted on the vertical axis.

Scatter diagram A plot of all the ordered pairs of bivariate data on a coordinate axis system.

To illustrate, let's work with data from Mr. Chamberlain's physical-fitness course in which several fitness scores were taken.

The following sample contains the numbers of push-ups and sit-ups done by 10 randomly selected students:

(27, 30) (22, 26) (15, 25) (35, 42) (30, 38)
(52, 40) (35, 32) (55, 54) (40, 50) (40, 43)

(continued on p. 60)

Americans Love Their Automobiles

America's love affair with the sport-utility vehicle (SUV) began in the late 1990s and early 2000s, but sales may be declining due to the vehicle's gasoline consumption, cost, and poor safety. SUVs are viewed as high-performance, rugged, four-wheel drive cars built on truck chassis that can go off road in any weather, have good towing abilities, can carry more than four passengers, and have greater safety because of their large size and heavy build.

This chart lists 16 of the 4-wheel drive 6-cylinder SUVs offered by auto manufacturers for 2009 and the values of four variables for each vehicle.

2009 4WD 6-Cylinder SUVs

Vehicle Manufacturer	Model	Type of Gas	Cost to Drive 25 miles	Cost to Fill up Tank	Tank Capacity (gallons)
Buick	Enclave	Reg.	2.51	37.82	22
Chevrolet	Trailblazer	Reg.	2.98	37.82	22
Chrysler	Aspen	Reg.	3.18	46.41	27
Dodge	Durango	Reg.	3.18	46.41	27
Ford	Escape	Reg.	2.39	28.36	16.5
GMC	Envoy	Reg.	2.98	37.82	22
Honda	Pilot	Reg.	2.65	36.1	21
Jeep	Grd Cherokee	Reg.	2.81	36.27	21.1
Kia	Sportage	Reg.	2.39	29.57	17.2
Lexus	RX 350	Prem.	2.83	37.15	19.2
Lincoln	MKX	Reg.	2.51	32.66	19
Mazda	CX-7	Prem.	2.99	35.22	18.2
Mercury	Mountaineer	Reg.	3.18	38.68	22.5
Mitsubishi	Outlander	Reg.	2.15	27.16	15.8
Nissan	Murano	Prem.	2.69	41.99	21.7
Toyota	RAV4	Reg.	2.27	27.33	15.9

SOURCE: http://www.fueleconomy.gov

Side-by-Side Plot of Cost to Dive 25 Miles by Grade of Gasoline

Cost of Gasoline to Drive 25 Miles, US$ — Grade of Gasoline (Premium, Regular)

This dotplot compares the cost of driving 25 miles to the grade of gasoline, revealing that three SUVs using regular gas still cost more than SUVs using premium gas.

This dotplot shows six of the SUVs that use regular gas have tanks with larger capacities than the three SUVs using premium. Why would some vehicles need 27-gallon gasoline tanks?

Side-by-Side Plot of Tank Capacity by Grade of Gasoline

Gas Tank Capacity (gallons) — Grade of Gasoline (Premium, Regular)

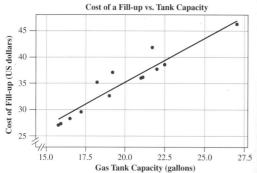

Cost of a Fill-up vs. Tank Capacity

Cost of Fill-up (US dollars) — Gas Tank Capacity (gallons)

This graph shows information you probably already knew; the larger the gas tank the more it costs to fill it. How could it be any other way? How do the distributions shown above appear on this graph?

Table 3.10 Data for Push-ups and Sit-ups

Student	1	2	3	4	5	6	7	8	9	10
Push-ups, x	27	22	15	35	30	52	35	55	40	40
Sit-ups, y	30	26	25	42	38	40	32	54	50	43

Table 3.10 shows these sample data, and Figure 3.3 shows a scatter diagram of the data.

The scatter diagram from Mr. Chamberlain's physical fitness course shows a definite pattern. Note that as the number of push-ups increased, so did the number of sit-ups.

Figure 3.3 Scatter Diagram

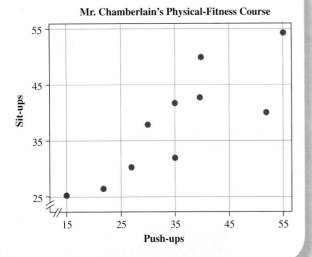

3.2 Linear Correlation

THE PRIMARY PURPOSE OF **LINEAR CORRELATION ANALYSIS** IS TO MEASURE THE STRENGTH OF A LINEAR RELATIONSHIP BETWEEN TWO VARIABLES.

Let's examine some scatter diagrams that demonstrate different relationships between input, or independent variables, x, and output, or dependent variables, y. If as x increases there is no definite shift in the values of y, we say there is **no correlation,** or no relationship between x and y. If as x increases there is a shift in the values of y, then there is a correlation. The correlation is **positive** when y tends to increase and **negative** when y tends to decrease. If the ordered pairs (x, y) tend to follow a straight-line path, there is a linear correlation. The preciseness of the shift in y as x increases determines the strength of the **linear correlation.** The scatter diagrams in Figure 3.4 demonstrate these ideas.

Perfect linear correlation occurs when all the points fall exactly along a straight line, as shown in the top two graphs of Figure 3.5. The correlation can be either positive or negative, depending on whether y increases or decreases as x increases. If the data form a straight horizontal or vertical line, there is no correlation, because one variable has no effect on the other, as also shown in the bottom two graphs of Figure 3.5.

Figure 3.4 Scatter Diagrams and Correlation

| No correlation | Positive | High positive | Negative | High negative |

Figure 3.5 Ordered Pairs Forming a Straight Line

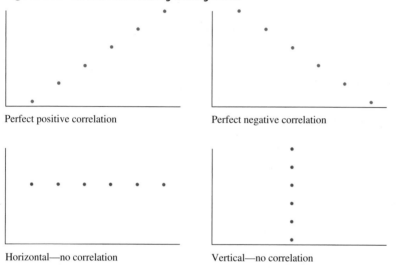

Perfect positive correlation   Perfect negative correlation

Horizontal—no correlation   Vertical—no correlation

Figure 3.6 No Linear Correlation

increase in the value of y, then r will be positive in value. For example, a positive value of r would be expected for the age and height of children because as children grow older, they grow taller. Also, consider the age, x, and resale value, y, of an automobile. As the car ages, its resale value decreases. Since as x increases, y decreases, the relationship results in a negative value for r.

Scatter diagrams do not always appear in one of the forms shown in Figures 3.4 and 3.5. Sometimes they suggest relationships other than linear, as in Figure 3.6. There appears to be a definite pattern; however, the two variables are not related linearly, and therefore there is no linear correlation.

Calculating the Linear Correlation Coefficient, *r*

The **coefficient of linear correlation, *r*,** is the numerical measure of the strength of the linear relationship between two variables. The coefficient reflects the consistency of the effect that a change in one variable has on the other. The value of the linear correlation coefficient helps us answer this question: Is there a linear correlation between the two variables under consideration? The linear correlation coefficient, r, always has a value between -1 and $+1$. A value of $+1$ signifies a perfect positive correlation, and a value of -1 shows a perfect negative correlation. If as x increases there is a general

y

r is positive for the age and height of children   **X**

The value of r is defined by **Pearson's product moment formula:**

Definition Formula

$$r = \frac{\Sigma(x - \bar{x})(y - \bar{y})}{(n-1)s_x s_y} \qquad (3.1)$$

NOTES:

1. s_x and s_y are the standard deviations of the x- and y-variables.

2. The development of this formula is discussed in Chapter 13.

To calculate r, we will use an alternative formula, formula (3.2), that is equivalent to formula (3.1). As preliminary calculations, we will separately calculate three sums of squares and then substitute them into formula (3.2) to obtain r.

Second, to complete the preliminary calculations, we substitute the five summations (the five column totals) from the extensions table into formulas (2.8), (3.3), and (3.4), and calculate the three sums of squares:

$$SS(x) = \Sigma x^2 - \frac{(\Sigma x)^2}{n} = 13{,}717 - \frac{(351)^2}{10} = 1{,}396.9$$

$$SS(y) = \Sigma y^2 - \frac{(\Sigma y)^2}{n} = 15{,}298 - \frac{(380)^2}{10} = 858.0$$

$$SS(xy) = \Sigma xy - \frac{\Sigma x \Sigma y}{n} = 14{,}257 - \frac{(351)(380)}{10} = 919.0$$

Third, we substitute the three sums of squares into formula (3.2) to find the value of the correlation coefficient:

$$r = \frac{SS(xy)}{\sqrt{SS(x)SS(y)}} = \frac{919.0}{\sqrt{(1{,}396.9)(858.0)}} = 0.8394 = \mathbf{0.84}$$

Computational Formula

$$\text{linear correlation coefficient} = \frac{\text{sum of squares for } xy}{\sqrt{(\text{sum of squares for } x)(\text{sum of squares for } y)}}$$

$$r = \frac{SS(xy)}{\sqrt{SS(x)SS(y)}} \qquad (3.2)$$

Recall the SS(x) calculation from formula (2.8) for sample variance:

$$\text{sum of squares for } x = \text{sum of } x^2 - \frac{(\text{sum of } x)^2}{n}$$

$$SS(x) = \Sigma x^2 - \frac{(\Sigma x)^2}{n} \qquad (2.8)$$

We can also calculate:

$$\text{sum of squares for } y = \text{sum of } y^2 - \frac{(\text{sum of } y)^2}{n}$$

$$SS(y) = \Sigma y^2 - \frac{(\Sigma y)^2}{n} \qquad (3.3)$$

$$\text{sum of squares for } xy = \text{sum of } xy - \frac{(\text{sum of } x)(\text{sum of } y)}{n}$$

$$SS(xy) = \Sigma xy - \frac{\Sigma x \Sigma y}{n} \qquad (3.4)$$

So let's apply these techniques and formulae to the push-up/sit-up data from Mr. Chamberlain's fitness course (page 58).

To find the linear correlation coefficient for the push-up/sit-up data, first we construct an extensions table (Table 3.11) listing all the pairs of values (x, y) to aid us in finding x^2, xy, and y^2 for each pair and the five column totals.

NOTE: TYPICALLY, r IS ROUNDED TO THE NEAREST HUNDREDTH.

The value of the linear correlation coefficient helps us answer the question: Is there a linear correlation between the two variables under consideration? When the calculated value of r is close to zero, we conclude that there is little or no linear correlation. As the calculated value of r changes from 0.0 toward either $+1.0$ or -1.0, it indicates an increasing linear correlation between the two variables. From a graphic viewpoint, when we calculate r, we are measuring how well a straight line describes the scatter diagram of ordered pairs. As the value of r changes from 0.0 toward $+1.0$ or -1.0, the data points create a pattern that moves closer to a straight line.

Table 3.11 **Extensions Table for Finding Five Summations**

Student	Push-ups, x	x^2	Sit-ups, y	y^2	xy
1	27	729	30	900	810
2	22	484	26	676	572
3	15	225	25	625	375
4	35	1,225	42	1,764	1,470
5	30	900	38	1,444	1,140
6	52	2,704	40	1,600	2,080
7	35	1,225	32	1,024	1,120
8	55	3,025	54	2,916	2,970
9	40	1,600	50	2,500	2,000
10	40	1,600	43	1,849	1,720
	$\Sigma x = 351$	$\Sigma x^2 = 13{,}717$	$\Sigma y = 380$	$\Sigma y^2 = 15{,}298$	$\Sigma xy = 14{,}257$
	sum of x	sum of x^2	sum of y	sum of y^2	sum of xy

Causation and Lurking Variables

As we try to explain the past, understand the present, and estimate the future, judgments about cause and effect are necessary because of our desire to impose order on our environment.

The **cause-and-effect relationship** is fairly straightforward. You may focus on the *effect* of a situation (e.g., a disease or social problem), and try to determine its *cause(s)*, or you may begin with a *cause* (unsanitary conditions or poverty) and discuss its *effect(s)*. To determine the cause of something, ask yourself *why* it happened. To determine the effect, ask yourself *what* happened. **Lurking variables** are also part of the cause-and-effect relationship being studied because even though they are not included in the study per se, they have an effect on the variables of the study and make it appear that those variables are related.

Lurking variable A variable that is not included in a study but has an effect on the variables of the study and makes it appear that those variables are related.

Here are some pitfalls to avoid:

1. In a direct cause-and-effect relationship, an increase (or decrease) in one variable causes an increase (or decrease) in another. Suppose there is a strong positive correlation between weight and height. Does an increase in weight *cause* an increase in height? Not necessarily. Or to put it another way, does a decrease in weight *cause* a decrease in height? Many other possible variables are involved, such as gender, age, and body type. These other variables are called *lurking variables*.

2. In the feature on page 59, a positive correlation existed between gas-tank capacity and the cost of a fill-up. If we had one of the SUVs with a smaller tank, would this cause us to save money on gas? Not necessarily. Other variables exist, including gas mileage.

3. Don't reason from *correlation* to *cause:* Just because all people who move to the city get old doesn't mean that the city *causes* aging. The city may be a factor, but you can't base your argument on the correlation.

Lurking Variables

A good example of lurking variables can be found in the strong positive relationship shown between the amount of damage caused by a fire and the number of firefighters who work the fire. The "size" of the fire is the lurking variable; it "causes" both the "amount" of damage and the "number" of firefighters. If there is a strong linear correlation between two variables, then one of the following situations may be true about the relationship between the two variables:

1 There is a direct cause-and-effect relationship between the two variables.

2 There is a reverse cause-and-effect relationship between the two variables.

3 Their relationship may be caused by a third variable.

4 Their relationship may be caused by the interactions of several other variables.

5 The apparent relationship may be strictly a coincidence.

REMEMBER that a strong correlation does not necessarily imply causation.

© mrPliskin/iStockphoto.com / © Alenate/iStockphoto.com

3.3 Linear Regression

ALTHOUGH THE CORRELATION COEFFICIENT MEASURES THE STRENGTH OF A LINEAR RELATIONSHIP, IT DOES NOT TELL US ABOUT THE MATHEMATICAL RELATIONSHIP BETWEEN THE TWO VARIABLES.

In Objective 3.2, the correlation coefficient for the push-up/sit-up data was found to be 0.84 (see page 62).

This and the pattern on the scatter diagram imply that there is a linear relationship between the number of push-ups and the number of sit-ups a student does. However, the correlation coefficient does not help us predict the number of sit-ups a person can do based on knowing that he or she can do 28 push-ups. **Regression analysis** finds the equation of the line that best describes the relationship between two variables. One use of this equation is to make predictions. We make use of these predictions regularly—for example, predicting the success a student will have in college based on high school results and predicting the distance required to stop a car based on its speed. Generally, the exact value of y is not predictable, and we are usually satisfied if the predictions are reasonably close.

Line of Best Fit

If a straight-line model seems appropriate, the best-fitting straight line is found by using the **method of least squares**. Suppose that $\hat{y} = b_0 + b_1 x$ is the equation of a straight line, where $\hat{y}$ (read "y-hat") represents the **predicted value of y** that corresponds to a particular value of x. The **least squares criterion** requires that we find the constants b_0 and b_1 such that $\Sigma(y - \hat{y})^2$ is as small as possible.

Figure 3.7 shows the distance of an observed value of y from a **predicted value of $\hat{y}$**. The length of this distance represents the value $(y - \hat{y})$ (shown as the red line segment in Figure 3.7). Note that $(y - \hat{y})$ is positive when the point (x, y) is above the line and negative when (x, y) is below the line.

Figure 3.8 shows a scatter diagram with what appears to be the **line of best fit**, along with 10 individual $(y - \hat{y})$ values. (Positive values are shown in red; negative, in green.) The sum of the squares of these differences is minimized (made as small as possible) if the line is indeed the line of best fit.

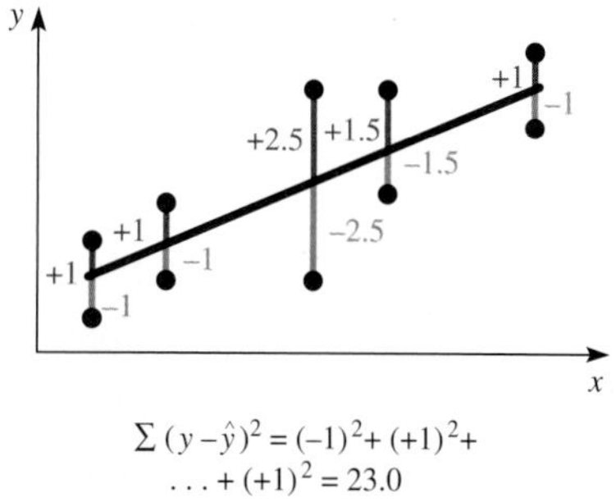

Figure 3.9 shows the same data points as Figure 3.8. The 10 individual values of $(y - \hat{y})$ are plotted with a line that is definitely not the line of best fit. [The value of $\Sigma(y - \hat{y})^2$ is 149, much larger than the 23 from Figure 3.8.] Every different line drawn through this set of 10 points will result in a different value for $\Sigma(y - \hat{y})^2$. Our job is to find the one line that will make $\Sigma(y - \hat{y})^2$ the smallest possible value.

The equation of the line of best fit is determined by its **slope (b_1)** and its **y-intercept (b_0)**. The values of the constants—slope and y-intercept—that satisfy the least squares criterion are found by using the formulas presented next:

Definition Formula

$$\text{slope:} \quad b_1 = \frac{\Sigma(x - \bar{x})(y - \bar{y})}{\Sigma(x - \bar{x})^2} \tag{3.5}$$

We will use a mathematical equivalent of formula (3.5) for the slope, b_1, that uses the sums of squares found in the preliminary calculations for correlation:

Computational Formula

$$\text{slope:} \quad b_1 = \frac{SS(xy)}{SS(x)} \tag{3.6}$$

Notice that the numerator of formula (3.6) is the $SS(xy)$ formula (3.4) (page 62) and the denominator is formula (2.8) (also on page 62) from the correlation coefficient calculations. Thus, if you have previously calculated the linear correlation coefficient using the procedure outlined on page 62, you can easily find the slope of the line of best

Figure 3.7 **Observed and Predicted Values of y**

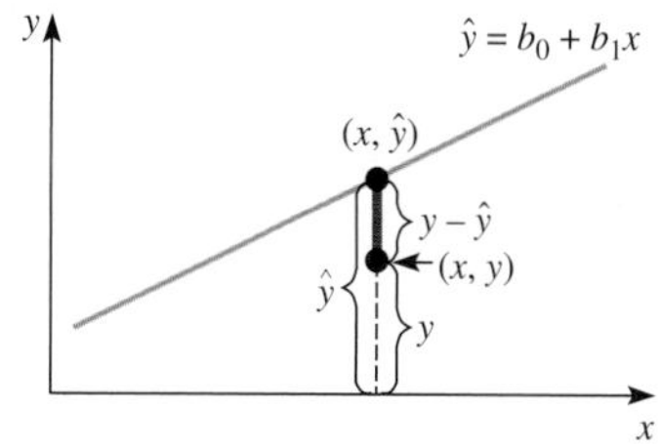

Figure 3.8 **The Line of Best Fit**

$$\Sigma(y - \hat{y})^2 = (-1)^2 + (+1)^2 + \ldots + (+1)^2 = 23.0$$

Figure 3.9 **Not the Line of Best Fit**

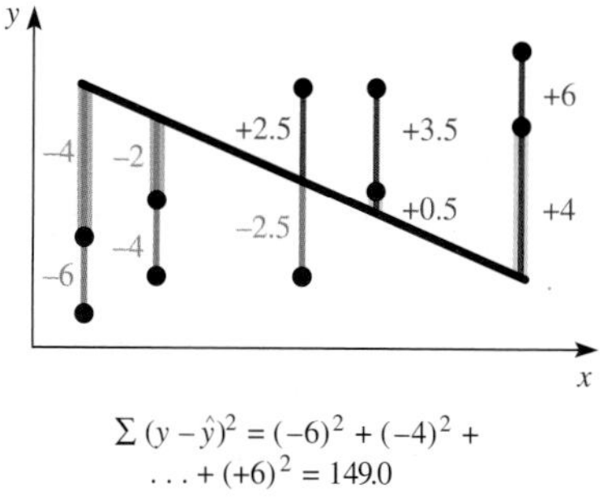

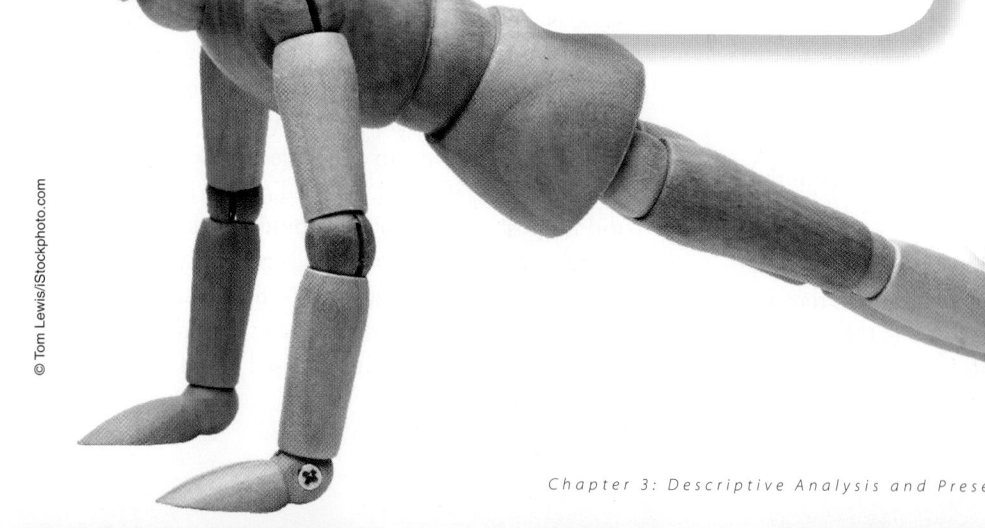

$$\Sigma(y - \hat{y})^2 = (-6)^2 + (-4)^2 + \ldots + (+6)^2 = 149.0$$

fit. If you did not previously calculate r, set up a table similar to Table 3.11 (page 63) and complete the necessary preliminary calculations.

For the y-intercept, we have:

Computational Formula

$$y\text{-intercept} = \frac{(\text{sum of } y) - [(\text{slope})(\text{sum of } x)]}{\text{number}}$$

$$b_0 = \frac{\Sigma y - (b_1 \cdot \Sigma x)}{n} \qquad (3.7)$$

Alternative Computational Formula

$$y\text{-intercept} = y\text{-bar} - (\text{slope} \cdot x\text{-bar})$$

$$b_0 = \bar{y} - (b_1 \cdot \bar{x}) \qquad (3.7a)$$

Now let's reconsider the data from Mr. Chamberlain's phys ed class (page 58) and the question of predicting a student's number of sit-ups based on the number of push-ups. We want to find the line of best fit, $\hat{y} = b_0 + b_1 x$. The preliminary calculations have already been completed in Table 3.11 (page 63). To calculate the slope, b_1, using formula (3.6), recall that $SS(xy) = 919.0$ and $SS(x) = 1,396.9$. Therefore,

$$\text{slope: } b_1 = \frac{SS(xy)}{SS(x)} = \frac{919.0}{1,396.9} = 0.6579 = \mathbf{0.66}$$

To calculate the y-intercept, b_0, using formula (3.7), recall that $\Sigma x = 351$ and $\Sigma y = 380$ from the extensions table. We have

$$y\text{-intercept: } b_0 = \frac{\Sigma y - (b_1 \cdot \Sigma x)}{n} = \frac{380 - (0.6579)(351)}{10}$$

$$= \frac{380 - 230.9229}{10} = 14.9077 = \mathbf{14.9}$$

By placing the two values just found into the model $\hat{y} = b_0 + b_1 x$, we get the equation of the line of best fit:

$$\hat{y} = \mathbf{14.9 + 0.66x}$$

NOTES:

1. Remember to keep at least three extra decimal places while doing the calculations to ensure an accurate answer.

2. When rounding off the calculated values of b_0 and b_1, always keep at least two significant digits in the final answer.

Now that we know the equation for the line of best fit, let's draw the line on the scatter diagram so that we can see the relationship between the line and the data. We need two points in order to draw the line on the diagram.

Select two convenient x values, one near each extreme of the domain ($x = 10$ and $x = 60$ are good choices for this illustration), and find their corresponding y values.

For $x = 10$: $\hat{y} = 14.9 + 0.66x = 14.9 + 0.66(10)$
$= 21.5$; **(10, 21.5)**

For $x = 60$: $\hat{y} = 14.9 + 0.66x = 14.9 + 0.66(60)$
$= 54.5$; **(60, 54.5)**

These two points, (10, 21.5) and (60, 54.5), are then located on the scatter diagram (we use a purple + to distinguish them from data points) and the line of best fit is drawn (shown in blue in Figure 3.10).

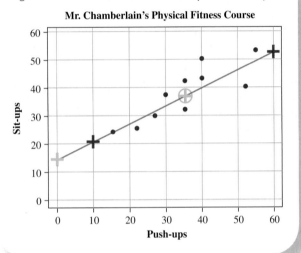

Figure 3.10 **Line of Best Fit for Push-ups versus Sit-ups**

There are some additional facts about the least squares method that we need to discuss.

1. The slope, b_1, represents the predicted change in y per unit increase in x. In our example, where $b_1 = 0.66$, if a student can do an additional 10 push-ups (x), we predict that he or she would be able to do approximately 7 (0.66×10) additional sit-ups (y).

2. The y-intercept is the value of y where the line of best fit intersects the y-axis. (When the vertical scale is located above $x = 0$, the y-intercept is easily seen on the scatter diagram, shown as a green + in Figure 3.10.) First, however, in interpreting b_0, you must consider whether $x = 0$ is a realistic x value before you can conclude that you would predict $\hat{y} = b_0$ if $x = 0$. To predict that if a student did no push-ups, he or she would still do approximately 15 sit-ups ($b_0 = 14.9$) is probably incorrect. Second, the x value of zero may be outside the domain of the data on which the regression line is based. In predicting y based on an x value,

est inch) and her weight (to the nearest 5 pounds). The data obtained are shown in Table 3.12. Find an equation to predict the weight of a college woman based on her height (the equation of the line of best fit), and draw it on the scatter diagram in Figure 3.11.

Before we start to find the equation for the line of best fit, it is often helpful to draw the scatter diagram, which provides visual insight into the relationship between the two variables. The scatter diagram for the data on the heights and weights of college women, shown in Figure 3.11, indicates that the linear model is appropriate.

To find the equation for the line of best fit, we first need to complete the preliminary calculations, as shown in Table 3.13. The other preliminary calculations include

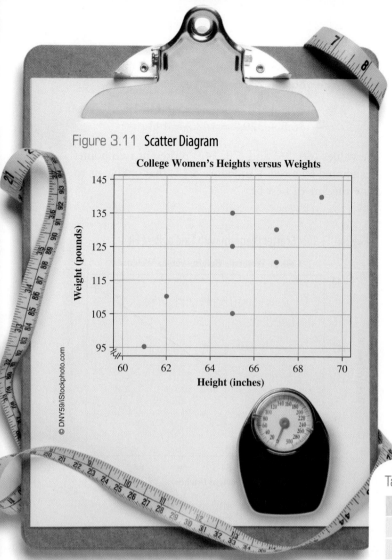

Figure 3.11 **Scatter Diagram**

College Women's Heights versus Weights

Table 3.12 **College Women's Heights and Weights**

	1	2	3	4	5	6	7	8
Height, x	65	65	62	67	69	65	61	67
Weight, y	105	125	110	120	140	135	95	130

Table 3.13 **Preliminary Calculations Needed to Find b_1 and b_0**

	Student Height, x	x^2	Weight, y	xy
1	65	4,225	105	6,825
2	65	4,225	125	8,125
3	62	3,844	110	6,820
4	67	4,489	120	8,040
5	69	4,761	140	9,660
6	65	4,225	135	8,775
7	61	3,721	95	5,795
8	67	4,489	130	8,710
	$\Sigma x = 521$	$\Sigma x^2 = 33,979$	$\Sigma y = 960$	$\Sigma xy = 62,750$

check to be sure that the x value is within the domain of the x values observed.

3. The line of best fit will always pass through the *centroid*, the point $(\bar{x}, \bar{y})$. When drawing the line of best fit on your scatter diagram, use this point as a check. For our illustration,

$$\bar{x} = \frac{\Sigma x}{n} = \frac{351}{10} = 35.1, \quad \bar{y} = \frac{\Sigma y}{n} = \frac{380}{10} = 38.0$$

We see that the line of best fit does pass through $(\bar{x}, \bar{y}) = (35.1, 38.0)$, as shown with the green ⊕ in Figure 3.10.

Let's work through another example to clarify the steps involved in regression analysis.

CALCULATING THE LINE OF BEST FIT EQUATION

In a random sample of eight college women, each woman was asked her height (to the near-

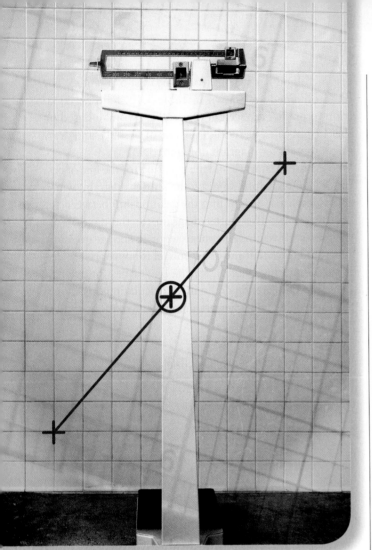

© PNC/Brand X Pictures/Jupiterimages / © AbleStock.com/Jupiterimages

$$\hat{y} = -186.5 + 4.71x = -186.5 + (4.71)(60)$$
$$= -186.5 + 282.6 = 96.1 \approx 96$$

$$\hat{y} = -186.5 + 4.71x = -186.5 + (4.71)(70)$$
$$= -186.5 + 329.7 = 143.2 \approx 143$$

The values (60, 96) and (70, 143) represent two points (designated by a red + in Figure 3.12) that enable us to draw the line of best fit.

Figure 3.12 **Scatter Diagram with Line of Best Fit**

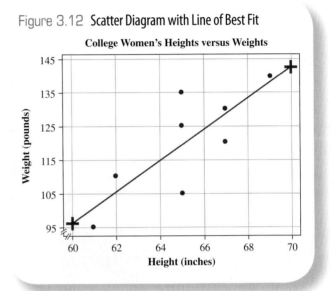

Making Predictions

One of the main reasons for finding a regression equation is to make predictions. Once a linear relationship has been established and the value of the input variable x is known, we can predict a value of y, $\hat{y}$. Consider the equation $\hat{y} = -186.5 + 4.71x$ relating the height and weight of college women. If a particular female college student is 66 inches tall, what do you predict her weight to be? The predicted value is

$$\hat{y} = -186.5 + 4.71x = -186.5 + (4.71)(66)$$
$$= -186.5 + 310.86$$
$$= 124.36 \approx 124 \text{ lb}$$

You should not expect this predicted value to occur exactly; rather, it is the average weight you would expect for all female college students who are 66 inches tall.

When you make predictions based on the line of best fit, observe the following restrictions:

1. The equation should be used to make predictions only about the population from which the sample was drawn. For example, using our relationship between the height and weight of college women

finding SS(x) from formula (2.8) and SS(xy) from formula (3.4):

$$SS(x) = \Sigma x^2 - \frac{(\Sigma x)^2}{n} = 33,979 - \frac{(521)^2}{8} = 48.875$$

$$SS(xy) = \Sigma xy - \frac{\Sigma x \Sigma y}{n} = 62,750 - \frac{(521)(960)}{8} = 230.0$$

Second, we need to find the slope and the y-intercept using formulas (3.6) and (3.7):

$$\text{slope: } b_1 = \frac{SS(xy)}{SS(x)} = \frac{230.0}{48.875} = 4.706 = \textbf{4.71}$$

$$y\text{-intercept: } b_0 = \frac{\Sigma y - (b_1 \cdot \Sigma x)}{n} = \frac{960 - (4.706)(521)}{8}$$
$$= -186.478 = \textbf{-186.5}$$

Thus, the equation of the line of best fit is

$$\hat{y} = \textbf{-186.5} + \textbf{4.71}x$$

To draw the line of best fit on the scatter diagram, we need to locate two points. Substitute two values for x—for example, 60 and 70—into the equation for the line of best fit to obtain two corresponding values for $\hat{y}$:

to predict the weight of professional athletes given their height would be questionable.

2. The equation should be used only within the sample domain of the input variable. We know that the data demonstrate a linear trend within the domain of the x data, but we do not know what the trend is outside this interval. Hence, predictions can be very dangerous outside the domain of the x data. For instance, in our current example, it is nonsense to predict that a college woman of height zero will weigh -186.5 pounds. Do not use a height outside the sample domain of 61 to 69 inches to predict weight. On occasion, you might wish to use the line of best fit to estimate values outside the domain interval of the sample. This can be done, but you should do it with caution and only for values close to the domain interval.

3. If the sample was taken in 2010, do not expect the results to have been valid in 1929 or to hold in 2020. The women of today may be different from the women of 1929 and the women in 2020.

Regression equations help you make accurate predictions.

UNLIKE

TEA LEAF READINGS

DID YOU KNOW

A Straight Regression Line

At the International Exposition in London in 1884, Sir Francis Galton set up a laboratory at which he paid people 3 pence to measure their heads. Galton was interested in predicting human intelligence and would give the person he paid his opinion of their intelligence. After the exposition, the laboratory moved to the London Museum, where Galton continued to collect data about human characteristics, such as height, weight, and strength. Galton made two-way plots of heights for parents and children, which eventually led to the slope of the regression line.

problems

Objective 3.1

*3.1 In a national survey of 500 business and 500 leisure travelers, each was asked where he or she would most like "more space."

	On Airplane	Hotel Room	All Other
Business	355	95	50
Leisure	250	165	85

a. Present the data as percentages of the total.

b. Present the data as percentages of the row totals. Why might one prefer the table to be expressed this way?

c. Present the data as percentages of the column totals. Why might one prefer the table to be expressed this way?

*3.2 What effect does the minimum amount have on the interest rate being offered on 3-month certificates of deposit (CDs)? The following are advertised rates of return, *y*, for a minimum deposit of $500, $1,000, $2,500, $5,000, or $10,000, *x*. (Note that *x* is in $100 and *y* is annual percentage rate of return.)

Min Deposit	Rate	Min Deposit	Rate	Min Deposit	Rate
100	0.95	25	1.00	25	0.75
100	1.24	50	1.00	10	0.75
10	1.24	100	1.00	100	0.70
10	1.15	5	1.00	5	0.64
100	1.10	10	1.00	10	0.50
50	1.09	10	0.80	100	0.35
100	1.07	10	0.75	25	0.35
5	1.00	10	0.75	5	0.99
25	0.75				

SOURCE: http://www. Bankrate.com, July 28, 2009

a. Prepare a dotplot of the six sets of data using a common scale.

b. Prepare a 5-number summary and a boxplot of the three sets of data. Use the same scale for the boxplots.

c. Describe any differences you see between the three sets of data.

*3.3 Can a woman's height be predicted from her mother's height? The heights of some mother-daughter pairs are listed; *x* is the mother's height and *y* is the daughter's height.

x	63	63	67	65	61	63	61	64	62	63
y	63	65	65	65	64	64	63	62	63	64

x	64	63	64	64	63	67	61	65	64	65	66
y	64	64	65	65	62	66	62	63	66	66	65

a. Draw two dotplots using the same scale and showing the two sets of data side by side.

b. What can you conclude from seeing the two sets of heights as separate sets in part (a)? Explain.

c. Draw a scatter diagram of these data as ordered pairs.

d. What can you conclude from seeing the data presented as ordered pairs? Explain.

3.4 Consider the two variables of a person's height and weight. Which variable, height or weight, would you use as the input variable when studying their relationship? Explain why.

3.5 Draw a coordinate axis and plot the points (0, 6), (3, 5), (3, 2), and (5, 0) to form a scatter diagram. Describe the pattern that the data show in this display.

3.6 Growth charts are commonly used by a child's pediatrician to monitor a child's growth. Consider the growth chart that follows.

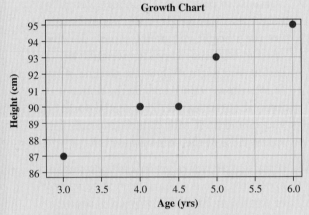

Growth Chart

a. What are the two variables shown in the graph?

b. What information does the ordered pair (3, 87) represent?

c. Describe how the pediatrician might use this chart and what types of conclusions might be based on the information displayed by it.

3.7 Does studying for an exam pay off?

a. Draw a scatter diagram of the number of hours studied, *x*, compared with the exam grade received, *y*.

x	2	5	1	4	2
y	80	80	70	90	60

b. Explain what you can conclude based on the pattern of data shown on the scatter diagram drawn in part (a).

*3.8 The accompanying data show the number of hours studied for an exam, *x*, and the grade received, *y* (*y* is measured in tens; that is, *y* = 8 means that the grade, rounded to the nearest 10 points, is 80). Draw the scatter diagram. (Retain this solution to use in problem 3.18, page 71.)

x	2	3	3	4	4	5	5	6	6	6	7	7	7	8	8
y	5	5	7	5	7	7	8	6	9	8	7	9	10	8	9

*3.9 An experimental psychologist asserts that the older a child is, the fewer irrelevant answers he or she will give during a controlled experiment. To investigate this claim, the following data were collected. Draw a scatter diagram. (Retain this solution to use in Exercise 3.19)

Age, x	2	4	5	6	6	7	9	9	10	12
Irrelevant Answers, y	12	13	9	7	12	8	6	9	7	5

*3.10 Refer to the information regarding 2009 4-wheel-drive, 6-cylinder SUVs on page 59. Specifically look at gas tank capacity, x, and the cost to fill it, y.

a. If you were to draw scatter diagrams of these two variables, on the same graph but separate, for the SUVs that use regular and premium gasoline, do you think the two sets of data would be distinguishable? Explain what you anticipate seeing.

b. Construct a scatter diagram of tank capacity, x, and fill-up cost, y, for the SUVs using regular gasoline.

c. Construct a scatter diagram of tank capacity, x, and fill-up cost, y, for the SUVs using premium gasoline on the scatter diagram for part (b).

d. Are the two sets of data distinguishable?

e. How does your answer in part (a) compare to your answer in part (d)? Explain any difference.

*3.11 Baseball stadiums vary in age, style, size, and in many other ways. Fans might think of the size of the stadium in terms of the number of seats; while the player might measure the size of the stadium by the distance from homeplate to the centerfield fence.

Seats	CF	Seats	CF	Seats	CF
38,805	420	36,331	434	40,950	435
41,118	400	43,405	405	38,496	400
56,000	400	48,911	400	41,900	400
45,030	400	50,449	415	42,271	404
34,077	400	50,091	400	43,647	401
40,793	400	43,772	404	42,600	396
56,144	408	49,033	407	46,200	400
50,516	400	47,447	405	41,222	403
40,615	400	40,120	422	52,355	408
48,190	406	41,503	404	45,000	408

CF = distance from homeplate to centerfield fence
SOURCE: http://mlb.mlb.com

Is there a relationship between these two measurements for the "size" of the 30 Major League Baseball stadiums?

a. What do you think you will find? Bigger fields have more seats? Smaller fields have more seats? No relationship exists between field size and number of seats? A strong relationship exists between field size and number of seats? Explain.

b. Construct a scatter diagram.

c. Describe what the scatter diagram tells you, including a reaction to your answer in (a).

Objective 3.2

3.12 How would you interpret the findings of a correlation study that reported a linear correlation coefficient of +0.3?

3.13 How would you interpret the finding of a correlation study that reported a linear correlation coefficient of −1.34?

3.14 Explain why it makes sense for a set of data to have a correlation coefficient of zero when the data show a very definite pattern, as in Figure 3.6 on page 61.

*3.15 Cell phones and iPods are necessities for this generation. Does the use of one indicate the use of the other? Seven junior high students who own both a cell phone and an iPod were randomly selected, resulting in the following data:

Cell, n(phone #s)	42	7	75	78	126	22	23
iPod, n(songs saved)	303	212	401	500	536	200	278

a. Complete the preliminary calculations: extensions, five sums, and SS(x), SS(y), SS(xy).

b. Find r.

3.16 Estimate the correlation coefficient for each of the following:

3.17 Manatees swim near the surface of the water. They often run into trouble with the many powerboats in Florida. Consider the graph that follows.

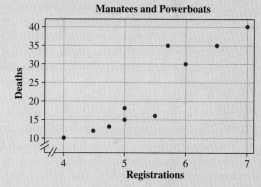

a. What two groups of subjects are being compared?

b. What two variables are being used to make the comparison?

c. What conclusion can one make based on this scatterplot?

d. What might you do if you were a wildlife official in Florida?

3.18 a. Use the scatter diagram you drew in problem 3.8 (page 70) to estimate r for the sample data on the number of hours studied and the exam grade.

b. Calculate r.

3.19 a. Use the scatter diagram you drew in Exercise 3.9 on page 70 to estimate r for the sample data on the number of irrelevant answers and the child's age.

b. Calculate r.

*3.20 Consider the following data for 2009 4WD 6-Cylinder SUVs.

2009 4WD 6-Cylinder SUVs			
Manufacturer	Model	Petro	Tons
Buick	Enclave	18.0	9.6
Chevrolet	Trailblazer	21.4	11.4
Chrysler	Aspen	22.8	12.2
Dodge	Durango	22.8	12.2

Continued on page 72

2009 4WD 6-Cylinder SUVs			
Manufacturer	Model	Petro	Tons
Ford	Escape	17.1	9.2
GMC	Envoy	21.4	11.4
Honda	Pilot	19.0	10.2
Jeep	Grand Cherokee	20.1	10.8
Kia	Sportage	17.1	9.2
Lexus	RX 350	18.0	9.6
Lincoln	MKX	18.0	9.6
Mazda	CX-7	19.0	10.2
Mercury	Mountaineer	22.8	12.2
Mitsubishi	Outlander	18.0	9.6
Nissan	Murano	17.1	9.2
Toyota	RAV4	16.3	8.7

a. What value do you anticipate for a correlation coefficient of the two variables, annual petroleum consumption in barrels, x, and annual tons of CO_2 emitted, y? Explain.

b. Calculate the linear correlation coefficient for the two variables, annual petroleum consumption in barrels, x, and annual tons of CO_2 emitted, y.

c. Is the value found in (b) approximately what you anticipated in (a)? Explain why or why not.

d. Does it make sense for the data to demonstrate such a high correlation? If the amount of consumption doubles, what do you think will happen to the tons of CO_2 emitted? Be specific in your explanation.

*3.21 The Children's Bureau of the U.S. Department of Health and Human Services has a monumental job. 510,000 children were in foster care in 2006. Of those, approximately 51,000 were adopted. Are more males or more females typically adopted? Is there a difference? The table lists the number of males and females adopted in each of sixteen randomly identified states.

State	Male	Female	State	Male	Female
Delaware	50	44	Wyoming	27	30
Nevada	231	213	New Jersey	689	636
Alabama	190	197	Arkansas	178	217
Michigan	1,296	1,296	Idaho	580	603
South Carolina	203	220	Hawaii	202	195
Iowa	512	472	Washington	586	610
Georgia	660	586	Tennessee	497	497
Vermont	90	74	Alaska	112	100

SOURCE: Children's Bureau, Administration for Children and Families, U.S. Department of Health and Human Services 2006

Is there a linear relationship between the number of males and females adopted from foster care during 2006? Use graphic and numerical statistics to support your answer.

*3.22 During the 2008 MLB All Star Game Home Run Derby, Josh Hamilton put on an amazing show with his 35 home runs. The recorded apex and distance of each home run he hit is listed here:

Apex—The highest point reached by the ball in flight above field level, in feet.

StdDist, Standard Distance—The estimated distance in feet the ball would have traveled if it flew uninterrupted all the way down to field level. Standard distance factors out the influence of wind, temperature, and altitude, and is thus the best way of comparing home runs hit under a variety of different conditions.

Apex	100	114	145	45	98	130	105	94	59
StdDist	459	474	404	378	479	443	393	410	356

Apex	112	50	144	154	153	132	126	123	118
StdDist	430	390	411	418	423	455	421	464	440

Apex	70	152	95	48	162	117	54	110	88
StdDist	432	435	447	386	364	447	379	423	442

Apex	125	47	119	111	84	155	153	116
StdDist	428	387	453	401	387	445	426	463

SOURCE: http://www.hittrackeronline.com/

a. Construct a scatter diagram using apex as x and standard distance as y.

b. Do the points seem to suggest a linear pattern? Explain.

c. Does it appear that the apex for the flight of a home run will be useful in predicting its length? Explain giving at least one reason that is non-statistical and at least one that is statistical.

d. What other factor about the flight of a home run ball might cause the pattern of points to be so varied?

e. Estimate the value of the linear correlation coefficient.

f. Calculate the correlation coefficient.

Objective 3.3

*3.23 Draw a scatter diagram for these data:

x	2	12	4	6	9	4	11	3	10	11	3	1	13	12	14	7	2	8
y	4	8	10	9	10	8	8	5	10	9	8	3	9	8	8	11	6	9

Would you be justified in using the techniques of linear regression on these data to find the line of best fit? Explain.

3.24 Does it pay to study for an exam? The number of hours studied, x, is compared to the exam grade received, y:

x	2	5	1	4	2
y	80	80	70	90	60

a. Find the equation for the line of best fit.

b. Draw the line of best fit on the scatter diagram of the data drawn in Exercise 3.7 on page 70.

c. Based on what you see in your answers to parts (a) and (b), does it pay to study for an exam? Explain.

3.25 Geoff is interested in purchasing a moderately priced SUV. He realizes that cars and trucks lose their value as soon as they are driven off the dealer's lot. Geoff used linear regression to get a better sense of how this

decline works. The regression line was
$\hat{y} = 34.03 - 3.04x$, where x is age of the car in years and y is the value of the car ($\times$ \$1,000). In terms of age and value:

a. Explain the meaning of the y-intercept, 34.03.
b. Explain the meaning of the slope, −3.04.

3.26 If all students from Mr. Chamberlain's physical fitness course (pages 58, 60, and 62–63) who can do 40 push-ups are asked to do as many sit-ups as possible:

a. How many sit-ups do you expect each can do?
b Will they all be able to do the same number?
c. Explain the meaning of the answer to part (a).

3.27 A study was conducted to investigate the relationship between the cost, y (in tens of thousands of dollars), per unit of equipment manufactured and the number of units produced per run, x. The resulting equation for the line of best fit was $\hat{y} = 7.31 - 0.01x$, with x being observed for values between 10 and 200. If a production run was scheduled to produce 50 units, what would you predict the cost per unit to be?

*3.28 "Now more than ever, a degree matters," according to an upstate New York college advertisement in the May 31, 2009, *Democrat and Chronicle*. The following statistics from the U.S. Bureau of Labor Statistics were presented on median usual weekly earnings.

Amount of Schooling	Median Usual Weekly Earnings	Years of Schooling
Less than a high school diploma	\$453	10
High school graduate, no college	\$618	12
Bachelor's degree	\$1,115	16
Advanced degree	\$1,287	18

a. Construct a scatter diagram with the years of schooling as the independent variable, x, and the median usual weekly earnings as the dependent variable, y.
b. Does there seem to be a linear relationship? Why?
c. Calculate the linear correlation coefficient.
d. Does the value of r seem reasonable compared with the pattern demonstrated in the scatter diagram? Explain.
e. Find the equation of the line of best fit.

f. Interpret the slope of the equation.
g. Plot the line of best fit on the scatter diagram.
h. What is the y-intercept for the equation? Interpret its meaning in this application.

*3.29 The following data are a sample of the ages and the asking prices for used Honda Accords that were listed on AutoTrader.com on March 10, 2005:

Age, x (years)	Price, y ($\times$ \$1000)	Age, x (years)	Price, y ($\times$ \$1000)
3	24.9	2	26.9
7	9.0	4	23.8
5	17.8	5	19.3
4	29.2	4	21.9
6	15.7	6	16.4
3	24.9	4	21.2
2	25.7	3	24.9
7	11.9	5	20.0
6	15.2	7	13.6
2	25.9	5	18.8

SOURCE: http://autotrader.com/

a. Draw a scatter diagram.
b. Calculate the equation of the line of best fit.
c. Graph the line of best fit on the scatter diagram.
d. Predict the average asking price for all Honda Accords that are 5 years old. Obtain this answer in two ways: Use the equation from part (b) and use the line drawn in part (c).
e. Can you think of any potential lurking variables for this situation? Explain any possible role they might play.

*remember

Problems marked with an asterisk have data sets available on the CourseMate for STAT2 site. Login at cengagebrain.com.

Probability

Have you ever opened a bag of M&M's to find only a couple of one color but a lot of another color? If you have, you may have wondered how one color compares to the rest of the colors. Percentages are a convenient way to make such a comparison. If you have 151 blue M&M's and there are 692 M&M's in the bag, you can figure out what percent of M&M's are blue with some simple math:

$$\frac{151}{692} = 0.218 \approx 0.22 \text{ or } 22\%$$

Thus, 22% of the M&M's in the bag are blue. Another way to look at this percentage is to imagine pulling one M&M from a thoroughly mixed container without looking. With the proportion above, you would have a 22% chance of picking a blue M&M.

objectives

4.1 Probability of Events

4.2 Conditional Probability of Events

4.3 Rules of Probability

4.4 Mutually Exclusive Events

4.5 Independent Events

4.6 Are Mutual Exclusiveness and Independence Related?

4.1 Probability of Events

WE ARE NOW READY TO DEFINE WHAT IS MEANT BY PROBABILITY.

Specifically, we talk about "the **probability of an event**" as the relative frequency with which that event can be expected to occur. The probability of an event may be obtained in three different ways: (1) *empirically*, (2) *theoretically*, and (3) *subjectively*.

The empirical method was just illustrated by the M&M's and their percentages and might be called **experimental** or **empirical probability.** This probability is the **observed relative frequency** with which an event occurs. In the M&M example, we observed that 151 of the 692 M&M's were blue. The observed empirical probability for the occurrence of blue is **151/692,** or 0.218.

The value assigned to the probability of event A as a result of experimentation can be found by means of the formula:

Empirical (Observed) Probability $P'(A)$

In words:

$$\text{empirical probability of A} = \frac{\text{number of times A occurred}}{\text{number of trials}}$$

In algebra: $P'(A) = \dfrac{n(A)}{n}$ (4.1)

The theoretical method for obtaining the probability of an event uses a *sample space*. A **sample space** is a listing of all possible outcomes from the experiment being considered (denoted by the capital letter S). When this method is used, the sample space must

Notation
for empirical probability

Notation for empirical probability: When the value assigned to the probability of an event results from experimental or empirical data, we will identify the probability of the event with the symbol $P'(\ )$.

contain **equally likely** sample points. For example, the sample space for the rolling of one die is S = {1, 2, 3, 4, 5, 6}. Each **outcome** (i.e., number) is equally likely. An **event** is a subset of the sample space (denoted by a capital letter other than S; A is commonly used for the first event). Therefore, the *probability of an event* A, P(A), is the ratio of the number of points that satisfy the definition of event A, $n(A)$, to the number of sample points in the entire sample space, $n(S)$. That is,

Theoretical (Expected) Probability P(A)

In words:

$$\text{theoretical probability of A} = \frac{\text{number of times A occurs in sample space}}{\text{number of elements in sample space}}$$

In algebra: $P(A) = \dfrac{n(A)}{n(S)}$ (4.2)

NOTES:

1. When the value assigned to the probability of an event results from a theoretical source, we will identify the probability of the event with the symbol P( ).

2. The prime symbol is *not used* with theoretical probabilities; it is used only for empirical probabilities.

Probability an event will occur The relative frequency with which that event can be expected to occur.

Sample space (S) A listing of all possible outcomes from the experiment being considered.

Event A subset of the sample space, denoted by a capital letter other than S.

Picturing the Sample Space

Consider one rolling of one die. In a single roll of a die, there are six possible outcomes, making $n(S) = 6$. Define event A as the occurrence of a

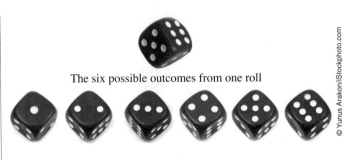

The six possible outcomes from one roll

© Yunus Arakon/iStockphoto.com

number "greater than 4." The event "greater than 4" is satisfied by the occurrence of either a 5 or a 6; thus, $n(A) = 2$. Assuming that the die is symmetrical and that each number has an equal likelihood of occurring, the probability of A is $\frac{2}{6}$, or $\frac{1}{3}$.

What happens when you role a pair of dice?

Chart Representation

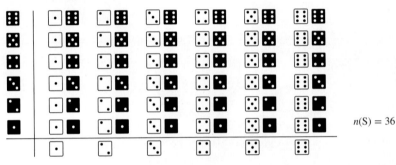

$n(S) = 36$

The sum of their dots is to be considered. A listing of the possible "sums" forms a sample space, S = {2, 3, 4, 5, 6, 7, 8, 9, 10, 11, 12} and $n(S) = 11$. However, the elements of this sample space are not equally likely; therefore, this sample space cannot be used to find theoretical probabilities—we must use the 36-point sample space shown in the above chart. By using the 36-point sample space, the sample space is entirely made up of equally likely sample points, and the probabilities for the sums of 2, 3, 4, and so on can be found quite easily. The sum of 2 represents {(1, 1)}, where the first element of the **ordered pair** is the outcome for the white die and the second element of the ordered pair is the outcome for the black die. The sum of 3 represents {(2, 1), (1, 2)}; and the sum of 4 represents {(1, 3), (3, 1), (2, 2)}; and so on. Thus, we can use formula (4.2) and the 36-point sample space to obtain the probabilities for each of the 11 sums.

$$P(2) = \frac{n(2)}{n(S)} = \frac{1}{36}; \quad P(3) = \frac{n(3)}{n(S)} = \frac{2}{36};$$

$$P(4) = \frac{n(4)}{n(S)} = \frac{3}{36}; \quad \text{and so forth.}$$

© Eliza Now/iStockphoto.com / © Wendy Idele/Nonstock/Getty Images

2. There are four branches; each branch starts at the "tree root" and continues to an "end" (made up of two branch segments each), showing a possible outcome.

Because the branch segments are equally likely, assuming equal likeliness of gender, the four branches are then equally likely. This means we need only the count of branches to use formula (4.2) to find the probability of the family having one child of each gender. The two middle branches, (B,G) and (G,B), represent the event of interest, so $n(A) = n(\text{one of each}) = 2$, whereas $n(S) = 4$ because there are a total of four branches. Thus,

$$P\left(\begin{array}{l}\text{one of each gender in} \\ \text{family of two children}\end{array}\right) = \frac{2}{4}$$
$$= \frac{1}{2} = 0.5$$

Now let's consider selecting a family of three children and finding the probability of "at least one boy" in that family. Again the family can be thought of as a sequence of three events—firstborn, second-born, and third-born. To create a tree diagram of this family, we need to add a third set of branch segments to our two-child family tree diagram. The green branch segments represent the third child (see Figure 4.2 on the next page).

Again, because the branch segments are equally likely, assuming equal likeliness of gender, the eight branches are then equally likely. This means we need only the count of branches to use formula (4.2) to find the probability of the family having at least one boy.

When a probability experiment can be thought of as a sequence of events, a **tree diagram** often is a very helpful way to picture the sample space. A family with two children is to be selected at random, and we want to find the probability that the family selected has one child of each gender. Because there will always be a firstborn and a second-born child, we will use a tree diagram to show the possible arrangements of gender, thus making it possible for us to determine the probability. Start by determining the sequence of events involved—firstborn and second-born in this case. Use the tree to show the possible outcomes of the first event (shown in brown in Figure 4.1) and then add branch segments to show the possible outcomes for the second event (shown in orange in Figure 4.1).

NOTES:

1. The two branch segments representing B and G for the second-born child must be drawn from each outcome for the firstborn child, thus creating the "tree" appearance.

Figure 4.1 **Tree Diagram Representation of Family with Two Children**

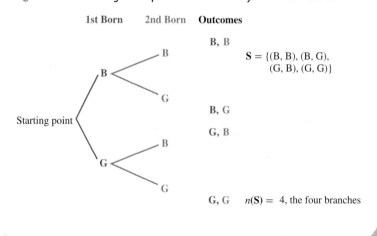

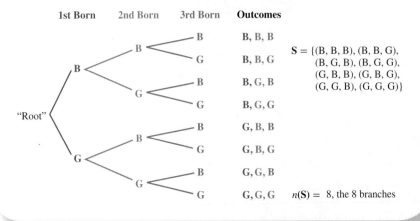

Figure 4.2 **Tree Diagram Representation of Family with Three Children**

1st Born	2nd Born	3rd Born	Outcomes
		B	B, B, B
	B	G	B, B, G
B		B	B, G, B
	G	G	B, G, G
		B	G, B, B
	B	G	G, B, G
G		B	G, G, B
	G	G	G, G, G

"Root"

$S = \{(B, B, B), (B, B, G),\ (B, G, B), (B, G, G),\ (G, B, B), (G, B, G),\ (G, G, B), (G, G, G)\}$

$n(S) = 8$, the 8 branches

The top seven branches all have one or more boys, the equivalent of "at least one."

$$P\left(\begin{array}{c}\text{at least one boy in a}\\\text{family of three children}\end{array}\right) = \frac{7}{8}$$
$$= 0.875$$

Let's consider one other question before we leave this example. What is the probability that the third child in this family of three children is a girl? The question is actually an easy one; the answer is 0.5, because we have assumed equal likelihood of either gender. However, if we look at the tree diagram in Figure 4.2, there are two ways to view the answer. First, if you look at only the branch segments for the third-born child, you see one of two is for a girl in each set, thus $\frac{1}{2}$, or 0.5. Also, if you look at the entire tree diagram, the last child is a girl on four of the eight branches; thus $\frac{4}{8}$, or 0.5.

Special attention should always be given to the sample space. Like the statistical population, the sample space must be well defined. Once the sample space is defined, you will find the remaining work much easier.

A **subjective probability** is a personal judgment determined by an observer with incomplete information. Your local weather forecaster often assigns a probability to the event "precipitation." For example, "There is a 20% chance of rain today," or "There is a 70% chance of snow tomorrow." The forecaster assigns a probability to the event based on past data about weather that followed after similar circumstances in the past, all the while knowing that all the factors that contribute to weather are not yet scientifically known. The more experience the observer (in this case, the forecaster) gains in relating current circumstances to past events, and the more types of data the observer takes into account, the more accurate the subjective probability becomes. Sub-

jective probabilities (also called Bayesian probabilities) are used increasingly in the natural sciences, the social sciences, medicine, and economics.

Properties of Probability Numbers

Whether the probability is *empirical*, *theoretical*, or *subjective*, the following two properties must hold. First, the probability is always a numerical value between zero and one. Second, the sum of the probabilities for all outcomes of an experiment is equal to exactly one.

PROPERTY 1 ("A probability is always a numerical value between zero and one.") can be expressed algebraically as follows:

$0 \le$ each $P(A) \le 1$ or $0 \le$ each $P'(A) \le 1$

Based on that formula, you can see that:

1. The probability is 0 if the event cannot occur.

2. The probability is 1 if the event occurs every time.

3. Otherwise, the probability is a fractional number between 0 and 1.

VS.

PROPERTY 2 ("The sum of the probabilities for all outcomes of an experiment is equal to exactly one.") can also be expressed algebraically:

$$\sum_{\text{all outcomes}} P(A) = 1 \quad \text{or} \quad \sum_{\text{all outcomes}} P'(A) = 1$$

For Property 2 to hold, the list of "all outcomes" must be a nonoverlapping set of events that includes all the possibilities (**all-inclusive**).

NOTES ABOUT PROBABILITY NUMBERS:

1. Probability represents a relative frequency.

2. $P(A)$ is the ratio of the number of times an event can be expected to occur divided by the number of trials. $P'(A)$ is the ratio of the number of times an event did occur divided by the number of data.

3. The numerator of the probability ratio must be a positive number or zero.

4. The denominator of the probability ratio must be a positive number (greater than zero).

5. As a result of Notes 1 through 4 above, the probability of an event, whether it be empirical, theoretical, or subjective, will always be a numerical value between zero and one, inclusively.

6. The rules for probability are the same for all three types of probability.

How Are Empirical and Theoretical Probabilities Related?

Consider the rolling of one die and define event A as the occurrence of a "1." An ordinary die has six equally likely sides, so the theoretical probability of event A is $P(A) = \frac{1}{6}$. What does this mean?

Do you expect to see one "1" in each trial of six rolls? Explain. If not, what results do you expect? If we were to roll the die several times and keep track of the proportion of the time event A occurs, we would observe an empirical probability for event A. What value would you expect to observe for $P'(A)$? Explain. How are the two probabilities $P(A)$ and $P'(A)$ related? Explain.

DEMONSTRATION—LAW OF LARGE NUMBERS

→ To gain some insight into this relationship, let's perform an experiment. The experiment will consist of 20 trials. Each trial of the experiment will consist of rolling a die six times and recording the number of times the "1" occurs. Perform 20 trials.

Each row of Table 4.1 shows the results of one trial; we conduct 20 trials, so there are 20 rows. Column 1 lists the number of 1s observed in each trial (set of six rolls); column 2 lists the observed relative frequency for each trial; and column 3 lists the cumulative relative frequency as each trial was completed.

Figure 4.3a on the next page shows the fluctuation (above and below) of the observed probability, $P'(A)$

Table 4.1 **Experimental Results of Rolling a Die Six Times in Each Trial**

Trial	Column 1: Number of 1s Observed	Column 2: Relative Frequency	Column 3: Cumulative Relative Frequency	Trial	Column 1: Number of 1s Observed	Column 2: Relative Frequency	Column 3: Cumulative Relative Frequency
1	1	1/6	1/6 = 0.17	11	1	1/6	10/66 = 0.15
2	2	2/6	3/12 = 0.25	12	0	0/6	10/72 = 0.14
3	0	0/6	3/18 = 0.17	13	2	2/6	12/78 = 0.15
4	1	1/6	4/24 = 0.17	14	1	1/6	13/84 = 0.15
5	0	0/6	4/30 = 0.13	15	1	1/6	14/90 = 0.16
6	1	1/6	5/36 = 0.14	16	3	3/6	17/96 = 0.18
7	2	2/6	7/42 = 0.17	17	0	0/6	17/102 = 0.17
8	2	2/6	9/48 = 0.19	18	1	1/6	18/108 = 0.17
9	0	0/6	9/54 = 0.17	19	0	0/6	18/114 = 0.16
10	0	0/6	9/60 = 0.15	20	1	1/6	19/120 = 0.16

Figure 4.3 Fluctuations Found in the Die-Tossing Experiment

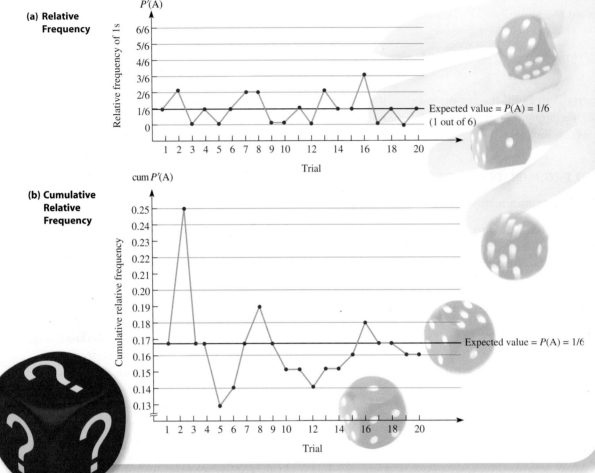

(a) Relative Frequency

P'(A)

Relative frequency of 1s

Expected value = *P*(A) = 1/6 (1 out of 6)

Trial

(b) Cumulative Relative Frequency

cum *P'(A)*

Cumulative relative frequency

Expected value = *P*(A) = 1/6

Trial

(Table 4.1, column 2), about the theoretical probability, $P(A) = \frac{1}{6}$, whereas Figure 4.3b shows the fluctuation of the cumulative relative frequency (Table 4.1, column 3) and how it becomes more stable. In fact, the cumulative relative frequency becomes relatively close to the theoretical or expected probability $\frac{1}{6}$, or $0.166\overline{6} = 0.167$.

A cumulative graph such as that shown in Figure 4.3b demonstrates the idea of a **long-term average.** Long-term average is often referred to as the **law of large numbers,** which states that as the number of times an experiment is repeated increases, the ratio of the number of successful occurrences to the number of trials will tend to approach the theoretical probability of the outcome for an individual trial.

The law of large numbers is telling us that the larger the number of experimental trials, n, the closer the empirical probability, $P'(A)$, is expected to be to the true or theoretical probability, $P(A)$. This concept has many applications. The preceding die-tossing experiment is an example in which we can easily compare actual results

against what we expected to happen; it gave us a chance to verify the claim of the law of large numbers.

Sometimes we live with the results obtained from large sets of data when the theoretical expectation is unknown. One such example occurs in the life insurance industry. The key to establishing proper life insurance rates is using the probability that those insured will live 1, 2, or 3 years, and so forth, from the time they purchase their policies. These probabilities are derived from actual life and death statistics and hence are empirical probabilities. They are published by the government and are extremely important to the life insurance industry.

Law of large numbers As the number of times an experiment is repeated increases, the ratio of the number of successful occurrences to the number of trials will tend to approach the theoretical probability of the outcome for an individual trial.

Probabilities as Odds

Probabilities can be and are expressed in many ways; we see and hear many of them in the news nearly every day (most of the time, they are subjective probabilities). **Odds** are a way of expressing probabilities by expressing the number of ways an event can happen compared to the number of ways it can't happen. The statement "It is four times more likely to rain tomorrow (R) than not rain (NR)" is a probability statement that can be expressed as odds: "The odds are 4 to 1 in favor of rain tomorrow" (also written 4:1).

The relationship between odds and probability is shown here.

> If the odds in favor of an event A are *a* to *b* (or *a:b*), then
>
> 1. The odds against event A are *b* to *a* (or *b:a*).
> 2. The probability of event A is $P(A) = \dfrac{a}{a+b}$.
> 3. The probability that event A will not occur is $P(\text{not A}) = \dfrac{b}{a+b}$.

To illustrate this relationship, consider the statement "The odds favoring rain tomorrow are 4 to 1." Using the preceding notation, $a = 4$ and $b = 1$.

Therefore, the probability of rain tomorrow is $\dfrac{4}{4+1}$, or $\dfrac{4}{5} = 0.8$. The odds against rain tomorrow are 1 to 4 (or 1:4), and the probability that there will be no rain tomorrow is $\dfrac{1}{4+1}$, or $\dfrac{1}{5} = 0.2$.

Let's consider a recognizable example of trying to beat the odds. Many young men aspire to become professional athletes. Only a few make it to the big time, as indicated in the following diagram. For every 13,600 college senior football players, only 250 are drafted by a professional team. That translates to a probability of only 0.018 (250/13,600).

There are many other interesting specifics hidden in this information. For example, many high school boys dream of becoming a professional football player, but according to these numbers, the probability of their dream being realized is only 0.000816 (250/306,200).

Once a player has made a college football team, he might be very interested in the odds that he will play as a senior. Of the 17,500 players making a college team as freshmen, 13,600 play as seniors, whereas 3,900 do not. Thus, if a player has made a college team, the odds he will play as a senior are 13,600 to 3,900, which reduces to 136 to 39.

If a college senior is fortunate enough to be playing, he will be interested in his chances of making it to the pros. We see that of the 13,600 college seniors, only 250 make the pros, whereas 13,350 do not; thus, the odds against him making it to the next level are 13,350 to 250, which reduces to 267 to 5. Odds are strongly against him making it.

Comparison of Probability and Statistics

Probability and **statistics** are two separate but related fields of mathematics. It has been said that "probability is the vehicle of statistics." That is, if it were not for the laws of probability, the theory of statistics would not be possible. Let's illustrate the relationship and the difference between these two branches of mathematics by looking at two sets of poker chips. On one hand, we know that the probability set contains twenty green, twenty red, and twenty blue poker chips. Probability tries to answer questions such as, "If one chip is randomly drawn from this set, what is the chance that it will be blue?" On the other hand, in the statistics set, we don't know what the combination of chips is. We draw a sample and, based on the

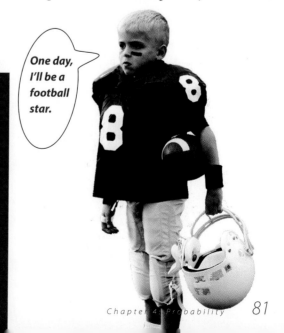

One day, I'll be a football star.

Players:
High school athletes: 306,200
College freshmen 17,500
College seniors 13,600
Drafted from college 250

findings in the sample, make conjectures about what we believe to be in the set. Note the difference: Probability asks you about the chance that something specific, such as drawing a blue chip, will happen when you know the possibilities (that is, you know the population). Statistics, in contrast, asks you to draw a sample, describe the sample (descriptive statistics), and then make inferences about the population based on the information found in the sample (inferential statistics).

Probability   Statistics

20G   20R   20B   ?   ?   ?

4.2 Conditional Probability of Events

MANY OF THE PROBABILITIES THAT WE SEE OR HEAR BEING USED ON A DAILY BASIS ARE THE RESULT OF CONDITIONS EXISTING AT THE TIME.

> **Conditional probability an event will occur** The relative frequency with which an event can be expected to occur under the condition that additional, preexisting information is known about some other event.

When the relative frequency with which an event can be expected to occur is based on the condition that additional preexisting information is known about some other event, we call this the **conditional probability an event will occur.** $P(A \mid B)$ is used to symbolize the probability of event A occurring under the condition that event B is known to already exist.

> **Some ways to say or express the conditional probability, $P(A \mid B)$, are:**
>
> - The *"probability of A, given B."*
> - The *"probability of A, knowing B."*
> - The *"probability of A happening, knowing B has already occurred."*

The concept of conditional probability is actually very familiar and occurs frequently without us even being aware of it. The news media often report many conditional probability values. However, they don't make the point that it is a conditional probability, and it passes for simple everyday arithmetic. Consider a poll of 13,660 voters in 250 precincts across the country during the 2008 presidential election (see Table 4.2).

➜ One person is to be selected at random from the sample of 13,660 voters. Using Table 4.2, find the answer to the following probability questions.

Table 4.2

Gender	Percentage of Voters	Percent for McCain	Percent for Obama	Percent for Others
Men	**48**	54	44	2
Women	52	**46**	56	1
Age				
18–29	**14**	36	63	1
30–44	27	55	44	1
45–64	39	44	45	1
65 and older	20	48	**52**	0

All of the percentages listed are to the nearest integer.

1. What is the probability that the person selected is a man? You answer: 0.48. Expressed in equation form:

P(voter selected is a man) = 0.48.

2. What is the probability that the person selected is 18 to 29 years old? You answer: 0.14. Expressed in equation form:

P(voter selected is of age 18 to 29) = 0.14.

3. Given that the voter selected was a woman, what is the probability she voted for McCain? You answer: 0.46. Expressed in equation form:

P(McCain | woman) = 0.46.

4. What is the probability that the person selected voted for Obama if the voter was 65 or older? Answer: 0.52. Expressed in equation form:

P(Obama | 65 and older) = 0.52.

NOTE: The first two are simple probabilities, whereas the last two are conditional probabilities.

The example above involved finding conditional probabilities from a table of percentages, but we can also find conditional probabilities from a table of count data. Let's stay with the field of politics but change our data set. This time we will consider a poll of 1,000 voters in 25 precincts across the country during the 2008 presidential election (see Table 4.3).

One person is to be selected at random from the sample of 1,000 voters. Using the table, find the answer to the following probability questions.

1. Given that the voter selected was a high school graduate, what is the probability the person voted for McCain? You Answer: 103/220 = 0.46818 = 0.47. Expressed in equation form:

P(McCain | HS graduate) = 103/220 = 0.46818 = 0.47.

2. Given that the voter selected had some college education, what is the probability the person voted for Obama? You Answer: 172/320 = 0.5375 = 0.54. Expressed in equation form:

P(Obama | some college) = 172/320 = 0.5375 = 0.54.

3. Given that the selected person voted for McCain, what is the probability the voter has a postgraduate education? You Answer: 88/477 = 0.1844 = 0.18. Expressed in equation form:

P(postgraduate | McCain) = 88/477 = 0.1844 = 0.18.

4. Given that the selected person voted for Obama, what is the probability the voter does not have a high school education? You Answer: 19/510 = 0.0372 = 0.04. Expressed in equation form:

P(no high school | Obama) = 19/510 = 0.0372 = 0.04.

NOTES:

1. The conditional probability notation is very informative and useful. When you express a conditional probability in equation form, it is to your advantage to use the most complete notation. That way, when you read the information back, all the information is there.

2. When finding a conditional probability, some possibilities will be eliminated as soon as the condition is known. Consider question 4 above. As soon as the conditional given that the selected person voted for Obama is stated, the 477 who voted for McCain and the 13 who voted for others are eliminated, leaving the 510 possible outcomes.

Table 4.3

Education	Number for Obama	Number for McCain	Number for Others	Number of Voters
No high school	19	20	1	40
High school graduate	114	103	3	220
Some college	172	147	1	320
College graduate	135	119	6	260
Postgraduate	70	88	2	160
Total	510	477	13	1,000

4.3 Rules of Probability

OFTEN, ONE WANTS TO KNOW THE PROBABILITY OF A **COMPOUND EVENT** BUT THE ONLY DATA AVAILABLE ARE THE PROBABILITIES OF THE RELATED SIMPLE EVENTS.

(Compound events are combinations of more than one simple event.) In the next few paragraphs, the relationship between these probabilities is summarized.

Finding the Probability of "Not A"

The concept of complementary events is fundamental to finding the probability of "not A." In **complementary events,** the complement of an event A, $\overline{A}$, is the set of all sample points in the sample space that do not belong to event A.

A few examples of complementary events are (1) the complement of the event "success" is "failure," (2) the complement of "selected voter is Republican" is "selected voter is not Republican," and (3) the complement of "no heads" on 10 tosses of a coin is "at least one head."

By combining the information in the definition of complement with Property 2 (page 79), we can say that

$$P(A) + P(\overline{A}) = 1.0 \text{ for any event A}$$

As a result of this relationship, we have the complement rule, which states that the probability of A complement = one minus the probability of A. This rule can be expressed algebraically this way:

$$P(\overline{A}) = 1 - P(A) \qquad (4.3)$$

Every event A has a complementary event $\overline{A}$. Complementary probabilities are very useful when the question asks for the probability of "at least one." Generally, this represents a combination of several events, but the complementary event "none" is a single outcome. It is easier to solve for the complementary event and get the answer by using formula (4.3).

USING COMPLEMENTS TO FIND PROBABILITIES

One way to find probabilities is to use complements. Two dice are rolled. What is the probability that the sum is at least 3 (that is, 3, 4, 5, ... , 12)? Suppose one of the dice is black and the other is white. (Recall the chart representation on page 76; it shows all 36 possible pairs of results when rolling a pair of dice.)

Rather than finding the probability for each of the sums 3, 4, 5, ... , 12 separately and adding, it is much simpler to find the probability that the sum is 2 ("less than 3") and then use formula (4.3) to find the probability of "at least 3," because "less than 3" and "at least 3" are complementary events. Using formula (4.3),

$$P(\text{sum of 2}) = P(A) = \frac{1}{36} \quad \left(\begin{array}{l} \text{"2" occurs only once in} \\ \text{the 36-point sample space} \end{array} \right)$$

$$P(\text{sum is at least 3}) = P(\overline{A}) = 1 - P(A) = 1 - \frac{1}{36} = \frac{35}{36}$$

Finding the Probability of "A B"

An hourly wage earner wants to estimate the chances of "receiving a promotion or getting a pay raise." The worker would be happy with either outcome. Historical information is available that will allow the worker to estimate the probability of "receiving a promotion" and "getting a pay raise" separately. In this section we will learn how to apply the **addition rule** to find the compound probability of interest.

> **Complementary event** The *complement of an event* A, $\overline{A}$, is the set of all sample points in the sample space that do not belong to event A.

The complement of event A is denoted by $\overline{A}$ (read "A complement").

GENERAL ADDITION RULE

→ Let A and B be two events defined in a sample space, S.

In words: *probability of A or B = probability of A + probability of B − probability of A and B*

In algebra: $P(A \text{ or } B) = P(A) + P(B) - P(A \text{ and } B)$ (4.4)

To see if the relationship expressed by the general addition rule works, let's look at it in action. A statewide poll of 800 registered voters in 25 precincts from across New York state was taken. Each voter was identified as being registered as Republican, Democrat, or other and then asked, "Are you in favor of or against the current budget proposal awaiting the governor's signature?" The resulting tallies are shown here.

	Number in Favor	Number Against	Number of Voters
Republican	136	88	224
Democrat	314	212	526
Other	14	36	50
Totals	464	336	800

Suppose one voter is to be selected at random from the 800 voters summarized in the preceding table. Let's consider the two events "The voter selected is in favor" and "The voter is a Republican." Find the four probabilities: $P(\text{in favor})$, $P(\text{Republican})$, $P(\text{in favor or Republican})$, and $P(\text{in favor and Republican})$. Then use the results to check the truth of the addition rule.

We find that the:

- Probability the voter selected is "in favor" = $P(\text{in favor}) = 464/800 = \underline{0.58}$.

- Probability the voter selected is "Republican" = $P(\text{Republican}) = 224/800 = \underline{0.28}$.

- Probability the voter selected is "in favor" or "Republican" = $P(\text{in favor or Republican})$ = $(136 + 314 + 14 + 88)/800 = 552/800 = \underline{0.69}$.

- Probability the voter selected is "in favor" and "Republican" = $P(\text{in favor and Republican}) = 136/800 = \underline{0.17}$.

NOTES ABOUT FINDING THE PRECEDING PROBABILITIES:

1. The connective "or" means "one or the other or both"; thus, "in favor or Republican" means all voters who satisfy either event.

2. The connective "and" means "both" or "in common"; thus, "in favor and Republican" means all voters who satisfy both events.

Now let's use the preceding probabilities to demonstrate the truth of the addition rule.

Let A = "in favor" and B = "Republican." The general addition rule then becomes:

$P(\text{in favor or Republican}) = P(\text{in favor}) + P(\text{Republican}) - P(\text{in favor and Republican})$

Remember, previously we found $P(\text{in favor or Republican}) = \underline{0.69}$. Using the other three probabilities, we see:

$P(\text{in favor}) + P(\text{Republican}) - P(\text{in favor and Republican}) = 0.58 + 0.28 - 0.17 = \underline{0.69}$

Thus, we obtain identical answers by applying the addition rule and by referring to the relevant cells in the table. You typically do not have the option of finding $P(A \text{ or } B)$ two ways, as we did here. You will be asked to find $P(A \text{ or } B)$ starting with $P(A)$ and $P(B)$. However, you will need a third piece of information. In the previous situation, we needed $P(A \text{ and } B)$. We will need to know either $P(A \text{ and } B)$ or some information that allows us to find it.

Finding the Probability of "A AND B"

Suppose a criminal justice professor wants his class to determine the likeliness of the event "a driver is ticketed for a speeding violation and the driver had previously attended a defensive driving class." The students are confident they can find the probabilities of "a driver being ticketed for speeding" and "a driver who has attended a defensive driving class" separately. In this section we will learn how to apply the **multiplication rule** to find the compound probability of interest.

© Brand X Pictures/Getty Images

GENERAL MULTIPLICATION RULE

Let A and B be two events defined in sample space S.

In words: *probability of A and B = probability of A × probability of B, knowing A*

In algebra: $P(A \text{ and } B) = P(A) \cdot P(B \mid A)$ (4.5)

NOTE: When two events are involved, either event can be identified as A, with the other identified as B. The general multiplication rule could also be written as $P(B \text{ and } A) = P(B) \cdot P(A \mid B)$.

To have a better understanding of the multiplication rule, let's go back to our statewide poll of 800 registered voters in 25 precincts from across New York state. Each voter was identified as being registered as Republican, Democrat, or other and then asked, "Are you are in favor of or against the current budget proposal awaiting the governor's signature?" The resulting tallies are shown here.

	Number in Favor	Number Against	Number of Voters
Republican	136	88	224
Democrat	314	212	526
Other	14	36	50
Totals	464	336	800

Suppose one voter is to be selected at random from the 800 voters summarized in the preceding table. Let's consider the two events: "The voter selected is in favor" and "The voter is a Republican." Find the three probabilities: $P(\text{in favor})$, $P(\text{Republican} \mid \text{in favor})$, and $P(\text{in favor and Republican})$. Then use the results to check the truth of the multiplication rule. We find that the:

- Probability the voter selected is "in favor"
  $= P(\text{in favor}) = 464/800 = \underline{0.58}$.

- Probability the voter selected is "Republican, given in favor"
  $= P(\text{Republican} \mid \text{in favor})$
  $= 136/464 = \underline{0.29}$.

- Probability the voter selected is "in favor" and "Republican"
  $= P(\text{in favor and Republican})$
  $= 136/800 = \frac{136}{800} = \underline{0.17}$.

NOTES ABOUT FINDING THE PRECEDING PROBABILITIES

1. The conditional "given" means there is a restriction; thus, "Republican | in favor" means we start with only those voters who are "in favor." In this case, this means we are looking only at 464 voters when determining this probability.

2. The connective "and" means "both" or "in common"; thus, "in favor and Republican" means all voters who satisfy both events.

Now let's use the previous probabilities to demonstrate the truth of the multiplication rule.

Let A = "in favor" and B = "Republican." The general multiplication rule then becomes:

$P(\text{in favor and Republican})$
$= P(\text{in favor}) \cdot P(\text{Republican} \mid \text{in favor})$

Previously we found:
$P(\text{in favor and Republican}) = \frac{136}{800} = \underline{0.17}$.

Using the other two probabilities, we see:

$P(\text{in favor}) \cdot P(\text{Republican} \mid \text{in favor})$
$= \frac{464}{800} \cdot \frac{136}{464} = \frac{136}{800} = \underline{0.17}$.

Figure 4.4 Tree Diagram—First Two Drawings, Carnival Game

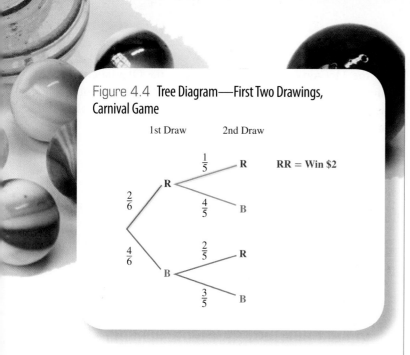

1st Draw   2nd Draw

R — $\frac{1}{5}$ — R   RR = Win \$2

R — $\frac{4}{5}$ — B

$\frac{2}{6}$ (R)

$\frac{4}{6}$ (B)

B — $\frac{2}{5}$ — R

B — $\frac{3}{5}$ — B

CONDITIONAL PROBABILITY AND DRAWING WITHOUT REPLACEMENT

You typically do not have the option of finding $P(A \text{ and } B)$ two ways, as we did here. When you are asked to find $P(A \text{ and } B)$, you will often be given $P(A)$ and $P(B)$. However, you will not always get the correct answer by just multiplying those two probabilities together. You will need a third piece of information: the conditional probability of one of the two events or information that will allow you to find it.

At a carnival game, the player blindly draws one colored marble at a time from a box containing two red and four blue marbles. The chosen marble is not returned to the box after being selected; that is, each drawing is done without replacement. The marbles are mixed before each drawing. It costs \$1 to play, and if the first two marbles drawn are red, the player receives a \$2 prize. If the first four marbles drawn are all blue, the player receives a \$5 prize. Otherwise, no prize is awarded. To find the probability of winning a prize, let's look first at the probability of drawing red or blue on consecutive drawings and organize the information on a tree diagram.

On the first draw (represented by the purple branch segments in Figure 4.4), the probability of red is two chances out of six, $\frac{2}{6}$, or $\frac{1}{3}$, whereas the probability of blue is $\frac{4}{6}$, or $\frac{2}{3}$. Because the marble is not replaced, only five marbles are left in the box; the number of each color remaining depends on the color of the first marble drawn. If the first marble was red, then the probabilities are $\frac{1}{5}$ and $\frac{4}{5}$, as shown on the tree diagram (green branch segments in Figure 4.4). If the first marble was blue, then the probabilities are $\frac{2}{5}$ and $\frac{3}{5}$,

as shown on the tree diagram (orange branch segments in Figure 4.4). The probabilities change with each drawing, because the number of marbles available keeps decreasing as each drawing takes place. The tree diagram is a marvelous pictorial aid in following the progression.

The probability of winning the \$2 prize can now be found using formula (4.5):

$$P(A \text{ and } B) = P(A) \cdot P(B \mid A)$$

$$P(\text{winning \$2}) = P(R_1 \text{ and } R_2) = P(R_1) \cdot P(R_2 \mid R_1)$$

$$= \frac{2}{6} \cdot \frac{1}{5} = \frac{1}{15} = 0.067$$

NOTE: The tree diagram, when labeled, has the probabilities needed for multiplying listed along the branch representing the winning effort.

4.4 Mutually Exclusive Events

TO FURTHER OUR DISCUSSION OF COMPOUND EVENTS, THE CONCEPT OF "MUTUALLY EXCLUSIVE" MUST BE INTRODUCED.

Mutually exclusive events are nonempty events defined on the same sample space with each event excluding the occurrence of the other. In other words, they are events that share no common elements. In algebra:

$$P(A \text{ and } B) = 0$$

> **Mutually exclusive events** Nonempty events defined on the same sample space with each event excluding the occurrence of the other. In other words, they are events that share no common elements.

There are several equivalent ways to express the concept of mutually exclusive:

1. If you know that either one of the events has occurred, then the other event is excluded or cannot have occurred.

2. If you are looking at the lists of the elements making up each event, none of the elements listed for either event will appear on the other event's list; there are "no shared elements."

3. The equation says, "the **intersection** of the two events has a probability of zero," meaning "the intersection is an empty set" or "there is no intersection."

NOTE: The concept of mutually exclusive events is based on the relationship between the sets of elements that satisfy the events. Mutually exclusive is not a probability concept by definition; it just happens to be easy to express the concept using a probability statement.

Understanding Mutually Exclusive (and Not Mutually Exclusive) Events

To help understand the difference between mutually exclusive and not mutually exclusive events, let's look at some examples.

Consider a poll of 1,000 voters in 25 precincts across the country during the 2008 presidential election, which provided the following:

Education	Number for McCain	Number for Obama	Number for Others	Number of Voters
No high school	19	20	1	40
High school graduate	114	103	3	220
Some college	172	147	1	320
College graduate	135	119	6	260
Postgraduate	70	88	2	160
Total	510	477	13	1,000

Suppose one voter is selected at random from the 1,000 voters summarized in the table. Consider the two events "The voter selected voted for McCain" and "The voter selected voted for Obama." In order for the event "the selected voter voted for McCain" to occur, the selected voter must be 1 of the 510 voters listed in the "Number for McCain" column. In order for the event "the selected voter voted for Obama" to occur, the voter selected must be 1 of the 477 voters listed in the "Number for Obama" column. Because no voter listed in the McCain column is also listed in the Obama column and because no voter listed in the Obama column is also listed in the McCain column, these two events are mutually exclusive.

In equation form:

P(voted for McCain and voted for Obama) = 0.

Now, let's look at the same situation but from a different angle to understand not mutually exclusive events. Consider the same poll of 1,000 voters in 25 precincts across the country during the 2008 presidential election, which provided the following:

Education	Number for McCain	Number for Obama	Number for Others	Number of Voters
No high school	19	20	1	40
High school graduate	114	103	3	220
Some college	172	147	1	**320**
College graduate	135	119	6	260
Postgraduate	70	88	2	160
Total	510	477	13	**1,000**

© PeJo29/iStockphoto.com

Suppose one voter is selected at random from the 1,000 voters summarized in the table. Consider the two events "The voter selected voted for McCain" and "The voter selected had some college education." In order for the event "the selected voter voted for Mc-Cain" to occur, the voter selected must be 1 of the 510 voters listed in the "Number for McCain" column. In order for the event "the selected voter had some college education" to occur, the selected voter must be 1 of the 320 voters listed in the "Some college" row. Because the 172 voters shown in the intersection of the "Number for McCain" column and the "Some college" row belong to both of the events ("the selected voter voted for McCain" and "the selected voter had some college education"), these two events are NOT mutually exclusive.

In equation form:

P(voted for McCain and some college education) = **172/1000** = 0.172, which is not equal to zero.

If you're having trouble visualizing these concepts in terms of politics, consider drawing one card from a regular deck of playing cards and the two events "card drawn is a queen" and "card drawn is an ace." The deck is to be shuffled and one card randomly drawn. In order for the event "card drawn is a queen" to occur, the card drawn must be one of the four queens: queen of hearts, queen of diamonds, queen of spades, or queen of clubs. In order for the event "card drawn is an ace" to occur, the card drawn must be one of the four aces: ace of hearts, ace of diamonds, ace of spades, or ace of clubs. Notice that there is no card that is both a queen and an ace. Therefore, these two events, "card drawn is a queen" and "card drawn is an ace," are mutually exclusive events.

In equation form:

P(queen and ace) = 0.

Similarly, we can demonstrate the concept of not mutually exclusive events with the same regular deck of playing cards and the two events "card drawn is a queen" and "card drawn is a heart." The deck is to be shuffled and one card randomly drawn. Are the events "queen" and "heart" mutually exclusive? The event "card drawn is a queen" is made up of the four queens: queen of hearts, queen of diamonds, queen of spades, and queen of clubs. The event "card drawn is a heart" is made up of the 13 hearts: ace of hearts, king of hearts, queen of hearts, jack of hearts, and the other nine hearts. Notice that the "queen of hearts" is on both lists, thereby making it possible for both events "card drawn is a queen" and "card drawn is a heart" to occur simultaneously. This means that when one of these two events occurs, it does not exclude the possibility of the other's occurrence. These events are not mutually exclusive events.

In equation form:

P(queen and heart) = 1/52, which is not equal to zero.

Visual Display and Understanding of Mutually Exclusive Events

How can we visually depict what happens with mutually exclusive events? Consider an experiment in which two dice are rolled. Three events are defined as follows:

A: The sum of the numbers on the two dice is 7.

B: The sum of the numbers on the two dice is 10.

C: Each of the two dice shows the same number.

Let's determine whether these three events are mutually exclusive.

We can show that three events are mutually exclusive by showing that each pair of events is mutually exclusive. Are events A and B mutually exclusive? Yes, they are, because the sum on the two dice cannot be both 7 and 10 at the same time. If a sum of 7 occurs, it is impossible for the sum to be 10.

Figure 4.5 presents the sample space for this experiment. This is the same sample space shown in the chart representation on page 76, except that ordered pairs are used in place of the pictures. The ovals, diamonds, and rectangles show the ordered pairs that are

Figure 4.5 **Sample Space for the Roll of Two Dice**

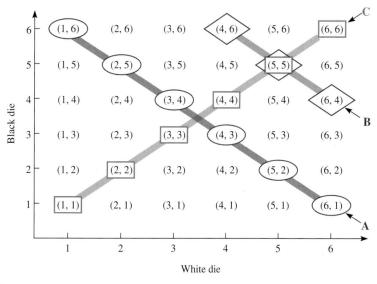

in events A, B, and C, respectively. We can see that events A and B do not intersect. Therefore, they are mutually exclusive. Point (5, 5) in Figure 4.5 satisfies both events B and C. Therefore, B and C are not mutually exclusive. Two dice can each show a 5, which satisfies C, and the total satisfies B. Since we found one pair of events that are not mutually exclusive, events A, B, and C are not mutually exclusive.

Special Addition Rule

The addition rule simplifies when the events involved are mutually exclusive. If we know that two events are mutually exclusive, then by applying $P(A \text{ and } B) = 0$ to the addition rule for probabilities, it follows that $P(A \text{ or } B) = P(A) + P(B) - P(A \text{ and } B)$ becomes $P(A \text{ or } B) = P(A) + P(B)$. In other words, when A and B are two mutually exclusive events defined in a sample space S, "the probability of A or B = probability of A + probability of B." This is known as the special addition rule. In basic algebraic terms:

$$P(A \text{ or } B) = P(A) + P(B) \qquad (4.6)$$

This formula can be expanded to consider more than two mutually exclusive events:

$$P(A \text{ or } B \text{ or} \ldots \text{ or } E) = P(A) + P(B) + P(C) + \ldots + P(E)$$

This equation is often convenient for calculating probabilities, but it does not help us understand the relationship between the events A and B. It is the *definition* that tells us how we should think about mutually exclusive events. Students who understand mutual exclusiveness this way gain insight into what mutual exclusiveness is all about. This should lead you to think more clearly about situations dealing with mutually exclusive events, thereby making you less likely to confuse the concept of mutually exclusive events with independent events (to be defined in Objective 4.5) or to make other common mistakes regarding the concept of mutual exclusivity.

NOTES:

1. Define mutually exclusive events in terms of the sets of elements satisfying the events and test for mutual exclusiveness in that manner.

2. Do not use $P(A \text{ and } B) = 0$ as the definition of mutually exclusive events. It is a property that results from the definition. It can be used as a test for mutually exclusive events; however, as a statement, it shows no meaning or insight into the concept of mutually exclusive events.

3. In equation form, the *definition* of mutually exclusive events states:

$P(A \text{ and } B) = 0$
(Both cannot happen at same time.)

$P(A \mid B) = 0$ and $P(B \mid A) = 0$
(If one is known to have occurred, then the other has not.)

Reconsider our mutually exclusive card event, with the two events "card drawn is a queen" and "card drawn is an ace" when drawing exactly one card from a deck of regular playing cards. The one card drawn is a queen, or the one card drawn is an ace. That one card cannot be both a queen and an ace at the same time, thereby making these two events mutually exclusive. The special addition rule therefore applies to the situation of finding $P(\text{queen or ace})$.

$$P(\text{queen or ace}) = P(\text{queen}) + P(\text{ace})$$
$$= \frac{4}{52} + \frac{4}{52} = \frac{8}{52} = \frac{2}{13}$$

4.5 Independent Events

THE CONCEPT OF **INDEPENDENT EVENTS** IS NECESSARY TO CONTINUE OUR DISCUSSION OF COMPOUND EVENTS.

Two events are *independent* if the occurrence (or nonoccurrence) of one gives us no information about the likeliness of occurrence of the other. In other words, if the probability of A remains unchanged after we know that B has happened (or has not happened), the events are independent. Algebraically, this situation is expressed as follows:

$$P(A) = P(A \mid B) = P(A \mid \text{not } B)$$

There are several equivalent ways to express the concept of independence:

1. The probability of event A is unaffected by knowledge that a second event, B, has occurred, knowledge that B has not occurred, or no knowledge about event B whatsoever.

> **Independent events** Two events are independent if the occurrence (or nonoccurrence) of one gives us no information about the likeliness of occurrence of the other.

2. The probability of event A is unaffected by knowledge, or by lack of knowledge, about a second event, B, having occurred or not occurred.

3. The probability of event A when we have no knowledge about event B is the same as the probability of event A when we know that event B has occurred, and both are the same as the probability of event A when we know that event B has not occurred.

Understanding Independent (and Not Independent) Events

Not all events are independent. Some events are **dependent**. That is, the occurrence of one event does have an effect on the probability of occurrence of the other event. To better illustrate this difference, let's continue with the contexts of elections and cards.

UNDERSTANDING INDEPENDENT EVENTS

→ A statewide poll of 750 registered Republicans and Democrats in 25 precincts from across New York state was taken. Each voter was identified as being registered as a Republican or a Democrat and then asked, "Are you are in favor of or against the current budget proposal awaiting the governor's signature?" The resulting tallies are shown here.

	Number in Favor	Number Against	Number of Voters
Republican	135	90	225
Democrat	315	210	525
Totals	450	300	750

Suppose one voter is to be selected at random from the 750 voters summarized in the preceding table. Let's consider the two events "The selected voter is in favor" and "The voter is a Republican." Are these two events independent?

To answer this, consider the following three probabilities: (1) probability the selected voter is in favor; (2) probability the selected voter is in favor, knowing the voter is a Republican; and (3) probability the selected voter is in favor, knowing the voter is not a Republican.

- Probability the selected voter is in favor
  = P(in favor) = **450/750** = 0.60.

- Probability the selected voter is in favor, knowing voter is a Republican
  = P(in favor | Republican) = **135/225** = 0.60.

- Probability the selected voter is in favor, knowing voter is not a Republican =
  Probability the selected voter is in favor, knowing voter is a Democrat
  = P(in favor | not Republican)
  = P(in favor | Democrat) = 315/**525** = 0.60.

Does knowing the voter's political affiliation have an influencing effect on the probability that the voter is in favor of the budget proposal? With no information about political affiliation, the probability of being in favor is 0.60. Information about the event "Republican" does not alter the probability of "in favor." They are all the value 0.60. Therefore, these two events are said to be *independent events*.

When checking the three probabilities, $P(A)$, $P(A \mid B)$, and $P(A \mid \text{not } B)$, we need to compare only two of them. If any two of the three probabilities are equal, the third will be the same value. Furthermore, if any two of the three probabilities are unequal, then all three will be different in value.

> **Dependent events**
> Events that are not independent, meaning the occurrence of one event does have an effect on the probability of occurrence of the other event.

NOTE: Determine all three values, using the third as a check. All will be the same, or all will be different—there is no other possible outcome.

Consider a regular deck of playing cards and the two events "card drawn is a queen" and "card drawn is a heart." Suppose that I shuffle the deck, randomly draw one card, and, before looking at the card, ask you the probability that it is a queen. You say $\frac{4}{52}$, or $\frac{1}{13}$. Then I peek at the card and tell you that it is a heart. Now, what is the probability that the card is a queen? You say it is $\frac{1}{13}$, the same as before knowing the card was a "heart."

The hint that the card was a heart provided you with additional information, but that information did not change the probability that it was a queen. Therefore, "queen" and "heart" are independent. Furthermore, suppose that after I drew the card and looked at it, I had told you the card was "not a heart." What would be the probability the card is a queen? You say $\frac{3}{39}$, or $\frac{1}{13}$. Again, notice that knowing the card was not a heart did provide additional information, but that information did not change the probability that it was a queen. This is what it means for the two events "card is a queen" and "card is a heart" to be independent.

In equation form:

$$P(\text{queen} \mid \text{heart}) = P(Q \mid H) = P(Q)$$

$$P(\text{queen} \mid \text{not heart}) = P(Q \mid \text{not H}) = P(Q)$$

Therefore, $P(Q) = P(Q \mid H) = P(Q \mid \text{not H})$, and the two events are independent.

UNDERSTANDING NOT INDEPENDENT EVENTS

With dependent (not independent) events, the occurrence of one event does have an effect on the probability of occurrence of the other event. Let's return to the election. Consider a poll of 13,660 voters in 250 precincts across the country during the 2008 presidential election, which provided the following:

	Percentage of Voters	Percent for Obama	Percent for McCain	Percent for Other
Men	48	**44**	54	2
Women	52	56	43	1

Suppose one voter is selected at random from the 13,660 voters summarized in the preceding table. Let's consider the two events "The voter is a woman" and "The voter voted for Obama." Are these two events independent? To answer this, consider this question: "Does knowing the voter is a woman have an influencing effect on the probability that the voter voted for Obama?" What is the probability of voting for Obama,

if the voter is a woman? You say, "0.56." Now compare this to the probability of voting for Obama if the voter is not a woman. You say that the probability is 0.44. So I ask you, "Did knowing the voter was a woman influence the probability of the voter having voted for Obama?" Yes, it did; it is 0.56 when the voter is a woman and 0.44 when the voter is not a woman. Information about the event "woman" does alter the probability of "voted for Obama." Therefore, these two events are *not independent* and are said to be *dependent events*.

In equation form:

$$P(\text{voted for Obama} \mid \text{knowing voter is a woman}) = P(O \mid W) = 0.56$$

$$P(\text{voted for Obama} \mid \text{knowing voter is not a woman}) = P(O \mid \overline{W}) = 0.44$$

Therefore, $P(O \mid W) \neq P(O \mid \overline{W})$, and the two events are not independent.

Now, let's consider the two events "card drawn is a heart" and "card drawn is red." Are the events "heart" and "red" independent? Following the same scenario as previously, I shuffle the deck of 52 cards, randomly draw one card, and before looking at it, you say the probability that the unknown card is red is $\frac{26}{52} = \frac{1}{2}$. However, when told the additional information that the card is a heart, you change your probability that the card is red to $\frac{13}{13}$, or 1. This additional information results in a different probability of red.

$P(\text{red} \mid \text{card is heart}) = P(R \mid H) = \frac{13}{13} = 1$, and $P(\text{red}) = P(\text{red} \mid \text{having no additional information}) = \frac{26}{52} = \frac{1}{2}$. Therefore, the additional information did change the probability of the event "red." These two events are not independent and therefore are said to be dependent events.

In equation form, the definition states:

A and B are independent if and only if $P(A \mid B) = P(A)$

Special Multiplication Rule

The multiplication rule simplifies when the events involved are independent.

If we know two events are independent, then by applying the definition of independence, $P(B \mid A) = P(B)$, to the multiplication rule, it follows that:

$$P(A \text{ and } B) = P(A) \cdot P(B \mid A)$$

becomes

$$P(A \text{ and } B) = P(A) \cdot P(B)$$

NOTE: Define independence in terms of conditional probability, and test for independence in that manner.

The result is then the **special multiplication rule.** Let A and B be two independent events defined in a sample space S. The special multiplication rule states that the "probability of A and B equals the probability of A times the probability of B." In algebraic terms, this means:

$$P(A \text{ and } B) = P(A) \cdot P(B) \qquad (4.7)$$

This formula can be expanded to consider more than two independent events:

$$P(A \text{ and } B \text{ and } C \text{ and } \ldots \text{ and } E)$$
$$= P(A) \cdot P(B) \cdot P(C) \cdot \ldots \cdot P(E)$$

To understand the independence relationship between the events A and B, you need to refer to the definition. If you think more clearly about situations dealing with independent events, you'll be less likely to confuse the concept of independent events with mutually exclusive events or to make other common mistakes regarding independence.

NOTE: Do not use $P(A \text{ and } B) = P(A) \cdot P(B)$ as the definition of independence. It is a property that results from the definition. It can be used as a test for independence, but as a statement, it shows no meaning or insight into the concept of independent events.

4.6 Are Mutual Exclusiveness and Independence Related?

MUTUALLY EXCLUSIVE EVENTS AND INDEPENDENT EVENTS ARE TWO VERY DIFFERENT CONCEPTS, BASED ON DEFINITIONS THAT START FROM VERY DIFFERENT ORIENTATIONS.

The two concepts can easily be confused because they interact with each other and are intertwined by the probability statements we use in describing these concepts.

To describe these two concepts and eventually understand the distinction between them as well as the relationship between them, we need to agree that the events being considered are two nonempty events defined on the same sample space and therefore each has nonzero probabilities.

The following examples give further practice with these probability concepts.

Calculating Probabilities and the Addition Rule

A pair of dice is rolled. Event T is defined as the occurrence of "a total of 10 or 11," and event D is the occurrence of "doubles." To find the probability $P(T \text{ or } D)$, you need to look at the sample space of 36 ordered pairs for the rolling of two dice in Figure 4.5 (page 89). Event T occurs if any one of 5 ordered pairs occurs: (4, 6), (5, 5), (6, 4), (5, 6), (6, 5). Therefore, $P(T) = \frac{5}{36}$. Event D occurs if any one of 6 ordered pairs occurs: (1, 1), (2, 2), (3, 3), (4, 4), (5, 5), (6, 6). Therefore, $P(D) = \frac{6}{36}$. Notice, however, that these two events are not mutually exclusive. The two events "share" the ordered pair (5, 5). Thus, the probability $P(T \text{ and } D) = \frac{1}{36}$. As a result, the probability $P(T \text{ or } D)$ will be found using formula (4.4).

$$P(T \text{ or } D) = P(T) + P(D) - P(T \text{ and } D)$$

$$= \frac{5}{36} + \frac{6}{36} - \frac{1}{36} = \frac{10}{36} = \frac{5}{18}$$

Look at the sample space in Figure 4.5 (page 89) and verify $P(T \text{ or } D) = \frac{5}{18}$.

Using Conditional Probabilities to Determine Independence

You can also use conditional probabilities to determine independence. In a sample of 150 residents, each person was asked if he or she favored the concept of having a single, countywide police agency. The county is composed of one large city and many suburban townships. The residences (city or outside the city) and the responses of the residents are summarized in Table 4.4 on the next page. If one of these residents was to be selected at random, what is the probability that the person would (a) favor the concept? (b) favor the concept if the person selected is a city resident? (c) favor the concept if the person selected resides outside the city? (d) Are the events F (favors the concept) and C (resides in city) independent?

Table 4.4 Sample Results

Residence	Favor (F)	Oppose (F̄)	Total
In city (C)	80	40	120
Outside of city (C̄)	20	10	30
Total	100	50	150

A series of simple calculations provides the solution:

(a) $P(F)$ is the proportion of the total sample that favors the concept. Therefore,

$$P(F) = \frac{n(F)}{n(S)} = \frac{100}{150} = \frac{2}{3}$$

(b) $P(F \mid C)$ is the probability that the person selected favors the concept, given that he or she lives in the city. The condition "is a city resident" reduces the sample space to the 120 city residents in the sample. Of these, 80 favored the concept; therefore,

$$P(F \mid C) = \frac{n(F \text{ and } C)}{n(C)} = \frac{80}{120} = \frac{2}{3}$$

(c) $P(F \mid \overline{C})$ is the probability that the person selected favors the concept, given that the person lives outside the city. The condition "lives outside the city" reduces the sample space to the 30 non-city residents; therefore,

$$P(F \mid \overline{C}) = \frac{n(F \text{ and } \overline{C})}{n(\overline{C})} = \frac{20}{30} = \frac{2}{3}$$

(d) All three probabilities have the same value, $\frac{2}{3}$. Therefore, we can say that the events F (favors) and C (resides in city) are independent. The location of residence did not affect $P(F)$.

In the next three chapters we will look at distributions associated with probabilistic events. This will prepare us for the statistics that follow. We must be able to predict the variability that the sample will show with respect to the population before we can be successful at "inferential statistics," in which we describe the population based on the sample statistics available.

problems

Objective 4.1

4.1 If you roll a die 40 times and 9 of the rolls result in a "5," what empirical probability was observed for the event?

4.2 A single die is rolled. What is the probability that the number on top is the following?
  a. A 3
  b. An odd number
  c. A number less than 5
  d. A number greater than 3

4.3 Mrs. Gordon wondered if her class was watching too much television on school nights. To find out, she did a quick poll of her seventh graders. Here are her results:

Hours	Number
0	2
1	3
2	2
3	0
4	3
5	2
6	1

  a. What percentage of the class is not watching television on school nights?
  b. What percentage of the class is watching, at most, 2 hours of television on school nights?
  c. What percentage of the class is watching at least 4 hours of television on school nights?

4.4 The table here shows the average number of births per day in the United States as reported by the CDC.

Day	Number
Sunday	7,563
Monday	11,733
Tuesday	13,001
Wednesday	12,598
Thursday	12,514
Friday	12,396
Saturday	8,605
Total	78,410

Based on this information, what is the probability that one baby identified at random was:
  a. Born on a Monday?
  b. Born on a weekend?
  c. Born on a Tuesday or Wednesday?
  d. Born on a Wednesday, Thursday, or Friday?

4.5 One single-digit number is to be selected randomly.
  a. List the sample space.
  b. What is the probability of each single digit?
  c. What is the probability of an even number?

4.6 Two dice are rolled. Find the probabilities in parts (b)–(e). Use the sample space and chart representation given on page 76.

  a. Why is the set {2, 3, 4, . . . , 12} not a useful sample space?
  b. P(white die is an odd number)
  c. P(sum is 6)
  d. P(both dice show odd numbers)
  e. P(number on black die is larger than number on white die)

4.7 Let x be the success rating of a new television show. The following table lists the subjective probabilities assigned to each x for a particular new show by three different media critics. Which of these sets of probabilities are inappropriate because they violate a basic rule of probability? Explain.

	Judge		
Success Rating, x	A	B	C
Highly Successful	0.5	0.6	0.3
Successful	0.4	0.5	0.3
Not Successful	0.3	−0.1	0.3

4.8 The odds for the Saints winning next year's Super Bowl are 1 to 6.
  a. What is the probability that the Saints will win next year's Super Bowl?
  b. What are the odds against the Saints winning next year's Super Bowl?

4.9 Alan Garole, a jockey at Saratoga Race Course had 195 starts between July and September 2008. Of those 195 starts, he finished with 39 first places, 17 second places, and 28 third places. If all the 2008 racing season conditions held for him at the beginning of the 2009 season, what would have been:
  a. the odds in favor of Alan Garole coming in first place during the 2009 race season at Saratoga?
  b. the probability of Alan Garole coming in first place during the 2009 race season at Saratoga?
  c. the odds in favor of Alan Garole placing (coming in 1st, 2nd, or 3rd) during the 2009 race season at Saratoga?
  d. the probability of Alan Garole placing during the 2009 race season at Saratoga?
  e. Based on the above statistics, should you bet for Alan Garole to come in first or to place? Why?

4.10 Many young women aspire to become professional athletes. Only a few make it to the big time as indicated in the table.

Student Athletes	Women's Basketball
High School Student Athletes	452,929
High School Senior Student Athletes	129,408
NCAA Student Athletes	15,096
NCAA Freshman Roster Positions	4,313
NCAA Senior Student Athletes	3,355
NCAA Student Athletes Drafted	32

a. What are the odds in favor of a high school female athlete being drafted by a pro basketball team?

b. What are the odds against a female basketball player who makes a freshman college roster playing as a senior?

c. What is the probability of a high school female athlete being drafted by a pro basketball team?

d. What is the probability of a NCAA senior female athlete being drafted by a pro basketball team?

4.11 Classify each of the following as a probability or a statistics problem:

a. Determining how long it takes to handle a typical telephone inquiry at a real estate office

b. Determining the length of life for the 100-watt lightbulbs a company produces

c. Determining the chance that a blue ball will be drawn from a bowl that contains 15 balls, of which 5 are blue

d. Determining the shearing strength of the rivets that your company just purchased for building airplanes

e. Determining the chance of getting "doubles" when you roll a pair of dice

Objective 4.2

4.12 Three hundred viewers were asked if they were satisfied with TV coverage of a recent disaster.

	Gender	
	Female	Male
Satisfied	80	55
Not Satisfied	120	45

One viewer is to be randomly selected from those surveyed.

a. Find P(satisfied)

b. Find P(satisfied | female)

c. Find P(satisfied | male)

4.13 Saturday mornings are busy times at the Webster Aquatic Center. Swim lessons ranging from Red Cross Level 2, Fundamental Aquatic Skills, through Red Cross Level 6, Swimming and Skill Proficiency, are offered during two sessions.

Level	Number of People in 10 A.M. Class	Number of People in 11 A.M. Class
2	12	12
3	15	10
4	8	8
5	2	0
6	2	0

Lauren, the program coordinator, is going to randomly select one swimmer to be interviewed for a local television spot on the center and its swim program. What is the probability that the selected swimmer is in the following?

a. A Level 3 class

b. The 10 A.M. class

c. A level 2 class, given that it is the 10 A.M. session

d. The 11 A.M. session, given that it is the Level 6 class

4.14 During the Spring 2009 semester at Monroe Community College, a random sample of students was questioned on their knowledge of the meaning of "sustainability." The primary motivation for the survey was to investigate how interested students might be in a Sustainability Certificate and to discover the best means of informing them of this option. The following table lists how much 224 students agreed with the statement: "Sustainability is important to me."

Level of agreement with statement "Sustainability is important to me"					
Generation (ages)	Strongly Agree	Agree	Disagree	Strongly Disagree	Total
Millennium Y (18–29)	74	109	11	1	195
Generation X (30–44)	14	8	1	0	23
Baby Boomers (45+)	2	3	0	1	6
All Respondents	90	120	12	2	224

SOURCE: Monroe Community College, Sustainability Certificate Survey

Find the probability that a randomly selected student

a. "strongly agrees" that sustainability is important to her.

b. is a member of Generation X.

c. "disagrees" with the importance of sustainability to them given she is a member of the Millennium Y generation.

d. is a member of the Baby Boomers given that she "agrees" with the importance of sustainability.

4.15 In 2007, data from two Youth Risk Behavior Surveys were analyzed to investigate seatbelt use among high school students aged 16 or older. The results were published in the September 2008 issue of *American Journal of Preventive Medicine*. Results (in percents) are included the table that followings:

	Always use when driving		Do not always use when driving	
Characteristic	Always use when passenger	Do not always use when passenger	Always use when passenger	Do not always use when passenger
Total	38.4	20.6	3.4	37.6
Age (years)				
16	38.2	22.5	3.2	36.1
17	38.1	19.9	3.6	38.4
≥18	39.4	18.4	3.6	38.6

SOURCE: http://www.ajpm-online.net/

If one student is selected at random from this population, what is the probability that the student selected:

a. Always uses a seatbelt when driving and always uses a seatbelt when a passenger?

b. Always uses a seatbelt when driving but not always when a passenger, given they are 18 or older?

c. Does not always use a seatbelt when driving and always does when a passenger, knowing they are 16?

d. Always wears a seatbelt when driving?

e. Does not always wear a seatbelt when driving and is 17 years old.

Objective 4.3

4.16 According to the American Pet Products Association 2007–2008 National Pet Owners Survey, about 63% of all American dog owners—some 60 million—are owners of one dog. Based on this information, find the probability that an American dog owner owns more than one dog.

4.17 According to the Sleep Disorder Channel (http://www. sleepdisorderchannel.com/), sleep apnea affects 18 million individuals in the United States. The sleep disorder interrupts breathing and can awaken sufferers as often as five times an hour. Many people do not recognize the condition even though it causes loud snoring. Assuming there are 304 million people in the United States, what is the probability that an individual chosen at random will not be affected by sleep apnea?

4.18 If $P(A) = 0.4$, $P(B) = 0.5$, and $P(A \text{ and } B) = 0.1$, find $P(A \text{ or } B)$.

4.19 Jason attends his high school reunion. Of the attendees, 50% are female. Common knowledge has it that 88% of people are right-handed. Being a left-handed male, Jason knows that of a given crowd, only approximately 6% are left-handed males. If Jason talks to the first person he meets at the reunion, what is the probability that the person is a male or left-handed?

4.20 A and B are events defined on a sample space, with $P(A) = 0.7$ and $P(B \mid A) = 0.4$. Find $P(A \text{ and } B)$.

4.21 A and B are events defined on a sample space, with $P(A) = 0.6$ and $P(A \text{ and } B) = 0.3$. Find $P(B \mid A)$.

4.22 Juan lives in a large city and commutes to work daily by subway or by taxi. He takes the subway 80% of the time because it costs less, and he takes a taxi the other 20% of the time. When taking the subway, he arrives at work on time 70% of the time, whereas he makes it on time 90% of the time when traveling by taxi.

a. What is the probability that Juan took the subway and is at work on time on any given day?

b. What is the probability that Juan took a taxi and is at work on time on any given day?

4.23 Nobody likes paying taxes, but cheating is not the way to get out of it! It is believed that 10% of all taxpayers intentionally claim some deductions to which they are not entitled. If 9% of all taxpayers both intentionally claim extra deductions and deny doing so when audited, find the probability that a taxpayer who does take extra deductions intentionally will deny it.

4.24 If you decide to play the carnival game on page 87, you would like to win the $5 prize, but what is the probability that you will?

a. Draw and completely label a tree diagram including the probabilities for all possible drawings.

b. What is the probability of drawing a red marble on the second drawing? What additional information is needed to find the probability? What "conditions" could exist?

c. Calculate the probability of winning the $5 prize.

d. Is the $2 prize or the $5 prize harder to win? Which is more likely? Justify your answer.

4.25 Suppose that A and B are events defined on a common sample space and that the following probabilities are known: $P(A) = 0.3$, $P(B) = 0.4$, and $P(A \mid B) = 0.2$. Find $P(A \text{ or } B)$.

4.26 Given $P(A \text{ or } B) = 1.0$, $P(\overline{A \text{ and } B}) = 0.7$, and $P(\overline{B}) = 0.4$, find:

a. $P(B)$ b. $P(A)$ c. $P(A \mid B)$

4.27 The probability of C is 0.4. The conditional probability that C occurs, given that D occurs is 0.5. The conditional probability that C occurs, given that D does not occur is 0.25.

a. What is the probability that D occurs?

b. What is the conditional probability that D occurs, given that C occurs?

Objective 4.4

4.28 Determine whether each of the following pairs of events is mutually exclusive.

a. Five coins are tossed: "one head is observed," "at least one head is observed."

b. A salesperson calls on a client and makes a sale: "the sale exceeds $100," "the sale exceeds $1,000."

c. One student is selected at random from a student body: the person selected is "male," the person selected is "older than 21 years of age."

d. Two dice are rolled: the total showing is "less than 7," the total showing is "more than 9."

4.29 Explain why $P(A \text{ and } B) = 0$ when events A and B are mutually exclusive.

4.30 If $P(A) = 0.3$, $P(B) = 0.4$, and if A and B are mutually exclusive events, find:

a. $P(\overline{A})$ b. $P(\overline{B})$

c. $P(A \text{ or } B)$ d. $P(A \text{ and } B)$

4.31 One student is selected from the student body of your college. Define the following events: M—the student selected is male, F—the student selected is female, S—the student selected is registered for statistics.

a. Are events M and F mutually exclusive? Explain.

b. Are events M and S mutually exclusive? Explain.

c. Are events F and S mutually exclusive? Explain.

d. Are events M and F complementary? Explain.

e. Are events M and S complementary? Explain.

f. Are complementary events also mutually exclusive events? Explain.

g. Are mutually exclusive events also complementary events? Explain.

4.32 Do people take indoor swimming lessons in the middle of the hot summer? They sure do at the Webster Aquatic Center. During the month of July 2009 alone, 283 people participated in various forms of lessons.

Swim Categories	Daytime	Evenings
Preschool	66	80
Levels	69	56
Adult and diving	10	2
Total	145	138

If one swimmer was selected at random from the July participants:

a. Are the events the selected participant is "daytime" and "evening" mutually exclusive? Explain.

b. Are the events the selected participant is "preschool" and "levels" mutually exclusive? Explain.

c. Are the events the selected participant is "daytime" and "preschool" mutually exclusive? Explain.

d. Find P(preschool).

e. Find P(daytime).

f. Find P(not levels).

g. Find P(preschool or evening).

h. Find P(preschool and daytime).

i. Find P(daytime | levels).

j. Find P(adult and diving | evening).

4.33 One student is selected at random from a student body. Suppose the probability that this student is female is 0.5 and the probability that this student works part time is 0.6. Are the two events "female" and "working" mutually exclusive? Explain.

4.34 Two dice are rolled. Define events as follows: A—sum of 7, C—doubles, E—sum of 8.

a. Which pairs of events, A and C, A and E, or C and E, are mutually exclusive? Explain.

b. Find the probabilities P(A or C), P(A or E), and P(C or E).

4.35 An aquarium at a pet store contains 40 orange swordfish (22 females and 18 males) and 28 green swordtails (12 females and 16 males). You randomly net one of the fish.

a. What is the probability that it is an orange swordfish?

b. What is the probability that it is a male fish?

c. What is the probability that it is an orange female swordfish?

d. What is the probability that it is a female or a green swordtail?

e. Are the events "male" and "female" mutually exclusive? Explain.

f. Are the events "male" and "swordfish" mutually exclusive? Explain.

Objective 4.5

4.36 Determine whether each of the following pairs of events is independent:

a. Rolling a pair of dice and observing a "1" on the first die and a "1" on the second die

b. Drawing a "spade" from a regular deck of playing cards and then drawing another "spade" from the same deck without replacing the first card

c. Same as part (b), except the first card is returned to the deck before the second drawing

d. Owning a red automobile and having blonde hair

e. Owning a red automobile and having a flat tire today

f. Studying for an exam and passing the exam

4.37 A and B are independent events, and P(A) = 0.7 and P(B) = 0.4. Find P(A and B).

4.38 Suppose P(A) = 0.3, P(B) = 0.4, and P(A and B) = 0.12

a. What is P(A|B)?

b. What is P(B|A)?

c. Are A and B independent?

4.39 A single card is drawn from a standard deck. Let A be the event "the card is a face card" (a jack, a queen, or a king), B is a "red card," and C is "the card is a heart." Determine whether the following pairs of events are independent or dependent:

a. A and B   b. A and C   c. B and C

4.40 A box contains four red and three blue poker chips. Three poker chips are to be randomly selected, one at a time.

a. What is the probability that all three chips will be red if the selection is done with replacement?

b. What is the probability that all three chips will be red if the selection is done without replacement?

c. Are the drawings independent in either part (a) or (b)? Justify your answer.

4.41 The U.S. space program has a history made up of many successes and some failures. Space flight reliability is of the utmost importance in the launching of space shuttles. The reliability of the complete mission is based on all of its components. Each of the six joints in the *Challenger* space shuttle's booster rocket had a 0.977 reliability. The six joints worked independently.

a. What does it mean to say that the six joints work independently?

b. What was the reliability (probability) for all six of the joints working together?

4.42 In a 2008 study by Experian Automotive, it was found that the average number of vehicles per household in the United States is 2.28 vehicles. The results also showed that nearly 35% of households have three or more vehicles (http://www.autospies.com/news/Study-Finds-Americans-Own-2-28-Vehicles-Per-Household-26437/).

a. If two U.S. households are randomly selected, find the probability that both will have three or more vehicles.

b. If two U.S. households are randomly selected, find the probability that neither of the two has three or more vehicles.

c. If four U.S. households are randomly selected, find the probability that all four will have three or more vehicles.

4.43 A *USA Today* Snapshot titled "Weighing Heavily" (February 5, 2009) provided the results from the National College Health Assessment 2007 Web Summary, in which 34% of the students said that "stress" was the health and mental-health issue that most often hampered their academic performance. If five college students are randomly selected, what is the probability that all five would say that "stress" is the health and mental-health issue that most often hampers their academic performance?

4.44 The owners of a two-person business make their decisions independently of each other and then compare their decisions. If they agree, the decision is made; if they do not agree, then further consideration is necessary before a decision is reached. If each has a history of making the right decision 60% of the time, what is the probability that together they:

a. Make the right decision on the first try

b. Make the wrong decision on the first try

c. Delay the decision for further study

Objective 4.6

4.45 a. Describe in your own words why two events cannot be independent if they are already known to be mutually exclusive.

b. Describe in your own words why two events cannot be mutually exclusive if they are already known to be independent.

4.46 One thousand employees at the Russell Microprocessor Company were polled about worker satisfaction. One employee is selected at random.

| | Male | | Female | | |
	Skilled	Unskilled	Skilled	Unskilled	Total
Satisfied	350	150	25	100	625
Unsatisfied	150	100	75	50	375
Total	500	250	100	150	1,000

a. Find the probability that an unskilled worker is satisfied with work.

b. Find the probability that a skilled female employee is satisfied with work.

c. Is satisfaction for female employees independent of their being skilled or unskilled?

4.47 $P(R) = 0.5$, $P(S) = 0.3$, and events R and S are independent.

a. Find $P(R$ and $S)$.   b. Find $P(R$ or $S)$.

c. Find $P(\bar{S})$.   d. Find $P(R \mid S)$.

e. Find $P(\bar{S} \mid R)$.

f. Are events R and S mutually exclusive? Explain.

4.48 $P(M) = 0.3$, $P(N) = 0.4$, and events M and N are mutually exclusive.

a. Find $P(M$ and $N)$.   b. Find $P(M$ or $N)$.

c. Find $P(M$ or $\bar{N})$.   d. Find $P(M \mid N)$.

e. Find $P(M \mid \bar{N})$.

f. Are events M and N independent? Explain.

4.49 Two flower seeds are randomly selected from a package that contains five seeds for red flowers and three seeds for white flowers.

a. What is the probability that both seeds will result in red flowers?

b. What is the probability that one of each color is selected?

c. What is the probability that both seeds are for white flowers?

HINT Draw a tree diagram.

4.50 A company that manufactures shoes has three factories. Factory 1 produces 25% of the company's shoes, Factory 2 produces 60%, and Factory 3 produces 15%. One percent of the shoes produced by Factory 1 are mislabeled, 0.5% of those produced by Factory 2 are mislabeled, and 2% of those produced by Factory 3 are mislabeled. If you purchase one pair of shoes manufactured by this company, what is the probability that the shoes are mislabeled?

Probability
Distributions (Discrete Variables)

Americans are very much in love with the automobile, and many have more than one available to them. The national average is 2.28 vehicles per household, with nearly 34 percent being single-vehicle households and 31 percent being two-vehicle households. However, nearly 35 percent of all households have three or more vehicles.

By pairing the number of vehicles per household as the variable x with the probability for each value of x, a probability distribution is created. This is much like the relative frequency distribution that we studied in Chapter 2.

Vehicles, x	1	2	3	4	5	6	7	8
$P(x)$	0.34	0.31	0.22	0.06	0.03	0.02	0.01	0.01

objectives

5.1 Random Variables

5.2 Probability Distributions of a Discrete Random Variable

5.3 Binomial Probability Distribution

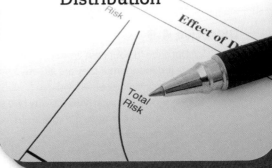

5.1 Random Variables

IF EACH OUTCOME OF A PROBABILITY **EXPERIMENT** IS ASSIGNED A NUMERICAL VALUE, THEN AS WE OBSERVE THE RESULTS OF THE EXPERIMENT WE ARE OBSERVING THE VALUES OF A RANDOM VARIABLE.

A **random variable** assumes a unique numerical value for each of the outcomes in the sample space of a probability experiment. This numerical value is the *random variable value*.

In other words, a random variable is used to denote the outcomes of a probability experiment. The random variable can take on any numerical value that belongs to the set of all possible outcomes of the experiment. (It is called "random" because the value it assumes is the result of a chance, or random event.) Each event in a probability experiment must also be defined in such a way that only one value of the random variable is assigned to it (**mutually exclusive events**), and every event must have a value assigned to it (**all-inclusive events**).

Numerical random variables can be subdivided into two classifications: *discrete random variables* and *continuous random variables*.

© Andreas Kuehn/Stone/Getty Images

→ The following illustrations demonstrate random variables.

- We toss five coins and observe the "number of heads" visible. The random variable x is the number of heads observed and may take on integer values from 0 to 5. (A discrete random variable.)

- Let the "number of phone calls received" per day by a company be the random variable. Integer values ranging from zero to some very large number are possible values. (A discrete random variable.)

- Let the "length of the cord" on an electrical appliance be a random variable. The random variable is a numerical value between 12 and 72 inches for most appliances. (A continuous random variable.)

© U.P.Images/iStockphoto.com

DISCRETE RANDOM VARIABLE
A quantitative random variable that can assume a countable number of values.

CONTINUOUS RANDOM VARIABLE
A quantitative random variable that can assume an uncountable number of values.

Recall from Chapter 1

- Let the "qualifying speed" for racecars trying to qualify for the Indianapolis 500 be a random variable. Depending on how fast the driver can go, the speeds are approximately 220 and faster and are measured in miles per hour (to the nearest thousandth of a mile). (A continuous random variable.)

Continuous random variables are covered in Chapter 6.

5.2 Probability Distributions of a Discrete Random Variable

CONSIDER A COIN-TOSSING EXPERIMENT WHERE TWO COINS ARE TOSSED AND NO HEADS, ONE HEAD, OR TWO HEADS ARE OBSERVED.

If we define the random variable x to be the number of heads observed when two coins are tossed, x can take on the value 0, 1, or 2. The probability of each of these three events can be calculated using techniques from Chapter 4:

$$P(x = 0) = P(0H) = P(TT)$$
$$= \frac{1}{2} \cdot \frac{1}{2} = \frac{1}{4} = 0.25$$

$$P(x = 1) = P(1H) = P(HT \text{ or } TH)$$
$$= \frac{1}{2} \cdot \frac{1}{2} + \frac{1}{2} \cdot \frac{1}{2} = \frac{1}{2} = 0.50$$

$$P(x = 2) = P(2H) = P(HH)$$
$$= \frac{1}{2} \cdot \frac{1}{2} = \frac{1}{4} = 0.25$$

These probabilities can be listed in any number of ways. One of the most convenient is a table format known as a *probability distribution* (see Table 5.1). A

Table 5.1 **Probability Distribution: Tossing Two Coins**

x	P(x)
0	0.25
1	0.50
2	0.25

Table 5.2 **Probability Distribution: Rolling a Die**

x	P(x)
1	$\frac{1}{6}$
2	$\frac{1}{6}$
3	$\frac{1}{6}$
4	$\frac{1}{6}$
5	$\frac{1}{6}$
6	$\frac{1}{6}$

© Anthony Rosenberg/iStockphoto.com

probability distribution is a distribution of the probabilities associated with each of the values of a random variable. The probability distribution is a theoretical distribution; it is used to represent populations.

In an experiment in which a single die is rolled and the number of dots on the top surface is observed, the random variable is the number observed. The probability distribution for this random variable is shown in Table 5.2.

Sometimes it is convenient to write a rule that algebraically expresses the probability of an event in terms of the value of the random variable. This expression is typically written in formula form and is called a **probability function**. A probability function can be as simple as a list that pairs the values of a random variable with their probabilities. Tables 5.1 and 5.2 show two such listings. However, a probability function is most often expressed in formula form.

Probability distribution A distribution of the probabilities associated with each of the values of a random variable. The probability distribution is a theoretical distribution; it is used to represent populations.

Probability function A rule that assigns probabilities to the values of the random variables.

Consider a die that has been modified so that it has one face with one dot, two faces with two dots, and three faces with three dots. Let x be the number of dots observed when this die is rolled. The probability distribution for this experiment is presented in Table 5.3.

Table 5.3 **Probability Distribution: Rolling the Modified Die**

x	$P(x)$
1	$\frac{1}{6}$
2	$\frac{2}{6}$
3	$\frac{3}{6}$

Each of the probabilities can be represented by the value of x divided by 6; that is, each $P(x)$ is equal to the value of x divided by 6, where $x = 1, 2,$ or 3. Thus,

$$P(x) = \frac{x}{6} \quad \text{for} \quad x = 1, 2, \text{or } 3$$

is the formula for the probability function of this experiment.

The probability function for the experiment of rolling one ordinary die is

$$P(x) = \frac{1}{6} \quad \text{for} \quad x = 1, 2, 3, 4, 5, \text{or } 6$$

This particular function is called a **constant function** because the value of $P(x)$ does not change as x changes.

Every probability function must display the two basic properties of probability (see pages 78–79). These two properties are: (1) the probability assigned to each value of the random variable must be between zero and one, inclusive, and (2) the sum of the probabilities assigned to all the values of the random variable must equal one—that is,

PROPERTY 1 $0 \leq$ each $P(x) \leq 1$

PROPERTY 2 $\sum_{\text{all } x} P(x) = 1$

Determining a Probability Function

How do you determine a probability function? For example, is $P(x) = \frac{x}{10}$ for $x = 1, 2, 3,$ or 4 a probability function? To answer this question we need only test the function in terms of the two basic properties. The probability distribution is shown in Table 5.4.

Property 1 is satisfied because 0.1, 0.2, 0.3, and 0.4 are all numerical values between zero and one. (See the ✓ showing that each value was checked.) Property 2 is also satisfied because the sum of all four probabilities is exactly one. (See the (ck) showing that the sum was

Table 5.4 **Probability Distribution for** $P(x) = \frac{x}{10}$ **for** $x = 1, 2, 3,$ or 4

x	$P(x)$	
1	$\frac{1}{10} = 0.1$	✓
2	$\frac{2}{10} = 0.2$	✓
3	$\frac{3}{10} = 0.3$	✓
4	$\frac{4}{10} = 0.4$	✓
	$\frac{10}{10} = 1.0$	(ck)

The values of the random variable are all-inclusive.

checked.) Since both properties are satisfied, we can conclude that $P(x) = \frac{x}{10}$ for $x = 1, 2, 3,$ or 4 is a probability function.

What about $P(x = 5)$ (or any value other than $x = 1, 2, 3,$ or 4) for the function $P(x) = \frac{x}{10}$ for $x = 1, 2, 3,$ or 4? $P(x = 5)$ is considered to be zero. That is, the probability function provides a probability of zero for all values of x other than the values specified as part of the domain.

Probability distributions can be presented graphically. Regardless of the specific graphic representation used, the values of the random variable are plotted on the horizontal scale, and the probability associated with each value of the random variable is plotted on the vertical scale.

A regular histogram is used frequently to present probability distributions. Figure 5.1 shows the probability distribution as a **probability histogram.** The histogram

Figure 5.1 **Histogram: Probability Distribution for** $P(x) = \frac{x}{10}$ **for** $x = 1, 2, 3, 4$

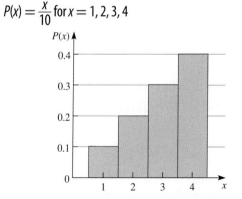

Colleges Strive to Fill Dorms

By Mary Beth Marklein, *USA Today*

Colleges and universities will mail their last batch of admission offers in the next few days, but the process is far from over.

Now, students have until May 1 to decide where they'll go this fall. And with lingering concerns about the economy and residual fears about travel and security since September 11, many admissions officials are less able this year to predict how students will respond.

Note the distribution depicted on the bar graph. It has the makings of a discrete probability distribution. The random variable "number of colleges applied to" is a discrete variable with values from zero to 11 or more. Each of the values has a corresponding probability, and the sum of the probabilities is equal to 1.

Students hedge their bets

Most students apply to more than one school, making it difficult for colleges to predict how many will actually enroll. Last fall's freshman class was asked:

To how many colleges, other than the one where you enrolled, did you apply for admission this year?

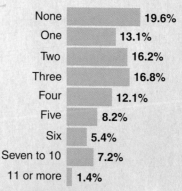

None	19.6%
One	13.1%
Two	16.2%
Three	16.8%
Four	12.1%
Five	8.2%
Six	5.4%
Seven to 10	7.2%
11 or more	1.4%

SOURCE: The American Freshman: National Norms for Fall 2001, survey of 281,064 freshmen entering 421 four-year colleges and universities.

Data from Julie Snider, © 2002 *USA Today*.

Atom tests 'affected IQ'

NUCLEAR FALLOUT from the first atomic bomb tests in New Mexico and, later, from tests in Nevada is directly responsible for a sharp decline in the intelligence of American youth, according to new research.

Radioactive iodine-131, which causes thyroid damage in the newborn, has impaired the mental development of a whole generation of American children. And the effects have been greatest in the far west of the United States, closest to the bomb test sites.

bomb test area, recorded a decline of 26 on the verbal test and 15 on the mathematical in 1974-6. Over the same period, Ohio - some 2,000 miles to the east of the test site and south of the fallout clouds - had a fall of just two points on the verbal test, while maths scores rose by four points.

The effect has persisted into the 1980s. In 1982, test scores for the US as a whole rose. Eight states went against the trend. "All were located in the west where iodine-131 was recorded from the first Chinese atomic test and where nuclear

WIFE-STEALING HORSE

By GEORGE EDWARDS

A HUSBAND gave his wife an ultimatum: "Either your horse goes, or I will."

The woman replied: "I'll never part with Fritz." Then she went out for a gallop.

And when she returned, her husband, Franz, had left for good.

Franz told a divorce court in Kaiserslautern, West Germany, that he gave the horse to his wife "I forgot to give Fritz his wedding anniversary"

Liselotte is no longer cooked proper meals and became too tired for sex.

Franz added: "One night, while we were making love, she suddenly cried 'I forgot to give Fritz his vitamin pill!'"

Franz got his divorce because of her "unreason-

about the animal. Often she stayed all day at the stable and sometimes spent the night with Fritz.

"It was a sad mistake," he said. "she became too tired

Pioneer heads for the unknown

AFTER 11 years in space, the American spacecraft Pioneer 10 will certainly become the first man-made object to leave the solar system. But when it will actually do so is far from certain and provides a nice opportunity for astronomical hair-splitting.

Last Tuesday Pioneer was as far from the sun as the so-called "outermost" planet, Pluto - a possible definition of leaving the solar system. But there is a catch: Pluto's orbit is highly eccentric and some of it

this will be the case for the next 17 years. And Pioneer will not cross Neptune's orbit until June 13.

But will it even then be out of the solar system? It will certainly not be clear of the heliosphere, the region of space affected by the million-mile-an-hour "wind" of electrified particles streaming out from the sun. The heliosphere extends to distances several times greater than Pioneer's present position, 2.8 billion miles from the sun. Farther still

---

of a probability distribution uses the physical area of each bar to represent its assigned probability. The bar for $x = 2$ is 1 unit wide (from 1.5 to 2.5) and 0.2 unit high. Therefore, its area (length × width) is $(0.2)(1) = 0.2$, the probability assigned to $x = 2$. The areas of the other bars can be determined in a similar fashion. This area representation will be an important concept in Chapter 6 when we begin to work with continuous random variables.

Mean and Variance of a Discrete Probability Distribution

Recall that in Chapter 2 we calculated several numerical sample statistics (mean, variance, standard deviation, and others) to describe empirical sets of data. Probability distributions may be used to represent theoretical pop-

ulations, the counterpart to samples. We use **population parameters** (mean, variance, and standard deviation) to describe these probability distributions just as we use **sample statistics** to describe samples.

The *mean of the probability distribution* of a discrete random variable, or the **mean of a discrete random variable,** is found in a manner that takes full advantage of the table format of a discrete probability distribution. The mean, μ, of a discrete random variable x is found

> **Mean of a discrete random variable (expected value)** The mean, μ, of a discrete random variable x is found by multiplying each possible value of x by its own probability and then adding all the products together.

notation notes

1. $\bar{x}$ is the mean of the sample.
2. s^2 and s are the variance and standard deviation of the sample, respectively.
3. $\bar{x}$, s^2, and s are called *sample statistics*.
4. μ (lowercase Greek letter mu) is the mean of the population.
5. σ^2 (sigma squared) is the variance of the population.
6. σ (lowercase Greek letter sigma) is the standard deviation of the population.
7. μ, σ^2, and σ are called *population parameters*. (A parameter is a constant; μ, σ^2, and σ are typically unknown values in real statistics problems. About the only time they are known is in a textbook problem setting for the purpose of learning and understanding.)

by multiplying each possible value of x by its own probability and then adding all the products together:

$$\text{mean of } x: \quad mu = sum \ of \begin{pmatrix} each \ x \ multiplied \\ by \ its \ own \ probability \end{pmatrix}$$

$$\mu = \Sigma[xP(x)] \tag{5.1}$$

The mean of a discrete random variable is often referred to as its *expected value*.

The **variance of a discrete random variable** is defined in much the same way as the variance of sample data, the mean of the squared deviations from the mean. To find the variance, σ^2, of a discrete random variable x, multiply each possible value of the squared deviation from the mean, $(x - \mu)^2$, by its own probability and then add all the products together:

$$\text{variance:} \quad \frac{sigma}{squared} = sum \ of \begin{pmatrix} squared \ deviation \\ times \ probability \end{pmatrix}$$

$$\sigma^2 = \Sigma[(x - \mu)^2 P(x)] \tag{5.2}$$

Formula (5.2) is often inconvenient to use; it can be reworked into the following form: "variance: sigma squared = sum of (x^2 times probability) − [sum of (x times probability)]²," or in the following algebraic form:

$$\sigma^2 = \Sigma[x^2 P(x)] - \{\Sigma[xP(x)]\}^2 \tag{5.3a}$$

or

$$\sigma^2 = \Sigma[x^2 P(x)] - \mu^2 \tag{5.3b}$$

Likewise, **standard deviation of a discrete random variable** is calculated in the same manner as the standard deviation of sample data—as the positive square root of variance:

$$\text{standard deviation:} \quad \sigma = \sqrt{\sigma^2} \tag{5.4}$$

To help you fully understand the application of these concepts, let's calculate the statistics for a probability function. Specifically, let's find the mean, variance, and standard deviation of the probability function

$$P(x) = \frac{x}{10} \quad \text{for} \quad x = 1, 2, 3, \text{ or } 4$$

First, we will find the mean using formula (5.1), the variance using formula (5.3a), and the standard deviation using formula (5.4). The most convenient way to organize the products and find the totals we need is to expand the probability distribution into an extensions table (see Table 5.5).

Table 5.5 **Extensions Table: Probability Distribution,** $P(x) = \frac{x}{10}$ for $x = 1, 2, 3,$ or 4

x	$P(x)$	$xP(x)$	x^2	$x^2P(x)$
1	$\frac{1}{10} = 0.1$ ✓	0.1	1	0.1
2	$\frac{2}{10} = 0.2$ ✓	0.4	4	0.8
3	$\frac{3}{10} = 0.3$ ✓	0.9	9	2.7
4	$\frac{4}{10} = 0.4$ ✓	1.6	16	6.4
	$\frac{10}{10} = 1.0$ (ck)	$\Sigma[xP(x)] = 3.0$		$\Sigma[x^2P(x)] = 10.0$

Variance of a discrete random variable The variance, σ^2, of a discrete random variable x is found by multiplying each possible value of the squared deviation from the mean, $(x - \mu)^2$, by its own probability and then adding all the products together.

Standard deviation of a discrete random variable The positive square root of variance.

To find the mean of x: The $xP(x)$ column contains each value of x multiplied by its corresponding probability, and the sum at the bottom is the value needed in formula (5.1):

$$\mu = \Sigma[xP(x)] = \mathbf{3.0}$$

To find the variance of x, the totals at the bottom of the $xP(x)$ and $x^2P(x)$ columns are substituted into formula (5.3a):

$$\sigma^2 = \Sigma[x^2P(x)] - \{\Sigma[xP(x)]\}^2$$
$$= 10.0 - (3.0)^2 = \mathbf{1.0}$$

To find the standard deviation of x, use formula (5.4):

$$\sigma = \sqrt{\sigma^2} = \sqrt{1.0} = \mathbf{1.0}$$

notes

1. The purpose of the extensions table is to organize the process of finding the three column totals: $\Sigma[P(x)]$, $\Sigma[xP(x)]$, and $\Sigma[x^2P(x)]$.

2. The other columns, x and x^2, should not be totaled; they are not used.

3. $\Sigma[P(x)]$ will always be 1.0; use this only as a check.

4. $\Sigma[xP(x)]$ and $\Sigma[x^2P(x)]$ are used to find the mean and variance of x.

© Sokolov Andrey/iStockphoto.com / © Stefan Klein/iStockphoto.com

5.3 Binomial Probability Distribution

MANY EXPERIMENTS ARE COMPOSED OF REPEATED TRIALS WHOSE OUTCOMES CAN BE CLASSIFIED INTO ONE OF TWO CATEGORIES: **SUCCESS** OR **FAILURE**.

Examples of such experiments are coin tosses, right/wrong quiz answers, and other more practical experiments such as determining whether a product did or did not do its prescribed job and whether a candidate gets elected or not. There are experiments in which the trials have many outcomes that, under the right conditions, may fit this general description of being classified in one of two categories. For example, when we roll a single die, we usually consider six possible outcomes. However, if we are interested only in knowing whether a "one" shows or not, there are really only two outcomes: the "one" shows or "something else" shows. Experiments like those just described are called **binomial probability experiments.**

A binomial probability experiment is made up of repeated trials that possess the following properties:

1. There are n repeated identical independent trials.

2. Each trial has two possible outcomes (success, failure).

3. $P(success) = p$, $P(failure) = q$, and $p + q = 1$.

4. The **binomial random variable** x is the count of the number of successful trials that occur; x may take on any integer value from zero to n.

→ To demonstrate the properties of a binomial probability experiment, let's look at the experiment of rolling a die 12 times and observing a "one" or "something else." At the end of all 12 rolls, the number of ones is reported. The random variable x is the number of times that a one is observed in the $n = 12$ trials. Since "one" is the outcome of concern, it is considered "success"; therefore, $p = P(\text{one}) = \frac{1}{6}$ and $q = P(\text{not one}) = \frac{5}{6}$. This experiment is binomial.

The key to working with any probability experiment is its probability distribution. All binomial probability experiments have the same properties, and therefore the same organization scheme can be used to represent all of them. The *binomial probability function* allows us to find the probability for each possible value of x.

BINOMIAL PROBABILITY FUNCTION

→ For a binomial experiment, let p represent the probability of a "success" and q represent the probability of a "failure" on a single trial. Then $P(x)$, the probability that there will be exactly x successes in n trials, is

$$P(x) = \binom{n}{x}(p^x)(q^{n-x}) \text{ for } x = 0, 1, 2, \ldots, n \quad (5.5)$$

When you look at the probability function, you notice that it is the product of three basic factors:

1. the number of ways that exactly x successes can occur in n trials, $\binom{n}{x}$,
2. the probability of exactly x successes, p^x, and
3. the probability that failure will occur on the remaining $(n - x)$ trials, q^{n-x}.

The number of ways that exactly x successes can occur in a set of n trials is represented by the symbol $\binom{n}{x}$, which must always be a positive integer. This term is called the **binomial coefficient** and is found by using the formula

$$\binom{n}{x} = \frac{n!}{x!(n-x)!} \tag{5.6}$$

NOTES:

1. $n!$ ("*n factorial*") is an abbreviation for the product of the sequence of integers starting with n and ending with one. For example, $3! = 3 \cdot 2 \cdot 1 = 6$ and $5! = 5 \cdot 4 \cdot 3 \cdot 2 \cdot 1 = 120$. There is one special case, $0!$, that is defined to be 1. For more information about **factorial notation,** go online to cengagebrain.com.

2. The values for $n!$ and $\binom{n}{x}$ can be found readily using most scientific calculators.

3. The binomial coefficient $\binom{n}{x}$ is equivalent to the number of combinations $_nC_x$, the symbol most likely on your calculator.

Binomial Probability Experiment

A four-question multiple choice (3 possible answers each) quiz qualifies as a binomial experiment made up of four trials when all four of the answers are obtained by random guessing.

1. **PROPERTY 1** A trial is the answering of one question, and it is repeated $n = 4$ times. The trials are independent because the probability of a correct answer on any one question is not affected by the answers on other questions.
2. **PROPERTY 2** The two possible outcomes on each trial are success = C, correct answer, and failure = W, wrong answer.
3. **PROPERTY 3** For each trial (each question): $p = P(correct) = \frac{1}{3}$ and $q = P(incorrect) = \frac{2}{3}$. $[p + q = 1 \checkmark]$
4. **PROPERTY 4** For the total experiment (the quiz): x = number of correct answers and can be any integer value from zero to $n = 4$.

- Properties 1 and 2 are the two basic properties of any binomial experiment.
- **Independent** trials means that the result of one trial does not affect the probability of success of any other trial in the experiment. In other words, the probability of success remains constant throughout the entire experiment.
- Property 3 gives the algebraic notation for each trial.
- Property 4 concerns the algebraic notation for the complete experiment.
- It is of utmost importance that both x and p be associated with "success."

4. Login to the CourseMate for STAT web site at cengagebrain.com for general information on the binomial coefficient.

A coin is tossed three times and we observe the number of heads that occurs in the three tosses. This is a binomial experiment because it displays all the properties of a binomial experiment:

1. There are $n = 3$ repeated independent trials (each coin toss is a separate trial, and the outcome of any one trial has no effect on the probability of another).

2. Each trial (each toss of the coin) has two outcomes: success = heads (what we are counting) and failure = tails.

3. The probability of success is $p = P(H) = 0.5$, and the probability of failure is $q = P(T) = 0.5$. $[p + q = 0.5 + 0.5 = 1 \ ⓒⓚ]$

4. The random variable x is the number of heads that occurs in the three trials; x will assume exactly one of the values 0, 1, 2, or 3 when the experiment is complete.

The binomial probability function for the tossing of three coins is

$$P(x) = \binom{n}{x}(p^x)(q^{n-x})$$
$$= \binom{3}{x}(0.5)^x(0.5)^{n-x} \quad \text{for} \quad x = 0, 1, 2, \text{ or } 3$$

Let's find the probability of $x = 1$ using the preceding binomial probability function:

$$P(x = 1) = \binom{3}{1}(0.5)^1(0.5)^2 = 3(0.5)(0.25) = \mathbf{0.375}$$

Determining a Binomial Experiment and Its Probabilities

We've already demonstrated the properties of a binomial probability experiment; now we can discuss how to determine a binomial experiment and its probabilities. Let's apply what we know about binomial experiments to a specific experiment that calls for drawing five cards, one at a time with replacement, from a well-shuffled deck of playing cards. The drawn card is identified as a spade or not a spade, it is returned to the deck, the deck is reshuffled, and so on. The random variable x is the number of spades observed in the set of five drawings. Is this a binomial experiment? Let's identify the four properties.

1. There are five repeated drawings: $n = 5$. These individual trials are independent because the drawn card is returned to the deck and the deck is reshuffled before the next drawing.

2. Each drawing is a trial, and each drawing has two outcomes: "spade" or "not spade."

3. $p = P(\text{spade}) = \frac{13}{52}$ and $q = P(\text{not spade}) = \frac{39}{52}$. $[p + q = 1 \ ⓒⓚ]$

4. x is the number of spades recorded upon completion of the five trials; the possible values are 0, 1, 2, ..., 5.

The binomial probability function is

$$P(x) = \binom{5}{x}\left(\frac{13}{52}\right)^x\left(\frac{39}{52}\right)^{5-x} = \binom{5}{x}\left(\frac{1}{4}\right)^x\left(\frac{3}{4}\right)^{5-x}$$
$$= \binom{5}{x}(0.25)^x(0.75)^{5-x} \quad \text{for} \quad x = 0, 1, \ldots, 5$$

$$P(0) = \binom{5}{0}(0.25)^0(0.75)^5 = (1)(1)(0.2373) = \mathbf{0.2373}$$

$$P(1) = \binom{5}{1}(0.25)^1(0.75)^4 = (5)(0.25)(0.3164)$$
$$= \mathbf{0.3955}$$

$$P(2) = \binom{5}{2}(0.25)^2(0.75)^3 = (10)(0.0625)(0.421875)$$
$$= \mathbf{0.2637}$$

$$P(3) = \binom{5}{3}(0.25)^3(0.75)^2 = (10)(0.015625)(0.5625)$$
$$= \mathbf{0.0879}$$

The two remaining probabilities are left for you to compute.

The preceding distribution of probabilities indicates that the single most likely value of x is one, the event of observing exactly one spade in a hand of five cards. What is the least likely number of spades that would be observed?

Binomial Probability of "Bad Eggs"

→ The manager of Steve's Food Market guarantees that none of his cartons of a dozen eggs will contain more than one bad egg. If a carton contains more than one bad egg, he will replace the whole dozen and allow the customer to keep the original eggs. If the probability that an individual egg is bad is 0.05, what is the probability that the manager will have to replace a given carton of eggs?

SOLUTION

At first glance, the manager's situation appears to fit the properties of a binomial experiment if we let x be the number of bad eggs found in a carton of a dozen eggs, let $p = P(\text{bad}) = 0.05$, and let the inspection of each egg be a trial that results in finding a "bad" or "not bad" egg. There will be $n = 12$ trials to account for the 12 eggs in a carton. However, trials of a binomial experiment must be independent; therefore, we will assume that the quality of one egg in a carton is independent of the quality of any of the other eggs. (This may be a big assumption! But with this assumption, we will be able to use the binomial probability distribution as our model.) Now, based on this assumption, we will be able to find/estimate the probability that the manager will have to make good on

© Mark Wragg/iStockphoto.com

his guarantee. The probability function associated with this experiment will be:

$$P(x) = \binom{12}{x}(0.05)^x(0.95)^{12-x} \text{ for } x = 0, 1, 2, \ldots, 12$$

The probability that the manager will replace a dozen eggs is the probability that $x = 2, 3, 4, \ldots, 12$. Recall that $\Sigma P(x) = 1$; that is,

$$P(0) + P(1) + P(2) + \cdots + P(12) = 1$$

$$P(\text{replacement}) = P(2) + P(3) + \cdots + P(12)$$

$$= 1 - [P(0) + P(1)]$$

It is easier to find the probability of replacement by finding $P(x = 0)$ and $P(x = 1)$ and subtracting their total from 1 than by finding all of the other probabilities. We have

$$P(x) = \binom{12}{x}(0.05)^x(0.95)^{12-x}$$

$$P(0) = \binom{12}{0}(0.05)^0(0.95)^{12} = \mathbf{0.540}$$

$$P(1) = \binom{12}{1}(0.05)^1(0.95)^{11} = \mathbf{0.341}$$

$$P(\text{replacement}) = 1 - (0.540 + 0.341) = \mathbf{0.119}$$

If $p = 0.05$ is correct, then the manager will be busy replacing cartons of eggs. If he replaces 11.9% of all the cartons of eggs he sells, he certainly will be giving away a substantial proportion of his eggs. This suggests that he should adjust his guarantee (or market better eggs). For example, if he were to replace a carton of eggs only when four or more were found to be bad, he would expect to replace only three out of 1,000 cartons $[1.0 - (0.540 + 0.341 + 0.099 + 0.017)]$, or 0.3% of the cartons sold. Notice that the manager will be able to control his "risk" (probability of replacement) if he adjusts the value of the random variable stated in his guarantee.

> NOTE: The values of many binomial probabilities for values of $n \leq 15$ and common values of p are found in Table 2 of Appendix B. In this example, we have $n = 12$ and $p = 0.05$, and we want the probabilities for $x = 0$ and 1. We need to locate the section of Table 2 where $n = 12$, find the column headed $p = 0.05$, and read the numbers across from $x = 0$ and $x = 1$. We find 0.540 and 0.341, as shown in Table 5.6. (Look up these values in Table 2 in Appendix B.)

A convenient notation to identify the binomial probability distribution for a binomial experiment with

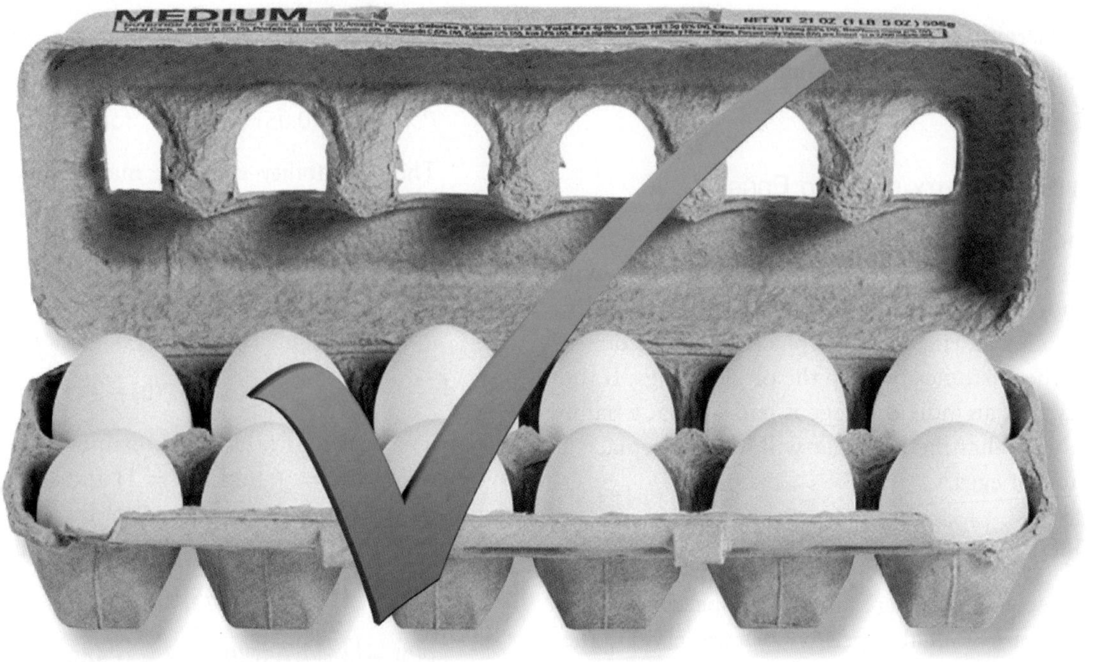

$n = 12$ and $p = 0.05$ is $B(12, 0.05)$. $B(12, 0.05)$, read as "*Binominal distribution for n* = 12 and $p = 0.05$," represents the entire distribution or "block" of probabilities shown in purple in Table 5.6. When used in combination with the $P(x)$ notation, $P[x = 1|B(12, 0.05)]$ indicates the probability of $x = 1$ from this distribution, or 0.341 as shown on Table 5.6.

Mean and Standard Deviation of the Binomial Distribution

The mean and standard deviation of a theoretical binomial probability distribution can be found by using the following two formulas:

Mean of Binomial Distribution

$$\mu = np \tag{5.7}$$

and

Standard Deviation of Binomial Distribution

$$\sigma = \sqrt{npq} \tag{5.8}$$

The formula for the mean, μ, seems appropriate: the number of trials multiplied by the probability of "success." The formula for the standard deviation, σ, is not

Table 5.6 **Excerpt of Table 2 in Appendix B, Binomial Probabilities**

							p								
n	x	0.01	0.05	0.10	0.20	0.30	0.40	0.50	0.60	0.70	0.80	0.90	0.95	0.99	x
	:														
	:														
12	0	.886	.540	.282	.069	.014	.002	0+	0+	0+	0+	0+	0+	0+	0
	1	.107	.341	.377	.206	.071	.017	.003	0+	0+	0+	0+	0+	0+	1
	2	.006	.099	.230	.283	.168	.064	.016	.002	0+	0+	0+	0+	0+	2
	3	0+	.017	.085	.236	.240	.142	.054	.012	.001	0+	0+	0+	0+	3
	4	0+	.002	.021	.133	.231	.213	.121	.042	.008	.001	0+	0+	0+	4
	:														

as easily understood. Thus, at this point it is appropriate to look at an example that demonstrates that formulas (5.7) and (5.8) yield the same results as formulas (5.1), (5.3a), and (5.4).

→ Imagine that you're going to toss a coin three times. Let x be the number of heads in three coin tosses, $n = 3$, and $p = \frac{1}{2} = 0.5$. Using formula (5.7), we find the mean of x to be

$$\mu = np = (3)(0.5) = \textbf{1.5}$$

Using formula (5.8), we find the standard deviation of x to be

$$\sigma = \sqrt{npq} = \sqrt{(3)(0.5)(0.5)}$$
$$= \sqrt{0.75} = 0.866 = \textbf{0.87}$$

The results would have been the same if we had used formulas (5.1), (5.3a), and (5.4). However, formulas (5.7) and (5.8) are much easier to use when x is a binomial random variable.

Using the formulas, we can calculate the mean and standard deviation of a binomial distribution (in other words, for any number n). Let's find the mean and standard deviation of the binomial distribution when $n = 20$ and $p = \frac{1}{5}$ (or 0.2, in decimal form). Recall that the "binomial distribution where $n = 20$ and $p = 0.2$" has the probability function

$$P(x) = \binom{20}{x}(0.2)^x(0.8)^{20-x} \quad \text{for } x = 0, 1, 2, \ldots, 20$$

and a corresponding distribution with 21 x values and 21 probabilities, as shown on the next page in the distribution chart, Table 5.7, and on the histogram in Figure 5.2.

Sir Francis Galton is credited with the "discovery" of fingerprints (that fingerprints are unique to each individual), and it was Galton who developed the methods used to identify them. It is the occurrence of irregular marks and cuts in the finger patterns that make each print unique. These marks are referred to as Galton marks. The Galton–Henry system of fingerprint classification was published in June 1900, began to be used at Scotland Yard in 1901, and was soon used throughout the world as an identifier in criminal investigations.

Sir Francis Galton

Table 5.7 Binomial Distribution: $n = 20, p = 0.2$

x	P(x)	x	P(x)
0	0.012	9	0.007
1	0.058	10	0.002
2	0.137	11	0+
3	0.205	12	0+
4	0.218	13	0+
5	0.175	⋮	⋮
6	0.109	⋮	⋮
7	0.055	⋮	⋮
8	0.022	20	0+

Let's find the mean and the standard deviation of this distribution of x using formulas (5.7) and (5.8):

$$\mu = np = (20)(0.2) = \mathbf{4.0}$$

$$\sigma = \sqrt{npq} = \sqrt{(20)(0.2)(0.8)} = \sqrt{3.2} = \mathbf{1.79}$$

Figure 5.3 shows the mean, $\mu = 4$ (shown by the location of the vertical blue line along the x-axis) relative to the variable x. This 4.0 is the mean value expected for x, the number of successes in each random sample of size 20 drawn from a population with $p = 0.2$. Figure 5.3 also shows the size of the standard deviation, $\sigma = 1.79$ (as shown by the length of the horizontal red line segment). It is the expected standard deviation for the values of the random variable x that occur in samples of size 20 drawn from this same population.

Figure 5.2 **Histogram of Binomial Distribution $B(20, 0.2)$**

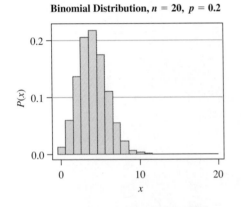

Figure 5.3 **Histogram of Binomial Distribution**

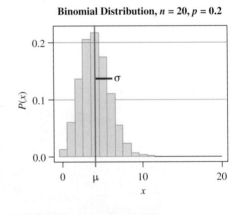

Recap:

In this chapter, we combined concepts of probability with some of the ideas presented in Chapter 2. We are now able to deal with distributions of probability values and find means, standard deviations, and other statistics and are prepared for extending those concepts to continuous random variables.

problems

Objective 5.1

5.1 Survey your classmates about the number of siblings they have and the length of the last conversation they had with their mothers. Identify the two random variables of interest and list their possible values.

5.2 a. The variables in problem 5.1 are either discrete or continuous. Which are they and why?
   b. Explain why the variable "number of dinner guests for Thanksgiving dinner" is discrete.
   c. Explain why the variable "number of miles to your grandmother's house" is continuous.

5.3 A social worker is involved in a study about family structure. She obtains information regarding the number of children per family for a certain community from the census data. Identify the random variable of interest, determine whether it is discrete or continuous, and list its possible values.

5.4 The staff at *Fortune* magazine recently isolated what they considered to be the 100 best companies in America to work for. Many of the companies on the list planned to hire during 2009. Those that planned to add the most employees included:

Company	New Jobs
Ernst & Young	2,800
Wegmans Food Markets	2,000
Edward Jones	1,040

SOURCE: "The 100 Best Companies to Work for 2009," *Fortune*

   a. What is the random variable involved in this study?
   b. Is the random variable discrete or continuous? Explain.

5.5 An archer shoots arrows at the bull's-eye of a target and measures the distance from the center of the target to the arrow. Identify the random variable of interest, determine whether it is discrete or continuous, and list its possible values.

*5.6 If you could stop time and live forever in good health, what age would you pick? Answers to this question were reported in a *USA Today* Snapshot. The average ideal age for each age group is listed in the following table; the average ideal age for all adults was found to be 41. Interestingly, those younger than 30 years want to be older, whereas those older than 30 years want to be younger.

Age Group	18–24	25–29	30–39	40–49	50–64	65+
Ideal Age	27	31	37	40	44	59

   Age is used as a variable twice in this application.
   a. The age of the person being interviewed is not the random variable in this situation. Explain why and describe how "age" is used with regard to age group.
   b. What is the random variable involved in this study? Describe its role in this situation.

   c. Is the random variable discrete or continuous? Explain.

Objective 5.2

5.7 Express the tossing of one coin as a probability distribution of x, the number of heads occurring (that is, $x = 1$ if a head occurs and $x = 0$ if a tail occurs).

5.8 a. Express $P(x) = \frac{1}{6}$; for $x = 1, 2, 3, 4, 5, 6$, in distribution form.
   b. Construct a histogram of the probability distribution $P(x) = \frac{1}{6}$; for $x = 1, 2, 3, 4, 5, 6$.
   c. Describe the shape of the histogram in part (b).

5.9 a. Explain how the various values of x in a probability distribution form a set of mutually exclusive events.
   b. Explain how the various values of x in a probability distribution form a set of "all-inclusive" events.

5.10 Test the following function to determine whether it is a probability function. If it is not, try to make it into a probability function.
$$R(x) = 0.2 \text{ for } x = 0, 1, 2, 3, 4$$
   a. List the distribution of probabilities.
   b. Sketch a histogram.

5.11 Test the following function to determine whether it is a probability function.
$$P(x) = \frac{x^2 + 5}{50}, \text{ for } x = 1, 2, 3, 4$$
   a. List the probability distribution.
   b. Sketch a histogram.

5.12 Census data are often used to obtain probability distributions for various random variables. Census data for families in a particular state with a combined income of $50,000 or more show that 20% of these families have no children, 30% have one child, 40% have two children, and 10% have three children. From this information, construct the probability distribution for x, where x represents the number of children per family for this income group.

5.13 In a *USA Today* Snapshot (June 1, 2009), the following statistics were reported on the number of hours of sleep that adults get.

Number of Hours	Percentage
5 or less	12
6	29
7	37
8 or more	24

SOURCE: StrategyOne survey for Tempur-Pedic of 1,004 adults in April

   a. Are there any other values the number of hours can attain?
   b. Explain why the total of the percentages is not 100%.
   c. Is this a discrete probability distribution? Is it a probability distribution? Explain.

5.14 a. Form the probability distribution table for $P(x) = \frac{x}{6}$, for $x = 1, 2, 3$.
   b. Find the extensions $xP(x)$ and $x^2P(x)$ for each x.
   c. Find $\Sigma[xP(x)]$ and $\Sigma[x^2P(x)]$.
   d. Find the mean for $P(x) = \frac{x}{6}$, for $x = 1, 2, 3$.
   e. Find the variance for $P(x) = \frac{x}{6}$, for $x = 1, 2, 3$.
   f. Find the standard deviation for $P(x) = \frac{x}{6}$, for $x = 1, 2, 3$.

5.15 If you find the sum of the x and the x^2 columns on the extensions table, exactly what have you found?

5.16 Given the probability function $P(x) = \frac{5 - x}{10}$ for $x = 1, 2, 3, 4$, find the mean and standard deviation.

5.17 Given the probability function $R(x) = 0.2$ for $x = 0, 1, 2, 3, 4$, find the mean and standard deviation.

5.18 The number of ships to arrive at a harbor on any given day is a random variable represented by x. The probability distribution for x is as follows:

x	10	11	12	13	14
$P(x)$	0.4	0.2	0.2	0.1	0.1

Find the mean and standard deviation of the number of ships that arrive at a harbor on a given day.

5.19 The number of children living per household, x, in the United States in 2008 is expressed as a probability distribution here.

x	0	1	2	3	4	5+
$P(x)$	0.209	0.384	0.249	0.106	0.032	0.020

SOURCE: U.S. Census Bureau

   a. Is this a discrete probability distribution? Explain.
   b. Draw a histogram for the distribution of x, the number of children per household.
   c. Replacing "5+" with exactly "5," find the mean and standard deviation.

5.20 Is a dog man's best friend? One would think so with 60 million pet dogs nationwide. But how many friends are needed? In the American Pet Products Association's 2007–2008 National Pet Owners Survey, the following statistics were reported.

Number of Pet Dogs	Percentage
One	63
Two	25
Three or more	12

SOURCE: APPA 2007–2008 National Pet Owners Survey

   a. Is this a discrete probability distribution? Explain.
   b. Draw a relative frequency histogram to depict the results shown in the table.
   c. Replacing the category "three or more" with exactly "three," find the mean and standard deviation of the number of pet dogs per household.
   d. How do you interpret the mean?
   e. Explain the effect that replacing the category "three or more" with "three" had on the mean and standard deviation.

5.21 As reported at the beginning of the chapter, Americans are in love with the automobile—the majority have more than one vehicle per household. In fact, the national average is 2.28 vehicles per household. The number of vehicles per household in the United States can be described as follows:

Vehicles, x	$P(x)$
1	0.34
2	0.31
3	0.22
4	0.06
5	0.03
6	0.02
7	0.01
8 or more	0.01

   a. Replacing the category "8 or more" with exactly "8," find the mean and standard deviation of the number of vehicles per household in the United States.
   b. How does the mean calculated in part (a) correspond to the national average of 2.28?
   c. Explain the effect that replacing the category "8 or more" with "8" had on the mean and standard deviation.

5.22 The random variable A has the following probability distribution:

A	1	2	3	4	5
$P(A)$	0.6	0.1	0.1	0.1	0.1

   a. Find the mean and standard deviation of A.
   b. How much of the probability distribution is within 2 standard deviations of the mean?
   c. What is the probability that A is between $\mu - 2\sigma$ and $\mu + 2\sigma$?

5.23 A *USA Today* Snapshot (March 4, 2009) presented a pie chart depicting how workers damage their laptops. Statistics were derived from a survey of 714 IT managers by Ponemon Institute for Dell. Is this a probability distribution? Explain.

Reason for Damage to Laptop	Percentage
Spilled food or liquids	34
Dropping them	28
Not protecting during travel	25
Worker anger	13

5.24 a. Use a computer (or random numbers table) to generate a random sample of 25 observations drawn from the discrete probability distribution.

x	1	2	3	4	5
$P(x)$	0.2	0.3	0.3	0.1	0.1

   Compare the resulting data to your expectations.
   b. Form a relative frequency distribution of the random data.

c. Construct a probability histogram of the given distribution and a relative frequency histogram of the observed data using class midpoints of 1, 2, 3, 4, and 5.

d. Compare the observed data with the theoretical distribution. Describe your conclusions.

e. Repeat parts (a) through (d) several times with $n = 25$. Describe the variability you observe between samples.

f. Repeat parts (a) through (d) several times with $n = 250$. Describe the variability you see between samples of this much larger size.

5.25 a. Use a computer (or random-number table) and generate a random sample of 100 observations drawn from the discrete probability population $P(x) = \frac{5 - x}{10}$, for $x = 1, 2, 3, 4$. List the resulting sample.

b. Form a relative frequency distribution of the random data.

c. Form a probability distribution of the expected probability distribution. Compare the resulting data with your expectations.

d. Construct a probability histogram of the given distribution and a relative frequency histogram of the observed data using class midpoints of 1, 2, 3, and 4.

e. Compare the observed data with the theoretical distribution. Describe your conclusions.

f. Repeat parts (a)–(d) several times with $n = 100$. Describe the variability you observe between samples.

5.26 Every Tuesday, Jason's Video has "roll-the-dice" day. A customer may roll two fair dice and rent a second movie for an amount (in cents) determined by the numbers showing on the dice, the larger number first. For example, if the customer rolls a one and a five, a second movie may be rented for $0.51. Let x represent the amount paid for a second movie on roll-the-dice Tuesday.

a. Use the sample space for the rolling of a pair of dice and express the rental cost of the second movie, x, as a probability distribution.

b. What is the expected mean rental cost (mean of x) of the second movie on roll-the-dice Tuesday?

c. What is the standard deviation of x?

d. Using a computer and the probability distribution found in part (a), generate a random sample of 30 values for x and determine the total cost of renting the second movie for 30 rentals.

e. Using a computer, obtain an estimate for the probability that the total amount paid for 30 second movies will exceed $15.00 by repeating part (d) 500 times and using the 500 results.

Objective 5.3

5.27 Identify the properties that make flipping a coin 50 times and keeping track of heads a binomial experiment.

5.28 State a very practical reason why the defective item in an industrial situation might be defined to be the "success" in a binomial experiment.

5.29 Evaluate each of the following.
   a. $4!$ b. $7!$ c. $0!$ d. $\frac{6!}{2!}$
   e. $\frac{5!}{2!3!}$ f. $\frac{6!}{4!(6 - 4)!}$ g. $(0.3)^4$ h. $\binom{7}{3}$
   i. $\binom{5}{2}$ j. $\binom{3}{0}$ k. $\binom{4}{1}(0.2)^1(0.8)^3$
   l. $\binom{5}{0}(0.3)^0(0.7)^5$

5.30 A carton containing 100 T-shirts is inspected. Each T-shirt is rated "first quality" or "irregular." After all 100 T-shirts have been inspected, the number of irregulars is reported as a random variable. Explain why x is a binomial random variable.

5.31 A die is rolled 20 times, and the number of "fives" that occurred is reported as being the random variable. Explain why x is a binomial random variable.

5.32 a. Calculate $P(4)$ and $P(5)$ for the drawing of a spade from the deck of cards on p. 108.

b. Verify that the 6 probabilities, $P(0)$, $P(1)$, $P(2)$, ... , $P(5)$ form a probability distribution.

5.33 If x is a binomial random variable, use Table 2 in Appendix B to determine the probability of x for each of the following:
   a. $n = 10, x = 8, p = 0.3$ b. $n = 8, x = 7, p = 0.95$
   c. $n = 15, x = 3, p = 0.05$ d. $n = 12, x = 12, p = 0.99$
   e. $n = 9, x = 0, p = 0.5$ f. $n = 6, x = 1, p = 0.01$
   g. Explain the meaning of the symbol 0+ that appears in Table 2.

5.34 According to a December 2008, *SELF Magazine* online poll, 66% responded "Yes" to the question, "Do you want to relive your college days?" What is the probability that exactly half of the next 10 randomly selected poll participants also respond "Yes" to this question?

5.35 According to a National Safety Council report, up to 78% of automobile collisions are a result of distractions such as text messaging, phoning a pal, or fumbling with the stereo. Consider a randomly selected group of 18 reported collisions.
   SOURCE: "Cruise Control," *SELF Magazine*, December 2008

a. What is the probability that all of the collisions will be due to the distractions mentioned?

b What is the probability that 15 of the collisions will be due to the distractions mentioned?

5.36 According to the "Season's Cleaning" article, the U.S. Department of Energy reports that 25% of people with two-car garages don't have room to park any cars inside. (January 1, 2009 Rochester D&C)

Assuming this to be true, what is the probability of the following?
a. Exactly 3 two-car garage households of a random sample of 5 two-car garage households do not have room to park any cars inside.
b. Exactly 7 two-car garage households of a random sample of 15 two-car garage households do not have room to park any cars inside.
c. Exactly 20 two-car garage households of a random sample of 30 two-car garage households do not have room to park any cars inside.

5.37 Can playing video games as a child and teenager lead to a gambling or substance addiction? According to the April 11, 2009, USA Today article, "Kids show addiction symptoms," research published in the journal Psychological Science found 8.5% of children and teens who play video games displayed behavioral signs that may indicate addiction.

Suppose a randomly selected group of 30 video-gaming eighth grade students is selected.
a. What is the probability that exactly 2 of the students will display addiction symptoms?
b. If the study also indicated that 12% of video-gaming boys display addiction symptoms, what is the probability that exactly 2 out of the 17 boys in the group will display addiction symptoms?
c. If the study also indicated that 3% of video-gaming girls displayed addiction symptoms, what is the probability that exactly 2 out of the 13 girls in the group will display addiction symptoms?

5.38 Of the parts produced by a particular machine, 0.5% are defective. If a random sample of 10 parts produced by this machine contains 2 or more defective parts, the machine is shut down for repairs. Find the probability that the machine will be shut down for repairs based on this sampling plan.

5.39 The survival rate during a risky operation for patients with no other hope of survival is 80%. What is the probability that exactly four of the next five patients survive this operation?

5.40 Of all the trees planted by a landscaping firm, 90% survive. What is the probability that 8 or more of the 10 trees they just planted will survive? (Find the answer by using a table.)

5.41 In the biathlon event of the Olympic Games, a participant skis cross-country and on four intermittent occasions stops at a rifle range and shoots a set of five shots. If the center of the target is hit, no penalty points are assessed. If a particular man has a history of hitting the center of the target with 90% of his shots, what is the probability of the following?
a. He will hit the center of the target with all five of his next set of five shots.
b. He will hit the center of the target with at least four of his next set of five shots. (Assume independence.)

5.42 The May 26, 2009, USA Today Snapshot "Overcoming Identity Theft" reported the results from a poll of identity theft victims. According to the source, Affinion Security Center, 20% of the victims stated that it took "one week to one month" to recover from identity theft.

A group of 14 identity theft victims are randomly selected in your hometown.
a. What is the probability none of them were able to recover from the theft in one week to one month?
b. What is the probability that exactly 3 were able to recover from the theft in one week to one month?
c. What is the probability that at least 5 were able to recover from the theft in one week to one month?
d. What is the probability that no more than 4 were able to recover from the theft in one week to one month?

5.43 If boys and girls are equally likely to be born, what is the probability that in a randomly selected family of six children, there will be at least one boy? (Find the answer using a formula.)

5.44 A January 2005 survey of bikers, commissioned by the Progressive Group of Insurance Companies, showed that 40% of bikers have body art, such as tattoos and piercings. A group of 10 bikers are in the process of buying motorcycle insurance.

SOURCE: http://www.syracuse.com/

a. What is the probability that none of the 10 has any body art?
b. What is the probability that exactly 3 have some body art?
c. What is the probability that at least 4 have some body art?
d. What is the probability that no more than 2 have some body art?

5.45 One-fourth of a certain breed of rabbits are born with long hair. What is the probability that in a litter of six rabbits, exactly three will have long hair? (Find the answer by using a formula.)

5.46 Find the mean and standard deviation for the binomial random variable x with $n = 30$ and $p = 0.6$, using formulas (5.7) and (5.8).

5.47 Consider the binomial distribution where $n = 11$ and $p = 0.05$.
a. Find the mean and standard deviation using formulas (5.7) and (5.8).
b. Using Table 2 in Appendix B, list the probability distribution and draw a histogram.
c. Locate μ and σ on the histogram.

5.48 Consider the binomial distribution where $n = 11$ and $p = 0.05$ (see problem 5.47).
a. Use the distribution [problem 5.47(b) or Table 2] and find the mean and standard deviation using formulas (5.1), (5.3a), and (5.4).
b. Compare the results of part (a) with the answers found in problem 5.47(a).

5.49 Given the binomial probability function

$$P(x) = \binom{5}{x} \cdot \left(\frac{1}{2}\right)^x \cdot \left(\frac{1}{2}\right)^{5-x} \quad \text{for } x = 0, 1, 2, 3, 4, 5$$

a. Calculate the mean and standard deviation of the random variable by using formulas (5.1), (5.3a), and (5.4).

b. Calculate the mean and standard deviation using formulas (5.7) and (5.8).

c. Compare the results of parts (a) and (b).

5.50 Find the mean and standard deviation of x for each of the following binomial random variables:

a. The number of tails seen in 50 tosses of a quarter

b. The number of left-handed students in a classroom of 40 students (assume that 11% of the population is left-handed)

c. The number of cars found to have unsafe tires among the 400 cars stopped at a roadblock for inspection (assume that 6% of all cars have one or more unsafe tires)

d. The number of melon seeds that germinate when a package of 50 seeds is planted (the package states that the probability of germination is 0.88)

5.51 Find the mean and standard deviation for each of the following binomial random variables in parts (a)–(c):

a. The number of sixes seen in 50 rolls of a die

b. The number of defective televisions in a shipment of 125 (the manufacturer claimed that 98% of the sets were operative)

c. The number of operative televisions in a shipment of 125 (the manufacturer claimed that 98% of the sets were operative)

d. How are parts (b) and (c) related? Explain.

5.52 According to United Mileage Plus Visa (November 22, 2004), 41% of passengers say they "put on the earphones" to avoid being bothered by their seatmates during flights. To show how important, or not important, the earphones are to people, consider the variable x to be the number of people in a sample of 12 who say they "put on the earphones" to avoid their seatmates. Assume the 41% is true for the whole population of airline travelers and that a random sample is selected.

a. Is x a binomial random variable? Justify your answer.

b. Find the probability that $x = 4$ or 5.

c. Find the mean and standard deviation of x.

d. Draw a histogram of the distribution of x: label it completely, highlight the area representing $x = 4$ and $x = 5$, draw a vertical line at the value of the mean, and mark the location of x that is 1 standard deviation larger than the mean.

5.53 If the binomial $(q + p)$ is squared, the result is $(q + p)^2 = q^2 + 2qp + p^2$. For the binomial experiment with $n = 2$, the probability of no successes in two trials is q^2 (the first term in the expansion), the probability of one success in two trials is $2qp$ (the second term in the expansion), and the probability of two successes in two trials is p^2 (the third term in the expansion). Find $(q + p)^3$ and compare its terms to the binomial probabilities for $n = 3$ trials.

5.54 Did you ever buy an incandescent light bulb that failed (either burned out or did not work) the first time you turned the switch on? When you put a new bulb into a light fixture, you expect it to light, and most of the time it does. Consider 8-packs of 60-watt bulbs and let x be the number of bulbs in a pack that "fail" the first time they are used. If 0.02 of all bulbs of this type fail on their first use and each 8-pack is considered a random sample:

a. List the probability distribution and draw the histogram of x.

b. What is the probability that any one 8-pack has no bulbs that fail on first use?

c. What is the probability that any one 8-pack has no more than one bulb that fails on first use?

d. Find the mean and standard deviation of x.

e. What proportion of the distribution is between $\mu - \sigma$ and $\mu + \sigma$?

f. What proportion of the distribution is between $\mu - 2\sigma$ and $\mu + 2\sigma$?

g. How does this information relate to the empirical rule and Chebyshev's theorem? Explain.

h. Use a computer to simulate testing 100 8-packs of bulbs and observing x, the number of failures per 8-pack. Describe how the information from the simulation compares with what was expected [answers to parts (a)–(g) describe the expected results].

i. Repeat part (h) several times. Describe how these results compare with those of parts (a)–(g) and with part (h).

*remember

Problems marked with an asterisk (*) have data sets available on the CourseMate for STAT2 site. Login at cengagebrain.com.

Normal Probability
Distributions

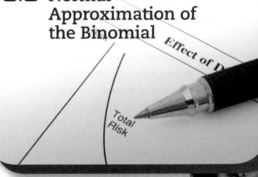

6.1 Normal Probability Distributions

THE **NORMAL PROBABILITY DISTRIBUTION**
IS CONSIDERED THE SINGLE MOST IMPORTANT
PROBABILITY DISTRIBUTION. AN UNLIMITED NUMBER
OF **CONTINUOUS RANDOM VARIABLES** HAVE
EITHER A NORMAL OR AN APPROXIMATELY **NORMAL**
DISTRIBUTION.

Several other probability distributions of both discrete and continuous random variables are also approximately normal under certain conditions.

Recall that in Chapter 5 we learned how to use a probability function to calculate the probabilities associated with **discrete random variables.** The normal probability distribution has a continuous random variable and uses two functions: one function to determine the ordinates (y values) of the graph displaying the distribution and a second to determine the probabilities. Formula (6.1) expresses the ordinate (y value) that corresponds to each abscissa (x value).

Normal Probability Distribution Function

$$y = f(x) = \frac{e^{-\frac{1}{2}\left(\frac{x-\mu}{\sigma}\right)^2}}{\sigma\sqrt{2\pi}} \text{ for all real } x \qquad (6.1)$$

When a graph of all such points is drawn, the normal (bell-shaped) curve will appear as shown in Figure 6.1

© Andreas Kuehn/Stone/Getty Images

Figure 6.1 **The Normal Probability Distribution**

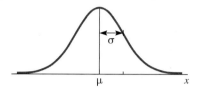

Figure 6.2 **Shaded Area:** $P(a \le x \le b)$

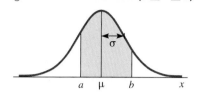

Note: Each different pair of values for the mean, μ, and standard deviation, σ, will result in a different normal probability distribution function.

Formula (6.2) yields the probability associated with the interval from $x = a$ to $x = b$:

$$P(a \le x \le b) = \int_a^b f(x)\, dx \qquad (6.2)$$

The probability that x is within the interval from $x = a$ to $x = b$ is shown as the shaded area in Figure 6.2.

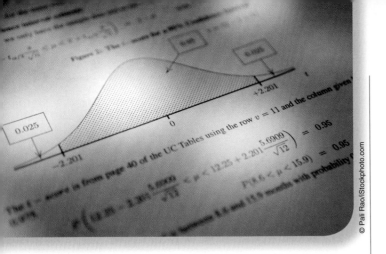

The definite integral of formula (6.2) is a calculus topic and is mathematically more advanced than what is expected in elementary statistics. Instead of using formulas (6.1) and (6.2), we will use a table to find probabilities for normal distributions. Formulas (6.1) and (6.2) were used to generate that table. Before we learn to use the table, however, it must be pointed out that the table is expressed in "standardized" form. It is standardized so that this one table can be used to find probabilities for all combinations of mean, μ, and standard deviation, σ, values. That is, the normal probability distribution with mean 38 and standard deviation 7 is similar to the normal probability distribution with mean 123 and standard deviation 32. Recall the empirical rule and the percentages of the distribution that fall within certain intervals of the mean (page 46). The same three percentages hold true for all normal distributions.

NOTE: Percentage, proportion, and probability

are basically the same concepts. Percentage (25%) is usually used when talking about a proportion $\left(\frac{1}{4}\right)$ of a population. Probability is usually used when talking about the chance that the next individual item will possess a certain property. Area is the graphic representation of all three when we draw a picture to illustrate the situation.

The empirical rule is a fairly crude measuring device; with it we are able to find probabilities associated only with whole-number multiples of the standard deviation (within 1, 2, or 3 standard deviations of the mean). We will often be interested in the probabilities associated with fractional parts of the standard deviation. For example, we might want to know the probability that x is within 1.37 standard deviations of the mean. Therefore, we must refine the empirical rule so that we can deal with more precise measurements. This refinement is discussed in the next section.

6.2 The Standard Normal Distribution

THERE ARE AN UNLIMITED NUMBER OF NORMAL PROBABILITY DISTRIBUTIONS, BUT FORTUNATELY THEY ARE ALL RELATED TO ONE DISTRIBUTION: THE **STANDARD NORMAL DISTRIBUTION.** THE STANDARD NORMAL DISTRIBUTION IS THE NORMAL DISTRIBUTION OF THE STANDARD VARIABLE z (CALLED THE **"STANDARD SCORE"** OR **"z-SCORE"**).

Table 3 in Appendix B lists the probabilities associated with the **cumulative area** to the left of a specified value of z. Probabilities of other intervals may be found by using the table entries and the operations of addition and subtraction, in accordance with the properties outlined in the box on the next page.

We have seen the standard normal distribution in earlier chapters where it appeared as the empirical rule. When using the empirical rule, the values of z were

Properties of the Standard Normal Distribution:

1. The total area under the **normal curve** is equal to 1.
2. The distribution is mounded and symmetrical; it extends indefinitely in both directions, approaching but never touching the horizontal axis.
3. The distribution has a mean of 0 and a standard deviation of 1.
4. The mean divides the area in half—0.5000 on each side.
5. Nearly all the area is between $z = -3.00$ and $z = 3.00$.

© Remigiusz Zaluck/iStockphoto.com

Figure 6.3 **Standard Normal Distribution According to Empirical Rule**

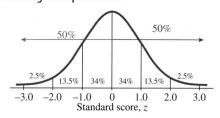

typically integer values, as in Figure 6.3. By using Table 3, the z-score will be measured to the nearest one-hundredth and allow increased accuracy.

Recall also that one of the basic properties of a probability distribution is that the sum of all probabilities is exactly 1.0. Since the area under the normal curve represents the measure of probability, the total area under the bell-shaped curve is exactly 1. Notice in Figure 6.3 that the distribution is also symmetrical with respect to a vertical line drawn through $z = 0$. That is, the area under the curve to the left of the mean is one-half, 0.5, and the area to the right is also one-half, 0.5. Note $z = 0.00$ in Table 3 in Appendix B. Areas (probabilities, percentages) not given directly by the table can be found with the aid of these properties. Let's look at several illustrations demonstrating how to use Table 3 to find probabilities of the standard normal score, z.

FINDING THE AREA TO THE LEFT OF A NEGATIVE z VALUE

Imagine you want to find the area under the standard normal curve to the left of $z = -1.52$ as shown in Figure 6.4.

Figure 6.4 **Area to the Left of $z = -1.52$**

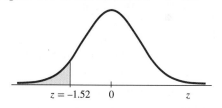

Table 3 in Appendix B is designed to give the area to the left of -1.52 directly. The z-score is located in the margins, with the units and tenths digits along the left side and the hundredths digit across the top. For $z = -1.52$, locate the row labeled -1.5 and the column labeled 0.02; at their intersection you will find 0.0643, the measure of the cumulative area to the left of $z = -1.52$ (see Table 6.1). Expressed as a probability:

$P(z < -1.52) = \mathbf{0.0643}$

Table 6.1 **A Portion of Table 3**

z	0.00	0.01	0.02	...
⋮				
-1.5			**0.0643**	...
⋮				

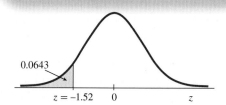

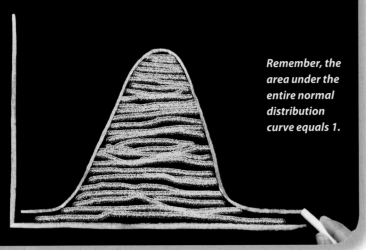

© Stephen Rees/iStockphoto.com

Remember, the area under the entire normal distribution curve equals 1.

FINDING THE AREA TO THE LEFT OF A POSITIVE z VALUE

Begin finding the area under the normal curve to the left of $z = 1.52$: $P(z < 1.52)$ by drawing and labeling a sketch. (Sketching and labeling the curve will help you visualize the calculations you are doing.) Finding the area to the left of a positive z value is much like finding it for a negative z value. Table 3 is designed to give the area to the left of positive z-values directly, with the z-score on the margins with the units and tenths digits on the left side, and the hundredths along the top. So, for $z = 1.52$, locate the row labeled 1.5 and the column labeled 0.02; as you can see in Table 6.2, at their intersection you will find 0.9357, the measure of the cumulative area to the left of $z = +1.52$.

$$P(z < 1.52) = \mathbf{0.9357}$$

Table 6.2 A Portion of Table 3

z	0.00	0.01	0.02	...
⋮				
1.5			**0.9357**	...
⋮				

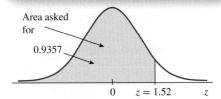

NOTES:

1. Probabilities associated with positive z-values are greater than 0.5000 since they include the entire left half of the normal curve.

2. Make it a habit to write the z-score with two decimal places and the areas (probabilities, percentages) with four decimal places, as in Table 3.

This will help with distinguishing between the two concepts.

FINDING THE AREA TO THE RIGHT OF A z VALUE

Knowing that the area under the entire normal distribution curve is equal to 1 is the key factor in determining probabilities associated with the values to the right of a z-value. To demonstrate why, let's look at how to find area that is not included in the shaded area. Find the area under the normal curve to the right of $z = -1.52$: $P(z > -1.52)$. The problem asks for the area that is not included in the 0.0643 shaded area. Since the area under the entire normal curve is 1, we subtract 0.0643 from 1:

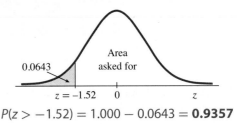

$$P(z > -1.52) = 1.000 - 0.0643 = \mathbf{0.9357}$$

This subtraction method applies when finding the area to the right of any z-score; using Table 3, find the area to the left and subtract the table value from 1.0. The total of the area to the left and the area to the right will always be 1.0.

FINDING THE AREA BETWEEN ANY TWO z VALUES

Whether both z-scores are negative, positive, or you have one of each, use the larger z-score to find the area between the two, as this example shows. The area between $z = -1.36$ and $z = 2.14$, $P(-1.36 < z < 2.14)$, is found using subtraction. The cumulative area to the left of the larger z, $z = 2.14$, includes both the area asked for and the area to the left of the smaller z, $z = -1.36$. Therefore, we subtract the area to the left of the smaller z, $z = -1.36$, from the area to the left of the larger z, $z = 2.14$:

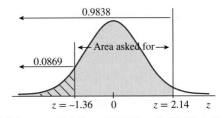

$$P(-1.36 < z < 2.14) = 0.9838 - 0.0869 = \mathbf{0.8969}$$

FINDING THE z-SCORE ASSOCIATED WITH A PERCENTILE

Table 3 can also be used to find the z-score that bounds a specified area; for example, find the z-score associated with the 75th percentile of a normal distribution. By finding the area or probability within the table, the z-score can be read from along the left side and top margins.

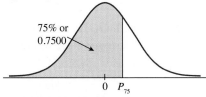

75% or
0.7500

0   P_{75}

Table 3's cumulative area matches the definition of a percentile. Recall that the 75th percentile means that 75% of the data are less than the percentile's value. To find the z-score for the 75th percentile, look in Table 3 and find the "area" entry that is closest to 0.7500; this area entry is 0.7486. Now read the z-score that corresponds to this area.

Table 6.3 A Portion of Table 3

z	...	0.07		0.08	...
⋮					
0.6	...	0.7486	*0.7500*	0.7518	...
⋮					

As you can see in Table 6.3, the z-score is found to be $z = 0.67$. This says that the 75th percentile in a normal distribution is 0.67 (approximately 2/3) standard deviations above the mean.

FINDING TWO z-SCORES THAT BOUND AN AREA

Finding z-scores around partial areas of a normal distribution is also possible. For example, what z-scores bound the middle 95% of a normal distribution? As shown in Figure 6.5, the 95% is

split into two equal parts by the mean, so 0.4750 is the area (percentage) of the z-score at the left boundary and $z = 0$, the mean (as well as the 0.4750 is the area between $z = 0$, the mean and the right boundary).

The area that is not included in either tail can be found by recalling that the area for each half of the normal curve is equal to 0.5000 and that the curve is symmetrical. Thus on the left side, $0.5000 - 0.4750 = 0.0250$ is needed; and on the right side, $0.5000 + 0.4750 = 0.9750$ is needed. To find the left boundary z-score, use the area 0.0250 in Table 3 and find the "area" entry that is closest to 0.0250; this entry is exactly 0.0250.

Reading Table 6.4, the z-score that corresponds to this area is found to be $z = -1.96$. Likewise, to find the right boundary z-score, use the area 0.9750 in Table 3 and find the "area" entry that is closest to 0.9750; this entry is exactly 0.9750. Reading this z-score gives $z = +1.96$.

Therefore, you can look up either one and utilize the symmetry of the normal distribution. $z = -1.96$ and $z = 1.96$ bound the middle 95% of a normal distribution.

IF YOU WANT TO DOUBLE CHECK YOUR WORK, DO THE PROBLEM BOTH WAYS.

Table 6.4 A Portion of Table 3

negative z side			and	positive z side		
z		0.06		z		0.06
⋮				⋮		
−1.9	...	0.0250	...	1.9	...	0.9750	...
⋮				⋮		

Figure 6.5

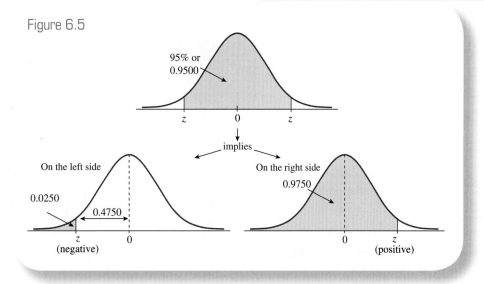

95% or
0.9500

z 0 z

implies

On the left side

0.0250

0.4750

z (negative)   0

On the right side

0.9750

0   z (positive)

6.3 Applications of Normal Distributions

IN OBJECTIVE 6.2 WE LEARNED HOW TO USE TABLE 3 IN APPENDIX B TO CONVERT INFORMATION ABOUT THE STANDARD NORMAL VARIABLE z INTO PROBABILITY AND HOW TO CONVERT PROBABILITY INFORMATION ABOUT THE STANDARD NORMAL DISTRIBUTION INTO z-SCORES.

Now we are ready to apply this methodology to all normal distributions. The key is the standard score, z. The information associated with a normal distribution will be in terms of x values or probabilities. We will use the z-score and Table 3 as the tools to "go between" the given information and the desired answer.

Probabilities and Normal Curves

To demonstrate the process of converting to a standard normal curve to find probabilities, let's consider IQ scores. IQ scores are normally distributed with a mean of 100 and a standard deviation of 16. If a person is picked at random, what is the probability that his or her IQ is between 100 and 115? That is, what is $P(100 < x < 115)$?

$P(100 < x < 115)$ is represented by the shaded area in the figure below.

The variable x must be standardized using formula (6.3). The z values are shown on the figure.

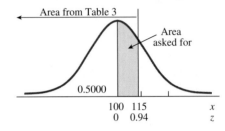

$$z = \frac{x - \mu}{\sigma}$$

when $x = 100$: $\quad z = \dfrac{100 - 100}{16} = 0.00$

when $x = 115$: $\quad z = \dfrac{115 - 100}{16} = 0.94$

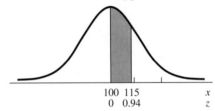

Therefore,

$$P(100 < x < 115) = P(0.00 < z < 0.94) = \textbf{0.3264.}$$

Thus, the probability is 0.3264 (found by using Table 3 in Appendix B) that a person picked at random has an IQ between 100 and 115.

What if we need to determine a probability for "any" normal curve? How do we calculate probability under "any" normal curve? Let's continue to use the example

Recall that the standard score, z, was defined in Chapter 2.

Standard Score

In words: $\quad z = \dfrac{x - (\text{mean of } x)}{\text{standard deviation of } x}$

In algebra: $\quad z = \dfrac{x - \mu}{\sigma}$ \hfill (6.3)

(Note that when $x = \mu$, the standard score $z = 0$.)

REMEMBER:

WHEN FINDING THE AREA BETWEEN TWO z-SCORES, SUBTRACT THE AREA CORRESPONDING TO THE SMALLER z FROM THE AREA CORRESPONDING TO THE LARGER z.

0.7357 chance that she has an IQ greater than 90

of IQ scores and try to find the probability that a person selected at random will have an IQ greater than 90.

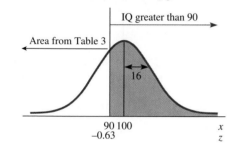

IQ greater than 90

Area from Table 3

16

90 100   *x*
−0.63   *z*

$$z = \frac{x - \mu}{\sigma} = \frac{90 - 100}{16} = \frac{-10}{16} = -0.625 = -0.63$$

$$P(x > 90) = P(z > -0.63)$$

$$= 1.0000 - 0.2643 = 0.7357$$

Thus, the probability is 0.7357 that a person selected at random will have an IQ greater than 90.

Using the Normal Curve and *z*

The normal table, Table 3, can be used to answer many kinds of questions that involve a normal distribution. Many problems call for the location of a "cut-off point," a particular value of *x* such that there is exactly a certain percentage in a specified area.

DETERMINE DATA VALUES

In a large class, suppose your instructor tells you that you need to obtain a grade in the top 10% of your class to get an A on a particular exam. From past experience, she is able to estimate that the mean and standard deviation on this exam will be 72 and 13,

respectively. What will be the minimum grade needed to obtain an A? (Assume that the grades will have an approximately normal distribution.) We can use the normal curve and *z* to determine that data value.

Start by converting the 10% to information that is compatible with Table 3 by subtracting:

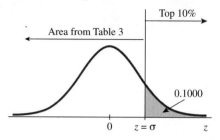

Top 10%

Area from Table 3

0.1000

0   *z* = σ   *z*

10% = 0.1000;
1.0000 − 0.1000 = 0.9000

Look in Table 3 to find the value of *z* associated with the area entry closest to 0.9000; it is *z* = 1.28. Thus,

$$P(z > 1.28) = 0.10$$

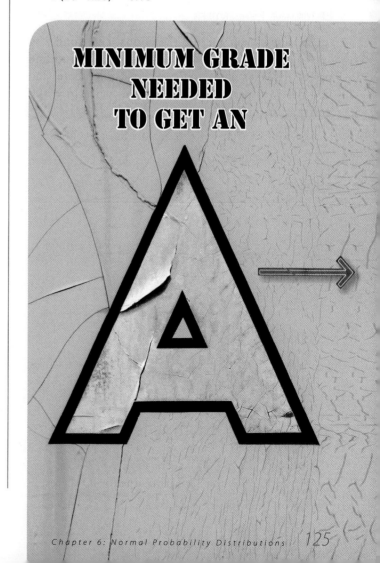

MINIMUM GRADE NEEDED TO GET AN A

Now find the x value that corresponds to $z = 1.28$ by using formula (6.3):

$$z = \frac{x - \mu}{\sigma}: \quad 1.28 = \frac{x - 72}{13}$$

$$x - 72 = (13)(1.28)$$

$$x = 72 + (13)(1.28) = 72 + 16.64 = 88.64, \text{ or } \mathbf{89}$$

Thus, if you receive an 89 or higher (the data value), you can expect to be in the top 10% (which means you can expect to receive an A).

DETERMINE PERCENTILES

Just as you can use the normal curve and z to find data values, you can also use them to find percentiles. Let's return to the example of IQ scores and find the 33rd percentile for IQ scores when $\mu = 100$ and $\sigma = 16$.

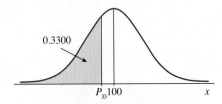

z	$\cdots$	0.04	$\cdots$
$\vdots$			
-0.4	$\cdots$	0.3300	$\cdots$

$$P(z < P_{33}) = 0.3300$$
33rd percentile is at $z = -0.44$

Now we convert the 33rd percentile of the z-scores, -0.44, to an x-score using formula (6.3):

$$z = \frac{x - \mu}{\sigma}: \quad -0.44 = \frac{x - 100}{16}$$

$$x - 100 = (16)(-0.44)$$

$$x = 100 - 7.04 = \mathbf{92.96}$$

Thus, 92.96 is the 33rd percentile for IQ scores.

DETERMINE POPULATION PARAMETERS

The normal curve and z can also be used to determine population parameters. That is, when given related information, you can find the standard deviation, σ. To illustrate how this can be done, let's look at the incomes of junior executives in a large corporation, which are approximately normally distributed. A pending cutback will not discharge those junior executives with earnings within $4,900 of the mean. If this represents the middle 80% of the incomes, what is the standard deviation for the salaries of this group of junior executives?

Table 3 indicates that the middle 80%, or 0.8000, of a normal distribution is bounded by -1.28 and 1.28. Consider point B shown in the figure below: 4,900 is the difference between the x-value at B and the value of the mean, the numerator of formula (6.3) [$x - \mu = 4,900$].

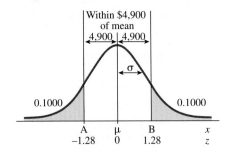

Using formula (6.3), we can find the value of σ:

$$z = \frac{x - \mu}{\sigma}: \quad 1.28 = \frac{4,900}{\sigma}$$

$$\sigma = \frac{4,900}{1.28}$$

$$\sigma = 3,828.125 = \mathbf{\$3,828}$$

That is, the current standard deviation for the salaries of junior executives is $3,828.

You are about to be unemployed.

6.4 Notation

THE *z*-SCORE IS USED THROUGHOUT STATISTICS IN A VARIETY OF WAYS; HOWEVER, THE RELATIONSHIP BETWEEN THE NUMERICAL VALUE OF *z* AND THE AREA UNDER THE STANDARD NORMAL DISTRIBUTION CURVE DOES NOT CHANGE. SINCE *z* IS USED WITH GREAT FREQUENCY, WE WANT A CONVENIENT NOTATION TO IDENTIFY THE NECESSARY INFORMATION.

The convention that we will use as an "algebraic name" for a specific z-score is $z(\alpha)$, where α represents the "area to the right" of the z being named.

Visual Interpretation of $z(\alpha)$

Figures 6.6 and 6.7 are both visual interpretations of $z(\alpha)$. Figure 6.6 depicts $z(0.05)$ (read "z of 0.05"), which is the algebraic name for z, such that the area to the right and under the standard normal curve is exactly 0.05. In a similar fashion, Figure 6.7 shows $z(0.90)$ (read "z of 0.90"), which is the value of z, such that 0.90 of the area lies to its right.

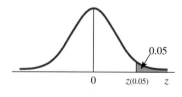

Figure 6.6 **Area Associated with** $z_{(0.05)}$

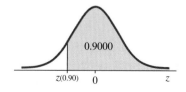

Figure 6.7 **Area Associated with** $z_{(0.90)}$

Determining Corresponding *z* Values for *z(α)*

Visual representations like those in Figures 6.6 and 6.7 also have corresponding numerical values. Now let's find the numerical values of $z(0.05)$, $z(0.90)$, and $z(0.95)$.

To find the numerical value of $z(0.05)$, we must convert the area information in the notation into information that we can use with Table 3 in Appendix B. By subtracting 0.05 from 1, we get 0.95, the area to the left of the $z(0.05)$, which you can see in Figure 6.8. When we look in Table 3, we look for an area as close as possible to 0.9500.

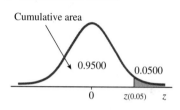

Figure 6.8 **Find the Value of** $z_{(0.05)}$

z	...	0.04	**0.05**	...
:				
1.6	...	0.9495	***0.9500***	0.9505	...
:				

We use the z that corresponds to the area closest in value. When the value happens to be exactly halfway between the table entries as above, always use the larger value of z. Therefore, $z(0.05) = \mathbf{1.65}$.

To find the numerical value of $z(0.90)$, we need to subtract the 0.90 area from 1, which results in an area of 0.10 to the left of $z(0.90)$. The 0.1000 area is the area we can use with Table 3 in Appendix B; as shown in the diagram below.

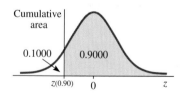

The closest values in Table 3 are 0.1003 and 0.0985, with 0.1003 being closer to 0.1000.

z	...	**0.08**		0.09
:				
−1.2	...	0.1003	0.1000	0.0985
:				

Therefore, $z(0.90)$ is related to -1.28. Since $z(0.90)$ is below the mean, it makes sense that $z(0.90) = -1.28$.

Because of the symmetrical nature of the normal distribution, $z(\alpha)$ and $z(1 - \alpha)$ are closely related, with the only difference being that one is positive and the other is negative. We already found the value of $z(0.05) = 1.65$. Now let's find $z(0.95)$.

$z(0.95)$ is located on the left-hand side of the normal distribution since the area to the right is 0.95. The area in the tail to the left then contains the other 0.05, as shown in Figure 6.9.

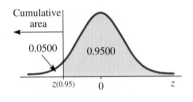

Figure 6.9   Area Associated with $z(0.95)$

Using Table 3, $z(0.95) = -1.65$.

Because of the symmetrical nature of the normal distribution, $z(0.95) = -1.65$ and $z(0.05) = 1.65$ differ only in sign and the side of the distribution to which they belong. Thus, $z(0.95) = -z(0.05) = -1.65$.

In many situations, it will be more convenient to refer to the area of the tail than to either the cumulative area or the area to the right, so we can use this difference as an alternative algebraic name for the z-values bounding a left-side tail situation. In general, when $1 - \alpha$ is larger than 0.5000, the notation convention we will use is $z(1 - \alpha) = -z(\alpha)$.

The $z(\alpha)$ notation is used regularly in connection with inferential situations involving the area of a tail (extreme ends of a distribution curve—either left or right) region. In later chapters this notation will be used on a regular basis. The values of z that will be used regularly come from one of the following situations: (1) the z-score such that there is a specified area in one tail of the normal distribution, or (2) the z-scores that bound a specified middle proportion of the normal distribution. When the middle proportion of a normal distribution is specified, we can also use the "area to the right" notation to identify the specific z-score involved.

Table 4 and Commonly Used z Values

We already solved two commonly used one-tail situations; $z(0.05) = 1.65$ is located so that 0.05 of the area

under the normal distribution curve is in the tail to the right and $z(0.90) = -1.28$ is located so that 0.10 of the area under the normal distribution curve is in the tail to the left.

Table 4, Critical Values of Standard Normal Distribution, was designed to provide only the most commonly used values of z when the area(s) of the tail regions are given. Part A, One-Tailed Situations, is used when the area of a tail is given. In order to examine this, let's find the values of $z(0.05)$ and $z(0.95)$ using Table 4. Table 4A, One-Tailed Situations shows us:

		Amount of α in One Tail			
α	$\cdots$	0.10	**0.05**	0.025	$\cdots$
$z(\alpha)$	$\cdots$	1.28	**1.65**	1.96	$\cdots$

$z(0.05) = 1.65$, and since the standard normal distribution is symmetrical, the value of $z(0.95) = -z(0.05) = -1.65$.

Determining z-Scores for Bounded Areas

z-scores can also be determined for bounded areas of a normal distribution. For example, we can find the z-scores that bound the middle 0.95 of the normal distribution. Given 0.95 as the area in the middle (see Figure 6.10), the two tails must contain a total of 0.05. Therefore, each tail contains $\frac{1}{2}$ of 0.05, or 0.025, as shown in Figure 6.11.

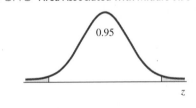

Figure 6.10   Area Associated with Middle 0.95

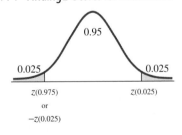

Figure 6.11   Finding z-Scores for Middle 0.95

The right tail value, $z(0.025)$, is found using Table 4, Part A, One-Tailed Situations, as shown previously.

A Portion of Table 4A, One-Tailed Situations					
	Amount of α in One Tail				
α	⋯	0.05	**0.025**	0.02	⋯
Z(α)	⋯	1.65	**1.96**	2.05	⋯

$z(0.025) = 1.96$, and since the standard normal distribution is symmetrical, the value of $z(0.975) = -z(0.025) = -1.96$.

USING TWO TAILS

You can use two tails to find the area as well. Given 0.95 as the area in the middle (Figure 6.11), the two tails must contain a total of 0.05. Table 4, Part B, Two-Tailed Situations, can be used when the combined area of both tails (or the area in the center) is given. Locate the column that corresponds to $\alpha = 0.05$ or $(1 - \alpha) = 0.95$.

A Portion of Table 4B, Two-Tailed Situations					
	Amount of α in Two Tails				
α	⋯	0.10	**0.05**	0.02	⋯
Z(α/2)	⋯	1.65	**1.96**	2.33	⋯
1 − α	⋯	0.90	**0.95**	0.98	⋯
Area in the "center"					

From Table 4B we find $z(0.05/2) = z(0.025) = 1.96$. Using the symmetry property of the distribution, we find $z(0.975) = -z(0.025) = -1.96$. Therefore, the middle 0.95 of the normal distribution is bounded by **−1.96** and **1.96**.

6.5 Normal Approximation of the Binomial

IN CHAPTER 5 WE INTRODUCED THE **BINOMIAL DISTRIBUTION**. RECALL THAT THE BINOMIAL DISTRIBUTION IS A PROBABILITY DISTRIBUTION OF THE DISCRETE RANDOM VARIABLE x, THE NUMBER OF SUCCESSES OBSERVED IN n REPEATED INDEPENDENT TRIALS.

We will now see how **binomial probabilities**—that is, probabilities associated with a binomial distribution—can be reasonably approximated by using the normal probability distribution.

Let's look first at a few specific binomial distributions. Figure 6.12 shows the probabilities of x for 0 to n for three situations: $n = 4, n = 8,$ and $n = 24$. For each of these distributions, the probability of success for one trial is 0.5. Notice that as n becomes larger, the distribution appears more and more like the normal distribution.

To make the desired approximation, we need to take into account one major difference between the binomial and the normal probability distribution. The binomial random variable is **discrete**, whereas the normal random variable is **continuous**. Recall that Chapter 5 demonstrated that the probability assigned to a particular value of x should be shown on a diagram by means of a straight-line segment whose length represents the probability (as in Figure 6.12). Chapter 5 suggested, however, that we can also use a histogram in which the area of each bar is equal to the probability of x.

Let's look at the distribution of the binomial variable x, when $n = 14$ and $p = 0.5$. The probabilities for each x value can be obtained from Table 2 in Appendix B. This distribution of x is shown in

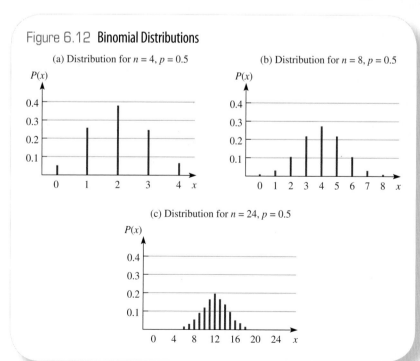

Figure 6.12 **Binomial Distributions**

(a) Distribution for $n = 4, p = 0.5$

(b) Distribution for $n = 8, p = 0.5$

(c) Distribution for $n = 24, p = 0.5$

Figure 6.13. We see the very same distribution in Figure 6.14 in histogram form.

Let's examine $P(x = 4)$ for $n = 14$ and $p = 0.5$ to study the approximation technique. $P(x = 4)$ is equal to 0.061 (see Table 2 in Appendix B), the area of the bar (rectangle) above $x = 4$ in Figure 6.15.

The area of a rectangle is the product of its width and height. In this case, the height is 0.061 and the width is 1.0, so the area is 0.061. Let's take a closer look at the width. For $x = 4$, the bar starts at 3.5 and ends at 4.5, so we are looking at an area bounded by $x = 3.5$ and $x = 4.5$. The addition and subtraction of 0.5 to the x value is commonly called the **continuity correction factor.** It is our method of converting a discrete variable into a continuous variable.

Now let's look at the normal distribution related to this situation. We will first need a normal distribution with a mean and a standard deviation equal to those of the binomial distribution we are discussing. Formulas (5.7) and (5.8) give us these values:

$$\mu = np = (14)(0.5) = \mathbf{7.0}$$

$$\sigma = \sqrt{npq} = \sqrt{(14)(0.5)(0.5)} = \sqrt{3.5} = \mathbf{1.87}$$

The probability that $x = 4$ is approximated by the area under the normal curve between $x = 3.5$ and $x = 4.5$, as shown in Figure 6.16. Figure 6.17 shows the entire distribution of the binomial variable x with a normal distribution of the same mean and standard deviation superimposed. Notice that the bars and the interval areas under the curve cover nearly the same area.

The probability that x is between 3.5 and 4.5 under this normal curve is found by using formula (6.3), Table 3, and the methods outlined in Objective 6.4:

$$z = \frac{x - \mu}{\sigma}:$$

$$P(3.5 < x < 4.5) = P\left(\frac{3.5 - 7.0}{1.87} < z < \frac{4.5 - 7.0}{1.87}\right)$$

$$= P(-1.87 < z < -1.34)$$

$$= 0.0901 - 0.0307 = 0.0594$$

Since the binomial probability of 0.061 and the normal probability of 0.0594 are reasonably close, the normal probability distribution seems to be a reasonable approximation of the binomial distribution.

The normal approximation of the binomial distribution is also useful for values of p that are not close to 0.5. The binomial probability distribution shown in Figure 6.18 suggests that binomial probabilities can be approximated using the normal distribution. Notice that as n increases, the binomial distribution begins to look like the normal distribution. As the value of p moves away from 0.5, a larger n is needed in order for the normal approximation to be reasonable. The following *rule of thumb* is generally used as a guideline:

RULE: The normal distribution provides a reasonable approximation to a binomial probability distribution whenever the values of np and $n(1 - p)$ both equal or exceed 5.

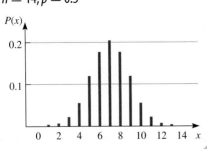

Figure 6.13 **The Distribution of x When $n = 14$, $p = 0.5$**

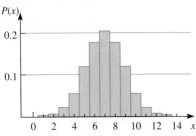

Figure 6.14 **Histogram for the Distribution of x When $n = 14$, $p = 0.5$**

Continuity Correction Factor:

$x \pm 0.5$

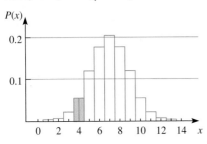

Figure 6.15 **Area of Bar above $x = 4$ is 0.061, for $B(n = 14, p = 0.5)$**

Figure 6.16 **Probability That $x = 4$ Is Approximated by Shaded Area**

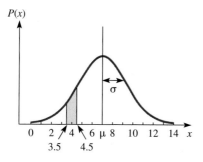

Figure 6.17 **Normal Distribution Superimposed over Distribution for Binomial Variable x**

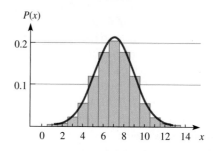

Figure 6.18 **Binomial Distributions**

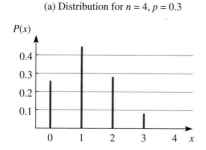

(a) Distribution for $n = 4$, $p = 0.3$

(b) Distribution for $n = 8$, $p = 0.3$

(c) Distribution for $n = 24$, $p = 0.3$

By now you may be thinking, "So what? I will just use the binomial table and find the probabilities directly and avoid all the extra work." That doesn't always work, though. Sometimes you must solve (and not find) a binomial probability problem with the normal distribution. For example, an unnoticed mechanical failure has caused $\frac{1}{3}$ of a machine shop's production of 5,000 coil springs to be defective. What is the probability that an inspector will find no more than 3 defective springs in a random sample of 25?

In this example of a binomial experiment, x is the number of defectives found in the sample, $n = 25$, and $p = P(\text{defective}) = \frac{1}{3}$. To answer the question using the binomial distribution, we will need to use the binomial probability function, formula (5.5):

$$P(x) = \binom{25}{x}\left(\frac{1}{3}\right)^x\left(\frac{2}{3}\right)^{25-x} \quad \text{for } x = 0, 1, 2, \ldots, 25$$

We must calculate the values for $P(0)$, $P(1)$, $P(2)$, and $P(3)$, because they do not appear in Table 2. This is a very tedious job because of the size of the exponent. In situations such as this, we can use the normal approximation method.

Now let's find $P(x \leq 3)$ by using the normal approximation method. We first need to find the mean and standard deviation of x using formulas (5.7) and (5.8):

$$\mu = np = (25)\left(\frac{1}{3}\right) = \textbf{8.333}$$

$$\sigma = \sqrt{npq} = \sqrt{(25)\left(\frac{1}{3}\right)\left(\frac{2}{3}\right)} = \sqrt{5.555} = \textbf{2.357}$$

These values are shown in the figure. The area of the shaded region ($x < 3.5$) represents the probability of $x = 0, 1, 2,$ or 3. Remember that $x = 3$, the discrete binomial variable, covers the continuous interval from 2.5 to 3.5.

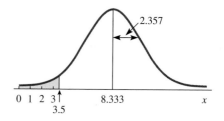

$$P(x \text{ is no more than } 3) = P(x \leq 3) \begin{pmatrix} \text{for a discrete} \\ \text{variable } x \end{pmatrix}$$

$$= P(x < 3.5) \begin{pmatrix} \text{for a continuous} \\ \text{variable } x \end{pmatrix}$$

$$z = \frac{x - \mu}{\sigma}:$$

$$P(x < 3.5) = P\left(z < \frac{3.5 - 8.333}{2.357}\right) = P(z < -2.05)$$

$$= \textbf{0.0202}$$

Thus, $P(\text{no more than 3 defectives})$ is approximately 0.02.

problems

Objective 6.1

6.1 Percentage, proportion, or probability—identify which is illustrated by each of the following statements.

   a. One-third of the crowd had a clear view of the event.

   b. Fifteen percent of the voters were polled as they left the voting precinct.

   c. The chance of rain during the day tomorrow is 0.2.

6.2 Percentage, proportion, or probability—in your own words, using between 25 and 50 words for each, describe the following:

   a. How percentage is different than the other two

   b. How proportion is different than the other two

   c. How probability is different than the other two

   d. How all three are basically the same thing

Objective 6.2

6.3 Find the area under the normal curve that lies to the left of the following z values.

   a. $z = -1.30$ b. $z = -2.56$

   c. $z = -3.20$ d. $z = -0.64$

6.4 Find the probability that a piece of data picked at random from a normal population will have a standard score (z) that lies to the left of the following z values.

   a. $z = 2.10$ b. $z = 1.20$

   c. $z = 3.26$ d. $z = 0.71$

6.5 Find the area under the standard normal curve to the right of $z = -2.35$, $P(z > -2.35)$.

6.6 Find the following areas under the standard normal curve.

   a. to the right of $z = -0.47$, $P(z > -0.47)$

   b. to the right of $z = -1.01$, $P(z > -1.01)$

   c. to the right of $z = -3.39$, $P(z > -3.39)$

6.7 Find the area under the standard normal curve to the right of $z = 2.03$, $P(z > 2.03)$.

6.8 Find the area under the standard normal curve between -1.39 and the mean, $P(-1.39 < z < 0.00)$.

6.9 Find the area under the standard normal curve between $z = -2.46$ and $z = 1.46$, $P(-2.46 < z < 1.46)$.

6.10 Find the area under the standard normal curve that corresponds to the following z values.

   a. between 0 and 1.55   b. to the right of 1.55

   c. to the left of 1.55   d. between -1.55 and 1.55

6.11 Find the following:

   a. $P(0.00 < z < 2.35)$ b. $P(-2.10 < z < 2.34)$

   c. $P(z > 0.13)$ d. $P(z < 1.48)$

6.12 Find the following:

   a. $P(-3.05 < z < 0.00)$ b. $P(-2.43 < z < 1.37)$

   c. $P(z < -2.17)$ d. $P(z > 2.43)$

6.13 Find the area under the standard normal curve between $z = 0.75$ and $z = 2.25$, $P(0.75 < z < 2.25)$.

6.14 Find the area under the normal curve that lies between the following pairs of z-values:

   a. $z = -1.20$ to $z = -0.22$

   b. $z = -1.75$ to $z = -1.54$

   c. $z = 1.30$ to $z = 2.58$

   d. $z = 0.35$ to $z = 3.50$

6.15 Find the z-score for the standard normal distribution shown on each of the following diagrams.

   a.

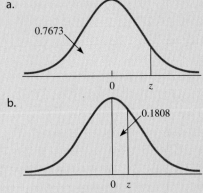

   b.

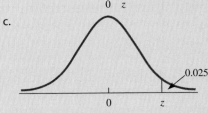

   c.

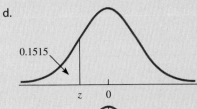

   d.

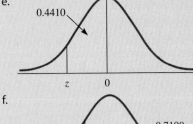

   e.

   f.

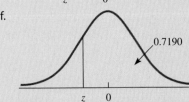

6.16 Assuming a normal distribution, what is the z-score associated with the 90th percentile? 95th percentile? 99th percentile?

6.17 Assuming a normal distribution, what is the z-score associated with the 1st quartile? 2nd quartile? 3rd quartile?

6.18 a. Find the standard z-score such that 80% of the distribution is below (to the left of) this value.

b. Find the standard z-score such that the area to the right of this value is 0.15.

c. Find the two z-scores that bound the middle 50% of a normal distribution.

6.19 a. Find the z-score for the 33rd percentile of the standard normal distribution.

b. Find the z-scores that bound the middle 40% of the standard normal distribution.

Objective 6.3

6.20 Given $x = 58$, $\mu = 43$, and $\sigma = 5.2$, find z.

6.21 Given that x is a normally distributed random variable with a mean of 60 and a standard deviation of 10, find the following probabilities.

a. $P(x > 60)$ b. $P(60 < x < 72)$

c. $P(57 < x < 83)$ d. $P(65 < x < 82)$

e. $P(38 < x < 78)$ f. $P(x < 38)$

6.22 Based on a survey conducted by Greenfield Online, 25–34-year-olds spend the most each week on fast food. The average weekly amount of $44 was reported in a May 2009 *USA Today* Snapshot. Assuming that weekly fast food expenditures are normally distributed with a standard deviation of $14.50, what is the probability that a 25- to 34-year-old will spend

a. less than $25 a week on fast food?

b. between $30 and $50 a week on fast food?

c. more than $75 a week on fast food?

6.23 Depending on where you live and on the quality of the daycare, costs of daycare can range from $3,000 to $15,000 a year (or $250 to $1,250 monthly) for one child, according to BabyCenter. Daycare centers in large cities such as New York and San Francisco are notoriously expensive. Suppose that daycare costs are normally distributed with a mean equal to $9,000 and a standard deviation equal to $1,800.

SOURCE: http://www.babycenter.com/

a. What percentage of daycare centers will cost between $7,200 and $10,800 annually?

b. What percentage of daycare centers will cost between $5,400 and $12,600 annually?

c. What percentage of daycare centers will cost between $3,600 and $14,400 annually?

d. Compare the results from parts (a) through (c) with the empirical rule. Explain the relationship.

6.24 As shown in the example on page 126, incomes of junior executives are normally distributed with a standard deviation of $3,828.

a. What is the mean for the salaries of junior executives if a salary of $62,900 is at the top end of the middle 80% of incomes?

b. With the additional information learned in part (a), what is the probability that a randomly selected junior executive earns less than $50,000?

6.25 According to ACT, results from 2008 ACT testing found that students had a mean reading score of 21.4 with a standard deviation of 6.0. Assuming that the scores are normally distributed,

a. Find the probability that a randomly selected student has a Reading ACT score less than 20.

b. Find the probability that a randomly selected student has a Reading ACT score between 18 and 24.

c. Find the probability that a randomly selected student has a Reading ACT score greater than 30.

d. Find the value of the 75th percentile for ACT scores.

6.26 A radar unit is used to measure the speed of automobiles on an expressway during rush-hour traffic. The speeds of individual automobiles are normally distributed with a mean of 62 mph.

a. Find the standard deviation of all speeds if 3% of the automobiles travel faster than 72 mph.

b. Using the standard deviation found in (a), find the percentage of these cars that are traveling less than 55 mph.

c. Using the standard deviation found in (a), find the 95th percentile for the variable "speed."

6.27 a. Generate a random sample of 100 data from a normal distribution with mean 50 and standard deviation 12.

b. Using the random sample of 100 data found in part (a) and the technology commands for calculating ordinate values on your Tech Card, find the 100 corresponding y-values for the normal distribution curve with mean 50 and standard deviation 12.

c. Use the 100 ordered pairs found in part (b) and draw the curve for the normal distribution with mean 50 and standard deviation 12. (Technology commands are included with part (b) commands on your Tech Card.)

d. Using the technology commands for cumulative probability on your Tech Card, find the probability that a randomly selected value from a normal distribution with mean 50 and standard deviation 12 will be between 55 and 65. Verify your results by using Table 3.

6.28 Generate 10 random samples, each of size 25, from a normal distribution with mean 75 and standard deviation 14. Answer parts (b) through (d) of question 6.25.

Objective 6.4

6.29 Using the $z(\alpha)$ notation (pass identify the value of α used within the parentheses), name each of the standard normal variable z's shown in the following diagrams.

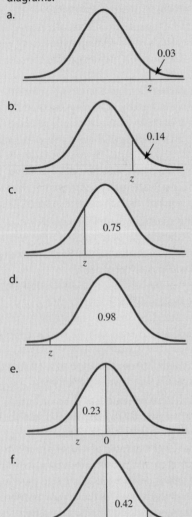

a.

0.03

z

b.

0.14

z

c.

0.75

z

d.

0.98

z

e.

0.23

z 0

f.

0.42

0   z

6.30 Draw a figure of the standard normal curve showing:

a. $z(0.15)$ b. $z(0.82)$

6.31 Draw a figure of the standard normal curve showing:

a. $z(0.04)$ b. $z(0.94)$

6.32 We are often interested in finding the value of z that bounds a given area in the right-hand tail of the normal distribution, as shown in the accompanying figure. The notation $z(\alpha)$ represents the value of z such that $P(z > z(\alpha)) = \alpha$.

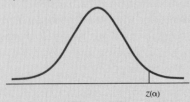

$z(\alpha)$

Find the following:

a. $z(0.025)$ b. $z(0.05)$ c. $z(0.01)$

6.33 Use Table 4A in Appendix B, and the symmetry property of normal distributions to find the following values of z.

a. $z(0.05)$ b. $z(0.01)$ c. $z(0.025)$

d. $z(0.975)$ e. $z(0.98)$

6.34 Using Table 4A and the symmetry property of the normal distribution, complete the following charts of z-scores. The area given in the tables is the area to the right under the normal distribution in the figures.

a. z-scores associated with the right-hand tail: Given the area A, find $z(A)$.

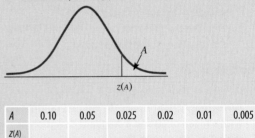

$z(A)$

A	0.10	0.05	0.025	0.02	0.01	0.005
$Z(A)$						

b. z-scores associated with the left-hand tail: Given the area B, find $z(B)$.

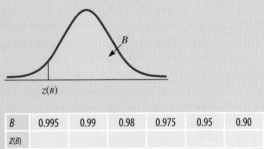

$z(B)$

B	0.995	0.99	0.98	0.975	0.95	0.90
$Z(B)$						

6.35 The z notation, $z(\alpha)$, combines two related concepts, the z-score and the area to the right, into a mathematical symbol. Identify the letter in each of the following as being a z-score or being an area, and then with the aid of a diagram explain what both the given number and the letter represent on the standard curve.

a. $z(A) = 0.10$ b. $z(0.10) = B$

c. $z(C) = -0.05$ d. $-z(0.05) = D$

Objective 6.5

6.36 Find the values np and nq (recall: $q = 1 - p$) for a binomial experiment with $n = 100$ and $p = 0.02$. Does this binomial distribution satisfy the rule for normal approximation? Explain.

6.37 In order to see what happens when the normal approximation is improperly used, consider the binomial distribution with $n = 15$ and $p = 0.05$. Since $np = 0.75$, the rule of thumb ($np > 5$ and $nq > 5$) is not satisfied. Using the binomial tables, find the probability of one or fewer successes and compare this with the normal approximation.

6.38 Find the normal approximation for the binomial probability $P(x = 6)$, where $n = 12$ and $p = 0.6$. Compare this to the value of $P(x = 6)$ obtained from Table 2 in Appendix B.

6.39 Find the normal approximation for the binomial probability $P(x = 4, 5)$, where $n = 14$ and $p = 0.5$. Compare this to the value of $P(x = 4, 5)$ obtained from Table 2 in Appendix B.

6.40 Find the normal approximation for the binomial probability $P(x \leq 8)$, where $n = 14$ and $p = 0.4$. Compare this to the value of $P(x \leq 8)$ obtained from Table 2 in Appendix B.

6.41 Find the normal approximation for the binomial probability $P(x \geq 9)$, where $n = 13$ and $p = 0.7$. Compare this to the value of $P(x \geq 9)$ obtained from Table 2 in Appendix B.

6.42 Melanoma is the most serious form of skin cancer and is increasing at a rate higher than any other cancer in the United States. If it is caught in its early stage, the five-year survival rate for patients on average is 98% in the United States. What is the probability that 235 or more of some group of 250 early-stage patients will survive five years or more after their melanoma diagnosis?
SOURCE: http://www.health.com/

6.43 According to a September 2008 Pew Internet & American Life Project report, 62% of employed adults use the Internet or e-mail on their jobs. What is the probability that more than 180 out of 250 employed adults use the Internet or e-mail on their jobs?
SOURCE: http://www.pewinternet.org/

6.44 A poll conducted in February 2007 by Gallup and reported by the Pew Research Center found that 88% of voters would vote for a female for president if she was qualified. Only 53% of voters felt this way in 1969. Assuming 88% is the true current proportion, what is the probability that another poll of 1,125 registered voters conducted randomly will show
SOURCE: http://pewresearch.org/
a. more than eight-ninths would vote for a female for president, if she was qualified?
b. less than 85% would vote for a female for president, if she was qualified?

6.45 Not all NBA coaches who enjoyed lengthy careers were consistently putting together winning seasons with the teams they coached. For example, Bill Fitch, who coached for 25 seasons of professional basketball after starting his coaching career at the University of Minnesota, won 944 games but lost 1,106 while working with the Cavaliers, Celtics, Rockets, Nets, and Clippers. If you were to randomly select 60 box scores from the historical records of games in which Bill Fitch coached one of the teams, what is the probability that less than half of them show his team winning? To obtain your answer, use the normal approximation to the binomial distribution.
SOURCE: http://www.basketball-reference.com

6.46 According to a December 2008 report from the Join Together Web site of the Boston University School of Public Health, approximately half (42%) of U.S. children are exposed to secondhand smoke on a weekly basis, with more than 25% of parents reporting that their child has been exposed to smoke in their homes. This statistic was one of many results from the Social Climate Survey of Tobacco Control. Use the normal approximation to the binomial distribution to find the probability that in a poll of 1,200 randomly selected parents, between 450 and 500 inclusive will report that their child has been exposed to smoke in their home.
SOURCE: http://www.socialclimate.org/

6.47 Technology is the key to our future. Apparently, students believe this also. According to an April 2009 poll of high school students by Ridgid, the top career choice for high school students was information technology. This top choice was selected by 25% of the surveyed students. Suppose you randomly select 200 students from your local high school. Use the normal approximation to the binomial distribution to find the probability that from within your sample:
a. More that 65 of the students pick information technology as their career choice
b. Less than 27 of the students pick information technology as their career choice
c. Between 45 to 60 of the students pick information technology as their career choice.

Sample Variability

7.1 Sampling Distributions

SAMPLES ARE TAKEN EVERY DAY FOR MANY REASONS; INDUSTRIES CONTINUALLY MONITOR THE QUALITY OF THEIR PRODUCTS, AGENCIES MONITOR THE ENVIRONMENT, AND MEDICAL PROFESSIONALS MONITOR OUR HEALTH—THE LIST IS LIMITLESS.

objectives

7.1 Sampling Distributions

7.2 The Sampling Distribution of Sample Means

7.3 Application of the Sampling Distribution of Sample Means

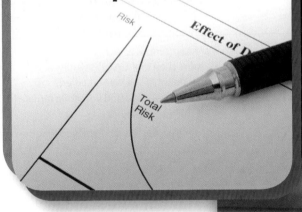

Many of these samples are one-time samples, while others are repeated for ongoing monitoring.

A census, which is a 100% survey or population sampling, is taken in the United States once every 10 years. It is an enormous and overwhelming job, but the information that is obtained is vital to our country's organization and structure. However, from census to census, new issues arise and times change. It is impractical to conduct a census more frequently, but new, up-to-date information is often needed. This is where a representative sample and everyday samples are useful.

In order to make inferences about a population, we need to discuss sample results a little more. A sample mean $\bar{x}$ is obtained from a sample. Do you expect this value, $\bar{x}$, to be exactly equal to the value of the population mean μ? Your answer should be no. We do not expect the means to be identical, but we will be satisfied with our sample results if the sample mean is "close" to the value of the population mean. Let's consider a second question: If a second sample is taken, will the second sample have a mean equal to the population mean? Equal to the first sample mean? Again, no, we do not expect the sample mean to be equal to the population mean, nor do we expect the second sample mean to be a repeat of the first one. We do, however, again expect the values to be "close." (This argument should hold for any other sample statistic and its corresponding population value.)

The next questions should already have come to mind: What is "close"? How do we determine (and measure) this closeness? Just how will **repeated sample statistics** be distributed? To answer these questions we must look at a *sampling distribution*. The **sampling distribution of a sample statistic** is the distribution of values for a sample statistic obtained from repeated samples, all of the same size and all drawn from the same population.

Let's start by investigating two different small theoretical sampling distributions. In the first, we'll introduce a basic sampling distribution of means; in the second, we'll examine sampling distribution of means in greater detail.

> **Sampling distribution of a sample statistic** The distribution of values for a sample statistic obtained from repeated samples, all of the same size and all drawn from the same population.

Forming a Sampling Distribution of Means

Let's consider a very small, finite population to illustrate the concept of a sampling distribution: the set of single-digit even integers {0, 2, 4, 6, 8} and all possible samples of size 2. We will look at a sampling distribution that might be formed: the sampling distribution of sample means.

First we need to list all possible samples of size 2; there are 25 possible samples:

{0, 0}	{2, 0}	{4, 0}	{6, 0}	{8, 0}
{0, 2}	{2, 2}	{4, 2}	{6, 2}	{8, 2}
{0, 4}	{2, 4}	{4, 4}	{6, 4}	{8, 4}
{0, 6}	{2, 6}	{4, 6}	{6, 6}	{8, 6}
{0, 8}	{2, 8}	{4, 8}	{6, 8}	{8, 8}

Each of these samples has a mean $\bar{x}$. These means are, respectively:

0	1	2	3	4
1	2	3	4	5
2	3	4	5	6
3	4	5	6	7
4	5	6	7	8

Each of these samples is equally likely, and thus each of the 25 sample means can be assigned a probability of $\frac{1}{25} = 0.04$. The **sampling distribution of sample means** is shown in Table 7.1 as a **probability distribution** and in Figure 7.1 as a histogram.

The example above is theoretical in nature and therefore expressed in probabilities. Since this population is small, it is easy to list all 25 possible samples of size 2 (a sample space) and assign probabilities. It is not always possible to do this.

Now, let's empirically (that is, by experimentation) investigate another sampling distribution.

Table 7.1 Probability Distribution: Sampling Distribution of Sample Means

$\bar{x}$	$P(\bar{x})$
0	0.04
1	0.08
2	0.12
3	0.16
4	0.20
5	0.16
6	0.12
7	0.08
8	0.04

Figure 7.1 **Histogram: Sampling Distribution of Sample Means**

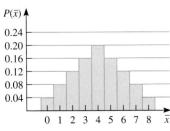

Creating a Sampling Distribution of Sample Means

Let's consider a population that consists of five equally likely integers: 1, 2, 3, 4, and 5. We can observe a portion of the sampling distribution of sample means when 30 samples of size 5 are randomly selected. Figure 7.2 shows a histogram representation of the population.

Table 7.2 shows 30 samples and their means. The resulting sampling distribution, a **frequency distribution**, of sample means is shown in Figure 7.3. Notice

Figure 7.2 **The Population: Theoretical Probability Distribution**

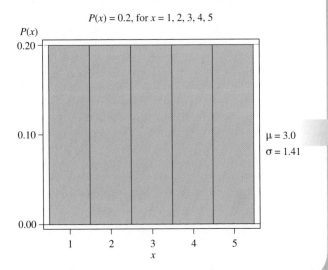

draw samples

Average Age of Urban Transit Rail Vehicles

There are many reasons for collecting data repeatedly. Not all repeated data collections are performed to form a sampling distribution. Consider the "Average Age of Urban Transit Rail Vehicles (Years)" statistics from the U.S. Department of Transportation. The table shows the average age for four different classifications of transit rail vehicles tracked over several years. By studying the pattern of change in the average age for each class of vehicle, a person can draw conclusions about what has been happening to the fleet over several years. Chances are the people involved in maintaining each fleet can also detect when a change in policies regarding replacement of older vehicles is needed. However useful this information is, there is no sampling distribution involved here.

SOURCE: U.S. Department of Transportation, Federal Transit Administration

Average Age of Urban Transit Rail Vehicles (Years)

Transit Rail	1985	1990	1995	2000	2003	2007
Commuter rail locomotives [a]	16.3	15.7	15.9	13.4	16.6	18.4
Commuter rail passenger coaches	19.1	17.6	21.4	16.9	20.5	18.9
Heavy-rail passenger cars	17.1	16.2	19.3	22.9	19.0	21.6
Light-rail vehicles (streetcars)	20.6	15.2	16.8	16.1	15.6	16.1

[a] Locomotives used in Amtrak intercity passenger services are not included.

that this distribution of sample means does not look like the population. Rather, it seems to display the characteristics of a normal distribution: it is mounded and nearly symmetric about its mean (approximately 3.0).

NOTE: The variable for the sampling distribution is $\bar{x}$; therefore, the mean of the $\bar{x}$'s is $\bar{\bar{x}}$ and the standard deviation of $\bar{x}$ is $s_{\bar{x}}$.

The theory involved with sampling distributions that will be described in the remainder of this chapter requires *random sampling*. Recall from Chapter 1 that a *random sample* is obtained in such a way that each possible sample of fixed size n has an equal probability of being selected (see page 14).

Figure 7.4 on the next page shows how the sampling distribution of sample means is formed.

*Table 7.2 **30 Samples of Size 5**

No.	Sample	$\bar{x}$	No.	Sample	$\bar{x}$
1	4,5,1,4,5	3.8	16	4,5,5,3,5	4.4
2	1,1,3,5,1	2.2	17	3,3,1,2,1	2.0
3	2,5,1,5,1	2.8	18	2,1,3,2,2	2.0
4	4,3,3,1,1	2.4	19	4,3,4,2,1	2.8
5	1,2,5,2,4	2.8	20	5,3,1,4,2	3.0
6	4,2,2,5,4	3.4	21	4,4,2,2,5	3.4
7	1,4,5,5,2	3.4	22	3,3,5,3,5	3.8
8	4,5,3,1,2	3.0	23	3,4,4,2,2	3.0
9	5,3,3,3,5	3.8	24	3,3,4,5,3	3.6
10	5,2,1,1,2	2.2	25	5,1,5,2,3	3.2
11	2,1,4,1,3	2.2	26	3,3,3,5,2	3.2
12	5,4,3,1,1	2.8	27	3,4,4,4,4	3.8
13	1,3,1,5,5	3.0	28	2,3,2,4,1	2.4
14	3,4,5,1,1	2.8	29	2,1,1,2,4	2.0
15	3,1,5,3,1	2.6	30	5,3,3,2,5	3.6

using the 30 means

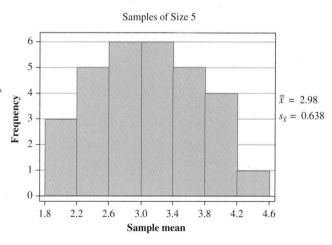

Figure 7.3 **Empirical Distribution of Sample Means**

Samples of Size 5

$\bar{\bar{x}} = 2.98$
$s_{\bar{x}} = 0.638$

SCALE
DRAWN BY
TRACED BY
CHECKED BY
SKETCH SHEET
DATE

figure 7.4 the sampling distribution of sample means

Statistical population being studied

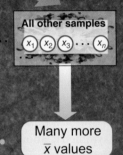

Statistical Population

Parameter of interest, μ

Repeated sampling is needed to form the sampling distribution

All possible samples of size n

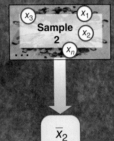

x_1 x_3
Sample 1 x_2
x_n

x_3 x_1
Sample 2 x_2
x_n

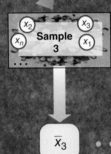

x_2 x_3
x_n **Sample 3** x_1

All other samples
x_1 x_2 x_3 $\cdots$ x_n

One value of the sample statistic ($\bar{x}$ in this case) corresponding to the parameter of interest (μ in this case) is obtained from each sample

$\bar{x}_1$

$\bar{x}_2$

$\bar{x}_3$ $\cdots$

Many more $\bar{x}$ values

Then all of these values of the sample statistic $\bar{x}$ are used to form the sampling distribution.

The Sampling Distribution of Sample Means

The elements of the sampling distribution:

$$\{\bar{x}_1, \bar{x}_2, \bar{x}_3, \ldots\}$$

Graphic description of sampling distribution:

Sampling Distribution of Sample Means

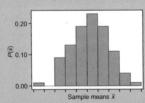

0.20

$P(\bar{x})$ 0.10

0.00

Sample means $\bar{x}$

Numerical description of sampling distribution:

$$\mu_{\bar{x}} = \mu \text{ and } \sigma_{\bar{x}} = \frac{\sigma}{\sqrt{n}}$$

7.2 The Sampling Distribution of Sample Means

ON THE PRECEDING PAGES WE DISCUSSED THE SAMPLING DISTRIBUTIONS OF SAMPLE MEANS, AND MANY OTHER SAMPLING DISTRIBUTIONS COULD BE DISCUSSED.

The only one of concern to us at this time, however, is the sampling distribution of sample means (SDSM):

> **I**f all possible random samples, each of size n, are taken from any population with mean μ and standard deviation σ, then the sampling distribution of sample means will have the following:
>
> 1. A mean $\mu_{\bar{x}}$ equal to μ
> 2. A standard deviation $\sigma_{\bar{x}}$ equal to $\dfrac{\sigma}{\sqrt{n}}$
>
> Furthermore, if the sampled population has a normal distribution, then the sampling distribution of $\bar{x}$ will also be normal for samples of all sizes.

The two-part statement in the box above is very interesting. The first part tells us about the relationship between the population mean and standard deviation, and the sampling distribution mean and standard deviation for all sampling distributions of sample means. The standard deviation of the sampling distribution is denoted by $\sigma_{\bar{x}}$ and given a specific name to avoid confusion with the population standard deviation, σ. We call $\sigma_{\bar{x}}$ the **standard error of the mean**.

> **Standard error of the mean ($\sigma_{\bar{x}}$)** The standard deviation of the sampling distribution of sample means.
>
> **Central limit theorem (CLT)** The sampling distribution of sample means will more closely resemble the normal distribution as the sample size increases.

The second part indicates that this information is not always useful. Stated differently, it says that the mean value of only a few observations will be normally distributed when samples are drawn from a normally distributed population but will not be normally distributed when the sampled population is uniform, skewed, or otherwise not normal. However, the central limit theorem gives us some additional and very important information about the sampling distribution of sample means. According to the **central limit theorem (CLT)**, the sampling distribution of sample means will more closely resemble the normal distribution as the sample size increases.

If the sampled distribution is normal, then the sampling distribution of sample means (SDSM) is normal, as stated above, and the central limit theorem (CLT) does not apply. But if the sampled population is not normal, the sampling distribution will still be

CENTRAL LIMIT THEOREM
and Games of Chance

Abraham De Moivre was a pioneer in the theory of probability and published *The Doctrine of Chance* in Latin in 1711 and then in expanded editions later in the century. The 1756 edition contained his most important contribution—the approximation of the binomial distributions for a large number of trials using the normal distribution. The definition of statistical independence also made its debut along with many dice games and other games of chance. De Moivre proved that the central limit theorem holds for numbers resulting from games of chance. With the use of mathematics, he also successfully predicted the date of his own death.

approximately normally distributed under the right conditions. If the sampled distribution is nearly normal, the $\bar{x}$ distribution is approximately normal for fairly small n (possibly as small as 15). When the sampled distribution lacks symmetry, n may have to be quite large (maybe 50 or more) before the normal distribution provides a satisfactory approximation.

By combining the preceding information, we can describe the sampling distribution of $\bar{x}$ completely: (1) the location of the center (mean), (2) a measure of spread indicating how widely the distribution is dispersed (standard deviation), and (3) an indication of how it is distributed.

1. $\mu_{\bar{x}} = \mu$; the mean of the sampling distribution ($\mu_{\bar{x}}$) is equal to the mean of the population (μ).

2. $\sigma_{\bar{x}} = \frac{\sigma}{\sqrt{n}}$; the standard error of the mean ($\sigma_{\bar{x}}$) is equal to the standard deviation of the population (σ) divided by the square root of the sample size, n.

3. The distribution of sample means is normal when the parent population is normally distributed, and the CLT tells us that the distribution of sample means becomes approximately normal (regardless of the shape of the parent population) when the sample size is large enough.

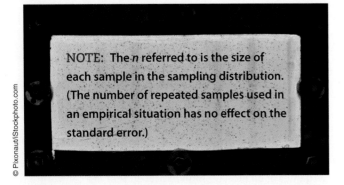

NOTE: The n referred to is the size of each sample in the sampling distribution. (The number of repeated samples used in an empirical situation has no effect on the standard error.)

We do not show the proof for the preceding three facts in this text; however, their validity will be demonstrated by examining two illustrations. For the first illustration, let's consider a population for which we can construct the theoretical sampling distribution of all possible samples.

Constructing a Sampling Distribution of Sample Means

Let's consider all possible samples of size 2 that could be drawn from a population that contains the three numbers 2, 4, and 6. First let's look at the population itself: Construct a histogram to picture its

distribution (Figure 7.5); calculate the mean μ and the standard deviation σ (Table 7.3). (Remember: We must use the techniques from Chapter 5 for discrete probability distributions.)

Figure 7.5 **Population**

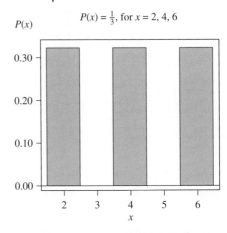

Table 7.3 **Extensions Table for x**

x	$P(x)$	$xP(x)$	$x^2P(x)$
2	$\frac{1}{3}$	$\frac{2}{3}$	$\frac{4}{3}$
4	$\frac{1}{3}$	$\frac{4}{3}$	$\frac{16}{3}$
6	$\frac{1}{3}$	$\frac{6}{3}$	$\frac{36}{3}$
Σ	$\frac{3}{3}$	$\frac{12}{3}$	$\frac{56}{3}$
	1.0 (ck)	4.0	18.66

$\mu = \mathbf{4.0}$

$\sigma = \sqrt{18.6\overline{6} - (4.0)^2} = \sqrt{2.6\overline{6}} = \mathbf{1.63}$

Table 7.4 lists all the possible samples of size 2 that can be drawn from this population. (One number is drawn, observed, and then returned to the population before the second number is drawn.) Table 7.4 also lists the means of these samples. The probability distribution for these means and the extensions are given in Table 7.5, along with the calculation of the mean and the standard error of the mean for the sampling distribution. The histogram for the sampling distribution of sample means is shown in Figure 7.6.

Table 7.4 All Nine Possible Samples of Size 2

Sample	$\bar{x}$	Sample	$\bar{x}$	Sample	$\bar{x}$
2, 2	2	4, 2	3	6, 2	4
2, 4	3	4, 4	4	6, 4	5
2, 6	4	4, 6	5	6, 6	6

Table 7.5 Extensions Table for $\bar{x}$

$\bar{x}$	$P(\bar{x})$	$\bar{x}P(\bar{x})$	$\bar{x}^2P(\bar{x})$
2	$\frac{1}{9}$	$\frac{2}{9}$	$\frac{4}{9}$
3	$\frac{2}{9}$	$\frac{6}{9}$	$\frac{18}{9}$
4	$\frac{3}{9}$	$\frac{12}{9}$	$\frac{48}{9}$
5	$\frac{2}{9}$	$\frac{10}{9}$	$\frac{50}{9}$
6	$\frac{1}{9}$	$\frac{6}{9}$	$\frac{36}{9}$
Σ	$\frac{9}{9}$	$\frac{36}{9}$	$\frac{156}{9}$
	1.0 (ck)	4.0	17.33

$$\mu_{\bar{x}} = 4.0$$
$$\sigma_{\bar{x}} = \sqrt{17.3\overline{3} - (4.0)^2} = \sqrt{1.3\overline{3}} = 1.15$$

Figure 7.6 Sampling Distribution of Sample Means

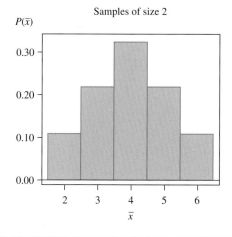

Samples of size 2

Let's now check the truth of the three facts about the sampling distribution of sample means:

3 FACTS

1. The mean $\mu_{\bar{x}}$ of the sampling distribution will equal the mean μ of the population: Both μ and $\mu_{\bar{x}}$ have the value **4.0**.

2. The standard error of the mean $\sigma_{\bar{x}}$ for the sampling distribution will equal the standard deviation σ of the population divided by the square root of the sample size, n: $\sigma_{\bar{x}} = $ **1.15** and $\sigma = 1.63$, $n = 2$, $\frac{\sigma}{\sqrt{n}} = \frac{1.63}{\sqrt{2}} = $ **1.15**; they are equal: $\sigma_{\bar{x}} = \frac{\sigma}{\sqrt{n}}$.

3. The distribution will become approximately normally distributed: The histogram in Figure 7.6 very strongly suggests normality.

Our example uses a theoretical situation and suggests that all three facts appear to hold true. Do these three facts hold when actual data are collected? Let's look back at the example using five equally likely integers (1, 2, 3, 4, 5) from page 138 and see if all three facts are supported by the empirical sampling distribution there.

First, let's look at the population, the theoretical probability distribution from which the samples were taken. Figure 7.2 is a histogram showing the probability distribution for randomly selected data from the population of equally likely integers 1, 2, 3, 4, 5. The population mean μ equals 3.0. The population standard deviation σ is $\sqrt{2}$, or 1.41. The population has a uniform distribution.

Now let's look at the empirical distribution of the 30 sample means found in our earlier example. From the 30 values of $\bar{x}$ in Table 7.2, the observed mean of the $\bar{x}$'s, $\bar{\bar{x}}$, is 2.98 and the observed standard error of the mean, $s_{\bar{x}}$, is 0.638. The histogram of the sampling distribution in Figure 7.3 appears to be mounded, approximately symmetrical, and centered near the value 3.0.

Now let's check the truth of the three specific properties:

1. $\mu_{\bar{x}}$ and μ will be equal: The mean of the population μ is 3.0, and the observed sampling distribution mean $\bar{\bar{x}}$ is 2.98; they are very close in value.

2. $\sigma_{\bar{x}}$ will equal $\frac{\sigma}{\sqrt{n}}$. $\sigma = 1.41$ and $n = 5$; therefore, $\frac{\sigma}{\sqrt{n}} = \frac{1.41}{\sqrt{5}} = 0.632$, and $s_{\bar{x}} = 0.638$; they are very close in value. (Remember that we have taken only 30 samples, not all possible samples, of size 5.)

3. The **sampling distribution** of $\bar{x}$ will be approximately normally distributed. Even though the population has a rectangular distribution, the histogram in Figure 7.3 suggests that the $\bar{x}$ distribution has some of the properties of normality (mounded, symmetric).

Although our examples do not constitute a proof, the evidence seems to strongly suggest that both statements, the sampling distribution of sample means and the central limit theorem, are true.

Having taken a look at these two specific illustrations, let's now look at a graphic illustration that presents the sampling distribution information and the CLT in a slightly different form. In Table 7.6, each row compares four distributions: The first row shows the distribution of the parent population, the distribution of the individual x values. Each of the other three rows shows a sampling distribution of sample means, $\bar{x}$'s, using three different sample sizes.

In the first column we have a uniform distribution, much like that in Figure 7.2 for the integer illustration, and the resulting distributions of sample means for samples of sizes 2, 5, and 30. The following two columns show U- and J-shaped distributions.

All three nonnormal distributions seem to verify the CLT; the sampling distributions of sample means appear to be approximately normal for all three when

Table 7.6 **Comparison of Sampling Distributions**

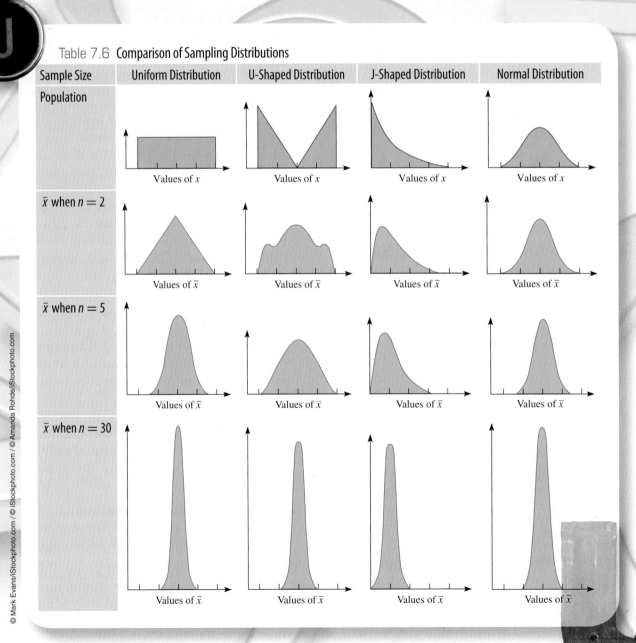

Sample Size	Uniform Distribution	U-Shaped Distribution	J-Shaped Distribution	Normal Distribution
Population	Values of x	Values of x	Values of x	Values of x
$\bar{x}$ when $n = 2$	Values of $\bar{x}$	Values of $\bar{x}$	Values of $\bar{x}$	Values of $\bar{x}$
$\bar{x}$ when $n = 5$	Values of $\bar{x}$	Values of $\bar{x}$	Values of $\bar{x}$	Values of $\bar{x}$
$\bar{x}$ when $n = 30$	Values of $\bar{x}$	Values of $\bar{x}$	Values of $\bar{x}$	Values of $\bar{x}$

samples of size 30 were used. With the normal population (the final column of Table 7.6), the sampling distributions for all sample sizes appear to be normal. Thus, you have seen an amazing phenomenon: No matter what the shape of a population, the sampling distribution of sample means either is normal or becomes approximately normal when n becomes sufficiently large.

You should notice one other point: The sample mean becomes less variable as the sample size increases. Notice that as n increases from 2 to 30, all the distributions become narrower.

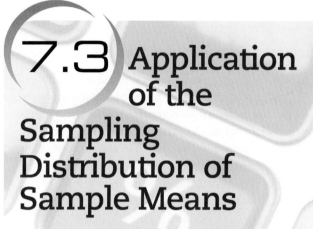

7.3 Application of the Sampling Distribution of Sample Means

WHEN THE SAMPLING DISTRIBUTION OF SAMPLE MEANS IS NORMALLY DISTRIBUTED, OR APPROXIMATELY NORMALLY DISTRIBUTED, WE WILL BE ABLE TO ANSWER PROBABILITY QUESTIONS WITH THE AID OF THE STANDARD NORMAL DISTRIBUTION (TABLE 3 OF APPENDIX B).

Converting $\bar{x}$ Information into z-Scores

➡️ When the population is normally distributed, the sampling distribution of $\bar{x}$'s is normally distributed. To determine probabilities associated with a normal distribution, we will need to format a probability statement involving the z-score in order to use Table 3 in Appendix B, the standard normal distribution table. Consider a normal population with $\mu = 100$ and $\sigma = 20$. If a random sample of size 16 is selected, what is the probability that this sample will have a mean value between 90 and 110? That is, what is $P(90 < \bar{x} < 110)$?

This population is normally distributed, so the sampling distribution of $\bar{x}$'s is normally distributed. We will need to convert the statement $P(90 < \bar{x} < 110)$ to a probability statement involving the z-score. The sampling distribution is shown in the figure, where the shaded area represents $P(90 < \bar{x} < 110)$.

The formula for finding the z-score corresponding to a known value of $\bar{x}$ is

$$z = \frac{\bar{x} - \mu_{\bar{x}}}{\sigma_{\bar{x}}} \qquad (7.1)$$

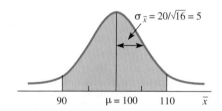

The mean and standard error of the mean are $\mu_{\bar{x}} = \mu$ and $\sigma_{\bar{x}} = \frac{\sigma}{\sqrt{n}}$. Therefore, we will rewrite formula (7.1) in terms of μ, σ, and n:

$$z = \frac{\bar{x} - \mu}{\sigma/\sqrt{n}} \qquad (7.2)$$

Returning to the illustration and applying formula (7.2), we find:

z-score for $\bar{x} = 90$: $z = \frac{\bar{x} - \mu}{\sigma/\sqrt{n}} = \frac{90 - 100}{20/\sqrt{16}} = \frac{-10}{5}$
$= -\mathbf{2.00}$

z-score for $\bar{x} = 110$: $z = \frac{\bar{x} - \mu}{\sigma/\sqrt{n}} = \frac{110 - 100}{20/\sqrt{16}} = \frac{10}{5}$
$= \mathbf{2.00}$

Therefore,

$P(90 < \bar{x} < 110) = P(-2.00 < z < 2.00)$

$= 0.9773 - 0.0228 = \mathbf{0.9545}$

Distribution of $\bar{x}$ and Increasing Individual Sample Size

Before we look at more illustrations, let's consider what is implied by $\sigma_{\bar{x}} = \frac{\sigma}{\sqrt{n}}$. To demonstrate, let's suppose that $\sigma = 20$ and let's use a sampling distribution of samples of size 4. Now $\sigma_{\bar{x}}$ is $20/\sqrt{4}$, or 10, and approximately 95% (0.9545) of all such sample means should be within the interval from 20 below to 20 above the population mean (within two standard deviations of the population mean). However, if the sample size is increased to 16, $\sigma_{\bar{x}}$ becomes $20/\sqrt{16} = 5$ and approximately 95% of the sampling distribution should be within 10 units of the mean, and so on. As the sample size increases, the size of $\sigma_{\bar{x}}$ becomes smaller and the distribution of sample means becomes much narrower. Figure 7.7 illustrates what happens to the distribution of $\bar{x}$'s as the size of the individual samples increases.

Recall that the area (probability) under the normal curve is always exactly one. So as the width of the curve narrows, the height has to increase in order to maintain this area.

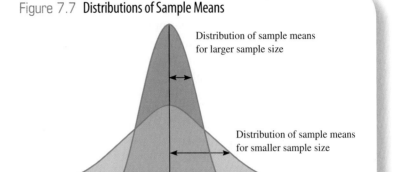

Figure 7.7 **Distributions of Sample Means**

Distribution of sample means for larger sample size

Distribution of sample means for smaller sample size

CALCULATING PROBABILITIES FOR THE MEAN

Calculating probabilities is one way we are able to make predictions about the corresponding population parameter we are looking at. Let's use the example of mean heights of kindergarteners to demonstrate. Kindergarten children have heights that are approximately normally distributed about a mean of 39 inches and a standard deviation of 2 inches. A random sample of size 25 is taken and the mean $\bar{x}$ is calculated. What is the

probability that this mean value will be between 38.5 and 40.0 inches?

To find out, we need to find $P(38.5 < \bar{x} < 40.0)$. The values of $\bar{x}$, 38.5 and 40.0, must be converted to z-scores (necessary for the use of Table 3), using $z = \frac{\bar{x} - \mu}{\sigma/\sqrt{n}}$:

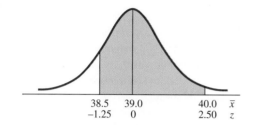

$\bar{x} = 38.5$:
$$z = \frac{\bar{x} - \mu}{\sigma/\sqrt{n}} = \frac{38.5 - 39.0}{2/\sqrt{25}}$$
$$= \frac{-0.5}{0.4} = -1.25$$

$\bar{x} = 40.0$:
$$z = \frac{\bar{x} - \mu}{\sigma/\sqrt{n}} = \frac{40.0 - 39.0}{2/\sqrt{25}}$$
$$= \frac{1.0}{0.4} = 2.50$$

Therefore,

$$P(38.5 < \bar{x} < 40.0) = P(-1.25 < z < 2.50)$$
$$= 0.9938 - 0.1057 = 0.8881$$

In the same vein, we can calculate mean height limits for a certain portion of our population. Using the heights of kindergarten children given in the previous example, we can figure out limits within which the middle 90% of the sampling distribution of sample means for samples of size 100 falls.

The two tools we have to work with are formula (7.2) and Table 3. The formula relates the key values of the population to the key values of the sampling distribution, and Table 3 relates areas to z-scores. First, using Table 3, we find that the middle 0.9000 is bounded by $z = \pm 1.65$.

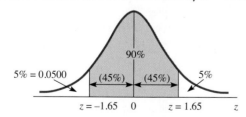

z	...	0.04		0.05	...
:					
−1.6	...	0.0505	*0.0500*	0.0495	...
:					

Second, we use formula (7.2), $z = \frac{\bar{x} - \mu}{\sigma/\sqrt{n}}$:

$z = -1.65$:
$$-1.65 = \frac{\bar{x} - 39.0}{2/\sqrt{100}}$$
$$\bar{x} - 39 = (-1.65)(0.2)$$
$$\bar{x} = 39 - 0.33$$
$$= 38.67$$

$z = 1.65$:
$$1.65 = \frac{\bar{x} - 39.0}{2/\sqrt{100}}$$
$$\bar{x} - 39 = (1.65)(0.2)$$
$$\bar{x} = 39 + 0.33$$
$$= 39.33$$

Thus,

$P(38.67 < \bar{x} < 39.33) = 0.90$

Therefore, 38.67 inches and 39.33 inches are the limits that capture the middle 90% of the sample means.

TO RECAP

The basic purpose for considering what happens when a population is repeatedly sampled, as discussed in this chapter, is to form sampling distributions. The sampling distribution is then used to describe the variability that occurs from one sample to the next. Once this pattern of variability is known and understood for a specific sample statistic, we are able to make predictions about the corresponding population parameter with a measure of how accurate the prediction is. The central limit theorem helps describe the distribution for sample means. We will begin to make inferences about population means and the values of population parameters in Chapter 8.

problems

Objective 7.1

7.1 Manufacturers use random samples to test whether or not their product is meeting specifications. These samples could be people, manufactured parts, or even samples during the manufacturing of potato chips.

    a. Do you think that all random samples taken from the same population will lead to the same result?

    b. What characteristic (or property) of random samples could be observed during the sampling process?

7.2 Consider the set of odd single-digit integers {1, 3, 5, 7, 9}.

    a. Make a list of all samples of size 2 that can be drawn from this set of integers. [Sample with replacement; that is, the first number is drawn, observed, and then replaced (returned to the sample set) before the next drawing.]

    b. Construct the sampling distribution of sample means for samples of size 2 selected from this set.

    c. Construct the sampling distributions of sample ranges for samples of size 2.

7.3 Consider the set of even single-digit integers {0, 2, 4, 6, 8}.

    a. Make a list of all the possible samples of size 3 that can be drawn from this set of integers. [Sample with replacement; that is, the first number is drawn, observed, and then replaced (returned to the sample set) before the next drawing.]

    b. Construct the sampling distribution of the sample means for samples of size 3.

    c. Construct the sampling distribution of the sample medians for samples of size 3.

7.4 Using the phone numbers listed in your directory as your population, obtain randomly 20 samples of size 3. From each phone number identified as a source, take the fourth, fifth, and sixth digits. (For example, for 245-8268, you would take the 8, the 2, and the 6 as your sample of size 3.)

    a. Calculate the mean of the 20 samples.

    b. Draw a histogram showing the 20 sample means. (Use classes −0.5 to 0.5, 0.5 to 1.5, 1.5 to 2.5, and so on.)

    c. Describe the distribution of $\bar{x}$'s that you see in (b) (shape of distribution, center, amount of dispersion).

    d. Draw 20 more samples and add the 20 new $\bar{x}$'s to the histogram in (b). Describe the distribution that seems to be developing.

7.5 From the table of random numbers in Table 1 in Appendix B, construct another table showing 20 sets of 5 randomly selected single-digit integers. Find the mean of each set and then the grand mean. Compare the grand mean with the theoretical population mean $\mu = 4.5$, using the absolute difference and the % error. Show all work.

7.6 a. Using a computer or a random-numbers table, simulate the drawing of 100 samples, each of size 5, from the uniform probability distribution of single-digit integers, 0 to 9.

    b. Find the mean for each sample.

    c. Construct a histogram of the sample means. (Use integer values as class midpoints.)

    d. Describe the sampling distribution shown in the histogram in part (c).

MINITAB

    a. Use the Integer RANDOM DATA commands on your Chapter 2 Tech Card, replacing *generate* with 100, *store in* with C1–C5, *minimum value* with 0, and *maximum value* with 9.

    b. *Choose:*   **Calc > Row Statistics**
         Select: **Mean**
         Enter: *Input variables:*   **C1–C5**
                     Store result in: **C6 > OK**

    c. Use the HISTOGRAM commands on your Chapter 2 Tech Card for the data in C6. To adjust the histogram, select Binning with midpoint and midpoint positions 0:9/1.

EXCEL

    a. Input 0 through 9 into column A and corresponding 0.1's into column B; then continue with:

         Choose: **Data > Data Analysis > Random Number Generation > OK**
         Enter: *Number of Variables:*   **5**
                   Number of Random Numbers: **100**
                   Distribution: **Discrete**
                   Value and Probability Input Range: **(A1:B10 or select cells)**
         Select: **Output Range:**
         Enter: **(C1 or select cell) > OK**

    b. Activate cell H1.

         Choose: **Insert Function, f_x > Statistical > AVERAGE > OK**
         Enter: *Number1:*   **(C1:G1 or select cells) > OK**
         Drag: **Bottom right corner of average value box down to give other averages**

    c. Use the HISTOGRAM commands on your Chapter 2 Tech Card with column H as the input range and column A as the bin range.

TI-83/84 PLUS

    a. Use the Integer RANDOM DATA and STO commands on your Chapter 2 Tech Card, replacing *Enter* with 0, 9, 100). Repeat preceding commands four more times, storing data in L2, L3, L4, and L5, respectively.

    b. *Choose:*   **STAT > EDIT > 1:Edit**
         Highlight: **L6** *(column heading)*
         Enter: **(L1+L2+L3+L4+L5)/5**
    c. *Choose:*   **2nd > STAT PLOT > 1:Plot1**
         Choose: **Window**
         Enter: **0, 9, 1, 0, 30, 5, 1**
         Choose: **Trace > > >**

7.7 a. Using a computer or a random-numbers table, simulate the drawing of 250 samples, each of size 18, from the uniform probability distribution of single-digit integers, 0 to 9.
   b. Find the mean for each sample.
   c. Construct a histogram of the sample means.
   d. Describe the sampling distribution shown in the histogram in part (c).

7.8 a. Use a computer to draw 500 random samples, each of size 20, from the normal probability distribution with mean 80 and standard deviation 15.
   b. Find the mean for each sample.
   c. Construct a frequency histogram of the 500 sample means.
   d. Describe the sampling distribution shown in the histogram in part (c), including the mean and standard deviation.

Objective 7.2

7.9 a. What is the total measure of the area for any probability distribution?
   b. Justify the statement "$\bar{x}$ becomes less variable as n increases."

7.10 If a population has a standard deviation σ of 25 units, what is the standard error of the mean if samples of size 16 are selected? Samples of size 36? Samples of size 100?

7.11 A certain population has a mean of 500 and a standard deviation of 30. Many samples of size 36 are randomly selected and the means calculated.
   a. What value would you expect to find for the mean of all these sample means?
   b. What value would you expect to find for the standard deviation of all these sample means?
   c. What shape would you expect the distribution of all these sample means to have?

7.12 According to Nielsen's Television Audience report, in 2009 the average American home had 2.86 television sets (more than the average number of people per household, at 2.5 people). If the standard deviation for the number of televisions in a U.S. household is 1.2 and a random sample of 80 U.S. households is selected, the mean of this sample belongs to a sampling distribution.
   a. What is the shape of this sampling distribution?
   b. What is the mean of this sampling distribution?
   c. What is the standard deviation of this sampling distribution?

7.13 According to *The World Factbook, 2009*, the total fertility rate (estimated mean number of children born per woman) for Uganda is 6.77. Suppose that the standard deviation of the total fertility rate is 2.6. The mean number of children for a sample of 200 randomly selected women is one value of many that form the sampling distribution of sample means.
   a. What is the mean value for this sampling distribution?
   b. What is the standard deviation of this sampling distribution?
   c. Describe the shape of this sampling distribution.

7.14 The American Meat Institute published the 2007 report U.S. Meat and Poultry Production & Consumption: An Overview. The 2007 fact sheet lists the annual consumption of chicken as 86.5 pounds per person.

   Suppose the standard deviation for the consumption of chicken per person is 29.3 pounds. The mean weight of chicken consumed for a sample of 150 randomly selected people is one value of many that form the sampling distribution of sample means.
   a. What is the mean value for this sampling distribution?
   b. What is the standard deviation of this sampling distribution?
   c. Describe the shape of this sampling distribution.

7.15 A researcher wants to take a simple random sample of about 5% of the student body at each of two schools. The university has approximately 20,000 students, and the college has about 5,000 students. Identify each of the following as true or false and justify your answer.
   a. The sampling variability is the same for both schools.
   b. The sampling variability for the university is higher than that for the college.
   c. The sampling variability for the university is lower than that for the college.
   d. No conclusion about the sampling variability can be stated without knowing the results of the study.

7.16 a. Use a computer to randomly select 100 samples of size 6 from a normal population with mean $\mu = 20$ and standard deviation $\sigma = 4.5$.
   b. Find mean $\bar{x}$ for each of the 100 samples.
   c. Using the 100 sample means, construct a histogram, find mean $\bar{\bar{x}}$, and find the standard deviation $s_{\bar{x}}$.
   d. Compare the results of part (c) with the three statements made in the SDSM.

MINITAB
   a. Use the Normal RANDOM DATA commands on your Chapter 2 Tech Card, replacing *generate* with 100, *store in* with C1–C6, *mean* with 20, and *standard deviation* with 4.5.
   b. Use the ROW STATISTICS commands from problem 7.6, replacing *input variables* with C1–C6 and *store result in* with C7.
   c. Use the HISTOGRAM commands on your Chapter 2 Tech Card for the data in C7. To adjust the histogram, select Binning with midpoint and midpoint positions 12.8:27.2/1.8. Use the MEAN and STANDARD DEVIATION commands on your Chapter 2 Tech Card for the data in C7.

EXCEL

a. Use the Normal RANDOM NUMBER GENERATION commands on your Chapter 2 Tech Card, replacing *number of variables* with 6, *number of random numbers* with 100, *mean* with 20, and *standard deviation* with 4.5.

b. Activate cell G1.

Choose: Insert Function, f_x > Statistical > AVERAGE > OK

Enter: Number1:   (A1:F1 or select cells) > OK

Drag: Bottom right corner of average value box down to give other averages

c. Use the following RANDOM NUMBER GENERATION Patterned Distribution commands:

Choose: Data > Data Analysis > Random Number Generation > OK

Enter: Number of Variables: 1

Distribution: Patterned

From: 12.8 to 27.2 in steps of 1.8

repeating each number: 1 times

repeating the sequence 1 times

Select: Output Range

Enter: (H1 or select cell) > OK

Use the HISTOGRAM commands on your Chapter 2 Tech Card with column G as the input range and column H as the bin range. Use the MEAN and STANDARD DEVIATION commands on your Chapter 2 Tech Card for the data in column G.

TI-83/84 PLUS

a. Use the Integer RANDOM DATA and STO commands on your Chapter 2 Tech Card, replacing *Enter* with 20,4.5,100). Repeat the preceding commands five more times, storing data in L2, L3, L4, L5, and L6, respectively.

b. Enter: (L1 + L2 + L3 + L4 + L5 + L6)/6

Choose: STO→ L7 (use ALPHA key for the 'L' or use 'MEAN')

c. Choose: 2nd > STAT

PLOT > 1:Plot1

Choose: Window

Enter: 12.8, 27.2, 1.8, 0, 40, 5, 1

Choose: Trace > > >

Choose: STAT > CALC > 1:1-VAR STATS > 2nd > LIST

Select: L7

7.17 a. Use a computer to randomly select 200 samples of size 24 from a normal population with mean $\mu = 20$ and standard deviation $\sigma = 4.5$.

b. Find mean $\bar{x}$ for each of the 200 samples.

c. Using the 200 sample means, construct a histogram, find mean $\bar{\bar{x}}$, and find the standard deviation $s_{\bar{x}}$.

d. Compare the results of part (c) with the three statements made for the SDSM and CLT on page 141.

e. Compare these results with the results obtained in problem 7.16. Specifically, what effect did the increase in sample size from 6 to 24 have? What effect did the increase from 100 samples to 200 samples have?

FYI If you use a computer, see problem 7.16.

Objective 7.3

7.18 Consider a normal population with $\mu = 43$ and $\sigma = 5.2$. Calculate the z-score for an $\bar{x}$ of 46.5 from a sample of size 16.

7.19 Consider a population with $\mu = 43$ and $\sigma = 5.2$.

a. Calculate the z-score for an $\bar{x}$ of 46.5 from a sample of size 35.

b. Could this z-score be used in calculating probabilities using Table 3 in Appendix B? Why or why not?

7.20 What is the probability that the sample of kindergarten children (pages 146–147) has a mean height of less than 39.75 inches?

7.21 A random sample of size 36 is to be selected from a population that has a mean $\mu = 50$ and a standard deviation σ of 10.

a. This sample of 36 has a mean value of $\bar{x}$, which belongs to a sampling distribution. Find the shape of this sampling distribution.

b. Find the mean of this sampling distribution.

c. Find the standard error of this sampling distribution.

d. What is the probability that this sample mean will be between 45 and 55?

e. What is the probability that the sample mean will have a value greater than 48?

f. What is the probability that the sample mean will be within 3 units of the mean?

7.22 The local bakery bakes more than a thousand 1-pound loaves of bread daily, and the weights of these loaves varies. The mean weight is 1 lb. and 1 oz., or 482 grams. Assume the standard deviation of the weights is 18 grams and a sample of 40 loaves is to be randomly selected.

a. This sample of 40 has a mean value of $\bar{x}$, which belongs to a sampling distribution. Find the shape of this sampling distribution.

b. Find the mean of this sampling distribution.

c. Find the standard error of this sampling distribution.

d. What is the probability that this sample mean will be between 475 and 495 grams?

e. What is the probability that the sample mean will have a value less than 478 grams?

f. What is the probability that the sample mean will be within 5 grams of the mean?

7.23 Consider the approximately normal population of heights of male college students with mean $\mu = 69$ inches and standard deviation $\sigma = 4$ inches. A random sample of 16 heights is obtained.

a. Describe the distribution of x, height of male college students.

b. Find the proportion of male college students whose height is greater than 70 inches.

c. Describe the distribution of $\bar{x}$, the mean of samples of size 16.

d. Find the mean and standard error of the $\bar{x}$ distribution.

e. Find $P(\bar{x} > 70)$.

f. Find $P(\bar{x} < 67)$.

7.24 The amount of fill (weight of contents) put into a glass jar of spaghetti sauce is normally distributed with mean $\mu = 850$ grams and standard deviation $\sigma = 8$ grams.

a. Describe the distribution of x, the amount of fill per jar.

b. Find the probability that one jar selected at random contains between 848 and 855 grams.

c. Describe the distribution of $\bar{x}$, the mean weight for a sample of 24 such jars of sauce.

d. Find the probability that a random sample of 24 jars has a mean weight between 848 and 855 grams.

7.25 The heights of the kindergarten children (pages 146–147) are approximately normally distributed with $\mu = 39$ and $\sigma = 2$.

a. If an individual kindergarten child is selected at random, what is the probability that he or she has a height between 38 and 40 inches?

b. A classroom of 30 of these children is used as a sample. What is the probability that the class mean $\bar{x}$ is between 38 and 40 inches?

c. If an individual kindergarten child is selected at random, what is the probability that he or she is taller than 40 inches?

d. A classroom of 30 of these kindergarten children is used as a sample. What is the probability that the class mean $\bar{x}$ is greater than 40 inches?

7.26 Salaries for various positions can vary significantly depending on whether the company is in the public or private sector. The U.S. Department of Labor posted the 2007 average salary for human resource managers employed by the federal government at $76,503. Assume that annual salaries for this job type are normally distributed and have a standard deviation of $8,850.

a. What is the probability that a randomly selected human resource manager received over $100,000 in 2007?

b. A sample of 20 labor relations managers is taken and annual salaries are reported. What is the probability that the sample mean annual salary falls between $70,000 and $80,000?

7.27 Based on data from 1996 through 2006 from the Western Regional Climate Center, the average speed of winds in Honolulu, Hawaii, equals 10.6 miles per hour.

Assume that wind speeds are approximately normally distributed with a standard deviation of 3.5 miles per hour.

a. Find the probability that the wind speed on any one reading will exceed 13.5 miles per hour.

b. Find the probability that the mean of a random sample of 9 readings exceeds 13.5 miles per hour.

c. Do you think the assumption of normality is reasonable? Explain.

d. What effect do you think the assumption of normality had on the answers to (a) and (b)? Explain.

7.28 The 2007 Trends in International Mathematics and Science Study (TIMSS), focused on the mathematics and science achievement of eighth-grade students throughout the world. A total of eight countries (including the United States) participated in the study. The mean math exam score for U.S. students was 509 with a standard deviation of 88.

SOURCE: http://nces.ed.gov/

Assuming the scores are normally distributed, find the following probabilities for a sample of 150 students.

a. Find the probability that the mean TIMSS score for a randomly selected group of eighth-grade students would be between 495 and 515.

b. Find the probability that the mean TIMSS score for a randomly selected group of eighth-grade students would be less than 520.

c. Do you think the assumption of normality is reasonable? Explain.

7.29 A popular flashlight that uses two D-size batteries was selected, and several of the same models were purchased to test the "continuous-use life" of D batteries. As fresh batteries were installed, each flashlight was turned on and the time noted. When the flashlight no longer produced light, the time was again noted. The resulting "life" data from Rayovac batteries had a mean of 21.0 hours

SOURCE: http://www.rayovac.com

Assume these values have a normal distribution with a standard deviation of 1.38 hours.

a. What is the probability that one randomly selected Rayovac battery will have a test life of between 20.5 and 21.5 hours?

b. What is the probability that a randomly selected sample of 4 Rayovac batteries will have a mean test life of between 20.5 and 21.5 hours?

c. What is the probability that a randomly selected sample of 16 Rayovac batteries will have a mean test life of between 20.5 and 21.5 hours?

d. What is the probability that a randomly selected sample of 64 Rayovac batteries will have a mean test life of between 20.5 and 21.5 hours?

e. Describe the effect that the increase in sample size had on the answers for parts (b)–(d).

Chapter 8

Introduction to
Statistical Inferences

8.1 The Nature of Estimation

RECENT DATA FROM THE NATIONAL CENTER FOR HEALTH STATISTICS (NCHS) GIVES THE AVERAGE HEIGHT OF FEMALES IN THE UNITED STATES TO BE 63.7 INCHES WITH A STANDARD DEVIATION OF 2.75 INCHES.

Now, suppose a sample of heights was gathered from 50 randomly selected American female health professionals. Do you expect the mean of this random sample of 50 females to be exactly equal to the population mean of 63.7 inches (an **estimation question**)? If the sample mean is greater than 63.7 inches, does it mean that female health professionals are taller than American females (a **hypothesis testing question**)?

As you recall, the central limit theorem gave us some very important information about the sampling distribution of sample means (SDSM). Specifically, it stated that in many realistic cases (when the random sample is large enough) a distribution of sample means is normally or approximately normally distributed about the mean of the population. With this information we were able to make probability statements about the likelihood of certain sample mean values occurring when samples are drawn from a population with a known mean and a known standard deviation. We are now ready to turn this situation around to the case in which the population mean is not known. We will draw one sample, calculate its mean value, and then make an inference about the value of the population mean based on the sample's mean value.

The objective of inferential statistics is to use the information contained in the sample data to increase our knowledge of the sampled population. We will learn about making two types of inferences: (1) estimating the value of a population parameter and (2) testing a hypothesis. The sampling distribution of sample means (SDSM) is the key to making these inferences.

In this chapter we deal with the questions about the population mean using two methods that assume the value of the population standard deviation is a known quantity. This assumption is

objectives

8.1 **The Nature of Estimation**

8.2 **Estimation of Mean μ (σ Known)**

8.3 **The Nature of Hypothesis Testing**

8.4 **Hypothesis Test of Mean μ (σ Known): A Probability-Value Approach**

8.5 **Hypothesis Test of Mean μ (σ Known): A Classical Approach**

seldom realized in real-life problems, but it will make our first look at the techniques of inference much simpler.

Let's start by looking at the concept of **estimation.** Estimations occur in two forms: a point estimate and an interval estimate. A **point estimate for a parameter** is a single number designed to estimate a quantitative parameter of a population, usually the value of the corresponding sample statistic.

Point estimate for a parameter A single number designed to estimate a quantitative parameter of a population, usually the value of the corresponding sample statistic.

To illustrate this, let's look at a company that manufactures rivets for use in building aircraft. One characteristic of extreme importance is the "shearing strength" of each rivet. The company's engineers must monitor production to be certain that the shearing strength of the rivets meets the required specs. To accomplish this, they take a sample and determine the mean shearing strength of the sample. Based on this sample information, the company can estimate the mean shearing strength for all the rivets it is manufacturing.

Shearing strength is the force required to break a material in a "cutting" action.

stat

A random sample of 36 rivets is selected, and each rivet is tested for shearing strength. The resulting sample mean is $\bar{x} = 924.23$ pounds. Based on this sample, we say, "We believe the mean shearing strength of all such rivets is 924.23 pounds." That is, the sample mean, $\bar{x}$, is the point estimate (single-number value) for the mean μ of the sampled population. For our rivet example, 924.23 is the point estimate for μ, the mean shearing strength of all rivets.

NOTE: Throughout Chapter 8 we will treat the standard deviation, σ, as a known, or given, quantity and concentrate on learning the procedures for making statistical inferences about the population mean μ. Therefore, to continue the explanation of statistical inferences, we will assume $\sigma = 18$ for the specific rivets described in our example.

The quality of this point estimate should be questioned. Is the estimate exact? Is the estimate likely to be high or low? Would another sample yield the same result? Would another sample yield an estimate of nearly the same value or a value that is very different? How is "nearly the same" or "very different" measured? The quality of an estimation procedure (or method) is greatly enhanced if the sample statistic is both *less variable* and *unbiased*. The variability of a statistic is measured by the standard error of its sampling distribution. The sample mean can be made less variable by reducing its standard error, $\sigma/\sqrt{n}$. That requires using a larger sample because as n increases, the standard error decreases.

Figure 8.1 illustrates the concept of being unbiased and the effect of variability on the point estimate. The value A is the parameter being estimated, and the dots represent possible sample statistic values from the sampling distribution of the statistic. If A represents the true

> **Unbiased statistic** A sample statistic whose sampling distribution has a mean value equal to the value of the population parameter being estimated. A statistic that is not unbiased is a biased statistic.

population mean, μ, then the dots represent possible sample means from the $\bar{x}$ sampling distribution.

Figures 8.1(a), (c), (d), and (f) show biased statistics. (a) and (d) show sampling distributions whose mean values are less than the value of the parameter, whereas (c) and (f) show sampling distributions whose mean values are greater than the parameter. Figures 8.1(b) and (e) show sampling distributions that appear to have a mean value equal to the value of the parameter; therefore, they are unbiased. Figures 8.1(a), (b), and (c) show more variability, whereas (d), (e), and (f) show less variability in the sampling distributions. Figure 8.1(e) represents the best situation; an estimator that is unbiased (on-target) and has low variability (all values close to the target).

The sample mean, $\bar{x}$, is an **unbiased statistic** because the mean value of the sampling distribution of sample means, $\mu_{\bar{x}}$, is equal to the population mean, μ. Therefore, the sample statistic $\bar{x} = 924.23$ is an unbiased point estimate for the mean strength of all rivets being manufactured in our example.

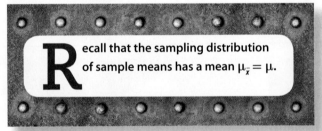

Recall that the sampling distribution of sample means has a mean $\mu_{\bar{x}} = \mu$.

Figure 8.1 **Effects of Variability and Bias**

	Negative Bias (under estimate)	Unbiased (on target estimate)	Positive Bias (over estimate)
High Variation	(a)	(b)	(c)
Low Variation	(d)	(e)	(f)

Sample means vary in value and form a sampling distribution in which not all samples result in $\bar{x}$ values equal to the population mean. Therefore, we should not expect this sample of 36 rivets to produce a point estimate (sample mean) that is exactly equal to the mean μ of the sampled population. We should, however, expect the point estimate to be fairly close in value to the population mean. The sampling distribution of sample means (SDSM) and the central limit theorem (CLT) provide the information needed to describe how close the point estimate, $\bar{x}$, is expected to be to the **population mean, μ**.

Recall that approximately 95% of a normal distribution is within two standard deviations of the mean and that the CLT describes the SDSM as being nearly normal when samples are large enough. Samples of size 36 from populations of variables such as rivet strengths are generally considered large enough. Therefore, we should anticipate that approximately 95% of all random samples selected from a population with unknown mean μ and standard deviation $\sigma = 18$ will have means $\bar{x}$ between

$$\mu - 2(\sigma_{\bar{x}}) \quad \text{and} \quad \mu + 2(\sigma_{\bar{x}})$$
$$\mu - 2\left(\frac{\sigma}{\sqrt{n}}\right) \quad \text{and} \quad \mu + 2\left(\frac{\sigma}{\sqrt{n}}\right)$$
$$\mu - 2\left(\frac{18}{\sqrt{36}}\right) \quad \text{and} \quad \mu + 2\left(\frac{18}{\sqrt{36}}\right)$$
$$\mu - 6 \quad \text{and} \quad \mu + 6$$

This suggests that 95% of all random samples of size 36 selected from the population of rivets should have a mean $\bar{x}$ between $\mu - 6$ and $\mu + 6$. Figure 8.2 shows the middle 95% of the distribution, the bounds of the interval covering the 95%, and the mean μ.

Figure 8.2 **Sampling Distribution of $\bar{x}$'s, Unknown μ**

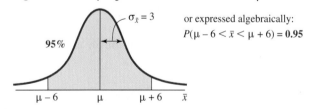

$\sigma_{\bar{x}} = 3$

95%

or expressed algebraically:
$P(\mu - 6 < \bar{x} < \mu + 6) = 0.95$

$\mu - 6$ μ $\mu + 6$ $\bar{x}$

The second form of estimate is an **interval estimate**, which is an interval bounded by two values and used to estimate the value of a population parameter. The values that bound this interval are statistics calculated from the sample that is being used as the basis for the estimation. Interval estimates involve a certain **level of confidence (1 − α)** which is the proportion of all interval estimates that include the parameter being estimated.

Combining an interval estimate with a specified level of confidence gives us a **confidence interval.** We

can pull all of the information from our rivet example together in the form of a confidence interval. To construct the confidence interval, we will use the point estimate $\bar{x}$ as the central value of an interval in much the same way as we used the mean μ as the central value to find the interval that captures the middle 95% of the $\bar{x}$ distribution in Figure 8.2.

For our rivet example, we can find the bounds to an interval centered at $\bar{x}$:

$$\bar{x} - 2(\sigma_{\bar{x}}) \quad \text{to} \quad \bar{x} + 2(\sigma_{\bar{x}})$$
$$924.23 - 6 \quad \text{to} \quad 924.23 + 6$$

The resulting interval is 918.23 to 930.23.

The level of confidence assigned to this interval is approximately 95%, or 0.95. The bounds of the interval are 2 multiples ($z = 2.0$) of the standard error from the sample mean, and by looking at Table 3 in Appendix B, we can more accurately determine the level of confidence as 0.9544. Putting all of this information together, we express the estimate as a confidence interval: **918.23 to 930.23** *is the 95.44% confidence interval for the mean shear strength of the rivets.* Or in an abbreviated form: **918.23 to 930.23,** *the 95.44% confidence interval for μ.*

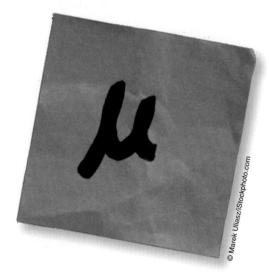

8.2 Estimation of Mean μ (σ Known)

IN OBJECTIVE 8.1 WE SURVEYED THE BASIC IDEAS OF ESTIMATION: POINT ESTIMATE, INTERVAL ESTIMATE, LEVEL OF CONFIDENCE, AND CONFIDENCE INTERVAL.

These basic ideas are interrelated and used throughout statistics when an inference calls for an estimate. In this section we formalize the interval estimation process as it applies to estimating the population mean μ based on a random sample under the restriction that the population standard deviation σ is a known value.

The SDSM and the CLT provide us with the information we need to ensure that the necessary *assumptions* for estimating a population mean are satisfied.

The assumption for estimating mean μ using a known σ:

The sampling distribution of $\bar{x}$ has a normal distribution.

The information needed to ensure that this assumption (or condition) is satisfied is contained in the SDSM and in the CLT. Recall from Chapter 7 that the SDSM $\bar{x}$ is distributed about a mean equal to μ with a standard error equal to $\sigma/\sqrt{n}$; and (1) if the randomly sampled

population is normally distributed, then $\bar{x}$ is normally distributed for all sample sizes, or (2) if the randomly sampled population is not normally distributed, then $\bar{x}$ is approximately normally distributed for sufficiently large sample sizes.

Therefore, we can satisfy the required assumption by either (1) knowing that the sampled population is normally distributed or (2) using a random sample that contains a sufficiently large number of data. The first possibility is obvious. We either know enough about the population to know that it is normally distributed or we don't. The second way to satisfy the assumption is by applying the CLT. Inspection of various graphic displays of the sample data should yield an indication of the type of distribution the population possesses. The CLT can be applied to smaller samples (say, $n = 15$ or larger) when the data provide a strong indication of a unimodal distribution that is approximately symmetric. If there is evidence of some skewness in the data, then the sample size needs to be much larger (perhaps $n \geq 50$). If the data provide evidence of an extremely skewed or J-shaped distribution, the CLT will still apply if the sample is large enough. In extreme cases, "large enough" may be unrealistically or impracticably large. There is no hard-and-fast rule defining "large enough"; the sample size that is "large enough" varies greatly according to the distribution of the population.

NOTE: The help of a professional statistician should be sought when treating extremely skewed data.

The $1 - \alpha$ confidence interval for the estimation of mean μ is found using the formula

Confidence Interval for Mean

$$\bar{x} - z(\alpha/2)\left(\frac{\sigma}{\sqrt{n}}\right) \quad \text{to} \quad \bar{x} + z(\alpha/2)\left(\frac{\sigma}{\sqrt{n}}\right) \tag{8.1}$$

Here are the parts of the confidence interval formula:

1. $\bar{x}$ is the point estimate and the center point of the confidence interval.

2. $z(\alpha/2)$ is the **confidence coefficient.** It is the number of multiples of the standard error needed to formulate an interval estimate of the correct width to have a level of confidence of $1 - \alpha$. Figure 8.3 shows the relationship among the level of confidence $1 - \alpha$

NOTE

The word *assumptions* is somewhat of a misnomer. It does not mean that we "assume" something to be the situation and continue, but that we must be sure the conditions expressed by the assumptions do exist before we apply a particular statistical method. If the assumptions are not met for estimating the mean μ, using a known σ, the level of confidence is most likely lower than stated.

(the middle portion of the distribution), α/2 (the "area to the right" used with the critical-value notation), and the confidence coefficient $z(\alpha/2)$ (whose value is found using Table 4B of Appendix B).

3. $\frac{\sigma}{\sqrt{n}}$ is the **standard error of the mean,** or the standard deviation of the sampling distribution of sample means.

4. $z(\alpha/2)\left(\frac{\sigma}{\sqrt{n}}\right)$ is one-half the width of the confidence interval (the product of the confidence coefficient and the standard error) and is called the **maximum error of estimate, E.**

5. $\bar{x} - z(\alpha/2)\left(\frac{\sigma}{\sqrt{n}}\right)$ is called the lower confidence limit (LCL), and $\bar{x} + z(\alpha/2)\left(\frac{\sigma}{\sqrt{n}}\right)$ is called the upper confidence limit (UCL) for the confidence interval.

The estimation procedure is organized into a five-step process that will take into account all of the above information and produce both the point estimate and the confidence interval.

Figure 8.3 **Confidence Coefficient $z(\alpha/2)$**

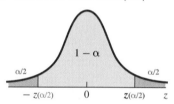

BASICALLY, THE CONFIDENCE INTERVAL IS "POINT ESTIMATE ± MAXIMUM ERROR."

Constructing a Confidence Interval

➡ Let's apply the confidence interval procedure to finding the mean for a one-way commute distance. The student body at many community colleges is considered a "commuter population." The student activities office wishes to obtain an answer to the question: How far (one way) does the average community college student commute to college each day? (Typically the "average student's commute distance" is meant to be the "mean distance" commuted by all students who commute.) A random sample of 100 commuting students was identified, and the one-way distance each commuted was obtained. The resulting sample mean distance was 10.22 miles. To estimate the mean one-way distance commuted by all commuting students, we'll use: (a) a point estimate and (b) a 95% confidence interval. (Use σ = 6 miles.) Our point estimate (a) for the mean one-way distance is **10.22** miles (the sample mean). Next we use the five-step procedure to find the 95% confidence interval (b).

The Confidence Interval: A Five-Step Procedure

STEP 1 **The Set-Up:**
Describe the population parameter of interest.

STEP 2 **The Confidence Interval Criteria:**
a. Check the assumptions.
b. Identify the probability distribution and the formula to be used.
c. State the level of confidence, $1 - \alpha$.

STEP 3 **The Sample Evidence:**
Collect the sample information.

STEP 4 **The Confidence Interval:**
a. Determine the confidence coefficient.
b. Find the maximum error of estimate.
c. Find the lower and upper confidence limits.

STEP 5 **The Results:**
State the confidence interval.

STEP 1 THE SET-UP:

Describe the population parameter of interest.
The mean μ of the one-way distances commuted by all commuting community college students is the parameter of interest.

STEP 2 THE CONFIDENCE INTERVAL CRITERIA:

a. **Check the assumptions.**
σ is known. The variable "distance commuted" most likely has a skewed distribution because the vast majority of the students commute between 0 and 25 miles, with fewer commuting more than 25 miles. A sample size of 100 should be large enough for the CLT to satisfy the assumption; the $\bar{x}$ sampling distribution is approximately normal.

b. **Identify the probability distribution and the formula to be used.**
The standard normal distribution, z, will be used to determine the confidence coefficient, and formula (8.1) will be used with $\sigma = 6$.

c. **State the level of confidence, $1 - \alpha$.**
The question asks for 95% confidence, or $1 - \alpha = 0.95$.

STEP 3 THE SAMPLE EVIDENCE:

Collect the sample information.
The sample information is given in the statement of the problem: $n = 100, \bar{x} = 10.22$.

STEP 4 THE CONFIDENCE INTERVAL:

a. **Determine the confidence coefficient.**
The confidence coefficient is found using Table 4B:

A Portion of Table 4B		
α	$\cdots$	0.05
$z(\alpha/2)$	$\cdots$	1.96
$1 - \alpha$	$\cdots$	0.95

Level of confidence:
$1 - \alpha = 0.95 \rightarrow$

Confidence coefficient:
$z(\alpha/2) = 1.96$

b. **Find the maximum error of estimate.**
Use the maximum error part of formula (8.1):

$$E = z(\alpha/2)\left(\frac{\sigma}{\sqrt{n}}\right) = (1.96)(0.6) = 1.176$$

c. **Find the lower and upper confidence limits.**
Using the point estimate, $\bar{x}$, from Step 3 and the maximum error, E, from Step 4b, we find the confidence interval limits:

$$\bar{x} - z(\alpha/2)\left(\frac{\sigma}{\sqrt{n}}\right) \quad \text{to} \quad \bar{x} + z(\alpha/2)\left(\frac{\sigma}{\sqrt{n}}\right)$$

$10.22 - 1.176$	to	$10.22 + 1.176$
9.044	to	11.396
9.04	to	11.40

STEP 5 THE RESULTS:

State the confidence interval.
9.04 to 11.40 is the 95% confidence interval for μ. That is, with 95% confidence we can say, "The mean one-way distance is between 9.04 and 11.40 miles."

Let's take another look at the concept "level of confidence." It was defined to be the probability that the sample to be selected will produce interval bounds that contain the parameter.

Demonstrating the Meaning of a Confidence Interval

Single-digit random numbers, like the ones in Table 1 in Appendix B, have a mean value $\mu = 4.5$ and a standard deviation $\sigma = 2.87$. Draw a sample of 40 single-digit numbers from Table 1 and construct the 90% confidence interval for the mean. Does the resulting interval contain the expected value of μ, 4.5? If we were to select another sample of 40 single-digit numbers from Table 1, would we get the same result? What might happen if we selected a total of 15 different samples and constructed the 90% confidence interval for each? Would the expected value for μ—namely, 4.5—be contained in all of them? Should we expect all 15 confidence intervals to contain 4.5? Think about the definition of "level of confidence"; it says that, in the long run, 90% of the samples will result in bounds that contain μ. In other words, 10% of the samples will not contain μ. Let's see what happens.

© Antenna/Getty Images / © pixhook/iStockphoto.com

First we need to address the assumptions; if the assumptions are not satisfied, we cannot expect the 90% and the 10% to occur.

We know: (1) the distribution of single-digit random numbers is rectangular (definitely not normal), (2) the distribution of single-digit random numbers is symmetric about their mean, (3) the $\bar{x}$ distribution for very small samples ($n = 5$) in our example from Chapter 7 displayed a distribution that appeared to be approximately normal, and (4) there should be no skewness involved. Therefore, it seems reasonable to assume that $n = 40$ is large enough for the CLT to apply.

The first random sample was drawn from Table 1 in Appendix B:

*Table 8.1 **Random Sample of Single-Digit Numbers**

2	8	2	1	5	5	4	0	9	1
0	4	6	1	5	1	1	3	8	0
3	6	8	4	8	6	8	9	5	0
1	4	1	2	1	7	1	7	9	3

The sample statistics are $n = 40$, $\Sigma x = 159$, and $\bar{x} = 3.975$. Here is the resulting 90% confidence interval:

$$\bar{x} \pm z_{(\alpha/2)}\left(\frac{\sigma}{\sqrt{n}}\right): \quad 3.975 \pm 1.65\left(\frac{2.87}{\sqrt{40}}\right)$$

$$3.975 \pm (1.65)(0.454)$$

$$3.975 \pm 0.749$$

$$3.975 - 0.749 = 3.23 \quad \text{to} \quad 3.975 + 0.749 = 4.72$$

3.23 to 4.72 is the 90% confidence interval for μ

Figure 8.4 shows this interval estimate, its bounds, and the expected mean μ.

Figure 8.4 **The 90% Confidence Interval**

With 90% confidence, we think μ is somewhere within this interval

$$3.23 \qquad \mu = 4.50 \quad 4.72 \quad \bar{x}$$

The expected value for the mean, 4.5, does fall within the bounds of the confidence interval for this sample. Let's now select 14 more random samples from Table 1 in Appendix B, each of size 40.

Table 8.2 on the next page lists the mean from the first sample and the means obtained from the 14 additional random samples of size 40. The 90% confidence intervals for the estimation of μ based on each of the 15 samples are listed in Table 8.2 and shown in Figure 8.5 on the next page.

We see that 86.7% (13 of the 15) of the intervals contain μ and two of the 15 samples (sample 7 and sample 12) do not contain μ. The results here are "typical"; repeated experimentation might result in any number of intervals that contain 4.5. However, in the long run we should expect approximately $1 - \alpha = 0.90$ (or 90%) of the samples to result in bounds that contain 4.5 and approximately 10% in bounds that do not contain 4.5.

Sample Size

The confidence interval has two basic characteristics that determine its quality: its level of confidence and its width. It is preferable for the interval to have a high level of confidence and be precise (narrow) at the same time. The higher the level of confidence, the more likely the interval is to contain the parameter, and the narrower the interval, the more precise the estimation. However, these two properties seem to work against each other, since it would seem that a narrower interval would tend to have a lower probability and a wider interval would be less

precise. The maximum error part of the confidence interval formula specifies the relationship involved.

Maximum Error of Estimate

$$E = z(\alpha/2)\left(\frac{\sigma}{\sqrt{n}}\right) \qquad (8.2)$$

This formula has four components: (1) the maximum error E, half of the width of the confidence interval; (2) the confidence coefficient, $z(\alpha/2)$, which is determined by the level of confidence; (3) the sample size, n; and (4) the standard deviation, σ. The standard deviation σ is not a concern in this discussion because it is a constant (the standard deviation of a population does not change in value). That leaves three factors. Inspection of formula (8.2) indicates the following: Increasing the level of confidence will make the confidence coefficient larger and will thereby require either the maximum error to increase or the sample size to increase; decreasing the maximum error will require the level of confidence to decrease or the sample size to increase; and decreasing the sample size will force the maximum error to increase or the level of confidence to decrease. We have a "three-way tug-of-war," as pictured in Figure 8.6.

An increase or decrease to any one of the three factors has an effect on one or both of the other two factors. The statistician's job is to "balance" the level of confidence, the sample size, and the maximum error so that an acceptable interval results.

Figure 8.5 **Confidence Intervals from Table 8.2**

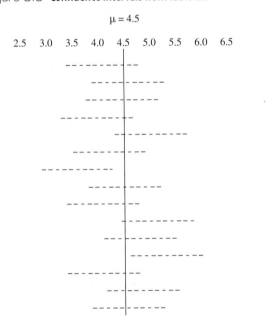

$\mu = 4.5$

2.5 3.0 3.5 4.0 4.5 5.0 5.5 6.0 6.5

Figure 8.6 **The "Three-Way Tug-of-War" between $1 - \alpha$, n, and E**

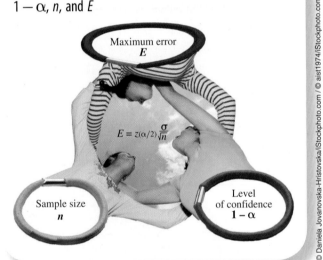

*Table 8.2 **Fifteen Samples of Size 40**

Sample Number	Sample Mean, $\bar{x}$	90% Confidence Interval Estimate for μ	Sample Number	Sample Mean, $\bar{x}$	90% Confidence Interval Estimate for μ
1	3.98	3.23 to 4.72	9	4.08	3.33 to 4.83
2	4.64	3.89 to 5.39	10	5.20	4.45 to 5.95
3	4.56	3.81 to 5.31	11	4.88	4.13 to 5.63
4	3.96	3.21 to 4.71	12	5.36	4.61 to 6.11
5	5.12	4.37 to 5.87	13	4.18	3.43 to 4.93
6	4.24	3.49 to 4.99	14	4.90	4.15 to 5.65
7	3.44	2.69 to 4.19	15	4.48	3.73 to 5.23
8	4.60	3.85 to 5.35			

SAMPLE SIZE AND CONFIDENCE INTERVALS

To get a better idea of how statisticians balance confidence, sample size, and error, let's look at the problem of determining the sample size needed to estimate the mean weight of all second-grade boys—and to be accurate within 1 pound with 95% confidence. We'll assume a normal distribution and that the standard deviation of the boys' weights is 3 pounds.

The desired level of confidence determines the confidence coefficient: The confidence coefficient is found using Table 4B: $z(\alpha/2) = z(0.025) = \textbf{1.96}$.

We know our desired maximum error is $E = 1.0$ (remember, 1 pound). Now we are ready to use the maximum error formula:

$$E = z(\alpha/2)\left(\frac{\sigma}{\sqrt{n}}\right): \qquad 1.0 = 1.96\left(\frac{3}{\sqrt{n}}\right)$$

$$\text{Solve for } n: \qquad 1.0 = \frac{5.88}{\sqrt{n}}$$

$$\sqrt{n} = 5.88$$

$$n = (5.88)^2 = 34.57 = \textbf{35}$$

Therefore, $n = 35$ is the sample size needed if you want a 95% confidence interval with a maximum error no greater than 1 pound.

NOTE: When we solve for the sample size n, it is customary to round up to the next larger integer, no matter what fraction (or decimal) results.

CALCULATING SAMPLE SIZE WITH UNKNOWN VALUE OF SIGMA (σ)

Using the maximum error formula (8.2) can be made a little easier by rewriting the formula in a form that expresses n in terms of the other values.

Sample Size

$$n = \left(\frac{z(\alpha/2)\cdot\sigma}{E}\right)^2 \tag{8.3}$$

If the maximum error is expressed as a multiple of the standard deviation σ, then the actual value of σ is not needed in order to calculate the sample size. If we wanted to find the sample size needed to estimate the population mean to within $\frac{1}{5}$ of a standard deviation with 99% confidence, we would first need to determine the confidence coefficient (using Table 4B): $1 - \alpha = 0.99$, $z(\alpha/2) = 2.58$. The desired maximum error is $E = \frac{\sigma}{5}$. Now we are ready to use the sample size formula (8.3):

$$n = \left(\frac{z(\alpha/2)\cdot\sigma}{E}\right)^2 \qquad n = \left(\frac{(2.58)\cdot\sigma}{\sigma/5}\right)^2 = \left(\frac{(2.58\sigma)(5)}{\sigma}\right)^2$$

$$= [(2.58)(5)]^2$$

$$= (12.90)^2 = 166.41 = \textbf{167}$$

8.3 The Nature of Hypothesis Testing

TO TEST A CLAIM, WE MUST FORMULATE A NULL HYPOTHESIS AND AN ALTERNATIVE HYPOTHESIS.

We all make decisions every day of our lives. Some of these decisions are of major importance; others are seemingly insignificant. All decisions follow the same basic pattern. We weigh the alternatives; then, based on our beliefs and preferences and whatever evidence is available, we arrive at a decision and take the appropriate action. The statistical hypothesis test follows much the same process, except that it involves statistical information. In this section we develop many of the concepts and attitudes of the hypothesis test while looking at several decision-making situations without using any statistics.

A friend is having a party (Super Bowl party, home-from-college party—you know the situation, any excuse will do), and you have been invited. You must make a decision: attend or not attend. That's simple; well maybe, except that you want to go only if you can be convinced the party is going to be more fun than your friend's typical parties; furthermore, you definitely do not want to go if the party is going to be just another dud. You have taken the position that "the party will be a dud" and you will not go unless you become convinced otherwise. Your friend assures you, "Guaranteed, the party will be a great time!" Do you go or not?

The decision-making process starts by identifying *something of concern* and then formulating *two* hypotheses about it. A **hypothesis** is a statement that something is true. Your friend's statement, "The party will be a great time," is a hypothesis. Your position, "The party will be a dud," is also a hypothesis. The process by which a decision is made between two opposing hypotheses is called a **statistical hypothesis test.** The two opposing hypotheses are formulated so that each hypothesis is the negation of the other. (That way one of them is always true, and the other one is always false.) Then one hypothesis is tested in hopes that it can be shown to be a very improbable occurrence, thereby implying that the other hypothesis is likely the truth.

The two hypotheses involved in making a decision are known as the *null hypothesis* and the *alternative hypothesis*. The **null hypothesis** is the hypothesis we will test. Generally this is a statement that a population parameter has a specific value. The null hypothesis is so named because it is the "starting point" for the investigation. (The phrase "there is no difference" is often used in its interpretation.) The **alternative hypothesis** is a statement about the same population parameter that is used in the null hypothesis. Generally this is a statement that specifies that the population parameter has a value that is somehow different from the value given in the null hypothesis. The rejection of the null hypothesis will imply the likely truth of this alternative hypothesis.

With regard to your friend's party, the two opposing viewpoints or hypotheses are "The party will be a great time" and "The party will be a dud." Which statement becomes the null hypothesis, and which becomes the alternative hypothesis?

Determining the statement of the null hypothesis and the statement of the alternative hypothesis is a very important step. The *basic idea* of the hypothesis test is for the evidence to have a chance to "disprove" the null hypothesis. The null hypothesis is the statement that the evidence might disprove. *Your concern* (belief or desired outcome), as the person doing the testing, is expressed in the alternative hypothesis. As the person making the decision, you believe that the evidence will

Hypothesis A statement that something is true.

Statistical hypothesis test A process by which a decision is made between two opposing hypotheses.

Null hypothesis, * H_o The hypothesis we will test.

Alternative hypothesis, H_a A statement about the same population parameter that is used in the null hypothesis.

demonstrate the feasibility of your "theory" by demonstrating the *unlikeliness* of the truth of the null hypothesis. The alternative hypothesis is sometimes referred to as the *research hypothesis* since it represents what the researcher hopes will be found to be "true." (If so, he or she will get a paper out of the research.)

Since the "evidence" (who's going to the party, what is going to be served, and so on) can only demonstrate the unlikeliness of the party being a dud, your initial position, "The party will be a dud," becomes the null hypothesis. Your friend's claim, "The party will be a great time," then becomes the alternative hypothesis.

$$H_o: \text{"Party will be a dud"}$$
$$\text{vs.}$$
$$H_a: \text{"Party will be a great time"}$$

Before returning to our example about the party, we need to look at the four possible outcomes that could result from the null hypothesis being either true or false and the decision being either to "reject H_o" or to "fail to reject H_o." Table 8.3 shows these four possible outcomes.

A **type A correct decision** occurs when the null hypothesis is true, and we decide in its favor. A **type B correct decision** occurs when the null hypothesis is

✱ We use the notation H_o for the null hypothesis to contrast it with H_a for the alternative hypothesis. Other texts may use H_0 (subscript zero) in place of H_o and H_1 in place of H_a.

© iStockphoto.com

Writing Hypotheses

Part of becoming a capable statistician is understanding how to write useful hypotheses. Here are a couple of examples to give you some ideas of how to do this in different situations.

Situation	Solution
You are testing a new design for airbags used in automobiles, and you are concerned that they might not open properly. State the null and alternative hypotheses.	The two opposing possibilities are "Bags open properly" and "Bags do not open properly." Testing can only produce evidence that discredits the hypothesis "Bags open properly." Therefore, the null hypothesis is "Bags open properly" and the alternative hypothesis is "Bags do not open properly." The alternative hypothesis can be the statement the experimenter wants to show to be true.
You suspect that a brand-name detergent outperforms the store's brand of detergent, and you wish to test the two detergents because you would prefer to buy the cheaper store brand. State the null and alternative hypotheses.	Your suspicion, "the brand-name detergent outperforms the store brand," is the reason for the test and therefore becomes the alternative hypothesis. H_o: "There is no difference in detergent performance." H_a: "The brand-name detergent performs better than the store brand." However, as a consumer, you are hoping not to reject the null hypothesis for budgetary reasons.

Table 8.3 **Four Possible Outcomes in a Hypothesis Test**

Decision	H_o True	H_o False
Fail to reject H_o	Type A correct decision	Type II error
Reject H_o	Type I error	Type B correct decision

false, and the decision is in opposition to the null hypothesis. A **type I error** is committed when a true null hypothesis is rejected—that is, when the null hypothesis is true but we decide against it. A **type II error** is committed when we decide in favor of a null hypothesis that is actually false.

When a decision is made, it would be nice to always make the correct decision. This, however, is not possible in statistics because we make our decisions on the basis of sample information. The best we can hope for is to control the probability with which an error occurs. The probability assigned to the type I error is **α** (called **"alpha"**; α is the first letter of the Greek alphabet). The probability of the type II error is **β** (called **"beta"**; β is the second letter of the Greek alphabet). See Table 8.4.

To control these errors, we will assign a small probability to each of them. The most frequently used probability values for α and β are 0.01 and 0.05. The probability assigned to each error depends on its seriousness. The more serious the error, the less willing we are to have it occur, and therefore a smaller probability will be assigned. α and β are probabilities of errors, each under separate conditions, and they cannot be combined. Therefore, we cannot determine a single probability for

Table 8.4 **Probability with Which Decisions Occur**

Error in Decision	Type	Probability	Correct Decision	Type	Probability
Rejection of a true H_o	I	α	Failure to reject a true H_o	A	$1 - α$
Failure to reject a false H_o	II	β	Rejection of a false H_o	B	$1 - β$

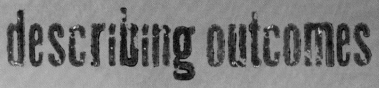

describing outcomes

How would we describe the four possible outcomes and the resulting actions that would occur for the hypothesis test about detergent described on page 163?

First, recall that we need a null hypothesis and an alternative hypothesis:

H_o: "There is no difference in detergent performance."

H_a: "The brand-name detergent performs better than the store brand."

NOTES:

1. The truth of the situation is not known before the decision is made, the conclusion reached, and the resulting actions take place. The truth of H_o may never be known.
2. The type II error often results in what represents a "lost opportunity"; lost in this situation is the chance to use a product that yields better results.

	H_o Is True	H_o Is False
Fail to Reject H_o	**Type A Correct Decision** **Truth of situation:** There is no difference between the detergents. **Conclusion:** It was determined that there was no difference. **Action:** The consumer bought the cheaper detergent, saving money and getting the same results.	**Type II Error** **Truth of situation:** The brand-name detergent is better. **Conclusion:** It was determined that there was no difference. **Action:** The consumer bought the cheaper detergent, saving money and getting inferior results.
Reject H_o	**Type I Error** **Truth of situation:** There is no difference between the detergents. **Conclusion:** It was determined that the brand-name detergent was better. **Action:** The consumer bought the brand-name detergent, spending extra money to attain no better results.	**Type B Correct Decision** **Truth of situation:** The brand-name detergent is better. **Conclusion:** It was determined that the brand-name detergent was better. **Action:** The consumer bought the brand-name detergent, spending more and getting better results.

making an incorrect decision. Likewise, the two correct decisions are distinctly separate and each has its own probability; $1 - \alpha$ is the probability of a correct decision when the null hypothesis is true, and $1 - \beta$ is the probability of a correct decision when the null hypothesis is false. $1 - \beta$ is called the *power of the statistical test* since it is the measure of the ability of a hypothesis test to reject a false null hypothesis, a very important characteristic.

NOTE: Regardless of the outcome of a hypothesis test, you can never be certain that a correct decision has been reached.

Let's look back at the two possible errors in decision that could occur in our example about the laundry detergent. Most people would become upset if they found out they were spending extra money for a detergent that performed no better than the cheaper brand. Likewise, most people would become upset if they found out they could have been buying a better detergent. Evaluating the relative seriousness of these errors requires knowing whether this is your personal laundry or a professional laundry business, how much extra the brand-name detergent costs, and so on.

There is an interrelationship among the probability of the type I error (α), the probability of the type II error (β), and the sample size (n). This is very much like the interrelationship among level of confidence, maximum error, and sample size discussed on page 160. Figure 8.7 shows the "three-way tug-of-war" among α, β, and n. If any one of the three is increased or decreased, it has an effect on one or both of the others.

Figure 8.7 **The "Three-Way Tug-of-War" between** α, β, and n

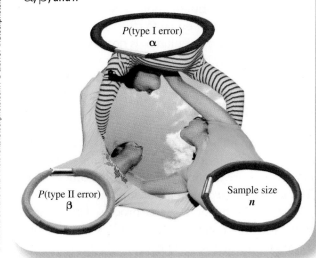

The statistician's job is to "balance" the three values of α, β, and n to achieve an acceptable testing situation.

If α is reduced, then either β must increase or n must be increased; if β is decreased, then either α increases or n must be increased; if n is decreased, then either α increases or β increases. The choices for α, β, and n are definitely not arbitrary. At this time in our study of statistics, α will be given in the statement of the problem, as will the sample size n. Further discussion on the role of β, P(type II error), is left for another time.

Sample size, n, is self-explanatory, so let's look at the role of α, or the **level of significance.** The level of significance α is the probability of committing the type I error.

Establishing the level of significance can be thought of as a "managerial decision." Typically, someone in charge determines the level of probability with which he or she is willing to risk a type I error.

At this point in the hypothesis test procedure, the evidence is collected and summarized and the value of a *test statistic* is calculated. A **test statistic** is a random variable whose value is calculated from the sample data and is used in making the decision "fail to reject H_o" or "reject H_o." The value of the calculated test statistic is used in conjunction with a decision rule to determine either "reject H_o" or "fail to reject H_o." This **decision rule** must be established prior to collecting the data; it specifies how you will reach the decision.

Back to your friend's party: You have to weigh the history of your friend's parties, the time and place, others going, and so on, against your own criteria and then make your decision. As a result of the decision about the null hypothesis ("The party will be a dud"), you will take the appropriate action; you will either go to or not go to the party.

To complete a hypothesis test, you will need to write a conclusion that carefully describes the meaning of the decision relative to the intent of the hypothesis test.

When writing the decision and the conclusion, remember that (1) the decision is about H_o, and (2) the conclusion is a statement about whether or not the contention of H_a was upheld. This is consistent with the "attitude" of the whole hypothesis test procedure. The null hypothesis is the statement that is "on trial," and therefore the decision must be

about it. The contention of the alternative hypothesis is the thought that brought about the need for a decision. Therefore, the question that led to the alternative hypothesis must be answered when the conclusion is written.

We must always remember that when the decision is made, nothing has been proved. Both decisions can lead to errors: "Fail to reject H_o" could be a type II error (the lack of sufficient evidence has led to great parties being missed more than once), and "reject H_o" could be a type I error (more than one person has decided to go to a party that was a dud).

THE CONCLUSION

a. If the decision is "reject H_o," then the conclusion should be worded something like, "There is sufficient evidence at the α level of significance to show that . . . (the meaning of the alternative hypothesis)."

b. If the decision is "fail to reject H_o," then the conclusion should be worded something like, "There is not sufficient evidence at the α level of significance to show that . . . (the meaning of the alternative hypothesis)."

8.4 Hypothesis Test of Mean μ (σ Known): A Probability-Value Approach

THE **PROBABILITY-VALUE APPROACH** MAKES FULL USE OF THE COMPUTER'S CAPABILITY IN DOING THE WORK OF THE DECISION-MAKING PROCESS.

In Objective 8.3, we surveyed the concepts and much of the reasoning behind a hypothesis test while looking at nonstatistical illustrations. In this section, we are going to formalize the hypothesis test procedure as it applies to statements concerning the mean μ of a population under the restriction that σ, the population standard deviation, is a known value.

> **The assumption for hypothesis tests about mean μ using a known σ:**
> The sampling distribution of $\bar{x}$ has a normal distribution.

The information we need to ensure that this assumption is satisfied is contained in the SDSM and in the CLT (see Chapter 7, pages 140–142):

The SDSM $\bar{x}$ is distributed about a mean equal to μ with a standard error equal to $\sigma/\sqrt{n}$; and (1) if the randomly sampled population is normally distributed, then $\bar{x}$ is normally distributed for all sample sizes, or (2) if the randomly sampled population is not normally distributed, then $\bar{x}$ is approximately normally distributed for sufficiently large sample sizes.

The hypothesis test is a well-organized, step-by-step procedure used to make a decision. Two different formats are commonly used for hypothesis testing. The *probability-value approach*, or simply p-*value approach*, is the hypothesis test process that has gained popularity in recent years, largely as a result of the convenience and the "number-crunching" ability of the

computer. This approach is organized as a five-step procedure outlined in the box below.

One-Tailed Hypothesis Test Using the *p*-Value Approach

To get a sense of how this procedure works, let's consider a commercial aircraft manufacturer that buys rivets to use in assembling airliners. Each rivet supplier that wants to sell rivets to the aircraft manufacturer must demonstrate that its rivets meet the required specifications. One of the specs is: "The mean shearing strength of all such rivets, μ, is at least 925 pounds." Each time the aircraft manufacturer buys rivets, it is concerned that the mean strength might be less than the specification of 925 pounds.

Each individual rivet has a shearing strength, which is determined by measuring the force required to shear ("break") the rivet. Clearly, not all the rivets can be tested.

Therefore, a sample of rivets will be tested, and a decision about the mean strength of all the untested rivets will be based on the mean from those sampled and tested.

STEP 1 THE SET-UP:

a. **Describe the population parameter of interest.**
   The population parameter of interest is the mean μ, the mean shearing strength of (or mean force required to shear) the rivets being considered for purchase.

b. **State the null hypothesis (H_o) and the alternative hypothesis (H_a).**
   The null hypothesis and the alternative hypothesis are formulated by inspecting the problem or statement to be investigated and first formulating two opposing statements about the mean μ. For our example, these two opposing statements are: (A) "The mean shearing strength is less than 925" (μ < 925, the aircraft manufacturer's concern),

> IF THE ASSUMPTIONS ARE NOT MET FOR HYPOTHESIS TESTS ABOUT THE MEAN μ USING A KNOWN σ, THAT CALCULATED *p*-VALUE COULD CAUSE A WRONG DECISION ABOUT H_o.

The Probability-Value Hypothesis Test: A Five-Step Procedure

STEP 1 The Set-Up:
a. Describe the population parameter of interest.
b. State the null hypothesis (H_o) and the alternative hypothesis (H_a).

STEP 2 The Hypothesis Test Criteria:
a. Check the assumptions.
b. Identify the probability distribution and the test statistic to be used.
c. Determine the level of significance, α.

STEP 3 The Sample Evidence:
a. Collect the sample information.
b. Calculate the value of the test statistic.

STEP 4 The Probability Distribution:
a. Calculate the *p*-value for the test statistic.
b. Determine whether or not the *p*-value is smaller than α.

STEP 5 The Results:
a. State the decision about H_o.
b. State the conclusion about H_a.

and (B) "The mean shearing strength is at least 925" ($\mu = 925$, the rivet supplier's claim and the aircraft manufacturer's spec).

NOTE: The trichotomy law from algebra states that two numerical values must be related in exactly one of three possible relationships: $<$, $=$, or $>$. All three of these possibilities must be accounted for in the two opposing hypotheses in order for the two hypotheses to be negations of each other. The three possible combinations of signs and hypotheses are shown in Table 8.5. Recall that the null hypothesis assigns a specific value to the parameter in question, and therefore "equals" will always be part of the null hypothesis.

Table 8.5 **The Three Possible Statements of Null and Alternative Hypotheses**

Null Hypothesis	Alternative Hypothesis
1. greater than or equal to ($\geq$)	less than ($<$)
2. less than or equal to ($\leq$)	greater than ($>$)
3. equal to ($=$)	not equal to ($\neq$)

The parameter of interest, the population mean μ, is related to the value 925. Statement (A) becomes the alternative hypothesis:

H_a: $\mu < 925$ (the mean is less than 925)

This statement represents the aircraft manufacturer's concern and says, "The rivets do not meet the required specs." Statement (B) becomes the null hypothesis:

H_o: $\mu = 925$ ($\geq$) (the mean is at least 925)

This hypothesis represents the negation of the aircraft manufacturer's concern and says, "The rivets do meet the required specs."

NOTE: We will write the null hypothesis with just the equal sign, thereby stating the exact value assigned. When "equal" is paired with "less than" or paired with "greater than," the combined symbol is written beside the null hypothesis as a reminder that all three signs have been accounted for in these two opposing statements.

Before continuing with our rivet example, let's look at two examples that demonstrate formulating the statistical null and alternative hypotheses involving the population mean μ.

WRITING NULL AND ALTERNATIVE HYPOTHESES (ONE-TAILED SITUATION)

Suppose the EPA was suing the city of Rochester for noncompliance with carbon monoxide standards. Specifically, the EPA would want to show that the mean level of carbon monoxide in downtown Rochester's air is dangerously high—higher than 4.9 parts per million. State the null and alternative hypotheses.

SOLUTION

To state the two hypotheses, we first need to identify the population parameter in question: the "mean level of carbon monoxide in Rochester." The parameter μ is being compared to the value 4.9 parts per million, the specific value of interest. The EPA is questioning the value of μ and wishes to show that it is higher than 4.9 (that is, $\mu > 4.9$). The three possible relationships—(1) $\mu < 4.9$, (2) $\mu = 4.9$, and (3) $\mu > 4.9$—must be arranged to form two opposing statements: One states the EPA's position, "The mean level is higher than 4.9 ($\mu > 4.9$)," and the other states the negation, "The mean level is not higher than 4.9 ($\mu \leq 4.9$)." One of these two statements will become the null hypothesis H_o, and the other will become the alternative hypothesis H_a.

NOTE: Recall that there are two rules for forming the hypotheses: (1) The null hypothesis states that the parameter in question has a specified value ("H_o must contain the equal sign"), and (2) the EPA's contention becomes the alternative hypothesis ("higher than"). Both rules indicate:

H_o: $\mu = 4.9$ ($\leq$)   and H_a: $\mu > 4.9$

WRITING NULL AND ALTERNATIVE HYPOTHESES (TWO-TAILED SITUATION)

Job satisfaction is very important to worker productivity. A standard job-satisfaction questionnaire was administered by union officers to a sample of assembly-line workers in a large plant in hopes of showing that the assembly workers' mean score on this questionnaire would be different from the established mean of 68. State the null and alternative hypotheses.

SOLUTION

Either the mean job-satisfaction score is different from 68 ($\mu \neq 68$) or the mean is equal to 68 ($\mu = 68$). Therefore,

H_o: $\mu = 68$ and H_a: $\mu \neq 68$

Table 8.6 lists some additional common phrases used in claims and indicates their negations and the hypothesis in which each phrase will be used. Again, notice that "equals" is always in the null hypothesis. Also notice that the negation of "less than" is "greater than or equal to." Think of negation as "all the others" from the set of three signs.

After the null and alternative hypotheses are established, we will work under the assumption that the null hypothesis is a true statement until there is sufficient evidence to reject it. This situation might be compared to a courtroom trial, where the accused is assumed to be innocent (H_o: Defendant is innocent vs. H_a: Defendant is not innocent) until sufficient evidence has been presented to show that innocence is totally unbelievable ("beyond reasonable doubt"). At the conclusion of the hypothesis test, we will make one of two possible decisions. We will decide in opposition to the null hypothesis and say that we "reject H_o" (this corresponds to "conviction" of the accused in a trial), or we will decide in agreement with the null hypothesis and say that we "fail to reject H_o" (this corresponds to "fail to convict" or an "acquittal" of the accused in a trial).

Let's return to the rivet example we interrupted on pages 167–168 and continue with Step 2. Recall that

H_o: $\mu = 925$ ($\geq$) (at least 925)
H_a: $\mu < 925$ (less than 925)

STEP 2 THE HYPOTHESIS TEST CRITERIA:

a. Check the assumptions.

σ is known. Variables like shearing strength typically have a mounded distribution; therefore, a sample of size 50 should be large enough for the CLT to apply and ensure that the SDSM will be normally distributed.

b. Identify the probability distribution and the test statistic to be used.

The standard normal probability distribution is used because $\bar{x}$ is expected to have a normal distribution.

For a hypothesis test of μ, we want to compare the value of the sample mean to the value of the population mean as stated in the null hypothesis. This comparison is accomplished using the test statistic in formula (8.4):

Test Statistic for Mean

$$z\star = \frac{\bar{x} - \mu}{\sigma/\sqrt{n}} \tag{8.4}$$

The resulting calculated value is identified as $z\star$ ("z star") because it is expected to have a standard normal distribution when the null hypothesis is true and the assumptions have been satisfied. The $\star$ ("star") is to remind us that this is the calculated value of the test statistic.

The test statistic to be used is $z\star = \frac{\bar{x} - \mu}{\sigma/\sqrt{n}}$ with $\sigma = 18$.

c. Determine the level of significance, α.

Setting α was described as a managerial decision in Objective 8.3. To see what is involved in determining α, the probability of the type I error, for our rivet example, we start by identifying the four possible outcomes, their meaning, and the action related to each.

The type I error occurs when a true null hypothesis is rejected. This would occur when the manufacturer tested rivets that in truth did meet the specs, and rejected them. Undoubtedly this would lead to the rivets not being purchased even though they did meet the specs. In order for the manager to set a level of significance, related information is needed—namely, how soon is the new supply of rivets needed? If they are needed tomorrow and this is the only vendor with an available supply, waiting a week to

Table 8.6 **Common Phrases and Their Negations**

H_o: ($\geq$)	H_a: ($<$)	H_o: ($\leq$)	H_a: ($>$)	H_o: ($=$)	H_a: ($\neq$)
at least	less than	at most	more than	is	is not
no less than	less than	no more than	more than	not different from	different from
not less than	less than	not greater than	greater than	the same as	not the same as

find acceptable rivets could be very expensive; therefore, rejecting good rivets could be considered a serious error. On the other hand, if the rivets are not needed until next month, then this error may not be very serious. Only the manager will know all the ramifications, and therefore the manager's input is important here.

After much consideration, the manager assigns the level of significance: $\alpha = 0.05$.

STEP 3 THE SAMPLE EVIDENCE:

a. **Collect the sample information.**
    We are ready for the data. The sample must be a random sample drawn from the population whose mean μ is being questioned. A random sample of 50 rivets is selected, each rivet is tested, and the sample mean shearing strength is calculated: $\bar{x} = 921.18$ and $n = 50$.

b. **Calculate the value of the test statistic.**
    The sample evidence ($\bar{x}$ and n found in Step 3a) is next converted into the **calculated value of the test statistic**, $z\star$, using formula (8.4). (μ is 925 from H_o; $\sigma = 18$ is the known quantity, as shown on the previous page.) We have

$$z\star = \frac{\bar{x} - \mu}{\sigma/\sqrt{n}}: \quad z\star = \frac{921.18 - 925.0}{18/\sqrt{50}}$$

$$= \frac{-3.82}{2.5456} = -1.50$$

STEP 4 THE PROBABILITY DISTRIBUTION:

a. **Calculate the p-value for the test statistic.**
    The **probability value**, or **p-value**, is the probability that the test statistic could be the value it is or a more extreme value (in the direction of the alternative hypothesis) when the null hypothesis is true. The p-value is represented by the area under the curve of the probability distribution for the test statistic that is more extreme than the calculated value of the test statistic. There are three separate cases, and the direction (or sign) of the alternative hypothesis is the key. Table 8.7 outlines the procedure for all three cases.

To apply this to Step 4 of our rivet example, draw a sketch of the standard normal distribution and locate $z\star$ (found in Step 3b) on it. To identify the area that represents the p-value, look at the sign in the alterna-

tive hypothesis. For this test, the alternative hypothesis indicates that we are interested in that part of the sampling distribution that is "*less than*" $z\star$. Therefore, the p-value is the area that lies to the *left* of $z\star$. Shade this area. You can see that for this example, we are dealing with a one-tailed, left tail hypothesis.

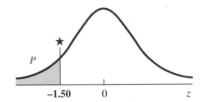

There are three ways to find the exact p-value:

Method 1: Use Table 3 in Appendix B to determine the tabled area related to the left of $z = -1.50$:

$$p\text{-value} = P(z < z\star) = P(z < -1.50) = \mathbf{0.0668}$$

Method 2: Use Table 5 in Appendix B and the symmetry property: Table 5 is set up to allow you to read the p-value directly from the table. Since $P(z < -1.50) = P(z > 1.50)$, simply locate $z\star = 1.50$ on Table 5 and read the p-value:

$$P(z < -1.50) = \mathbf{0.0668}$$

Method 3: Use the cumulative probability function on a computer or calculator to find the p-value:

$$P(z < -1.50) = \mathbf{0.0668}$$

b. **Determine whether or not the p-value is smaller than α.**
    In our example, the p-value (0.0668) is not smaller than α (0.05).

STEP 5 THE RESULTS:

a. **State the decision about H_o.**
    Is the p-value small enough to indicate that the sample evidence is highly unlikely in the event that the null hypothesis is true? In order to make the decision, we need to know the *decision rule*.

Decision Rule

a. If the p-value is *less than or equal to* the level of significance α, then the decision must be to **reject H_o**.
b. If the p-value is *greater than* the level of significance α, then the decision must be to **fail to reject H_o**.

Table 8.7 Finding *p*-Values Using the Cumulative Distribution

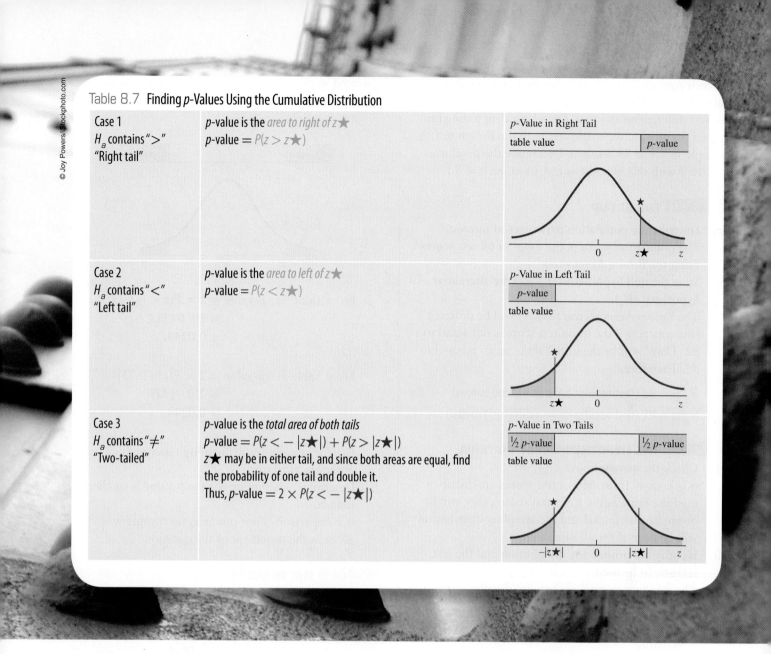

Case 1 H_a contains ">" "Right tail"	*p*-value is the *area to right of z★* *p*-value $= P(z > z★)$	*p*-Value in Right Tail
Case 2 H_a contains "<" "Left tail"	*p*-value is the *area to left of z★* *p*-value $= P(z < z★)$	*p*-Value in Left Tail
Case 3 H_a contains "≠" "Two-tailed"	*p*-value is the *total area of both tails* *p*-value $= P(z < -|z★|) + P(z > |z★|)$ z★ may be in either tail, and since both areas are equal, find the probability of one tail and double it. Thus, *p*-value $= 2 \times P(z < -|z★|)$	*p*-Value in Two Tails

Decision about H_o: Fail to reject H_o.

b. **State the conclusion about H_a.**

Review page 166 for specific information on writing the conclusion. In our case, there is not sufficient evidence at the 0.05 level of significance to show that the mean shearing strength of the rivets is less than 925. We "failed to convict" the null hypothesis. In other words, a sample mean as small as 921.18 is likely to occur (as defined by α) when the true population mean value is 925.0 and $\bar{x}$ is normally distributed. The resulting action by the manager would be to buy the rivets.

NOTE: When the decision reached is "fail to reject H_o" (or "accept H_o," as many say improperly), it simply means "for the lack of better information, act as if the null hypothesis is true."

Two-Tailed Hypothesis Test Using the *p*-Value Approach

Our rivet example involved a one-tailed procedure. Let's now look at an illustration involving the two-tailed procedure. To do this, we'll use the example of an employee selection test.

Many large companies in a certain city have for years used the Kelly Employment Agency for testing prospective employees. The employment selection test used has historically resulted in scores normally distributed about a mean of 82 with a standard deviation of 8. The Brown Agency has developed a new test that is quicker and easier to administer and therefore less expensive. Brown claims that its test results are the same as those obtained on the Kelly test. Many of the companies are considering a change from the Kelly Agency to the

Brown Agency in order to cut costs. However, they are unwilling to make the change if the Brown test results have a different mean value. An independent testing firm tested 36 prospective employees with the Brown test. A sample mean of 79 resulted. Determine the *p*-value associated with this hypothesis test. (Assume $\sigma = 8$.)

STEP 1 THE SET-UP:

a. **Describe the population parameter of interest.**
The population mean μ, the mean of all test scores using the Brown Agency test.

b. **State the null hypothesis (H_o) and the alternative hypothesis (H_a).**
The Brown Agency's test results "will be different" (the concern) if the mean test score is not equal to 82. They "will be the same" if the mean is equal to 82. Therefore,

H_o: $\mu = 82$ (test results have the same mean)

H_a: $\mu \neq 82$ (test results have a different mean)

STEP 2 THE HYPOTHESIS TEST CRITERIA:

a. **Check the assumptions.**
σ is known. If the Brown test scores are distributed the same as the Kelly test scores, they will be normally distributed and the sampling distribution will be normal for all sample sizes.

b. **Identify the probability distribution and the test statistic to be used.**
The standard normal probability distribution and the test statistic

$$z\star = \frac{\bar{x} - \mu}{\sigma/\sqrt{n}}$$

will be used with $\sigma = 8$.

c. **Determine the level of significance, α.**
The level of significance is omitted because the question asks for the *p*-value and not a decision.

STEP 3 THE SAMPLE EVIDENCE:

a. **Collect the sample information.** $n = 36, \bar{x} = 79$.

b. **Calculate the value of the test statistic.**
μ is 82 from H_o; $\sigma = 8$ is the known quantity. We have

$$z\star = \frac{\bar{x} - \mu}{\sigma/\sqrt{n}}: \quad z\star = \frac{79 - 82}{8/\sqrt{36}} = \frac{-3}{1.3333} = -2.25$$

STEP 4 THE PROBABILITY DISTRIBUTION:

a. **Calculate the *p*-value for the test statistic.**
Since the alternative hypothesis indicates a two-tailed test, we must find the probability associ-

ated with both tails. The *p*-value is found by doubling the area of one tail (see Table 8.7).
$z\star = -2.25$

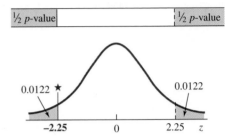

From Table 3: $p\text{-value} = 2 \times P(z < -2.25)$
$= 2(0.0122)$
$= 0.0244.$

or

From Table 5: $p\text{-value} = 2 \times P(z > 2.25)$
$= 2(0.0122)$
$= 0.0244.$

or

Use the cumulative probability function on a computer or calculator.

b. **Determine whether or not the *p*-value is smaller than α.**
A comparison is not possible; no α value was given in the statement of the question.

STEP 5 THE RESULTS:

The *p*-value for this hypothesis test is 0.0244. Each individual company now will decide whether to continue to use the Kelly Agency's services or change to the Brown Agency. Each will need to establish the level of significance that best fits its own situation and then make a decision using the decision rule described previously.

Evaluating the *p*-Value Approach

The fundamental idea of the *p*-value is to express the degree of belief in the null hypothesis:

- When the *p*-value is minuscule (something like 0.0003), the null hypothesis would be rejected by everybody because the sample results are very unlikely for a true H_o.

- When the *p*-value is fairly small (like 0.012), the evidence against H_o is quite strong and H_o will be rejected by many.

- When the *p*-value begins to get larger (say, 0.02 to 0.08), there is too much probability that data like the sample involved could have occurred even if H_o were true, and the rejection of H_o is not an easy decision.

- When the *p*-value is large (like 0.15 or more), the data are not at all unlikely if the H_o is true, and no one will reject H_o.

The advantages of the *p*-value approach are as follows: (1) The results of the test procedure are expressed in terms of a continuous probability scale from 0.0 to 1.0, rather than simply on a "reject" or "fail to reject" basis. (2) A *p*-value can be reported and the user of the information can decide on the strength of the evidence as it applies to his or her own situation. (3) Computers can do all the calculations and report the *p*-value, thus eliminating the need for tables.

The disadvantage of the *p*-value approach is the tendency for people to put off determining the level of significance. This should not be allowed to happen, because it is then possible for someone to set the level of significance after the fact, leaving open the possibility that the "preferred" decision will result. This is probably important only when the reported *p*-value falls in the "hard choice" range (say, 0.02 to 0.08), as described previously.

Let's return to

8.5 Hypothesis Test of Mean μ (σ Known): A Classical Approach

THE CLASSICAL APPROACH USES CRITICAL VALUES IN DOING THE WORK OF THE DECISION-MAKING PROCESS.

In Objective 8.4, we explored the *p*-value approach to hypothesis testing. Now, we'll examine the classical approach, which has enjoyed popularity for many years. Like the *p*-value approach, the classical approach is a well-organized, step-by-step procedure used to make a

The Classical Hypothesis Test: A Five-Step Procedure

STEP 1 **The Set-Up:**
   a. Describe the population parameter of interest.
   b. State the null hypothesis (H_o) and the alternative hypothesis (H_a).

STEP 2 **The Hypothesis Test Criteria:**
   a. Check the assumptions.
   b. Identify the probability distribution and the test statistic to be used.
   c. Determine the level of significance, α.

STEP 3 **The Sample Evidence:**
   a. Collect the sample information.
   b. Calculate the value of the test statistic.

STEP 4 **The Probability Distribution:**
   a. Determine the critical region and critical value(s).
   b. Determine whether or not the calculated test statistic is in the critical region.

STEP 5 **The Results:**
   a. State the decision about H_o.
   b. State the conclusion about H_a.

decision. The classical hypothesis test is also organized as a five-step procedure.

With the classical procedure, we still assume that about mean μ using a known σ, the sampling distribution of $\bar{x}$ has a normal distribution. (Review page 166.)

To keep things simple, let's return to our rivet example. Recall the set-up: A commercial aircraft manufacturer buys rivets to use in assembling airliners. Each rivet supplier that wants to sell rivets to the aircraft manufacturer must demonstrate that its rivets meet the required specifications. One of the specs is "The mean shearing strength of all such rivets, μ, is at least 925 pounds." Each time the aircraft manufacturer buys rivets, it is concerned that the mean strength might be less than the 925-pound specification. The same stipulations about shearing strength and sampling apply here as well.

STEP 1 THE SET-UP:

As with the p-value approach, our basic problem set up is the same (see page 167). Note also that the trichotomy law from algebra is in force for the classical hypothesis test as well. The three possible combinations of signs and hypotheses were shown in Table 8.5. Recall that the null hypothesis assigns a specific value to the parameter in question, and therefore "equals" will always be part of the null hypothesis.

Again, the parameter of interest, the population mean μ, is related to the value 925. Statement (A) becomes the alternative hypothesis:

H_a: μ < 925 (the mean is less than 925)

This statement represents the aircraft manufacturer's concern and says, "The rivets do not meet the required specs." Statement (B) becomes the null hypothesis:

H_o: μ = 925 (≥) (the mean is at least 925)

This hypothesis represents the negation of the aircraft manufacturer's concern and says, "The rivets do meet the required specs."

NOTE: We use the same notation for writing the null hypothesis as we did with the p-value approach. (See page 168 for notes.)

Keep in mind that our example here is of a one-tailed situation. Writing for a two-tailed situation using the classical approach is comparable to writing it using the p-value approach (see page 168).

Whether you use the p-value or classical approach for the rivet example, Steps 2 and 3 are the same. To refresh your memory on how those were conducted, review pages 169–170. When we get to Step 4, however, things are done a bit differently.

STEP 4 THE PROBABILITY DISTRIBUTION:

a. **Determine the critical region and critical value(s).** The standard normal variable z is our test statistic for this hypothesis test; therefore, we draw a sketch of the standard normal distribution, label the scale as z, and locate its mean value, 0. The **critical region** is the set of values for the test statistic that will cause us to reject the null hypothesis. The set of values that are not in the critical region is called the **noncritical region** (sometimes called the *acceptance region*).

Recall that we are working under the assumption that the null hypothesis is true. Thus, we are assuming that the mean shearing strength of all rivets in the sampled population is 925. If this is the case, then when we select a random sample of 50 rivets, we can expect this sample mean, $\bar{x}$, to be part of a normal distribution that is centered at 925 and to have a standard error of $\sigma/\sqrt{n} = 18/\sqrt{50}$, or approximately 2.55. Approximately 95% of the sample mean values will be greater than 920.8 [a value 1.65 standard errors below the mean: $925 - (1.65)(2.55) = 920.8$]. Thus, if H_o is true and μ = 925, then we expect $\bar{x}$ to be greater than 920.8 approximately 95% of the time and less than 920.8 only 5% of the time.

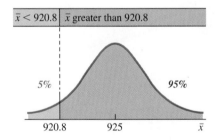

If, however, the value of $\bar{x}$ that we obtain from our sample is less than 920.8—say, 919.5—we will have

Critical region The set of values for the test statistic that will cause us to reject the null hypothesis.

Noncritical region (acceptance region) The set of values that are not in the critical region.

MPG

Writing Null and Alternative Hypotheses (One-Tailed Situation)

A consumer advocate group would like to disprove a car manufacturer's claim that a specific model will average 24 miles per gallon of gasoline. Specifically, the group would like to show that the mean miles per gallon is considerably less than 24. State the null and alternative hypotheses.

Solution

To state the two hypotheses, we first need to identify the population parameter in question: the "mean mileage attained by this car model." The parameter μ is being compared to the value 24 miles per gallon, the specific value of interest. The advocates are questioning the value of μ and wish to show it to be less than 24 (that is, $\mu < 24$).

There are three possible relationships: (1) $\mu < 24$, (2) $\mu = 24$, and (3) $\mu > 24$. These three cases must be arranged to form two opposing statements: One states what the advocates are trying to show, "The mean level is less than 24 ($\mu < 24$)," whereas the "negation" is "The mean level is not less than 24 ($\mu \geq 24$)." One of these two statements will become the null hypothesis H_o, and the other will become the alternative hypothesis H_a.

Note: Recall that there are two rules for forming the hypotheses: (1) The null hypothesis states that the parameter in question has a specified value ("H_o must contain the equal sign"), and (2) the consumer advocate group's contention becomes the alternative hypothesis ("less than"). Both rules indicate:

$$H_o: \mu = 24 \, (\geq) \quad \text{and} \quad H_a: \mu < 24$$

---

to make a choice. It could be that either (A) such an $\bar{x}$ value (919.5) is a member of the sampling distribution with mean 925, although it has a very low probability of occurrence (less than 0.05), or (B) $\bar{x} = 919.5$ is a member of a sampling distribution whose mean is less than 925, which would make it a value that is more likely to occur.

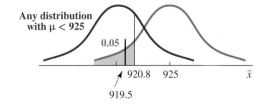

Any distribution with $\mu < 925$

0.05

919.5 → 920.8   925 $\bar{x}$

In statistics, we "bet" on the "more probable to occur" and consider the second choice (B) to be the right one. Thus, the left-hand tail of the z-distribution becomes the critical region. And the level of significance α becomes the measure of its area.

We also need to identify the **critical value,** or first or boundary value, of the critical region. The critical value for our example is $-z(0.05)$ and has the value of -1.65, as found in Table 4A in Appendix B.

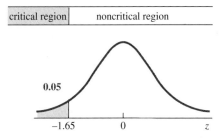

critical region	noncritical region

0.05

−1.65   0 z

b. Determine whether or not the calculated test statistic is in the critical region.

> **Critical value(s)** The "first" or "boundary" value(s) of the critical region(s).

Graphically this determination is shown by locating the value for $z\star$ on the sketch in Step 4a. The calculated value of z, $z\star = -1.50$, is **not in the critical region** (it is in the unshaded portion of the figure).

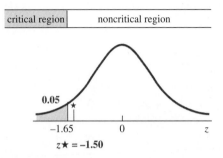

STEP 5 **THE RESULTS:**

a. State the decision about H_o.
    In order to make the decision, we need to know the *decision rule*.

De**Cis**i**on** **Ru**le

a. If the test statistic falls *within the critical region*, then the decision must be to **reject** H_o. (The critical value is part of the critical region.)
b. If the test statistic is *not in the critical region*, then the decision must be to **fail to reject** H_o.

The decision is: Fail to reject H_o.

b. State the conclusion about H_a.
    There is not sufficient evidence at the 0.05 level of significance to show that the rivets have a mean shearing strength less than 925. We "failed to convict" the null hypothesis. In other words, a sample mean as small as 921.18 is likely to occur (as defined by α) when the true population mean value is 925.0. Therefore, the resulting action would be to buy the rivets.

The value assigned to α is called the *significance level* of the hypothesis test. Alpha cannot be interpreted to be anything other than the risk (or probability) of rejecting the null hypothesis when it is actually true. We will seldom be able to determine whether the null hypothesis is true or false; we will decide only to "reject H_o" or to "fail to reject H_o." The relative frequency with which we reject a true hypothesis is α, but we will never know the relative frequency with which we make an error in decision. The two ideas are quite different; that is, a type I error and an error

in decision are two different things altogether. Remember that there are two types of errors: type I and type II.

Two-Tailed Hypothesis Test Using the Classical Approach

Let's look at one last hypothesis test involving the two-tailed procedure. For this example, we'll look at the mean weight of female college students. It has been claimed that the mean weight of women students at a certain college is 54.4 kg. Professor Hart does not believe the claim and sets out to show that the mean weight is not 54.4 kg. To test the claim, he collects a random sample of 100 weights from among the women students. A sample mean of 53.75 kg results. To determine if this is sufficient evidence for Professor Hart to reject the statement, let $\alpha = 0.05$ and $\sigma = 5.4$ kg. Once again, we'll apply our five-step procedure.

STEP 1 **THE SET-UP:**

a. Describe the population parameter of interest.
    The population parameter of interest is the mean μ, the mean weight of all women students at the college.

SOME OF THE DETAILS WE HAVE SEEN THUS FAR:

1. **The null hypothesis specifies a particular value of a population parameter.**
2. **The alternative hypothesis can take three forms. Each form dictates a specific location of the critical region(s), as shown in the following table.**
3. **For many hypothesis tests, the sign in the alternative hypothesis "points" in the direction in which the critical region is located. [Think of the not equal to sign ($\neq$) as being both less than ($<$) and greater than ($>$), thus pointing in both directions.]**

	Sign in the Alternative Hypothesis		
	<	**≠**	**>**
	One region	Two regions	One region
Critical Region	Left side	Half on each side	Right side
	One-tailed test	**Two-tailed test**	**One-tailed test**

b. State the null hypothesis (H_o) and the alternative hypothesis (H_a).
The mean weight is equal to 54.4 kg, or the mean weight is not equal to 54.4 kg.

H_o: $\mu = 54.4$ (mean weight is 54.4)

H_a: $\mu \neq 54.4$ (mean weight is not 54.4)
(Remember: $\neq$ is $<$ and $>$ together.)

STEP 2 **THE HYPOTHESIS TEST CRITERIA:**

a. **Check the assumptions.**
σ is known. The weights of an adult group of women are generally approximately normally distributed; therefore, a sample of $n = 100$ is large enough to allow the CLT to apply.

b. **Identify the probability distribution and the test statistic to be used.**
The standard normal probability distribution and the test statistic

$$z\star = \frac{\bar{x} - \mu}{\sigma / \sqrt{n}}$$

will be used; $\sigma = 5.4$.

c. **Determine the level of significance, α.**
$\alpha = 0.05$ (given in the statement of the problem).

STEP 3 **THE SAMPLE EVIDENCE:**

a. **Collect the sample information.**
$\bar{x} = 53.75$ and $n = 100$.

b. **Calculate the value of the test statistic.**
Use formula (8.4), information from H_o—
$\mu = 54.4$, and $\sigma = 5.4$ (known):

$$z\star = \frac{\bar{x} - \mu}{\sigma / \sqrt{n}}: \quad z\star = \frac{53.75 - 54.4}{5.4 / \sqrt{100}} = \frac{-0.65}{0.54}$$
$$= -1.204 = -1.20$$

STEP 4 **THE PROBABILITY DISTRIBUTION:**

a. **Determine the critical region and critical value(s).**
The critical region is both the left tail and the right tail because both smaller and larger values of the sample mean suggest that the null hypothesis is wrong. The level of significance will be split in half, with 0.025 being the measure of each tail. The critical values are found in Table 4B in Appendix B: $\pm z(0.025) = \pm 1.96$. (Table 4B instructions are on page 158.)

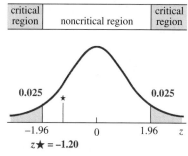

critical region	noncritical region	critical region

0.025   0.025

−1.96   0   1.96   z

$z\star = -1.20$

b. **Determine whether or not the calculated test statistic is in the critical region.**
The calculated value of z, $z\star = -1.20$, is not in the critical region (shown in red on the preceding figure).

STEP 5 **THE RESULTS:**

a. State the decision about H_o: Fail to reject H_o.
b. State the conclusion about H_a.
There is not sufficient evidence at the 0.05 level of significance to show that the women students have a mean weight different from the 54.4 kg claimed. In other words, there is no statistical evidence to support Professor Hart's contentions.

In this chapter we have restricted our discussion of inferences to the mean of a population for which the standard deviation is known. In Chapters 9 and 10 we will discuss inferences about the population mean and remove the restriction about the known value for standard deviation. We will also look at inferences about the parameters proportion, variance, and standard deviation.

problems

Objective 8.1

*8.1 The number of engines owned per fire department was obtained from a random sample taken from the profiles of fire departments from across the United States.

SOURCE: *Firehouse*

29	8	7	33	21	26	6	11	4	54	7	4

Use the data to find a point estimate for each of the following parameters:

a. Mean   b. Variance   c. Standard deviation

8.2 In each diagram below, I and II represent sampling distributions of two statistics that might be used to estimate a parameter. In each case, identify the statistic that you think would be the better estimator, or neither, and describe your choice.

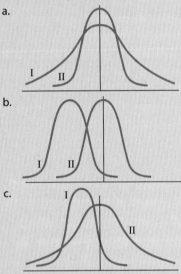

a.

b.

c.

8.3 Being unbiased and having a small variability are two desirable characteristics of a statistic if it is going to be used as an estimator. Describe how the SDSM addresses both of these properties when estimating the mean of a population.

8.4 The U.S. Census Bureau reports the estimated mean U.S. married-couple family income is $90,835 ± $101. They describe the margin of error as providing a 90% probability that the interval defined by the estimate minus the margin of error and the estimate plus the margin of error (the lower and upper confidence bounds) contains the true value.

SOURCE: U.S. Census Bureau, 2005–2007 American Community Survey

a. What is the population and variable of interest?
b. What parameter is being estimated? What is its estimated value?
c. How is the margin of error related to the maximum error of estimate?
d. What value is being reported as the margin of error?
e. What level of confidence is being reported?
f. Find the confidence interval and state exactly what it represents.

8.5 The Consumer Reports National Research Center reported that 76% of women respond, "daily or more often" when asked, "How often is your bed made?" This additional information is included as a footnote: margin of error ± 3.2 percentage points.

a. What is the population and variable of interest?
b. What parameter is being estimated? What is its estimated value?
c. What value is being reported as the margin of error?
d. Find the interval and state exactly what it represents.
e. What additional information might you want to know about this confidence interval?

8.6 Explain why the standard error of sample means is 3 for the rivet example on page 153.

8.7 Find the level of confidence assigned to an interval estimate of the mean formed using the following intervals:

a. $\bar{x} - 1.28 \cdot \sigma_{\bar{x}}$ to $\bar{x} + 1.28 \cdot \sigma_{\bar{x}}$
b. $\bar{x} - 1.44 \cdot \sigma_{\bar{x}}$ to $\bar{x} + 1.44 \cdot \sigma_{\bar{x}}$
c. $\bar{x} - 1.96 \cdot \sigma_{\bar{x}}$ to $\bar{x} + 1.96 \cdot \sigma_{\bar{x}}$
d. $\bar{x} - 2.33 \cdot \sigma_{\bar{x}}$ to $\bar{x} + 2.33 \cdot \sigma_{\bar{x}}$

8.8 "Population Requirement for Primary Hip-Replacement Surgery: A Cross-Sectional Study" was conducted by the University of Bristol in the United Kingdom. The findings resulted in the following statement: "The prevalence of self-reported hip pain was 107 per 1,000 (95% CI 101–113) for men and 173 per 1,000 (166–180) for women."

a. Explain the meaning of the confidence interval, (95% CI 101–113).
b. Find the standard error for the men's self-reported hip pain 95% confidence interval.
c. Assuming the women's data was also from a 95% confidence interval, find the standard error.

8.9 A recruiter estimates that if you are hired to work for her company and you put in a full week at the commissioned sales representative position she is offering, you will make "$525 plus or minus $250, 80% of the time." She adds, "It all depends on you!"

a. What does the "$525 plus or minus $250" mean?
b. What does the "80% of the time" mean?
c. If you make $300 to the nearest $10 most weeks, will she have told you the truth? Explain.

Objective 8.2

8.10 Determine the value of the confidence coefficient $z_{(\alpha/2)}$ for each situation described:

a. $1 - \alpha = 0.90$ b. $1 - \alpha = 0.95$

8.11 Given the information, the sampled population is normally distributed, $n = 16$, $\bar{x} = 28.7$, and $\sigma = 6$:

a. Find the 0.95 confidence interval for μ.
b. Are the assumptions satisfied? Explain

8.12 Given the information, the sampled population is normally distributed, $n = 55$, $\bar{x} = 78.2$, and $\sigma = 12$:
a. Find the 0.98 confidence interval for μ.
b. Are the assumptions satisfied? Explain.

8.13 In your own words, describe the relationship between the following:
a. Sample mean and point estimate
b. Sample size, sample standard deviation, and standard error
c. Standard error and maximum error

8.14 A sample of 60 night-school students' ages is obtained in order to estimate the mean age of night-school students. $\bar{x} = 25.3$ years. The population variance is 16.
a. Give a point estimate for μ.
b. Find the 95% confidence interval for μ.
c. Find the 99% confidence interval for μ.

8.15 The Eurostar was Europe's first international train, designed to take advantage of the Channel Tunnel that connects England with Continental Europe. It carries nearly 800 passengers and occasionally reaches a peak speed of more than 190 mph.

SOURCE: http://www.o-keating.com/

Assume the standard deviation of train speed is 19 mph in the course of all the journeys back and forth and that the train's speed is normally distributed. Suppose speed readings are made during the next 20 trips of the Eurostar and the mean speed of these measurements is 184 mph.
a. What is the variable being studied?
b. Find the 90% confidence interval estimate for the mean speed.
c. Find the 95% confidence interval estimate for the mean speed.

8.16 Based on a survey conducted by Greenfield Online, 25–34-year-olds spend the most each week on fast food. The average weekly amount of $44 (based on 115 participants) was reported in a May 2009 *USA Today* Snapshot. Assuming that weekly fast-food expenditures are normally distributed with a known standard deviation of $14.50, construct a 90% confidence interval for the mean weekly amount that 25–34-year-olds spend each week on fast food.

8.17 "College Costs Rise" (October 29, 2008), an article on the CNN Money website, gave the latest figures from the College Board on annual tuition, fees, and room and board. The average total figure for private colleges is $34,132 and $14,333 for public colleges.

SOURCE: http://money.cnn.com/

In an effort to compare those same costs in New York State, a sample of 32 college juniors is randomly selected statewide from the private colleges and 32 more from the public colleges. The private college sample resulted in a mean of $34,020, and the public college sample mean was $14,045.
a. Assume the annual college fees for private colleges have a mounded distribution and the standard deviation is $2,200. Find the 95% confidence interval for the mean college costs.
b. Assume the annual college fees for public colleges have a mounded distribution and the standard deviation is $1,500. Find the 95% confidence interval for the mean college costs.
c. How do the New York State college costs compare to the College Board's values? Explain.
d. Compare the confidence intervals found in (a) and (b), and describe the effect the two different sample means had on the resulting answers.
e. Compare the confidence intervals found in (a) and (b), and describe the effect the two different sample standard deviations had on the resulting answers.

8.18 How large a sample should be taken if the population mean is to be estimated with 99% confidence to within $75? The population has a standard deviation of $900.

8.19 By measuring the amount of time it takes a component of a product to move from one workstation to the next, an engineer has estimated that the standard deviation is 5 seconds.
a. How many measurements should be made to be 95% certain that the maximum error of estimation will not exceed 1 second?
b. What sample size is required for a maximum error of 2 seconds?

8.20 The new mini-laptop computers can deliver as much computing power as machines several times their size, but they weigh in at less than 3 pounds. How large a sample would be needed to estimate the population mean weight if the maximum error of estimate is to be 0.4 of 1 standard deviation with 95% confidence?

Objective 8.3

8.21 State the null and alternative hypotheses for each of the following:
a. You want to show an increase in buying and selling of single-family homes this year when compared with last year's rate.
b. You are testing a new recipe for "low-fat" cheesecake and expect to find that its taste is not as good as traditional cheesecake.
c. You are trying to show that music lessons have a positive effect on a child's self-esteem.
d. You are investigating the relationship between a person's gender and the automobile he or she drives—specifically you want to show that more males than females drive truck-type vehicles.

8.22 State the null and alternative hypotheses for each of the following:
a. You are investigating a complaint that "special delivery mail takes too much time" to be delivered.
b. You want to show that people find the new design for a recliner chair more comfortable than the old design.
c. You are trying to show that cigarette smoke affects the quality of a person's life.

d. You are testing a new formula for hair conditioner and hope to show that it is effective on "split ends."

8.23 When a medic at the scene of a serious accident inspects each victim, she administers the appropriate medical assistance to all victims, unless she is certain the victim is dead.
   a. State the null and alternative hypotheses.
   b. Describe the four possible outcomes that can result depending on the truth of the null hypothesis and the decision reached.
   c. Describe the seriousness of the two possible errors.

8.24 A supplier of highway construction materials claims he can supply an asphalt mixture that will make roads that are paved with his materials less slippery when wet. A general contractor who builds roads wishes to test the supplier's claim. The null hypothesis is "Roads paved with this asphalt mixture are no less slippery than roads paved with other asphalt." The alternative hypothesis is "Roads paved with this asphalt mixture are less slippery than roads paved with other asphalt."
   a. Describe the meaning of the two possible types of errors that can occur in the decision when this hypothesis test is completed.
   b. Describe how the null hypothesis, as stated previously, is a "starting point" for the decision to be made about the asphalt.

8.25 Describe the actions that would result in a type I error and a type II error if each of the following null hypotheses were tested. (Remember, the alternative hypothesis is the negation of the null hypothesis.)
   a. H_o: The majority of Americans favor laws against assault weapons.
   b. H_o: The choices on the fast-food menu are not low in salt.
   c. H_o: This building must not be demolished.
   d. H_o: There is no waste in government spending.

8.26 a. If α is assigned the value 0.001, what are we saying about the type I error?
   b. If α is assigned the value 0.05, what are we saying about the type I error?
   c. If α is assigned the value 0.10, what are we saying about the type I error?

8.27 a. If the null hypothesis is true, the probability of a decision error is identified by what name?
   b. If the null hypothesis is false, the probability of a decision error is identified by what name?

8.28 Suppose that a hypothesis test is to be carried out by using $\alpha = 0.05$. What is the probability of committing a type I error?

8.29 Explain why α is not always the probability of rejecting the null hypothesis.

8.30 Explain how assigning a small probability to an error controls the likelihood of its occurrence.

8.31 The conclusion is the part of the hypothesis test that communicates the findings of the test to the reader.

As such, it needs special attention so that the reader receives an accurate picture of the findings.
   a. Carefully describe the "attitude" of the statistician and the statement of the conclusion when the decision is "reject H_o."
   b. Carefully describe the "attitude" and the statement of the conclusion when the decision is "fail to reject H_o."

8.32 A normally distributed population is known to have a standard deviation of 5, but its mean is in question. It has been argued to be either $\mu = 80$ or $\mu = 90$, and the following hypothesis test has been devised to settle the argument. The null hypothesis, H_o: $\mu = 80$, will be tested using one randomly selected data value and comparing it with the critical value of 86. If the data value is greater than or equal to 86, the null hypothesis will be rejected.
   a. Find α, the probability of the type I error.
   b. Find β, the probability of the type II error.

Objective 8.4

8.33 State the null hypothesis H_o and the alternative hypothesis H_a that would be used for a hypothesis test related to each of the following statements:
   a. The mean age of the students enrolled in evening classes at a certain college is greater than 26 years.
   b. The mean weight of packages shipped on Air Express during the past month was less than 36.7 pounds.
   c. The mean life of fluorescent light bulbs is at least 1,600 hours.
   d. The mean strength of welds by a new process is different from 570 pounds per unit area, the mean strength of welds by the old process.

8.34 A manufacturer wishes to test the hypothesis that "by changing the formula of its toothpaste, it will give its users improved protection." The null hypothesis represents the idea that "the change will not improve the protection," and the alternative hypothesis is "the change will improve the protection." Describe the meaning of the two possible types of errors that can occur in the decision when the test of the hypothesis is conducted.

8.35 Assume that z is the test statistic and calculate the value of $z\star$ for each of the following:
   a. H_o: $\mu = 51$, $\sigma = 4.5$, $n = 40$, $\bar{x} = 49.6$
   b. H_o: $\mu = 20$, $\sigma = 4.3$, $n = 75$, $\bar{x} = 21.2$
   c. H_o: $\mu = 138.5$, $\sigma = 3.7$, $n = 14$, $\bar{x} = 142.93$
   d. H_o: $\mu = 815$, $\sigma = 43.3$, $n = 60$, $\bar{x} = 799.6$

8.36 Find the test statistic $z\star$ and the p-value for each of the following situations.
   a. H_o: $\mu = 22.5$, H_a: $\mu > 22.5$; $\bar{x} = 24.5$, $\sigma = 6$, $n = 36$
   b. H_o: $\mu = 200$, H_a: $\mu < 200$; $\bar{x} = 192.5$, $\sigma = 40$, $n = 50$
   c. H_o: $\mu = 12.4$, H_a: $\mu \neq 2.4$; $\bar{x} = 11.52$, $\sigma = 2.2$, $n = 16$

8.37 Calculate the p-value for each of the following:
   a. H_o: $\mu = 10$, H_a: $\mu > 10$, $z\star = 1.48$
   b. H_o: $\mu = 105$, H_a: $\mu < 105$, $z\star = -0.85$
   c. H_o: $\mu = 13.4$, H_a: $\mu \neq 13.4$, $z\star = 1.17$
   d. H_o: $\mu = 8.56$, H_a: $\mu < 8.56$, $z\star = -2.11$
   e. H_o: $\mu = 110$, H_a: $\mu \neq 110$, $z\star = -0.93$

8.38 Ponemon Institute, along with Intel, published "The Cost of a Lost Laptop" study in April 2009. With an increasingly mobile workforce carrying around more sensitive data on their laptops, the loss involves much more than the laptop itself. The average cost of a lost laptop based on cases from various industries is $49,246. This average includes laptop replacement, data breach cost, lost productivity cost, and other legal and forensic costs. A separate study of 30 cases from the health care industry produced a mean of $67,873. Given these figures, is there sufficient evidence to support the claim that health care laptop replacement costs are higher in general? Use a 0.001 level of significance.

SOURCE: http://communities.intel.com/

8.39 From candy to jewelry to flowers, the average consumer was expected to spend $123.89 for Mother's Day 2009, according to an April 2009 National Retail Federation survey. Local merchants felt this average was too high for their area. They contracted an agency to conduct a study. A random sample of 60 consumers was taken at a local shopping mall the Saturday before Mother's Day and produced a sample mean amount of $106.27. If $\sigma = 39.50, does the sample provide sufficient evidence to support the merchants' claim at the 0.05 level of significance?

*8.40 Nationally, the ratio of nurses to students falls short of the recommended federal standard according to the *USA Today* article "School Nurses in Short Supply" (August 11, 2009). The recommendation from the Centers for Disease Control and Prevention (CDC) is one nurse per 750 students. Use the sample below from 38 randomly selected New York State schools to test the statement, "The New York mean number of students per school nurse is significantly higher than the CDC standard of 750." Assume $\sigma = 540$.

1,062	1,070	353	675	1,557	1,374	459	302	1,946	487	295
1,047	1,751	784	480	377	883	1,035	332	330	989	1,098
1,241	778	1,691	963	1,645	1,594	2,125	338	1,380	885	707
1,267	1,412	1,037	1,603	915						

a. Describe the parameter of interest.
b. State the null and alternative hypotheses.
c. Calculate the value for $z\star$ and find the *p*-value.
d. State your decision and conclusion using $\alpha = 0.01$.

Objective 8.5

8.41 State the null hypothesis, H_o, and the alternative hypothesis, H_a, that would be used for a hypothesis test for each of the following statements:
a. The mean age of the youths who hang out at the mall is less than 16 years.
b. The mean height of professional basketball players is greater than 6'6".

c. The mean elevation drop for ski trails at eastern ski centers is at least 285 feet.
d. The mean diameter of the rivets is no more than 0.375 inch.
e. The mean cholesterol level of male college students is different from 200 mg/dL.

8.42 Determine the critical region and the critical values used to test the following null hypotheses:
a. $H_o: \mu = 55 (\geq), H_a: \mu < 55, \alpha = 0.02$
b. $H_o: \mu = -86 (\geq), H_a: \mu < -86, \alpha = 0.01$
c. $H_o: \mu = 107, H_a: \mu \neq 107, \alpha = 0.05$
d. $H_o: \mu = 17.4 (\leq), H_a: \mu > 17.4, \alpha = 0.10$

8.43 Women own an average of 15 pairs of shoes. This is based on a survey of female adults by Kelton Research for New York City-based Enslow, The Foot Comfort Center.

Suppose a random sample of 35 newly hired female college graduates was taken, and the sample mean was 18.37 pairs of shoes. If $\sigma = 6.12$, does this sample provide sufficient evidence that the mean number of shoes for young female college graduates is greater than the overall mean number for all female adults? Use a 0.10 level of significance.

8.44 According to the Center on Budget and Policy Priorities article "Curbing Flexible Spending Accounts Could Help Pay for Health Care Reform" (revised June 10, 2009), flexible spending accounts encourage the over-consumption of health care; people buy things they do not need, otherwise they lose the money. In 2007, for those that did not use all of their account (about one out of every seven), the average amount lost was $723.

SOURCE: http://www.cbpp.org/

Suppose a random sample of 150 employees who did not use all of their funds in 2009 is taken and an average amount of $683 was lost. Test the hypothesis that there is no significant difference in the average amount forfeited. Assume that $\sigma = 307 per year. Use $\alpha = 0.05$.
a. Define the parameter.
b. State the null and alternative hypotheses.
c. Specify the hypothesis test criteria.
d. Present the sample evidence.
e. Find the probability distribution information.
f. Determine the results.

*remember

Problems marked with an asterisk (*) have data sets available on the CourseMate for STAT2 site. Login at cengagebrain.com.

Inferences
Involving One Population

From the time you get out of bed until you walk out the door, how long does it take you to get ready in the morning? Some will tell you as little as 5 minutes, but when timed, it is tough to shower, fully groom, eat, and dress in less than 15 minutes. If you were given the job of estimating "floor-to-door" time for the typical college female, what information would you need? How would you use it to determine the estimate? Examine the timeline below, and see if you think it would provide sufficient information.

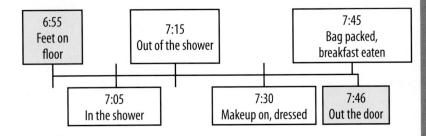

| 6:55 Feet on floor | 7:15 Out of the shower | 7:45 Bag packed, breakfast eaten |
| 7:05 In the shower | 7:30 Makeup on, dressed | 7:46 Out the door |

objectives

9.1 Inferences about the Mean μ (σ Unknown)

9.2 Inferences about the Binomial Probability of Success

9.3 Inferences about the Variance and Standard Deviation

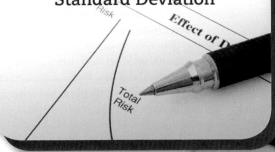

9.1 Inferences about the Mean μ (σ Unknown)

INFERENCES ABOUT THE POPULATION MEAN μ ARE BASED ON THE SAMPLE MEAN $\bar{x}$ AND INFORMATION OBTAINED FROM THE SAMPLING DISTRIBUTION OF SAMPLE MEANS.

Recall that the sampling distribution of sample means has a mean μ and a **standard error** of $\sigma/\sqrt{n}$ for all samples of size n, and it is normally distributed when the sampled population has a normal distribution or approximately normally distributed when the **sample size** is sufficiently large. This means that the test statistic $z\bigstar = \frac{\bar{x} - \mu}{\sigma/\sqrt{n}}$ has a standard normal distribution. However, when σ **is unknown,** the standard error $\sigma/\sqrt{n}$ is also unknown. Therefore, the sample standard deviation s will be used as the point estimate for σ. As a result, an estimated standard error of the mean, $s/\sqrt{n}$, will be used and our test statistic will become $\dfrac{\bar{x} - \mu}{s/\sqrt{n}}$.

When a **known** σ is being used to make an inference about the mean μ, a sample provides one value for use in the formulas; that one value is $\bar{x}$. When the sample standard deviation s is also used, the sample provides two values: the sample mean $\bar{x}$ and the estimated standard error $s/\sqrt{n}$. As a result, the z-statistic will be replaced with a statistic that accounts for the use of an estimated standard error. This new statistic is known as the **Student's t-statistic.**

In 1908 W. S. Gosset, an Irish brewery employee, published a paper about this t-distribution under the pseudonym "Student." In deriving the t-distribution, Gosset assumed that the samples were taken from normal

z OR t

Figure 9.1 Do I Use the z-Statistic or the t-Statistic?

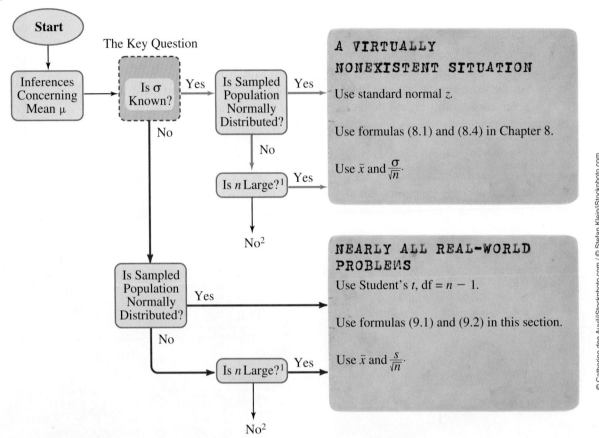

1. Is *n* large? Samples as small as *n* = 15 or 20 may be considered large enough for the central limit theorem to hold if the sample data are unimodal, nearly symmetrical, short-tailed, and without outliers. Samples that are not symmetrical require larger sample sizes, with 50 sufficing except for extremely skewed samples. See the discussion in Chapter 8 (objective 8.2).

2. Requires the use of a nonparametric technique; see Chapter 14.

populations. Although this might seem to be restrictive, satisfactory results are obtained when large samples are selected from many nonnormal populations.

Figure 9.1 presents a diagrammatic organization for the inferences about the population mean as discussed in Chapter 8 and in this first section of Chapter 9. Two situations exist: σ is known, or σ is unknown. As stated before, σ is almost never a known quantity in real-world problems; therefore, the standard error will almost always be estimated by $s/\sqrt{n}$. The use of an estimated standard error of the mean requires the use of the *t*-distribution. Almost all real-world inferences about the population mean will be made with the Student's *t*-statistic.

Figure 9.2 shows some *t*-distributions. The box on the following page explains in detail the specific **properties of the *t*-distribution**.

The number of degrees of freedom associated with s^2 is the divisor $(n-1)$ used to calculate the sample

Figure 9.2 Student's *t*-Distributions

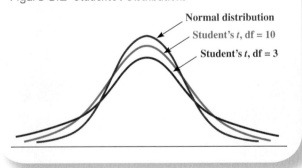

variance s^2 [formula (2.5), page 40]; that is, df $= n - 1$. The sample variance is the mean of the squared deviations. The number of degrees of freedom is the "number of unrelated deviations" available for use in estimating σ^2. Recall that the sum of the deviations, $\Sigma(x - \bar{x})$, must be zero. From a sample of size *n*, only the first

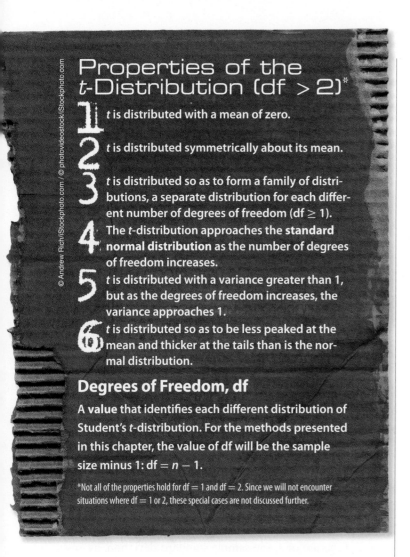

Properties of the t-Distribution (df > 2)*

1 *t* is distributed with a mean of zero.

2 *t* is distributed symmetrically about its mean.

3 *t* is distributed so as to form a family of distributions, a separate distribution for each different number of degrees of freedom (df ≥ 1).

4 The t-distribution approaches the **standard normal distribution** as the number of degrees of freedom increases.

5 *t* is distributed with a variance greater than 1, but as the degrees of freedom increases, the variance approaches 1.

6 *t* is distributed so as to be less peaked at the mean and thicker at the tails than is the normal distribution.

Degrees of Freedom, df

A **value** that identifies each different distribution of Student's *t*-distribution. For the methods presented in this chapter, the value of df will be the sample size minus 1: df = $n - 1$.

*Not all of the properties hold for df = 1 and df = 2. Since we will not encounter situations where df = 1 or 2, these special cases are not discussed further.

$n - 1$ of these deviations has freedom of value. That is, the last, or *n*th, value of $(x - \overline{x})$ must make the sum of the *n* deviations total exactly zero. As a result, variance is said to average $n - 1$ unrelated squared deviation values, and this number, $n - 1$, was named "degrees of freedom."

Although there is a separate *t*-distribution for each degree of freedom, df = 1, df = 2, . . . , df = 20, . . . , df = 40, and so on, only certain key **critical values of t** will be necessary for our work. Consequently, the table for the Student's *t*-distribution (Table 6 in Appendix B) is a table of critical values rather than a complete table, such as Table 3 is for the standard normal distribution for *z*. As you look at Table 6, you will note that the left side of the table is identified by "df," degrees of freedom. This left-hand column starts at 3 at the top and lists consecutive df values to 30, then jumps to 35, . . . , to "df > 100" at the bottom. As we stated, as the degrees of freedom increase, the *t*-distribution approaches the characteristics of the standard normal *z*-distribution. Once df is "greater than 100," the critical values of the

t-distribution are the same as the corresponding critical values of the standard normal distribution as given in Table 4A in Appendix B.

Using the t-Distribution Table (Table 6, Appendix B)

The critical values of the Student's *t*-distribution that are to be used both for constructing a confidence interval and for hypothesis testing will be obtained from Table 6 in Appendix B. To find the value of *t*, you will need to know two identifying values: (1) df, the number of degrees of freedom (identifying the distribution of interest), and (2) α, the area under the curve to the right of the right-hand critical value. A notation much like that used with *z* will be used to identify a critical value. $t(\text{df}, \alpha)$, read as "*t* of df, α," is the symbol for the value of *t* with df degrees of freedom and an area of α in the right-hand tail, as shown in Figure 9.3.

Figure 9.3 *t*-Distribution Showing $t(\text{df}, \alpha)$

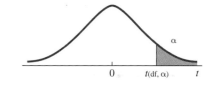

FINDING *t* IN RELATION TO THE MEAN

There are three relationships of *t* to the mean: *t* can be on the right side of the mean, on the left side, or have values that bound a certain percentage. Let's start by finding the value of *t* on the right side of the mean, specifically finding the value of $t(10, 0.05)$ (see the diagram below).

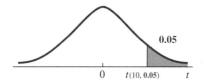

There are 10 degrees of freedom, and 0.05 is the area to the right of the critical value. In Table 6 of Appendix B, we look for the row df = 10 and the column marked "Area in One Tail," $\alpha = 0.05$. At their intersection, we see that $t(10, 0.05) = \textbf{1.81.}$

Portion of Table 6			
	Area in One Tail		
df	...	0.05	...
:			
10		**1.81**	

$t(10, 0.05) = 1.81$

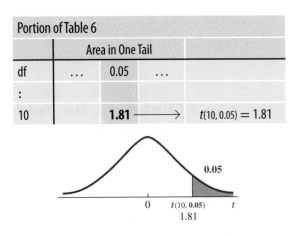

For the values of t on the left side of the mean, we can use one of two notations. The t-value shown in Figure 9.4 could be named $t(df, 0.95)$, since the area to the right of it is 0.95, or it could be identified by $-t(df, 0.05)$, since the t-distribution is symmetric about its mean, zero.

Figure 9.4 t-Value on the Left Side

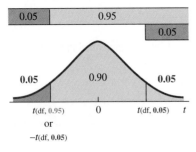

Let's find the value of $t(15, 0.95)$. There are 15 degrees of freedom. In Table 6 we look for the column marked $\alpha = 0.05$ (one tail) and its intersection with the row df = 15. The table gives us $t(15, 0.05) = 1.75$; therefore, $t(15, 0.95) = -t(15, 0.05) = -1.75$. The value is negative because it is to the left of the mean, zero; see the figure below.

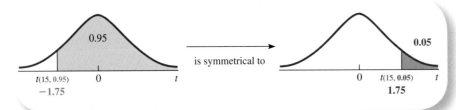

Let's look at another example: finding t-values that bound a middle percentage. We can find the values of the t-distribution that bound the middle 0.90 of the area under the curve for the distribution with df = 17.

The middle 0.90 leaves 0.05 for the area of each tail. The value of t that bounds the right-hand tail is $t(17, 0.05) = \mathbf{1.74}$, as found in Table 6. The value

that bounds the left-hand tail is $-\mathbf{1.74}$ because the t-distribution is symmetric about its mean, zero.

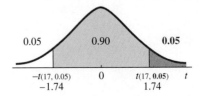

If the df needed is not listed in the left-hand column of Table 6, then use the next smaller value of df that is listed. For example, $t(72, 0.05)$ is estimated using $t(70, 0.05) = 1.67$.

Confidence Interval Procedure

We are now ready to make inferences about the population mean μ using the sample standard deviation. As we mentioned earlier, use of the t-distribution has a condition:

> #### The assumption for inferences about the mean μ when σ is unknown:
> **The sampled population is normally distributed.**

The procedure to make confidence intervals using the sample standard deviation is very similar to that used when σ is known (see pages 154–158). The difference is the use of the Student's t in place of the standard normal z and the use of s, the sample standard deviation, as an estimate of σ. The central limit theorem implies that this technique can also be applied to nonnormal populations when the sample size is sufficiently large.

Confidence Interval for Mean:

$$\bar{x} - t(df, \alpha/2)\left(\frac{s}{\sqrt{n}}\right) \quad \text{to} \quad \bar{x} + t(df, \alpha/2)\left(\frac{s}{\sqrt{n}}\right)$$
$$\text{with df} = n - 1 \tag{9.1}$$

CONFIDENCE INTERVAL FOR μ WITH σ UNKNOWN

To illustrate how confidence intervals can be formed utilizing the t-distribution, consider a random sample of 20 weights taken from 1-year-olds born at Northside Hospital last year. A mean of 20.73 pounds and a standard deviation of 2.17 pounds were found for the sample. Based on past information, it is assumed that weights of 1-year-olds are normally distributed. Using the 5-step process, we can estimate with 95% confidence the mean weight of all 1-year-olds born in this hospital.

19.72-21.74

STEP 1 THE SET-UP:

Describe the population parameter of interest.

μ, the mean weight of 1-year-olds born at Northside Hospital last year.

STEP 2 THE CONFIDENCE INTERVAL CRITERIA:

a. Check the assumptions.

σ is unknown, and past information indicates that the sampled population is normal.

b. Identify the probability distribution and the formula to be used.

The Student's t-distribution will be used with formula (9.1).

c. State the level of confidence.

$1 - \alpha = 0.95$.

STEP 3 THE SAMPLE EVIDENCE:

Collect the sample information.

$n = 20, \bar{x} = 20.73$, and $s = 2.17$.

STEP 4 THE CONFIDENCE INTERVAL:

a. Determine the confidence coefficients.

Since $1 - \alpha = 0.95$, $\alpha = 0.05$; therefore, $\alpha/2 = 0.025$. Also, since $n = 20$, df $= 19$. At the intersection of row df $= 19$ and one-tailed column $\alpha = 0.025$ in Table 6, we find $t(df, \alpha/2) = t(19, 0.025) = 2.09$. See the figure below.

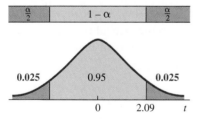

Information about the confidence coefficient and using Table 6 is on pages 185–186.

b. Find the maximum error of estimate.

$E = t(df, \alpha/2)\left(\dfrac{s}{\sqrt{n}}\right)$:

$E = t(19, 0.025)\left(\dfrac{s}{\sqrt{n}}\right)$

$= 2.09\left(\dfrac{2.17}{\sqrt{20}}\right) = (2.09)(0.485) = 1.01$

c. Find the lower and upper confidence limits.

$$\bar{x} - E \quad \text{to} \quad \bar{x} + E$$
$$20.73 - 1.01 \quad \text{to} \quad 20.73 + 1.01$$
$$19.72 \quad \text{to} \quad 21.74$$

STEP 5 THE RESULTS:

State the confidence interval.

19.72 to 21.74 is the 95% confidence interval for μ. That is, with 95% confidence we estimate the mean weight to be between 19.72 to 21.74 pounds.

Hypothesis-Testing Procedure

The t-statistic is used to complete a hypothesis test about the population mean μ in much the same manner z was used in Chapter 8. In hypothesis-testing situations, we use formula (9.2) to calculate the value of the **test statistic $t\bigstar$**:

Test Statistic for Mean:

$$t\bigstar = \frac{\bar{x} - \mu}{s/\sqrt{n}} \quad \text{with df} = n - 1 \qquad (9.2)$$

The **calculated t** is the number of estimated standard errors that $\bar{x}$ is from the hypothesized mean μ. As with confidence intervals, the central limit theorem indicates that the t-distribution can also be applied to nonnormal populations when the **sample size** is sufficiently large.

ONE-TAILED HYPOTHESIS TEST FOR μ WITH σ UNKNOWN

To conduct a one-tailed hypothesis test for μ with σ unknown, let's return to the hypothesis of the example from Chapter 8 (page 168) where the EPA wanted to show that the mean carbon monoxide level is higher than 4.9 parts per million. Does a random sample of 22 readings (sample results: $\bar{x} = 5.1$ and $s = 1.17$) present sufficient evidence to support the EPA's claim? Use $\alpha = 0.05$. Previous studies have indicated that such readings have an approximately normal distribution. Again we'll follow the five-step procedure:

STEP 1 THE SET-UP:

a. **Describe the population parameter of interest.**
μ, the mean carbon monoxide level of air in downtown Rochester.

b. **State the null hypothesis (H_o) and the alternative hypothesis (H_a).**

$$H_o: \mu = 4.9 \ (\leq) \text{ (no higher than)}$$

$$H_a: \mu > 4.9 \text{ (higher than)}$$

STEP 2 THE HYPOTHESIS TEST CRITERIA:

a. **Check the assumptions.**
The assumptions are satisfied because the sampled population is approximately normal and the sample size is large enough for the CLT to apply (see pages 183–185).

b. **Identify the probability distribution and the test statistic to be used.**
σ is unknown; therefore, we use the t-distribution with df $= n - 1 = 21$, and the test statistic is $t\star$, formula (9.2).

c. **Determine the level of significance.**
$\alpha = 0.05$.

STEP 3 THE SAMPLE EVIDENCE:

a. **Collect the sample information.**
$n = 22, \bar{x} = 5.1$, and $s = 1.17$.

b. **Calculate the value of the test statistic.**
Use formula (9.2):

$$t\star = \frac{\bar{x} - \mu}{s/\sqrt{n}}: \quad t\star = \frac{5.1 - 4.9}{1.17/\sqrt{22}} = \frac{0.20}{0.2494} = 0.8018 = \mathbf{0.80}$$

STEP 4 THE PROBABILITY DISTRIBUTION:

As always, we can use either the p-value procedure or the classical procedure.

Using the p-value procedure:

a. **Calculate the p-value for the test statistic.**
Use the right-hand tail because H_a expresses concern for values related to "higher than."
$P = P(t\star > 0.80$, with df $= 21)$ as shown on the figure.

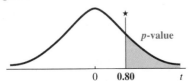

To find the p-value, use one of three methods:
1. Use Table 6 in Appendix B to place bounds on the p-value: $0.10 < P < 0.25$.
2. Use Table 7 in Appendix B to read the value directly: $P = 0.216$.
3. Use a computer or calculator to calculate the p-value: $P = 0.2163$.

b. **Determine whether or not the p-value is smaller than α.**
The p-value is not smaller than α, the level of significance.

Using the classical procedure:

a. **Determine the critical region and critical value(s).**
The critical region is the right-hand tail because H_a expresses concern for values related to "higher than." The critical value is found at the intersection of the df $= 21$ row and the one-tailed 0.05 column of Table 6: $t(21, 0.05) = 1.72$.

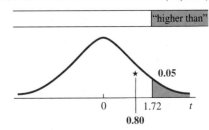

b. **Determine whether or not the calculated test statistic is in the critical region.**
$t\star$ is not in the critical region, as shown in red in the figure above.

STEP 5 THE RESULTS:

a. **State the decision about H_o.**
Fail to reject H_o.

b. **State the conclusion about H_a.**
At the 0.05 level of significance, the EPA does not have sufficient evidence to show that the mean carbon monoxide level is higher than 4.9.

Calculating the *p*-value when using the *t*-distribution

Method 1: Use Table 6 in Appendix B to place bounds on the p-value. By inspecting the df = 21 row of Table 6, you can determine an interval within which the *p*-value lies. Locate *t*★ along the row labeled df = 21. If the *t*★ value is not listed, locate the two table values it falls between and read the bounds for the *p*-value from the top of the table. In this case, *t*★ = 0.80 is between 0.686 and 1.32; therefore, **P** is between 0.10 and 0.25. Use the one-tailed heading, since H_a is one-tailed in this illustration. (Use the two-tailed heading when H_a is two-tailed.)

The 0.686 entry in the table tells us that $P(t > 0.686) = 0.25$, as shown in purple on the figure at the bottom of the page. The 1.32 entry in the table tells us that $P(t > 1.32) = 0.10$, as shown in green. You can see that the *p*-value **P** (shown in blue) is between 0.10 and 0.25. Therefore, $0.10 < P < 0.25$, and we say that 0.10 and 0.25 are the "bounds" for the *p*-value.

Method 2: Use Table 7 in Appendix B to read the p-value or to "place bounds" on the p-value. Table 7 is designed to yield *p*-values given the *t*★ and df values or to produce bounds on **P** that are narrower than those produced by Table 6.

In the preceding example, *t*★ = 0.80 and df = 21. These happen to be row and column headings, so the *p*-value can be read directly from the table. Locate the *p*-value at the intersection of the *t*★ = 0.80 row and the df = 21 column. The *p*-value for *t*★ = 0.80 with df = 21 is **0.216**.

Portion of Table 7				
t★	df	...	21	
⋮				
0.80			**0.216** ⟶ **P** = $P(t★ > 0.80$, with df = 21)$ = 0.216	

To illustrate how to place bounds on the *p*-value when *t*★ and df are not the heading values, let's consider the situation where *t*★ = 2.43 with df = 16. The *t*★ = 2.43 is between rows *t* = 2.4 and *t* = 2.5, while df = 16 is between columns df = 15 and df = 18. These two rows and two columns intersect a total of four times, namely at 0.015 and 0.014 in the row *t*★ = 2.4 and at 0.012 and 0.011 in the row *t*★ = 2.5. The *p*-value we are looking for is bounded by the smallest and largest of these four values, namely, 0.011 (lower right) and 0.015 (upper left). Therefore, the bounds for the *p*-value are $0.011 < P < 0.015$.

Portion of Table 7					
t★	df ...	15	**16**	18	
⋮					
2.4		0.015		0.014	**P** = $P(t★ > 2.43$, with df = 16)
2.43			**P**	⟶	$0.011 < P < 0.015$
2.5		0.012		0.011	

Method 3: If you are doing the hypothesis test with the aid of a computer or calculator, most likely it will calculate the *p*-value for you.

TWO-TAILED HYPOTHESIS TEST FOR μ WITH σ UNKNOWN

Let's look at a two-tailed hypothesis-testing situation, which we can also do for μ with σ unknown. This time, we'll examine data from a popular self-image test that results in normally distributed scores. The mean score for public-assistance recipients is expected to be 65. A random sample of 28 public-assistance recipients in Emerson County is given the test. They achieve a mean score of 62.1, and their scores have a standard deviation of 5.83. Do the Emerson County public-assistance recipients test differently, on the average, than what is expected, at the 0.02 level of significance? To find out, we again turn to our five-step procedure.

Finding **P** = $P(t★ > 0.80$, with df = 21)

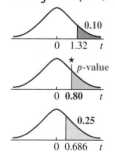

Portion of Table 6			
	Area in One Tail		
df	0.25	P	0.10
⋮	⋮	↑	⋮
21	0.686	0.80	1.32

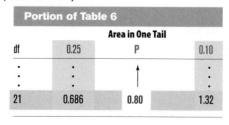

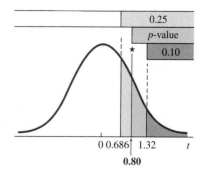

STEP 1 — THE SET-UP:

a. Describe the population parameter of interest.

μ, the mean self-image test score for all Emerson County public-assistance recipients.

b. State the null hypothesis (H_o) and the alternative hypothesis (H_a).

$H_o: \mu = 65$ (mean is 65)

$H_a: \mu \neq 65$ (mean is different from 65)

STEP 2 — THE HYPOTHESIS TEST CRITERIA:

a. Check the assumptions.

The test is expected to produce normally distributed scores; therefore, the assumption has been satisfied; σ is unknown.

b. Identify the probability distribution and the test statistic to be used.

The t-distribution with df $= n - 1 = 27$, and the test statistic is $t\star$, formula (9.2).

c. Determine the level of significance.

$\alpha = 0.02$ (given in statement of problem).

STEP 3 — THE SAMPLE EVIDENCE:

a. Collect the sample information.

$n = 28, \bar{x} = 62.1$, and $s = 5.83$.

b. Calculate the value of the test statistic.

Use formula (9.2):

$$t\star = \frac{\bar{x} - \mu}{s/\sqrt{n}}:$$

$$t\star = \frac{62.1 - 65.0}{5.83/\sqrt{28}}$$

$$= \frac{-2.9}{1.1018}$$

$$= -2.632 = \mathbf{-2.63}$$

STEP 4 — THE PROBABILITY DISTRIBUTION:

Again, we can choose either the p-value or classical procedure.

Using the *p*-value procedure:

a. Calculate the p-value for the test statistic.

Use both tails because H_a expresses concern for values related to "different from."

$P = P(t < -2.63) + P(t > 2.63)$

$= 2 \cdot P(t > 2.63)$, with df $= 27$ as shown in the figure.

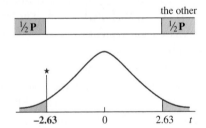

To find the p-value, use one of three methods:

1. Use Table 6 in Appendix B to place bounds on the p-value: $0.01 < P < 0.02$.
2. Use Table 7 in Appendix B to place bounds on the p-value: $0.012 < P < 0.016$.
3. Use a computer or calculator to calculate the p-value: $P = 0.0140$.

Specific details follow this example.

b. Determine whether or not the p-value is smaller than α.

The p-value is smaller than the level of significance, α.

Using the classical procedure:

a. Determine the critical region and critical value(s).

The critical region is both tails because H_a expresses concern for values related to "different from." The critical value is found at the intersection of the df $= 27$ row and the one-tailed 0.01 column of Table 6: $t(27, 0.01) = 2.47$.

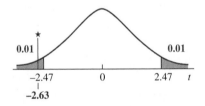

b. Determine whether or not the calculated test statistic is in the critical region.

$t\star$ is in the critical region, as shown in red in the preceding figure.

THE RESULTS:

a. State the decision about H_o.
   Reject H_o.

b. State the conclusion about H_a.
   At the 0.02 level of significance, we do have sufficient evidence to conclude that the Emerson County
   assistance recipients' test results are significantly
   different, on the average, from the expected 65.

Calculating the p-value when using the t-distribution

Method 1: Using Table 6, find 2.63 between two entries in the df = 27 row and read the bounds for **P** from
the two-tailed heading at the top of the table:

$$0.01 < \textbf{P} < 0.02$$

Method 2: Generally, bounds found using Table 7 will
be narrower than bounds found using Table 6. The table
below shows you how to read the bounds from Table 7;
find $t\star = 2.63$ between two rows and df = 27 between
two columns, and locate the four intersections of these
columns and rows. The value of $\frac{1}{2}\textbf{P}$ is bounded by the
upper left and the lower right of these table entries.

Portion of Table 7				
	Degrees of Freedom			
$t\star$	25	**27**	29	$P = 2P(t\star > 2.63,$ with df = 27$)$
$\vdots$		$\vdots$		
2.6	0.008		0.007	
2.63	$\longrightarrow$	$\frac{1}{2}\textbf{P}$	$\longrightarrow$	$0.006 < \frac{1}{2}\textbf{P} < 0.008$
2.7	0.006		0.006	$0.012 < \textbf{P} < 0.016$

Method 3: If you are doing the hypothesis test with the
aid of a computer or calculator, most likely it will calculate the p-value for you (do not double it).

9.2 Inferences about the Binomial Probability of Success

PERHAPS THE MOST COMMON INFERENCE
INVOLVES THE **BINOMIAL PARAMETER p,**
THE "PROBABILITY OF SUCCESS."

Yes, every one of us uses this inference, even if only
casually. In thousands of situations we are concerned
about something either "happening" or "not happening." There are only two possible outcomes of concern,
and that is the fundamental property of a **binomial
experiment.** The other necessary ingredient is multiple
independent trials. Asking five people whether they are
"for" or "against" some issue can create five independent trials; if 200 people are asked the same question,
200 independent trials may be involved; if 30 items are
inspected to see if each "exhibits a particular property"
or "not," there will be 30 repeated trials; these are the
makings of a binomial inference.

The binomial parameter p is defined to be the
probability of success on a single trial in a binomial
experiment.

Sample Binomial Probability

$$p' = \frac{x}{n} \tag{9.3}$$

where the **random variable x** represents the number of
successes that occur in a sample consisting of n trials.

© asiseeit/iStockphoto.com / © Image Source/Getty Images

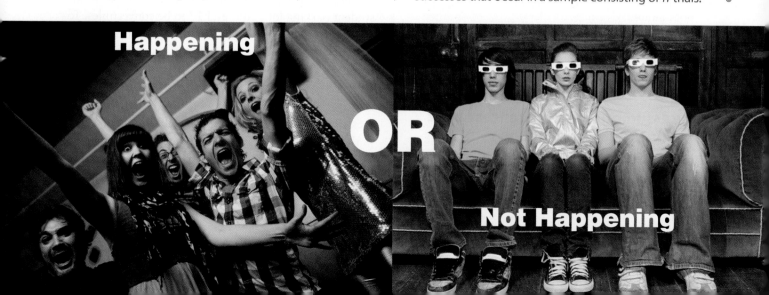

Happening OR Not Happening

Recall that the mean and standard deviation of the binomial random variable x are found by using formula (5.7), $\mu = np$, and formula (5.8), $\sigma = \sqrt{npq}$, where $q = 1 - p$. The distribution of x is considered to be approximately normal if n is greater than 20 and if np and nq are both greater than 5. This commonly accepted *rule of thumb* allows us to use the **standard normal distribution** to estimate probabilities for the binomial random variable x, the number of successes in n trials, and to make inferences concerning the binomial parameter p, the probability of success on an individual trial.

Generally, it is easier and more meaningful to work with the distribution of p' (the observed probability of occurrence) than with x (the number of occurrences). Consequently, we will convert formulas (5.7) and (5.8) from units of x (integers) to units of **proportions** (percentages expressed as decimals) by dividing each formula by n, as shown in Figure 9.5.

Recall that $\mu_{p'} = p$ and that the *sample statistic p'* is an **unbiased estimator for p**. Therefore, the information about the sampling distribution of p' is summarized as follows:

> **I**f a random sample of size n is selected from a large population with $p = P(\text{success})$, then the sampling distribution of p' has:
>
> 1. A mean $\mu_{p'}$ equal to p (9.4)
> 2. A standard error $\sigma_{p'}$ equal to $\sqrt{\dfrac{pq}{n}}$ (9.5)
> 3. An approximately normal distribution if n is sufficiently large

In practice, using these guidelines will ensure normality:

1. The sample size is greater than 20.
2. The products np and nq are both greater than 5.
3. The sample consists of less than 10% of the population.

We are now ready to make inferences about the population parameter p. Use of the z-distribution involves an assumption:

The assumption for inferences about the binomial parameter p:

The n random observations that form the sample are selected independently from a population that does not change during the sampling.

Confidence Interval Procedure

Inferences concerning the population binomial parameter p, $P(\text{success})$, are made using procedures that closely parallel the inference procedures used for the population mean μ. When we estimate the **population proportion p**, we will base our estimations on the **point estimate p'**. The point estimate, p', becomes the center of the confidence interval, and the maximum error of estimate is a multiple of the **standard error**. The **level of confidence** determines the confidence coefficient, the number of multiples of the standard error.

Confidence Interval for a Proportion

$$p' - z(\alpha/2)\left(\sqrt{\frac{p'q'}{n}}\right)$$

to

$$p' + z(\alpha/2)\left(\sqrt{\frac{p'q'}{n}}\right) \tag{9.6}$$

where $p' = \frac{x}{n}$ and $q' = 1 - p'$.

Notice that the standard error, $\sqrt{\frac{pq}{n}}$, has been replaced by $\sqrt{\frac{p'q'}{n}}$. Since we are estimating p, we do not know its value and therefore we must use the best replacement available. That replacement is p', the observed value or the point estimate for p. This replacement will cause little change in the standard error or the width of our confidence interval provided n is sufficiently large.

CONFIDENCE INTERVAL FOR p

→ We can illustrate the formation of a confidence interval for the binomial parameter p with the following example. In a discussion about the cars that fellow students drive, several statements were made about types, ages, makes, colors, and so on. Dana decided he wanted to estimate the proportion of convertibles students drive, so he randomly identified 200 cars in the student parking lot and found 17 to be convertibles. To find the 90% confidence interval for the proportion of convertibles driven by students, we once again follow the five-step process.

STEP 1 THE SET-UP:
Describe the population parameter of interest.
p, the proportion (percentage) of convertibles driven by students.

STEP 2 THE CONFIDENCE INTERVAL CRITERIA:
a. Check the assumptions.

Figure 9.5 Formulas (9.4) and (9.5)

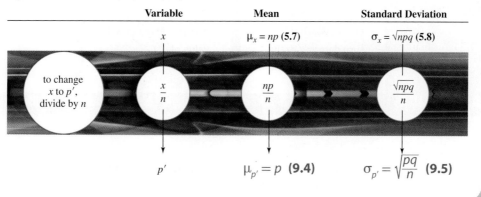

Variable	Mean	Standard Deviation

x $\mu_x = np$ **(5.7)** $\sigma_x = \sqrt{npq}$ **(5.8)**

to change x to p', divide by n $\dfrac{x}{n}$ $\dfrac{np}{n}$ $\dfrac{\sqrt{npq}}{n}$

p' $\mu_{p'} = p$ **(9.4)** $\sigma_{p'} = \sqrt{\dfrac{pq}{n}}$ **(9.5)**

The sample was randomly selected, and each student's response is independent of those of the others surveyed.

b. Identify the probability distribution and the formula to be used.

The standard normal distribution will be used with formula (9.6) as the test statistic. p' is expected to be approximately normal because: (1) $n = 200$ is greater than 20, and (2) both np [approximated by $np' = 200(17/200) = 17$] and nq [approximated by $nq' = 200(183/200) = 183$] are greater than 5.

c. State the level of confidence.

$1 - \alpha = 0.90$.

STEP 3 THE SAMPLE EVIDENCE:

Collect the sample information.
$n = 200$ cars were identified, and $x = 17$ were convertibles:

$$p' = \frac{x}{n} = \frac{17}{200} = 0.085$$

STEP 4 THE CONFIDENCE INTERVAL:

a. Determine the confidence coefficient.

This is the z-score [$z(\alpha/2)$, "z of one-half of alpha"] identifying the number of standard errors needed to attain the level of confidence and is found using Table 4 in Appendix B; $z(\alpha/2) = z(0.05) = 1.65$ (see the diagram below).

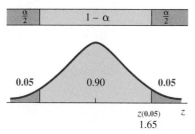

b. Find the maximum error of estimate.

Use the maximum error part of formula (9.6):

$$E = z(\alpha/2)\left(\sqrt{\frac{p'q'}{n}}\right) = 1.65\left(\sqrt{\frac{(0.085)(0.915)}{200}}\right)$$

$$= (1.65)\sqrt{0.000389} = (1.65)(0.020) = \mathbf{0.033}$$

c. Find the lower and upper confidence limits.

$$p' - E \text{ to } p' + E$$

$$0.085 - 0.033 \text{ to } 0.085 + 0.033$$

$$0.052 \text{ to } 0.118$$

STEP 5 THE RESULTS:

State the confidence interval.
0.052 to 0.118 is the 90% confidence interval for $p = P(\text{drives convertible})$.

That is, the true proportion of students who drive convertibles is between 0.052 and 0.118, with 90% confidence.

Determining Sample Size

By using the maximum error part of the confidence interval formula, it is possible to determine the **size of the sample** that must be taken in order to estimate p with a desired accuracy. Here is the formula for the **maximum error of estimate for a proportion:**

$$E = z(\alpha/2)\left(\sqrt{\frac{pq}{n}}\right) \tag{9.7}$$

To determine the sample size from this formula, we must decide on the quality we want for our final confidence interval. This quality is measured in two ways: the level of confidence and the preciseness (narrowness) of the interval. The level of confidence we establish will

| 3 | the level of confidence $(1 - \alpha$, which in turn determines the confidence coefficient) | the provisional value of p (p^* determines the value of q^*) | the maximum error, E |

components determine the sample size:

in turn determine the confidence coefficient, $z(\alpha/2)$. The desired preciseness will determine the maximum error of estimate, E. (Remember that we are estimating p, the binomial probability; therefore, E will typically be expressed in hundredths.)

For ease of use, we can solve formula (9.7) for n as follows:

Sample Size for $1 - \alpha$ Confidence Interval for p

$$n = \frac{[z(\alpha/2)]^2 \cdot p^* \cdot q^*}{E^2} \tag{9.8}$$

where p^* and q^* are provisional values of p and q used for planning.

An increase or decrease in one of the three components shown in the film strip above affects the sample size. If the level of confidence is increased or decreased (while the other components are held constant), then the sample size will increase or decrease, respectively. If the product of p^* and q^* is increased or decreased (with other components held constant), then the sample size will increase or decrease, respectively. (The product $p^* \cdot q^*$ is largest when $p^* = 0.5$ and decreases as the value of p^* moves farther from 0.5.) An increase or decrease in the desired maximum error will have the opposite effect on the sample size, since E appears in the denominator of the formula. If no provisional values for p and q are available, then use $p^* = 0.5$ and $q^* = 0.5$. Using $p^* = 0.5$ is safe because it gives the largest sample size of any possible value of p. Using $p^* = 0.5$ works reasonably well when the true value is "near 0.5" (say, between 0.3 and 0.7); however, as p gets nearer to either zero or one, a sizable overestimate in sample size will occur.

SAMPLE SIZE FOR ESTIMATING p (NO PRIOR INFORMATION)

Using confidence intervals, we can determine the sample size required for estimating p with no prior information.

For example, to find the sample size required to estimate the true proportion of blue-eyed community college students if you want your estimate to be within 0.02 with 90% confidence, we would do the following:

STEP 1 The level of confidence is $1 - \alpha = 0.90$; therefore, the confidence coefficient is $z(\alpha/2) = z(0.05) = 1.65$ from Table 4 in Appendix B; see the diagram below.

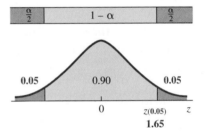

STEP 2 The desired maximum error is $E = 0.02$.

STEP 3 Since no estimate was given for p, use $p^* = 0.5$ and $q^* = 1 - p^* = 0.5$.

STEP 4 Use formula (9.8) to find n:

$$n = \frac{[z(\alpha/2)]^2 \cdot p^* \cdot q^*}{E^2};$$

$$n = \frac{(1.65)^2 \cdot 0.5 \cdot 0.5}{(0.02)^2}$$

$$= \frac{0.680625}{0.0004} = 1{,}701.56 = \mathbf{1{,}702}$$

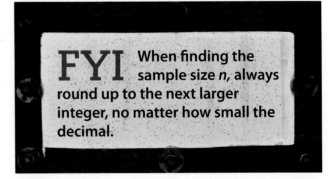

FYI When finding the sample size n, always round up to the next larger integer, no matter how small the decimal.

Sample Size for Estimating p (Prior Information)

 We can also determine the sample size for estimating p when we do have prior information. Consider an automobile manufacturer that purchases bolts from a supplier who claims the bolts are approximately 5% defective. To determine the sample size required to estimate the true proportion of defective bolts if we want our estimate to be within ± 0.02 with 90% confidence, we would do the following:

STEP 1 The level of confidence is $1 - \alpha = 0.90$; the confidence coefficient is $z(\alpha/2) = z(0.05) = 1.65$.

STEP 2 The desired maximum error is $E = 0.02$.

STEP 3 Since there is an estimate for p (supplier's claim is "5% defective"), use $p^* = 0.05$ and $q^* = 1 - p^* = 0.95$.

STEP 4 Use formula (9.8) to find n:

$$n = \frac{[z(\alpha/2)]^2 \cdot p^* \cdot q^*}{E^2}:$$

$$n = \frac{(1.65)^2 \cdot 0.05 \cdot 0.95}{(0.02)^2}$$

$$= 0.12931875/0.0004 = 323.3 = \mathbf{324}$$

Notice the difference in the sample sizes required in the two previous examples (with and without prior information). The only mathematical difference between the problems is the value used for p^*. In the first example, we used $p^* = 0.5$, and in the second example we used $p^* = 0.05$. Recall that the use of the provisional value $p^* = 0.5$ gives the maximum sample size. As you can see, it will be an advantage to have some indication of the value expected for p, especially as p moves increasingly farther from 0.5.

Hypothesis-Testing Procedure

When the binomial parameter p is to be tested using a hypothesis-testing procedure, we will use a test statistic that represents the difference between the observed proportion and the hypothesized proportion, divided by the standard error. This test statistic is assumed to be normally distributed when the null hypothesis is true, when the assumptions for the test have been satisfied, and when n is sufficiently large ($n > 20$, $np > 5$, and $nq > 5$).

Test Statistic for a Proportion

$$z \bigstar = \frac{p' - p}{\sqrt{\frac{pq}{n}}} \quad \text{with } p' = \frac{x}{n} \tag{9.9}$$

To demonstrate the use of this formula, we'll use two examples: one- and two-tailed hypothesis tests for proportion p.

One-Tailed Hypothesis Test for Proportion p

Many people sleep in on the weekends to make up for "short nights" during the workweek. The Better Sleep Council reports that 61% of us get more than seven hours of sleep per night on the weekend. A random sample of 350 adults found that 235 had more than seven hours of sleep each night last weekend. At the 0.05 level of significance, does this evidence show that more than 61% sleep seven or more hours per night on the weekend?

STEP 1 THE SET-UP:

a. Describe the population parameter of interest.
   p, the proportion of adults who get more than seven hours of sleep per night on weekends.

b. State the null hypothesis (H_o) and the alternative hypothesis (H_a).

$$H_o: p = P(7+ \text{ hours of sleep}) = 0.61 \ (\leq)$$
$$\text{(no more than 61\%)}$$

$$H_a: p > 0.61 \ (\text{more than 61\%})$$

STEP 2 THE HYPOTHESIS TEST CRITERIA:

a. Check the assumptions.
   The random sample of 350 adults was independently surveyed.

b. Identify the probability distribution and the test statistic to be used.
   The standard normal z will be used with formula (9.9). Since $n = 350$ is greater than 20 and both $np = (350)(0.61) = 213.5$ and $nq = (350)(0.39) = 136.5$ are greater than 5, p' is expected to be approximately normally distributed.

c. Determine the level of significance: $\alpha = 0.05$.

STEP 3 THE SAMPLE EVIDENCE:

a. Collect the sample information: $n = 350$ and $x = 235$:
$$p' = \frac{x}{n} = \frac{235}{350} = 0.671$$

b. Calculate the value of the test statistic.
   Use formula (9.9):

$$z \bigstar = \frac{p' - p}{\sqrt{\frac{pq}{n}}}: \quad z \bigstar = \frac{0.671 - 0.61}{\sqrt{\frac{(0.61)(0.39)}{350}}}$$

$$= \frac{0.061}{\sqrt{0.0006797}}$$

$$= \frac{0.061}{0.0261} = \mathbf{2.34}$$

Again, we can choose either the *p*-value or classical procedure.

Using the *p*-value procedure:

a. **Calculate the *p*-value for the test statistic.**
Use the right-hand tail because H_a expresses concern for values related to "more than."
P = *p*-value = $P(z > 2.34)$, as shown on the figure below.

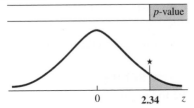

To find the *p*-value, use one of three methods:
1. Use Table 3 in Appendix B to calculate the *p*-value:

   P = $1.000 - 0.9904 =$ **0.0096.**

2. Use Table 5 in Appendix B to place bounds on the *p*-value:

   $0.0094 <$ **P** $< 0.0107.$

3. Use a computer or calculator to calculate the *p*-value:

   P = 0.0096.

   For specific instructions, see *Method 3* at the end of Step 5.

b. **Determine whether or not the *p*-value is smaller than α.**
The *p*-value is smaller than α.

Using the classical procedure:

a. **Determine the critical region and critical value(s).**
The critical region is the right-hand tail because H_a expresses concern for values related to "more than." The critical value is obtained from Table 4A: $z(0.05) =$ **1.65.**

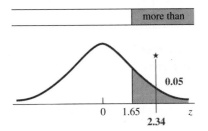

Specific instructions for finding critical values are given on pages 174–176.

b. **Determine whether or not the calculated test statistic is in the critical region.**
$z\star$ is in the critical region, as shown in red on the figure above.

---

a. **State the decision about H_o.**
Reject H_o.

b. **State the conclusion about H_a.**
There is sufficient reason to conclude that the proportion of adults in the sampled population who are getting more than seven hours of sleep nightly on weekends is significantly higher than 61% at the 0.05 level of significance.

Method 3: If you are doing the hypothesis test with the aid of a computer or calculator, most likely it will calculate the *p*-value for you.

TWO-TAILED HYPOTHESIS TEST FOR PROPORTION *p*

Now let's work through a two-tailed hypothesis test for proportion *p* by picking up the example on page 192

about cars students drive. While talking about the cars that fellow students drive, Tom claimed that 15% of the students drive convertibles. Jody finds this hard to believe, and she wants to check the validity of Tom's claim using Dana's random sample. At a level of significance of 0.10, we want to determine if there is sufficient evidence to reject Tom's claim if there were 17 convertibles in his sample of 200 cars.

a. **Describe the population parameter of interest.**
$p = P(\text{student drives convertible})$.

b. State the null hypothesis (H_o) and the alternative hypothesis (H_a).

H_o: $p = 0.15$ (15% do drive convertibles)

H_a: $p \neq 0.15$ (the percentage is different from 15%)

STEP 2 THE HYPOTHESIS TEST CRITERIA:

a. **Check the assumptions.**

The sample was randomly selected, and each subject's response is independent of other responses.

b. **Identify the probability distribution and the test statistic to be used.**

The standard normal z and formula (9.9) will be used. Since $n = 200$ is greater than 20 and both np and nq are greater than 5, p' is expected to be approximately normally distributed.

c. **Determine the level of significance.**

$\alpha = 0.10$.

Using the *p*-value procedure:

a. **Calculate the *p*-value for the test statistic.**

Use both tails because H_a expresses concern for values related to "different from."

$$P = p\text{-value} = P(z < -2.57) + P(z > 2.57)$$

$$= 2 \times P(|z| > 2.57)$$

as shown on the figure below.

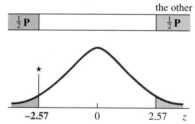

To find the *p*-value, use one of three methods:

1. Use Table 3 in Appendix B to calculate the *p*-value:

$$\mathbf{P} = 2 \times 0.0051 = 0.0102.$$

2. Use Table 5 in Appendix B to place bounds on the *p*-value:

$$0.0094 < \mathbf{P} < 0.0108.$$

3. Use a computer or calculator to calculate the *p*-value:

$$\mathbf{P} = 0.0102.$$

For specific instructions, see pages 170–171.

b. **Determine whether or not the *p*-value is smaller than α.**

The *p*-value is smaller than α.

STEP 3 THE SAMPLE EVIDENCE:

a. **Collect the sample information.**

$n = 200$ and $x = 17$:

$$p' = \frac{x}{n} = \frac{17}{200} = 0.085$$

b. **Calculate the value of the test statistic.**

Use formula (9.9):

$$z\bigstar = \frac{p' - p}{\sqrt{\dfrac{pq}{n}}}:$$

$$z\bigstar = \frac{0.085 - 0.150}{\sqrt{\dfrac{(0.15)(0.85)}{200}}}$$

$$= \frac{-0.065}{\sqrt{0.00064}}$$

$$= \frac{-0.065}{0.02525} = \mathbf{-2.57}$$

STEP 4 THE PROBABILITY DISTRIBUTION:

Again, we can choose either the *p*-value or classical procedure.

Using the classical procedure:

a. **Determine the critical region and critical value(s).**

The critical region is two-tailed because H_a expresses concern for values related to "different from." The critical value is obtained from Table 4B: $z(0.05) = 1.65$.

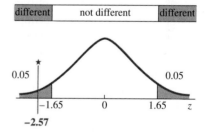

For specific instructions, see pages 174–176.

b. **Determine whether or not the calculated test statistic is in the critical region.**

$z\bigstar$ is in the critical region, as shown in red on the figure above.

a. State the decision about H_o.
   Reject H_o.
b. State the conclusion about H_a.
   There is sufficient evidence to reject Tom's claim and conclude that the percentage of students who drive convertibles is different from 15% at the 0.10 level of significance.

9.3 Inferences about the Variance and Standard Deviation

PROBLEMS OFTEN ARISE THAT REQUIRE US TO MAKE INFERENCES ABOUT VARIABILITY.

For example, a soft-drink bottling company has a machine that fills 16-ounce bottles. The company needs to control the standard deviation σ (or variance σ^2) in the amount of soft drink, x, put into each bottle. The mean amount placed in each bottle is important, but a correct mean amount does not ensure that the filling machine is working correctly. If the variance is too large, many bottles will be overfilled and many underfilled. Thus, the bottling company wants to maintain as small a standard deviation (or variance) as possible.

When discussing inferences about the spread of data, we usually talk about variance instead of standard deviation because the techniques (the formulas used) employ the sample variance rather than the standard deviation. However, remember that the standard deviation is the positive square root of the variance; thus, talking about the variance of a population is comparable to talking about the standard deviation.

Inferences about the variance of a normally distributed population use the **chi-square, χ^2**, distributions ("ki-square"—that's "ki" as in "kite"; χ is the Greek lowercase letter chi). The chi-square distributions, like Student's t-distributions, are a family of probability distributions, each one identified by the parameter number of degrees of freedom. In order to use the chi-square distribution, we must be aware of its properties (see Figure 9.6).

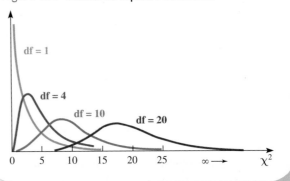

Figure 9.6 **Various Chi-Square Distributions**

NOTE: When df > 2, the mean value of the chi-square distribution is df. The mean is located to the right of the mode (the value where the curve reaches its high point) and just to the right of the median (the value that splits the distribution, 50% on either side). By locating zero at the left extreme and the value of df on your sketch of the χ^2 distribution, you will establish an approximate scale so that other values can be located in their respective positions. See Figure 9.7.

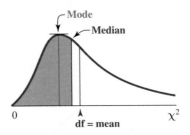

Figure 9.7 **Location of Mean, Median, and Mode for χ^2 Distribution**

Properties of the Chi-Square Distribution

1. χ^2 is nonnegative in value; it is zero or positively valued.
2. χ^2 is not symmetrical; it is skewed to the right.
3. χ^2 is distributed so as to form a family of distributions, a separate distribution for each different number of degrees of freedom.

Critical Values of Chi-Square

The **critical values for chi-square** are obtained from Table 8 in Appendix B. Each critical value is identified by two pieces of information: degrees of freedom (df) and area under the curve to the right of the critical value being sought. Thus, $\chi^2(df, \alpha)$ (read "*chi-square of df, alpha*") is the symbol used to identify the critical value of chi-square with df degrees of freedom and with α area to the right, as shown in Figure 9.8. Since the chi-square distribution is not symmetrical, the critical values associated with the right and left tails are given separately in Table 8.

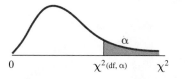

Figure 9.8 **Chi-Square Distribution Showing** $\chi^2(df, \alpha)$

To illustrate finding χ^2 associated with the right tail, let's find $\chi^2(20, 0.05)$. See the figure below. Use Table 8 in Appendix B to find the value of $\chi^2(20, 0.05)$ at the intersection of row df = 20 and the column for an area of 0.05 to the right, as shown in the portion of the table below:

Portion of Table 8			
	Area to the Right		
df	...	0.05	...
⋮			
20		**3.14**	

$\chi^2(20, 0.05) = 3.14$

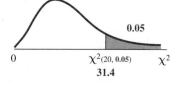

We can also find χ^2 associated with the left tail. To do so, let's find $\chi^2(14, 0.90)$.

We use Table 8 in Appendix B to find the value of $\chi^2(14, 0.90)$ at the intersection of row df = 14 and column $\alpha = 0.90$, as shown in the portion of the table below:

Portion of Table 8			
	Area to the Right		
df	...	0.90	...
⋮			
14		7.79	

$\chi^2(14, 0.90) = 7.79$

Applying that number to our curve produces the corresponding figure.

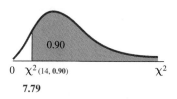

Most computer software packages or statistical calculators will calculate the area related to a specified χ^2-value. The accompanying figure shows the relationship between the cumulative probability distribution and a specific χ^2-value for a χ^2-distribution with df degrees of freedom.

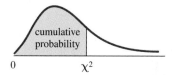

Hypothesis-Testing Procedure

We are now ready to use chi-square to make inferences about the population variance or standard deviation.

The assumption for inferences about the variance σ^2 or standard deviation σ:

The sampled population is normally distributed.

The *t* procedures for inferences about the mean (see Objective 9.1) were based on the assumption of normality, but they are generally useful even when the sampled population is nonnormal, especially for larger samples. However, the same is not true about the inference procedures for the standard deviation. The statistical procedures for the standard deviation are very sensitive to nonnormal distributions (skewness in particular), and this makes it difficult to determine whether an apparent significant result is the result of the sample evidence or a violation of the assumptions. Therefore, the only inference procedure to be presented here is the hypothesis test for the standard deviation of a normal population.

The **test statistic** that will be used in testing hypotheses about the population variance or standard deviation is obtained by using the following formula.

Test Statistic for Variance and Standard Deviation

$$\chi^2\star = \frac{(n-1)s^2}{\sigma^2} \quad \text{with df} = n - 1 \qquad (9.10)$$

When random samples are drawn from a normal population with a known variance σ^2, the quantity $\frac{(n-1)s^2}{\sigma^2}$ possesses a probability distribution that is known as the chi-square distribution with $n - 1$ degrees of freedom.

ONE-TAILED HYPOTHESIS TEST FOR VARIANCE σ^2

Let's return to the illustration about the bottling company that wishes to detect when the variability in the amount of soft drink placed into each bottle gets out of control. A variance of 0.0004 is considered acceptable, and the company wants to adjust the bottle-filling machine when the variance, σ^2, becomes larger than this value. The decision will be made using the hypothesis-testing procedure. In this scenario, we're going to conduct a one-tailed hypothesis test for variance, σ^2.

The soft-drink bottling company wants to control the variability in the amount of fill by not allowing the variance to exceed 0.0004. We need to know if a sample of size 28 with a variance of 0.0007 indicates that the bottling process is out of control (with regard to variance) at the 0.05 level of significance.

STEP 1 THE SET-UP:

a. Describe the population parameter of interest.
σ^2, the variance in the amount of fill of a soft drink during a bottling process.

b. State the null hypothesis (H_o) and the alternative hypothesis (H_a).

H_o: $\sigma^2 = 0.0004$ ($\leq$) (variance is not larger than 0.0004)

H_a: $\sigma^2 > 0.0004$ (variance is larger than 0.0004)

STEP 2 THE HYPOTHESIS TEST CRITERIA:

a. Check the assumptions.
The amount of fill put into a bottle is generally normally distributed. By checking the distribution of the sample, we could verify this.

b. Identify the probability distribution and the test statistic to be used.
Use the chi-square distribution and formula (9.10), with df $= n - 1 = 28 - 1 = 27$.

c. Determine the level of significance: $\alpha = 0.05$.

STEP 3 THE SAMPLE EVIDENCE:

a. Collect the sample information.
$n = 28$ and $s^2 = 0.0007$.

b. Calculate the value of the test statistic.
Use formula (9.10):

$$\chi^2\star = \frac{(n-1)s^2}{\sigma^2}:$$

$$\chi^2\star = \frac{(28-1)(0.0007)}{0.0004}$$

$$= \frac{(27)(0.0007)}{0.0004}$$

$$= \textbf{47.25}$$

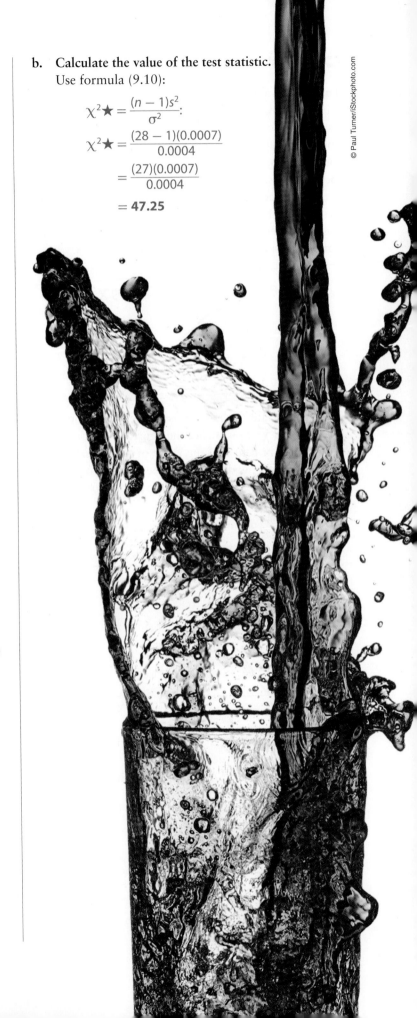

 **STEP 4** **THE PROBABILITY DISTRIBUTION:**
Again, we can choose either the *p*-value or classical procedure.

Using the *p*-value procedure:

a. **Calculate the *p*-value for the test statistic.**
Use the right-hand tail because H_a expresses concern for values related to "larger than."
$P = P(\chi^2\star > 47.25,$ with df $= 27)$, as shown on the figure.

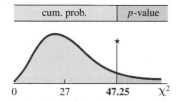

To find the *p*-value, use one of two methods:
1. Use Table 8 in Appendix B to place bounds on the *p*-value: $0.005 < P < 0.01$.
2. Use a computer or calculator to calculate the *p*-value: $P = 0.0093$.
Specific instructions follow this five-step procedure.

b. **Determine whether or not the *p*-value is smaller than α.**
The *p*-value is smaller than the level of significance, $\alpha(0.05)$.

Using the classical procedure:

a. **Determine the critical region and critical value(s).**
The critical region is the right-hand tail because H_a expresses concern for values related to "larger than." The critical value is obtained from Table 8, at the intersection of row df $= 27$ and column $\alpha = 0.05$: $\chi^2(27, 0.05) = 40.1$.

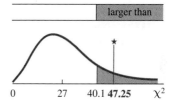

For specific instructions, see page 199.

b. **Determine whether or not the calculated test statistic is in the critical region.**
$\chi^2\star$ is in the critical region, as shown in red on the figure above.

STEP 5 **THE RESULTS:**

a. **State the decision about H_o.**
Reject H_o.

b. **State the conclusion about H_a.**
At the 0.05 level of significance, we conclude that the bottling process is out of control with regard to the variance.

Calculating the *p*-value when using the χ^2-distribution

Method 1: Use Table 8 in Appendix B to place bounds on the p-value. By inspecting the df $= 27$ row of Table 8, you can determine an interval within which the *p*-value lies. Locate $\chi^2\star$ along the row labeled df $= 27$. If $\chi^2\star$ is not listed, locate the two values that $\chi^2\star$ falls between, and then read the bounds for the *p*-value from the top of the table. In this case, $\chi^2\star = 47.25$ is between 47.0 and 49.6; therefore, **P** is between 0.005 and 0.01.

Portion of Table 8

df	...	Area to the Right				
		0.01	**P**	0.005	→	$0.005 < P < 0.01$
⋮		↑		↑		
27		47.0	**47.25**	49.6		

Two-Tailed Hypothesis Test for Standard Deviation, σ

Decisions can also be made using the two-tailed hypothesis-testing procedure. Let's look at another scenario. The manufacturer of a photographic chemical claims that its product has a shelf life that is normally distributed about a mean of 180 days with a standard deviation of no more than 10 days. As a user of this chemical, Fast Photo is concerned that the standard deviation might be different from 10 days; otherwise, it will buy a larger quantity while the chemical is part of a special promotion. Twelve random samples were selected and tested, with a standard deviation of 14 days resulting. The managers at Fast Photo want to know if, at the 0.05 level of significance, this sample presents sufficient evidence to show that the standard deviation is different from 10 days.

STEP 1 THE SET-UP:

a. Describe the population parameter of interest.
 σ, the standard deviation for the shelf life of the chemical.

b. State the null hypothesis (H_o) and the alternative hypothesis (H_a).

H_o: $\sigma = 10$ (standard deviation is 10 days)

H_a: $\sigma \neq 10$ (standard deviation is different from 10 days)

STEP 2 THE HYPOTHESIS TEST CRITERIA:

a. Check the assumptions.

The manufacturer claims shelf life is normally distributed; this could be verified by checking the distribution of the sample.

b. Identify the probability distribution and the test statistic to be used.
 The chi-square distribution will be used and formula (9.10), with df = $n - 1 = 12 - 1 = 11$.

c. Determine the level of significance.
 $\alpha = 0.05$.

STEP 3 THE SAMPLE EVIDENCE:

a. Collect the sample information.
 $n = 12$ and $s = 14$.

b. Calculate the value of the test statistic.
 Use formula (9.10):

$$\chi^2\star = \frac{(n-1)s^2}{\sigma^2}; \quad \chi^2\star = \frac{(12-1)(14)^2}{(10)^2} = \frac{2156}{100} = \mathbf{21.56}$$

STEP 4 THE PROBABILITY DISTRIBUTION:

Again, we can choose either the p-value or classical procedure.

Using the p-value procedure:

a. Calculate the p-value for the test statistic.
 Since the concern is for values "different from" 10, the p-value is the area of both tails. The area of each tail will represent $\frac{1}{2}$P. Since $\chi^2\star = 21.56$ is in the right tail, the area of the right tail is $\frac{1}{2}$P:

$\frac{1}{2}$P $= P(\chi^2 > 21.56$, with df = 11)

as shown on the figure.

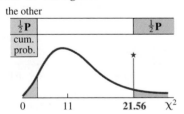

To find $\frac{1}{2}$P, use one of two methods:

1. Use Table 8 in App. B to place bounds on $\frac{1}{2}$P: $0.025 < \frac{1}{2}$P$ < 0.05$. Double both bounds to find the bounds for P: $2 \times (0.025 < \frac{1}{2}P < 0.05)$ becomes $0.05 < P < 0.10$.

2. Use a computer or calculator to find $\frac{1}{2}$P: $\frac{1}{2}$P $= 0.0280$; therefore, P $= 0.0560$.

 Specific instructions follow this five-step procedure.

b. Determine whether or not the p-value is smaller than α.
 The p-value is not smaller than the level of significance, $\alpha(0.05)$.

Using the classical procedure:

a. Determine the critical region and critical value(s).
 The critical region is split into two equal parts because H_a expresses concern for values related to "different from." The critical values are obtained from Table 8 at the intersections of row df = 11 with columns $\alpha = 0.975$ and 0.025 (area to right): $\chi^2(11, 0.975) = 3.82$ and $\chi^2(11, 0.025) = 21.9$.

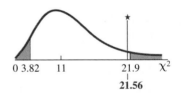

For specific instructions, see page 199.

b. Determine whether or not the calculated test statistic is in the critical region.
 $\chi^2\star$ is not in the critical region; see the figure above.

STEP 5 THE RESULTS:

a. State the decision about H_o.
    Fail to reject H_o.
b. State the conclusion about H_a.
    There is not sufficient evidence at the 0.05 significance level to conclude that the shelf life of this chemical has a standard deviation different from 10 days. Therefore, Fast Photo should purchase the chemical accordingly.

Calculating the *p*-value when using the χ^2-distribution

Method 1: Use Table 8 in Appendix B to place bounds on the p-*value.* Inspect the df = 11 row to locate $\chi^2\bigstar$ = 21.56. Notice that 21.56 is between two table entries. The bounds for $\frac{1}{2}P$ are read from the Right-Hand Tail heading at the top of the table.

Portion of Table 8				
		Area to the right		
df	...	0.05	$\frac{1}{2}$P	0.025
⋮		↑		↑
11		19.7	**21.56**	21.9

$\longrightarrow 0.025 < \frac{1}{2}P < 0.05$

Double both bounds to find the bounds for **P**:
$2 \times (0.025 < \frac{1}{2}P < 0.05)$ becomes $\mathbf{0.05 < P < 0.10}$.

NOTE: When sample data are skewed, just one outlier can greatly affect the standard deviation. It is very important, especially when using small samples, that the sampled population be normal; otherwise, the procedures are not reliable.

recap

We have been studying inferences, both confidence intervals and hypothesis tests, for the three basic population parameters (μ, p, and σ) of a single population. In the next chapter we will discuss inferences about two populations whose respective means, proportions, and standard deviations are to be compared.

problems

Objective 9.1

9.1 Find:
    a. $t(12, 0.01)$ b. $t(22, 0.025)$ c. $t(50, 0.10)$ d. $t(8, 0.005)$

9.2 Find these critical values using Table 6 in Appendix B:
    a. $t(21, 0.95)$ b. $t(26, 0.975)$ c. $t(27, 0.99)$ d. $t(60, 0.025)$

9.3 Construct a 95% confidence interval estimate for the mean μ using the sample information $n = 24$, $\bar{x} = 16.7$, and $s = 2.6$.

9.4 The National Highway Traffic Safety Administration found the average U.S. EMS response time from EMS notification to arrival at the crash scene in urban areas to be 6.85 minutes. A random sample of 20 reported fatal crashes in South Dakota had a mean notification-to-arrival time of 5.25 minutes with a standard deviation of 2.78 minutes. Find the 98% confidence interval for the true mean notification-to-arrival time in South Dakota if response times are considered nearly symmetrical.

9.5 Based on a survey of 1,000 adults by Greenfield Online and reported in a May 2009 *USA Today* Snapshot, adults 24 years of age and under spend a weekly average of $35 on fast food. If 200 of the adults surveyed were in the age category of 24 and under and they provided a standard deviation of $14.50, construct a 95% confidence interval for the weekly average expenditure on fast food for adults 24 years of age and under. Assume fast food weekly expenditures are normally distributed.

9.6 While writing an article on the high cost of college education, a reporter took a random sample of the cost of new textbooks for a semester. The random variable *x* is the cost of one book. Her sample data can be summarized by $n = 41$, $\Sigma x = 3{,}582.17$, and $\Sigma(x - \bar{x})^2 = 9{,}960.336$.
    a. Find the sample mean, $\bar{x}$.
    b. Find the sample standard deviation, s.
    c. Find the 90% confidence interval to estimate the true mean textbook cost for the semester based on this sample.

*9.7 Lunch breaks are often considered too short, and employees frequently develop a habit of "stretching" them. The manager at Giant Mart randomly identified 22 employees and observed the length of their lunch breaks (in minutes) for one randomly selected day during the week:

| 30 | 24 | 38 | 35 | 27 | 35 | 23 | 28 | 28 | 22 | 26 |
| 34 | 29 | 25 | 28 | 34 | 24 | 26 | 28 | 32 | 29 | 40 |

    a. Show evidence that the normality assumptions are satisfied.
    b. Find the 95% confidence interval for "mean length of lunch breaks" at Giant Mart.

*9.8 Many studies have revealed to us that we need to exercise to lower various health risks such as high blood pressure, heart disease, and high cholesterol. But knowing so and doing so are not the same thing. People in the health profession should be even more aware of the need for exercise. The data below is from a study surveying cardiovascular technicians (individuals who perform various cardiovascular diagnostic procedures) as to their own physical exercise per week, measured in minutes.

60	40	50	30	60	50	90	30	60	60
60	80	90	90	60	30	20	120	60	50
20	60	30	120	50	30	90	20	30	40
50	40	30	40	20	30	60	50	60	80

a. Determine whether an assumption of normality is reasonable. Explain.

b. Estimate the mean amount of weekly exercise time for all cardiovascular technicians using a point estimate and a 95% confidence interval.

9.9 State the null hypothesis, H_o, and the alternative hypothesis, H_a, that would be used to test each of the following claims:

a. A chicken farmer at Best Broilers claims that his chickens have a mean weight of 56 ounces.

b. The mean age of U.S. commercial jets is less than 18 years.

c. The mean monthly unpaid balance on credit card accounts is more than $400.

9.10 Calculate the value of $t\star$ for the hypothesis test: $H_o: \mu = 32, H_a: \mu > 32, n = 16, \bar{x} = 32.93, s = 3.1$.

9.11 Determine the p-value for the following hypothesis tests involving Student's t-distribution with 10 degrees of freedom.

a. $H_o: \mu = 15.5, H_a: \mu < 15.5, t\star = -2.01$

b. $H_o: \mu = 15.5, H_a: \mu > 15.5, t\star = 2.01$

c. $H_o: \mu = 15.5, H_a: \mu \neq 15.5, t\star = 2.01$

9.12 Determine the critical region and critical value(s) that would be used in the classical approach to test the following null hypotheses:

a. $H_o: \mu = 10, H_a: \mu \neq 10$ ($\alpha = 0.05, n = 15$)

b. $H_o: \mu = 37.2, H_a: \mu > 37.2$ ($\alpha = 0.01, n = 25$)

c. $H_o: \mu = -20.5, H_a: \mu < -20.5$ ($\alpha = 0.05, n = 18$)

9.13 Homes in a nearby college town have a mean value of $88,950. It is assumed that homes in the vicinity of the college have a higher mean value. To test this theory, a random sample of 12 homes is chosen from the college area. Their mean value is $92,460, and the standard deviation is $5,200. Complete a hypothesis test using $\alpha = 0.05$. Assume prices are normally distributed.

a. Solve using the p-value approach.

b. Solve using the classical approach.

9.14 According to the August 2009 *Readers Digest* article "Where Our Garbage Goes," the average American tosses 4.6 pounds of garbage every day. A small town in Vermont initiated a Going Green campaign and asked residents to work on recycling more and reducing their generation of garbage each day. To estimate the average amount of trash discarded by people in their town, 18 households were randomly selected and all were asked on the same day to carefully weigh their trash for that day. The average amount for the sample was 3.89 pounds, with a standard deviation of 1.322 pounds. Is there sufficient evidence that the Vermont town now has significantly lower average daily garbage amounts than the average American household? Use a 0.05 level of significance and assume weights are normally distributed.

*9.15 The recommended number of hours of sleep per night is 8 hours, but everybody "knows" that the average college student sleeps less than 7 hours. The number of hours slept last night by 10 randomly selected college students is listed here:

5.2	6.8	6.2	5.5	7.8	5.8	7.1	8.1	6.9	5.6

Use a computer or calculator to complete the hypothesis test: $H_o: \mu = 7, H_a: \mu < 7, \alpha = 0.05$.

*9.16 According to statements from The National Women's Health Information Center and the Centers for Disease Control and Prevention, people should exercise at least 60 minutes per week to lower various health risks.

a. Based on the data from Exercise 9.8, determine if the technicians exercise at least 60 minutes a week. Use a 0.05 level of significance.

b. Did the statistical decision reached in (a) result in the same conclusion as you expressed in answering part (b) of Exercise 9.6 for this same data?

*9.17 Use a computer or calculator to complete the calculations and the hypothesis test for this problem. Delco Products, a division of General Motors, produces commutators designed to be 18.810 mm in overall length. (A commutator is a device used in the electrical system of an automobile.) The following data are the lengths of a sample of 35 commutators taken while monitoring the manufacturing process:

18.802	18.810	18.780	18.757	18.824	18.827	18.825
18.809	18.794	18.787	18.844	18.824	18.829	18.817
18.785	18.747	18.802	18.826	18.810	18.802	18.780
18.830	18.874	18.836	18.758	18.813	18.844	18.861
18.824	18.835	18.794	18.853	18.823	18.863	18.808

SOURCE: With permission of Delco Products Division, GMC

Is there sufficient evidence to reject the claim that these parts meet the design requirements "mean length is 18.810" at the $\alpha = 0.01$ level of significance?

9.18 Acetaminophen is an active ingredient found in more than 600 over-the-counter and prescription medicines, such as pain relievers, cough suppressants, and cold medications. It is safe and effective when used correctly, but taking too much can lead to liver damage.

SOURCE: http://www.keepkidshealthy.com/

A researcher believes the amount of acetaminophen in a particular brand of cold tablets contains a mean amount of acetaminophen per tablet different from the 600 mg claimed by the manufacturer. A random sample of 30 tablets had a mean acetaminophen content of 596.3 mg with a standard deviation of 4.7 mg.

a. Is the assumption of normality reasonable? Explain.

b. Construct a 99% confidence interval for the estimate of the mean acetaminophen content.

c. What does the confidence interval found in part (b) suggest about the mean acetaminophen content of one pill? Do you believe there is 600 mg per tablet? Explain.

Objective 9.2

9.19 Of the 150 elements in a random sample, 45 are classified as "success."

a. Explain why x and n are assigned the values 45 and 150, respectively.

b. Determine the value of p'. Explain how p' is found and the meaning of p'.

For each of the following situations, find p'.

c. $x = 24$ and $n = 250$

d. $x = 640$ and $n = 2,050$

e. 892 of 1,280 responded "yes"

9.20 a. What is the relationship between $p = P$ (success) and $q = P$ (failure)? Explain.

b. Explain why the relationship between p and q can be expressed by the formula $q = 1 - p$.

c. If $p = 0.6$, what is the value of q?

d. If the value of $q' = 0.273$, what is the value of p'?

9.21 A bank randomly selected 250 checking account customers and found that 110 of them also had savings accounts at the same bank. Construct a 95% confidence interval for the true proportion of checking account customers who also have savings accounts.

9.22 In a poll conducted by Harris Interactive of 1,179 U.S. video-gaming youngsters, 8.5% displayed behavioral signs that may indicate addiction. Using a 99% confidence interval for the true binomial proportion based on a random sample of 1,179 binomial trials and an observed proportion of 0.085, estimate the proportion of video-gaming youngsters that may go on to have an addiction.

SOURCE: "Kids Show Addiction Symptoms," *USA Today*, April 21, 2009

9.23 Just one serving a month of kale or collard greens or more than 2 servings of carrots a week can reduce the risk of glaucoma by more than 60%, according to a UCLA study of 1,000 women. Using a 90% confidence interval for the true binomial proportion based on a random sample of 1,000 binomial trials and an observed proportion of 0.60, estimate the proportion of glaucoma risk reduction in women who eat the

recommended servings of kale, collard greens, or carrots.

SOURCE: "Tasty Sight Savers," *Readers Digest*, February 2009

9.24 Three nationwide poll results are described below.

> *USA Today* Snapshot/Rent.com, August 18, 2009; $n = 1,000$ adults 18 and over. MoE ± 3.
>
> What renters look for the most when seeking an apartment: Washer/dryer-39%, Air conditioning-30%, Fitness center-10%, Pool-10%
>
> *USA Today*/Harris Interactive Poll, February 10–15, 2009; $n = 1,010$ adults. MoE ± 3.
>
> Americans who say people on Wall Street are "as honest and moral as other people": Disagree-70%, Agree-26%, Not sure/refused-4%
>
> American Association of Retired Persons Bulletin/AARP survey, July 22–August 2, 2009; $n = 1,006$ adults age 50 and older. MoE ± 3.
>
> 16% of adults age 50 and older said they are likely to return to school.
>
> MoE stands for Margin of Error

Each of the polls is based on approximately 1,005 randomly selected adults.

a. Calculate the 95% confidence maximum error of estimate for the true binomial proportion based on binomial experiments with the same sample size and observed proportion as listed first in each article.

b. Explain what caused the values of the maximum errors to vary.

c. The margin of error being reported is typically the value of the maximum error rounded to the next larger whole percentage. Do your results in (a) verify this?

d. Explain why the round-up practice is considered "conservative"?

e. What value of p should be used to calculate the standard error if the most conservative margin of error is desired?

9.25 According to a May 2009 Harris Poll, 72% of those who drive and own cell phones say they use them to talk while they are driving. You wish to conduct a survey in your city to determine what percent of the drivers with cell phones use them to talk while driving. Use the national figure of 72% for your initial estimate of p.

a. Find the sample size if you want your estimate to be within 0.02 with 90% confidence.

b. Find the sample size if you want your estimate to be within 0.04 with 90% confidence.

c. Find the sample size if you want your estimate to be within 0.02 with 98% confidence.

d. What effect does changing the maximum error have on the sample size? Explain.

e. What effect does changing the level of confidence have on the sample size? Explain.

9.26 Lung cancer is the leading cause of cancer deaths in both women and men in the United States. According to the Center for Disease Control and Prevention 2005 statistics, lung cancer accounted for more deaths than breast cancer, prostate cancer, and colon cancer combined.

Overall, only about 16% of all people who develop lung cancer survive for 5 years.

SOURCE: http://www.cdc.gov/

Suppose you wanted to see if this survival rate were still true. How large a sample would you need to take to estimate the true proportion surviving for five years after diagnosis to within 1% with 95% confidence? (Use 16% as the initial value of p.)

9.27 State the null hypothesis, H_o, and the alternative hypothesis, H_a, that would be used to test these claims:
 a. More than 60% of all students at our college work part-time jobs during the academic year.
 b. No more than one-third of cigarette smokers are interested in quitting.
 c. A majority of the voters will vote for the school budget this year.
 d. At least three-fourths of the trees in our county were seriously damaged by the storm.
 e. The results show the coin was not tossed fairly.

9.28 Calculate the test statistic $z\star$ used in testing the following:
 a. $H_o: p = 0.70$ vs. $H_a: p > 0.70$, with the sample $n = 300$ and $x = 224$
 b. $H_o: p = 0.35$ vs. $H_a: p \neq 0.35$, with the sample $n = 280$ and $x = 94$

9.29 Determine the test criteria that would be used to test the following hypotheses when z is used as the test statistic and the classical approach is used.
 a. $H_o: p = 0.5$ and $H_a: p > 0.5$, with $\alpha = 0.05$
 b. $H_o: p = 0.5$ and $H_a: p \neq 0.5$, with $\alpha = 0.05$
 c. $H_o: p = 0.4$ and $H_a: p < 0.4$, with $\alpha = 0.10$

9.30 Determine the p-value for each of the following hypothesis-testing situations.
 a. $H_o: p = 0.5, H_a: p \neq 0.5, z\star = 1.48$
 b. $H_o: p = 0.4, H_a: p > 0.4, z\star = 0.98$
 c. $H_o: p = 0.2, H_a: p < 0.2, z\star = -1.59$

9.31 An insurance company states that 90% of its claims are settled within 30 days. A consumer group selected a random sample of 75 of the company's claims to test this statement. If the consumer group found that 55 of the claims were settled within 30 days, do they have sufficient reason to support their contention that less than 90% of the claims are settled within 30 days? Use $\alpha = 0.05$.
 a. Solve using the p-value approach.
 b. Solve using the classical approach.

9.32 An April 21, 2009 USA Today article titled "On Road, It's 'Do as I Say, Not as I Do'" reported that 58% of U.S. adults sped up to beat a yellow light. Suppose you conduct a survey of 150 randomly selected adults in your hometown and find that 71 out of the 150

admit to speeding up to beat a yellow light. Does your hometown have a lower rate for speeding up to beat a yellow light than the nation as a whole? Use a 0.05 level of significance.

9.33 The popularity of personal watercraft (PWCs, also known as jet skis) continues to increase, despite the apparent danger associated with their use. In fact, a sample of 54 watercraft accidents reported to the Game and Parks Commission in the state of Nebraska revealed that 85% of them involved PWCs even though only 8% of the motorized boats registered in the state are PWCs.

SOURCE: "Officer's Notebook: The Personal Problem," NEBRASKAland

Suppose the national average proportion of watercraft accidents involving PWCs was 78%. Does the watercraft accident rate for PWCs in the state of Nebraska exceed that of the nation as a whole? Use a 0.01 level of significance.
 a. Solve using the p-value approach
 b. Solve using the classical approach.

Objective 9.3

9.34 Find these critical values by using Table 8 of Appendix B.
 a. $\chi^2(18, 0.01)$ b. $\chi^2(16, 0.025)$ c. $\chi^2(8, 0.10)$
 d. $\chi^2(28, 0.01)$ e. $\chi^2(22, 0.95)$ f. $\chi^2(10, 0.975)$
 g. $\chi^2(50, 0.90)$ h. $\chi^2(24, 0.99)$

9.35 Using the notation of problem 9.28, name and find the critical values of χ^2.

a.
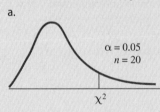

b.
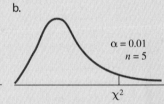

c.
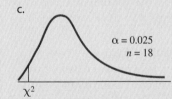

d.
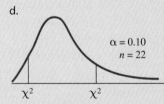

9.36 State the null hypothesis, H_o, and the alternative hypothesis, H_a, that would be used to test these claims:
 a. The standard deviation has increased from its previous value of 24.
 b. The standard deviation is no larger than 0.5 ounce.
 c. The standard deviation is not equal to 10.
 d. The variance is no less than 18.
 e. The variance is different from the value of 0.025, the value called for in the specs.

9.37 Calculate the value for the test statistic, $\chi^2\star$, for each of these situations:
 a. $H_o: \sigma^2 = 20, n = 15, s^2 = 17.8$
 b. $H_o: \sigma^2 = 30, n = 18, 's = 5.7$

c. H_o: '$\sigma = 42$, $n = 25$, '$s = 37.8$

d. H_o: '$\sigma = 12$, $n = 37$, $s^2 = 163$

9.38 Calculate the p-value for each of the following hypothesis tests.

   a. H_a: $\sigma^2 \neq 20$, $n = 15$, $\chi^2\star = 27.8$

   b. H_a: $\sigma^2 > 30$, $n = 18$, $\chi^2\star = 33.4$

   c. H_a: $\sigma^2 < 12$, df $= 40$, $\chi^2\star = 26.3$

9.39 Determine the critical region and critical value(s) that would be used to test the following using the classical approach:

   a. H_o: $\sigma = 0.5$ and H_a: $\sigma > 0.5$, with $n = 18$ and $\alpha = 0.05$

   b. H_o: $\sigma^2 = 8.5$ and H_a: $\sigma^2 < 8.5$, with $n = 15$ and $\alpha = 0.01$

   c. H_o: $\sigma = 20.3$ and H_a: $\sigma \neq 20.3$, with $n = 10$ and $\alpha = 0.10$

   d. H_o: $\sigma^2 = 0.05$ and H_a: $\sigma^2 \neq 0.05$, with $n = 8$ and $\alpha = 0.02$

   e. H_o: $\sigma = 0.5$ and H_a: $\sigma < 0.5$, with $n = 12$ and $\alpha = 0.10$

9.40 A random sample of 51 observations was selected from a normally distributed population. The sample mean was $\bar{x} = 98.2$, and the sample variance was $s^2 = 37.5$. Does this sample show sufficient reason to conclude that the population standard deviation is not equal to 8 at the 0.05 level of significance?

   a. Solve using the p-value approach.

   b. Solve using the classical approach.

9.41 Variation in the life of a battery is expected, but too much variation would be of concern to the consumer, never knowing if the purchased battery would have a very short life. A random sample of 30 AA batteries of a particular brand produced a standard deviation of 350 hours. If a standard deviation of 288 hours (12 days) is considered acceptable, does this sample show sufficient evidence that this brand of battery has greater variation than what is acceptable at the 0.05 level of significance? Assume battery life is normally distributed.

9.42 A commercial farmer harvests his entire field of a vegetable crop at one time. Therefore, he would like to plant a small variety of green beans that mature all at one time (small standard deviation between maturity times of individual plants). A seed company has developed a new hybrid strain of green beans that it believes to be better for the commercial farmer. The maturity time of the standard variety has an average of 50 days and a standard deviation of 2.1 days. A random sample of 30 plants of the new hybrid showed a standard deviation of 1.65 days. Does this sample show

a significant lowering of the standard deviation at the 0.05 level of significance? Assume that maturity time is normally distributed.

   a. Solve using the p-value approach.

   b. Solve using the classical approach.

*9.43 The dry weight of a cork is another quality that does not affect the ability of the cork to seal the bottle, but it is a variable that is monitored regularly. The weights of the no. 9 natural corks (24 mm in diameter by 45 mm in length) have a normal distribution. Ten randomly selected corks were weighed to the nearest hundredth of a gram.

Dry Weight (in grams)									
3.26	3.58	3.07	3.09	3.16	3.02	3.64	3.61	3.02	2.79

   a. Does the preceding sample show sufficient reason to show the standard deviation of the dry weights is different from 0.3275 gram at the 0.02 level of significance?

A different random sample of 20 is taken from the same batch.

Dry Weight (in grams)									
3.53	3.77	3.49	3.24	3.00	3.41	3.33	3.51	3.02	3.46
2.80	3.58	3.05	3.51	3.61	2.90	3.69	3.62	3.26	3.58

   b. Does the preceding sample show sufficient reason to show the standard deviation of the dry weights is different from 0.3275 gram at the 0.02 level of significance?

   c. What effect did the two different sample standard deviations have on the calculated test statistic in parts (a) and (b)? What effect did they have on the p-value or critical value? Explain.

   d. What effect did the two different sample sizes have on the calculated test statistic in parts (a) and (b)? What effect did they have on the p-value or critical value? Explain.

*remember

Problems marked with an asterisk (*) have data sets available on the CourseMate for STAT2 site. Login at <u>cengagebrain.com</u>.

Inferences Involving
Two Populations

10.1 Dependent and Independent Samples

IN THIS CHAPTER WE ARE GOING TO STUDY THE PROCEDURES FOR MAKING INFERENCES ABOUT TWO POPULATIONS.

Figure 10.1 gives you an overview of how the procedures will unfold over the course of the chapter.

When comparing two populations, we need two samples, one from each population. Two basic kinds of samples can be used: independent and dependent. The dependence or independence of two samples is determined by the sources of the data. A **source** can be a person, an object, or anything else that yields a piece of data. If the same set of sources or related sets are used to obtain the data representing both populations, we have **dependent samples.** If two unrelated sets of sources are used, one set from each population, we have **independent samples.**

Source A person, an object, or anything that yields a piece of data.

Dependent samples If the same set of sources or related sets are used to obtain the data representing both populations, we have dependent samples.

Independent samples If two unrelated sets of sources are used, one set from each population, we have independent samples.

Figure 10.1 "Road Map" to Two Population Inferences

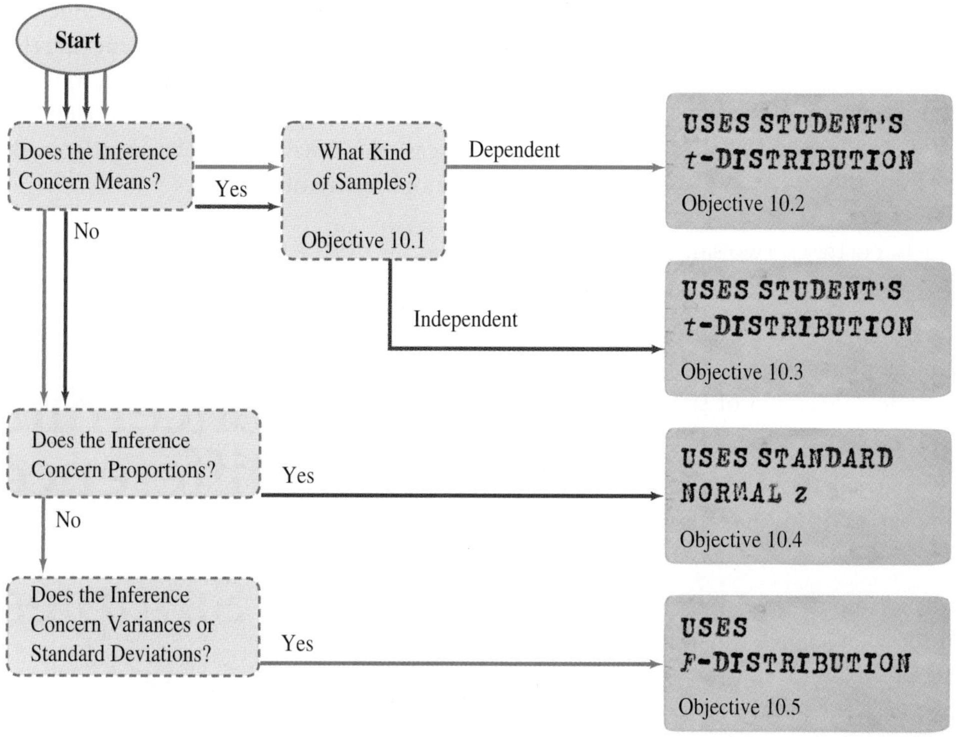

```
Start

Does the Inference        What Kind         Dependent      USES STUDENT'S
Concern Means?    ─Yes─→  of Samples?    ──────────────→   t-DISTRIBUTION
                                                            Objective 10.2
     │No               Objective 10.1
                                                            USES STUDENT'S
                           Independent     ──────────────→  t-DISTRIBUTION
                                                            Objective 10.3

Does the Inference                                          USES STANDARD
Concern Proportions?      ─Yes─────────────────────────→    NORMAL z
                                                            Objective 10.4
     │No

Does the Inference                                          USES
Concern Variances or      ─Yes─────────────────────────→    F-DISTRIBUTION
Standard Deviations?                                        Objective 10.5
```

To get a better sense of the differences between dependent and independent samples, imagine that you are conducting a test to see whether the participants in a physical fitness class actually improve in their level of fitness. It is anticipated that approximately 500 people will sign up for this course. You decide to give 50 of the participants a set of tests before the course begins (a pretest), and then give another set of tests to 50 participants at the end of the course (a posttest). Two sampling procedures are proposed:

PLAN A: Randomly select 50 participants from the list of those enrolled and give them the pretest. At the end of the course, make a second random selection of size 50 and give them the posttest.

PLAN B: Randomly select 50 participants and give them the pretest; give the same set of 50 the posttest when they complete the course.

Plan A illustrates independent sampling; the sources (the class participants) used for each sample (pretest and posttest) were selected separately. Plan B illustrates dependent sampling; the sources used for both samples (pretest and posttest) are the same.

Typically, when both a pretest and a posttest are used, the same subjects participate in the study. Thus, pretest versus posttest (before versus after) studies usually use dependent samples. Studies can, however, also use before versus after examinations in conjunction with independent samples. For example, a test is being designed to compare the wearing quality of two brands of automobile tires. The automobiles will be selected and equipped with the new tires and then driven under "normal" conditions for one month. Then a measurement will be taken to determine how much wear took place. Two plans are proposed:

PLAN C: A sample of cars will be selected randomly, equipped with brand A tires, and driven for the month. Another sample of cars will be selected, equipped with brand B tires, and driven for the month.

PLAN D: A sample of cars will be selected randomly, equipped with one tire of brand A and one tire of brand B (the other two tires are not part of the test), and driven for the month.

We suspect that many other factors must be taken into account when testing automobile tires—such as age, weight, and mechanical condition of the car; driving habits of drivers; location of the tire on the car; and

where and how much the car is driven. However, at this time we are trying only to illustrate dependent and independent samples. Plan C is independent (unrelated sources), and plan D is dependent (common sources).

Independent and dependent samples each have their advantages; these will be emphasized later. Both methods of sampling are often used.

10.2 Inferences Concerning the Mean Difference Using Two Dependent Samples

THE PROCEDURES FOR COMPARING TWO POPULATION MEANS ARE BASED ON THE RELATIONSHIP BETWEEN TWO SETS OF SAMPLE DATA, BOTH SAMPLES FROM THE SAME OR RELATED SOURCES.

When dependent samples are involved, the data are thought of as "paired data." The data may be paired as a result of being obtained from "before" and "after" studies; from pairs of identical twins; from a "common" source, as with the amounts of tire wear for each brand in plan D in Objective 10.1; or from matching two subjects with similar traits to form "matched pairs." The pairs of data values are compared directly to each other by using the difference in their numerical values. The resulting difference is called a **paired difference.**

Paired Difference

$$d = x_1 - x_2 \qquad (10.1)$$

Using paired data this way has a built-in ability to remove the effect of otherwise uncontrolled factors. The tire-wear problem using plan C and plan D is an excellent example of such additional factors. The wearing ability of a tire is greatly affected by a multitude of factors: the size, weight, age, and condition of the car, the driving habits of the driver, the number of miles driven, the condition and types of roads driven on, the quality of the material used to make the tire, and so on. With plan D, we create paired data by mounting one tire from each brand on the same car. Since one tire of each brand will be tested under the same conditions, same car, same driver, and so on, the extraneous causes of wear are neutralized.

Procedures and Assumptions for Inferences Involving Paired Data

The test comparing the wear of tires from two different tire companies uses plan D as described in Objective 10.1. All the aforementioned factors will have an equal effect on both brands of tires, car by car. The test places one tire of each brand on each of the six test cars. The position (left or right side, front or back) was determined with the aid of a random-number table. Table 10.1 lists the resulting amounts of wear (in thousandths of an inch).

*Table 10.1 **Amount of Tire Wear**

Car	1	2	3	4	5	6
Brand A	125	64	94	38	90	106
Brand B	133	65	103	37	102	115

Since the various cars, drivers, and conditions are the same for each tire of a paired set of data, it makes sense to use a third variable, the paired difference d. Our two dependent samples of data may be combined into one set of d values, where $d = B - A$.

Car	1	2	3	4	5	6
$d = B - A$	8	1	9	−1	12	9

The difference between the two population means, when dependent samples are used (often called **dependent means**), is equivalent to the **mean of the paired differences**. Therefore, when an inference is to be made about the difference of two means and paired differences are used, the inference will in fact be about the mean of the paired differences. The sample mean of the

paired differences will be used as the point estimate for these inferences.

In order to make inferences about the mean μ_d of all possible paired differences, we need to know about the *sampling distribution* of $\overline{d}$.

When paired observations are randomly selected from normal populations:

The paired difference, $d = x_1 - x_2$, will be approximately normally distributed about a mean μ_d with a standard deviation of σ_d.

This is another situation in which the *t*-test for one mean is applied; namely, we wish to make inferences about an unknown mean (μ_d) where the random variable (d) involved has an approximately normal distribution with an unknown standard deviation (σ_d).

Inferences about the mean of all possible paired differences μ_d are based on samples of n dependent pairs of data and the **t-distribution** with $n - 1$ degrees of freedom, under the following assumption:

Assumption for inferences about the mean of paired differences μ_d:

The paired data are randomly selected from normally distributed populations.

Confidence Interval Procedure

The $1 - \alpha$ confidence interval for estimating the mean difference μ_d is found using this formula:

Confidence Interval for μ_d

$$\overline{d} - t(\text{df}, \alpha/2) \cdot \frac{s_d}{\sqrt{n}} \quad \text{to} \quad \overline{d} + t(\text{df}, \alpha/2) \cdot \frac{s_d}{\sqrt{n}},$$

where df $= n - 1$ (10.2)

$\overline{d}$ is the mean of the sample differences:

$$\overline{d} = \frac{\Sigma d}{n}$$ (10.3)

and s_d is the standard deviation of the sample differences:

$$s_d = \sqrt{\frac{\Sigma d^2 - \frac{(\Sigma d)^2}{n}}{n - 1}}$$ (10.4)

CONSTRUCTING A CONFIDENCE INTERVAL FOR μ_d

To illustrate how the confidence interval for μ_d can be formed, we will use the paired data on tire wear as reported in Table 10.1 and assume the amounts of wear are approximately normally distributed for both brands of tires. Using the five-step process, we can construct the 95% confidence interval for the mean difference in the paired data. The sample information is $n = 6$ pieces of paired data, $\bar{d} = 6.3$, and $s_d = 5.1$.

STEP 1 Parameter of interest:
μ_d, the mean difference in the amounts of wear between the two brands of tires.

STEP 2 a. Assumptions:
Both sampled populations are approximately normal.
b. Probability distribution:
The t-distribution with df $= 6 - 1 = 5$ and formula (10.2) will be used.
c. State the level of confidence:
$1 - \alpha = 0.95$.

STEP 3 Sample information:
$$n = 6, \bar{d} = 6.3, \text{ and } s_d = 5.1.$$

STEP 4 a. Confidence coefficient:
This is a two-tailed situation with $\alpha/2 = 0.025$ in one tail. From Table 6 in Appendix B,
$t(\text{df}, \alpha/2) = t(5, 0.025) = 2.57$.
b. Maximum error of estimate:
Using the maximum error part of formula (10.2), we have
$$E = t(\text{df}, \alpha/2) \cdot \frac{s_d}{\sqrt{n}}:$$
$$E = 2.57 \cdot \left(\frac{5.1}{\sqrt{6}} \right) = (2.57)(2.082) = 5.351 = \textbf{5.4}$$
c. Lower/upper confidence limits:
$$\bar{d} \pm E$$
$$6.3 \pm 5.4$$
$$6.3 - 5.4 = \textbf{0.9} \text{ to } 6.3 + 5.4 = \textbf{11.7}$$

STEP 5 a. Confidence interval:
0.9 to 11.7 is the 95% confidence interval for μ_d.
b. That is, with 95% confidence we can say that the mean difference in the amounts of wear is between 0.9 and 11.7 thousandths of an inch. Or, in other words, the population mean tire wear from Brand B is between 0.9 and 11.7 thousandth of an inch greater than the population mean tire wear for Brand A.

This is quite a wide confidence interval, in part because of the small sample size. Recall from the central limit theorem that as the sample size increases, the standard error (estimated by $s_d/\sqrt{n}$) decreases.

Hypothesis-Testing Procedure

When we test a null **hypothesis about the mean difference,** the test statistic used will be the difference between the sample mean $\bar{d}$ and the hypothesized value of μ_d, divided by the estimated **standard error.** This statistic is assumed to have a t-distribution when the null hypothesis is true and the assumptions for the test are satisfied. The value of the **test statistic $t\bigstar$** is calculated as follows:

Test Statistic for μ_d

$$t\bigstar = \frac{\bar{d} - \mu_d}{s_d/\sqrt{n}} \text{ where df} = n - 1 \qquad (10.5)$$

NOTE: A hypothesized mean difference, μ_d, can be any specified value. The most common value specified is zero; however, the difference can be nonzero.

ONE-TAILED HYPOTHESIS TEST FOR μ_d

Let's conduct a one-tailed hypothesis test for the mean difference by looking at a study on high blood pressure and the drugs used to control it. The effect of calcium channel blockers on pulse rate was one of many specific concerns in the study. Twenty-six patients were randomly selected from a large pool of potential subjects, and their pulse rates were recorded. A calcium channel blocker was administered to each patient for a fixed period of time, and then each patient's pulse rate was again determined. The two resulting sets of data appeared to have approximately normal distributions, and the statistics were $\bar{d} = 1.07$ and $s_d = 1.74$ ($d =$ before $-$ after). Using the five-step method, can we determine if the sample information provides sufficient evidence to show that this calcium channel blocker lowered the pulse rate? In other words, if "lower pulse rate" means that "after" is less than "before," then "before $-$ after" should be positive. Does the sample information provide confirming evidence? Use $\alpha = 0.05$.

STEP 1 **a. Parameter of interest:**
μ_d, the mean difference (reduction) in pulse rate from before to after using the calcium channel blocker for the time period of the test.

b. Statement of hypotheses:

$$H_o: \mu_d = 0 \ (\leq) \ \text{(did not lower rate)}$$

Remember: $d = $ before $-$ after.

$$H_a: \mu_d > 0 \ \text{(did lower rate)}$$

STEP 2 **a. Assumptions:**
Since the data in both sets are approximately normal, it seems reasonable to assume that the two populations are approximately normally distributed.

b. Test statistic:
The t-distribution with df $= n - 1 = 25$, and the test statistic is $t\star$ from formula (10.5).

c. Level of significance: $\alpha = 0.05$.

STEP 3 **a. Sample information:**
$n = 26, \bar{d} = 1.07$, and $s_d = 1.74$.

b. Calculate the value of the test statistic.

$$t\star = \frac{\bar{d} - \mu_d}{s_d/\sqrt{n}}: \quad t\star = \frac{1.07 - 0.0}{1.74/\sqrt{26}} = \frac{1.07}{0.34} = \mathbf{3.14}$$

STEP 4 **The probability distribution:**
As always, we can use either the p-value or the classical procedure:

Using the *p*-value procedure:

a. Use the right-hand tail because H_a expresses concern for values related to "greater than."
$\mathbf{P} = P(t\star > 3.14, \text{ with df} = 25)$ as shown on the figure.

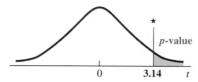

To find the p-value, you have three options:
1. Use Table 6 in Appendix B to place bounds on the p-value: **P < 0.005.**
2. Use Table 7 in Appendix B to read the value directly: **P = 0.002.**
3. Use a computer or calculator to find the p-value: **P = 0.0022.**

Specific instructions are on page 189.

b. The p-value is smaller than the level of significance, α.

Using the classical procedure:

a. The critical region is the right-hand tail because H_a expresses concern for values related to "greater than." The critical value is obtained from Table 6: $t(25, 0.05) = \mathbf{1.71}$.

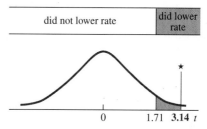

Specific instructions are on pages 185–186.

b. $t\star$ is in the critical region, as shown in red on the figure.

STEP 5 **a. Decision:** Reject H_o.

b. Conclusion:
At the 0.05 level of significance, we can conclude that the average pulse rate is lower after the administration of the calcium channel blocker.

In the preceding detailed hypothesis test, the results showed a statistical significance with a p-value of 0.002—that is, 2 chances in 1,000. We can see that *statistical significance* does not always have the same meaning when the "practical" application of the results is considered. A more practical question might be: Is lowering the pulse rate by this small average amount, estimated to be 1.07 beats per minute, worth the risks of possible side effects of this medication? Actually, the whole issue is much broader than just this one issue of pulse rate.

TWO-TAILED HYPOTHESIS TEST FOR μ_d

To conduct this two-tailed hypothesis test, let's return to the data we collected from two different brands of tires. Suppose the sample data in Table 10.1 (page 211) were collected with the hope of showing that the two brands do not wear equally. Assuming the amounts of wear are approximately normally distributed, we will use the five-step process to determine whether the data provide sufficient evidence to conclude that the two brands show unequal wear at the 0.05 level of significance.

STEP 1 a. **Parameter of interest:**

μ_d, the mean difference in the amounts of wear between the two brands.

b. **State the null hypothesis (H_o) and the alternative hypothesis (H_a):**

$H_o: \mu_d = 0$ (no difference)

$H_a: \mu_d \neq 0$ (difference)

Remember: $d = B - A$.

STEP 2 a. **Assumptions:**

The assumption of normality is included in the statement of this problem.

b. **Test statistic:**

The t-distribution with df $= n - 1 = 6 - 1 = 5$, and $t\bigstar = (\bar{d} - \mu_d)/(s_d/\sqrt{n})$.

c. **Level of significance:** $\alpha = 0.05$.

STEP 3 a. **Sample information:**

$n = 6, \bar{d} = 6.3$, and $s_d = 5.1$.

b. **Calculated test statistic:**

$$t\bigstar = \frac{\bar{d} - \mu_d}{s_d/\sqrt{n}}:$$

$$t\bigstar = \frac{6.3 - 0.0}{5.1/\sqrt{6}} = \frac{6.3}{2.08} = \mathbf{3.03}$$

STEP 4 The probability distribution:

Again, we can use either the p-value or the classical procedure:

Using the p-value procedure

a. Use both tails because H_a expresses concern for values related to "different from."

$\mathbf{P} = p\text{-value} = P(t\bigstar < -3.03) + P(t\bigstar > 3.03)$
$= 2 \times P(t\bigstar > 3.03)$

as shown on the figure below.

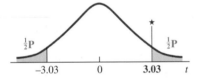

To find the p-value, you have three options:
1. Use Table 6 in Appendix B: $\mathbf{0.02 < P < 0.05}$.
2. Use Table 7 in Appendix B to place bounds on the p-value: $\mathbf{0.026 < P < 0.030}$.
3. Use a computer or calculator to find the p-value: $\mathbf{P} = 2 \times 0.0145 = \mathbf{0.0290}$.
For specific instructions, see page 189.

b. The p-value is smaller than α.

Using the classical procedure

a. The critical region is two-tailed because H_a expresses concern for values related to "different than." The critical value is obtained from Table 6: $t(5, 0.025) = \mathbf{2.57}$.

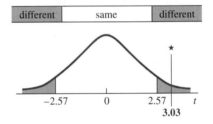

For specific instructions, see pages 185–186.

b. $t\bigstar$ is in the critical region, as shown in red on the figure.

STEP 5 a. **Decision:** Reject H_o.

b. **Conclusion:**
There is a significant difference in the mean amounts of wear at the 0.05 level of significance.

10.3 Inferences Concerning the Difference between Means Using Two Independent Samples

WHEN COMPARING THE MEANS OF TWO POPULATIONS, WE TYPICALLY CONSIDER THE DIFFERENCE BETWEEN THEIR MEANS, $\mu_1 - \mu_2$ (OFTEN CALLED **INDEPENDENT MEANS**).

The inferences about $\mu_1 - \mu_2$ will be based on the difference between the observed sample means, $\bar{x}_1 - \bar{x}_2$. This observed difference, $\bar{x}_1 - \bar{x}_2$, belongs to a sampling distribution with the characteristics described in the following statement.

IF independent samples of sizes n_1 and n_2 are drawn randomly from large populations with means μ_1 and μ_2 and variances σ_1^2 and σ_2^2, respectively,

THEN the sampling distribution of $\bar{x}_1 - \bar{x}_2$, the difference between the sample means, has

1. mean $\mu_{\bar{x}_1 - \bar{x}_2} = \mu_1 - \mu_2$ and

2. standard error $\sigma_{\bar{x}_1 - \bar{x}_2} = \sqrt{\left(\dfrac{\sigma_1^2}{n_1}\right) + \left(\dfrac{\sigma_2^2}{n_2}\right)}$. (10.6)

If both populations have normal distributions, then the sampling distribution of $\bar{x}_1 - \bar{x}_2$ will also be normally distributed.

The preceding statement is true for all sample sizes, given that the populations involved are normal and the population variances σ_1^2 and σ_2^2 are known quantities. However, as with inferences about one mean, the variance of a population is generally an unknown quantity. Therefore, it will be necessary to estimate the standard

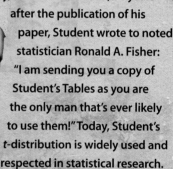

Guinness Beer and the "t-Distribution"

As head brewer at Guinness Brewing Company, William Gosset (aka Student) was faced with many small sets of data—small by necessity because a 24-hour period often resulted in only one data value. Thus, he developed the t-test to handle these small samples for quality control in brewing. In his paper *The Probable Error of a Mean*, he set out to find the distribution of the amount of error in the sample mean, $(\bar{x} - \mu)$ divided by s, where s was from a sample of any known size. He then found the probable error of a mean, $\bar{x}$, for any size sample, by using the distribution of $(\bar{x} - \mu)/(s/\sqrt{n})$. Student's t-distribution did not immediately gain popularity, and in fact in 1922, 14 years after the publication of his paper, Student wrote to noted statistician Ronald A. Fisher: "I am sending you a copy of Student's Tables as you are the only man that's ever likely to use them!" Today, Student's t-distribution is widely used and respected in statistical research.

error by replacing the variances, σ_1^2 and σ_2^2, in formula (10.6) with the best estimates available—namely, the sample variances, s_1^2 and s_2^2. The *estimated standard error* will be found using the following formula:

$$\text{estimated standard error} = \sqrt{\left(\frac{s_1^2}{n_1}\right) + \left(\frac{s_2^2}{n_2}\right)} \quad (10.7)$$

Inferences about the difference between two population means, $\mu_1 - \mu_2$, will be based on the following assumptions:

Assumptions for inferences about the difference between two means, $\mu_1 - \mu_2$:

The samples are randomly selected from normally distributed populations, and the samples are selected in an independent manner.

NO ASSUMPTIONS ARE MADE ABOUT THE POPULATION VARIANCES.

Since the samples provide the information for determining the standard error, the *t*-distribution will be used as the test statistic. The inferences are divided into two cases.

CASE 1: The *t*-distribution will be used, and the number of degrees of freedom will be calculated.

CASE 2: The *t*-distribution will be used, and the number of degrees of freedom will be approximated.

Case 1 will occur when you are completing the inference *using a computer or statistical calculator and the statistical software or program calculates the number of degrees of freedom* for you. The calculated value for df is a function of both sample sizes and their relative sizes, and both sample variances and their relative sizes. The value of df will be a number between the smaller of $df_1 = n_1 - 1$ or $df_2 = n_2 - 1$ and the sum of the degrees of freedom, $df_1 + df_2 = [(n_1 - 1) + (n_2 - 1)] = n_1 + n_2 - 2$.

Case 2 will occur when you are completing the inference *without the aid of a computer or calculator and its statistical software package.* Use of the *t*-distribution with the smaller of $df_1 = n_1 - 1$ or $df_2 = n_2 - 1$ will give conservative results. Because of this approximation, the true level of confidence for an interval estimate will be slightly higher than the reported level of confidence, or the true *p*-value and the true level of significance for a hypothesis test will be slightly less than reported. The gap between these reported values and the true values will be quite small, unless the sample sizes are quite small and unequal or the sample variances are very different. The gap will decrease as the samples increase in size or as the sample variances are more alike.

Since the only difference between the two cases is the number of degrees of freedom used to identify the *t*-distribution involved, we will study case 2 first.

NOTE $A > B$ ("A is greater than B") is equivalent to $B < A$ ("B is less than A"). When the difference between A and B is being discussed, it is customary to express the difference as "larger − smaller" so that the resulting difference is positive: $A - B > 0$. To express the difference as "smaller − larger" results in $B - A < 0$ (the difference is negative), which is usually unnecessarily confusing. Therefore, it is recommended that the difference be expressed as "larger − smaller."

Confidence Interval Procedure

We will use the following formula for calculating the endpoints of the $1 - \alpha$ confidence interval.

Confidence Interval for the Difference between Two Means (Independent Samples)

$$(\bar{x}_1 - \bar{x}_2) - t(df, \alpha/2) \cdot \sqrt{\left(\frac{s_1^2}{n_1}\right) + \left(\frac{s_2^2}{n_2}\right)} \quad \text{to}$$

$$(\bar{x}_1 - \bar{x}_2) + t(df, \alpha/2) \cdot \sqrt{\left(\frac{s_1^2}{n_1}\right) + \left(\frac{s_2^2}{n_2}\right)} \tag{10.8}$$

where df equals the smaller of df_1 or df_2.

In order to construct a confidence interval for the difference between two means, let's look at some sample information on student heights from a certain college campus. The sample information, given in Table 10.2, contains the heights (in inches) of 20 randomly selected women and 30 randomly selected men, taken in order to estimate the difference in their mean heights. We will assume that the heights are approximately normally distributed for both populations as we begin the five-step process to find the 95% confidence interval for the difference between the mean heights, $\mu_m - \mu_f$.

Table 10.2 **Sample Information on Student Heights**

Sample	Number	Mean	Standard Deviation
Female (*f*)	20	63.8	2.18
Male (*m*)	30	69.8	1.92

STEP 1 **Parameter of interest:**
$\mu_m - \mu_f$, the difference between the mean height of male students and the mean height of female students.

STEP 2 **a. Assumptions:**
Both populations are approximately normally distributed, and the samples were random and independently selected.

b. **Probability distribution:**
The *t*-distribution with df = 19, the smaller of $n_m - 1 = 30 - 1 = 29$ or $n_f - 1 = 20 - 1 = 19$, and formula (10.8).

c. **Level of confidence:**
$1 - \alpha = 0.95$.

STEP 3 **Sample information:** See Table 10.2.

STEP 4 **a. Confidence coefficient:**
We have a two-tailed situation with $\alpha/2 = 0.025$ in one tail and df = 19. From

Table 6 in Appendix B, $t(df, \alpha/2) = t(19, 0.025) = 2.09$. See the figure below.

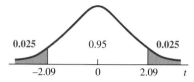

See pages 185–186 for instructions on using Table 6.

b. **Maximum error of estimate:**
Using the maximum error part of formula (10.8), we have

$$E = t(df, \alpha/2) \cdot \sqrt{\left(\frac{s_1^2}{n_1}\right) + \left(\frac{s_2^2}{n_2}\right)}:$$

$$E = 2.09 \cdot \sqrt{\left(\frac{1.92^2}{30}\right) + \left(\frac{2.18^2}{20}\right)}$$

$$= (2.09)(0.60) = \mathbf{1.25}$$

c. **Lower and upper confidence limits:**

$$(\bar{x}_1 - \bar{x}_2) \pm E$$

$$6.00 \pm 1.25$$

$$6.00 - 1.25 = \mathbf{4.75} \quad \text{to} \quad 6.00 + 1.25 = \mathbf{7.25}$$

STEP 5 a. **Confidence interval:**
4.75 to 7.25 is the 95% confidence interval for $\mu_m - \mu_f$.

b. That is, with 95% confidence, we can say that the difference between the mean heights of the male and female students is between 4.75 and 7.25 inches; that is, the mean height of male students is between 4.75 and 7.25 inches greater than the mean height of female students.

Hypothesis-Testing Procedure

When we test a null **hypothesis about the difference between two population means,** the test statistic used will be the difference between the observed difference of the sample means and the hypothesized difference of the population means, divided by the estimated standard error. The test statistic is assumed to have approximately a t-distribution when the null hypothesis is true and the normality assumption has been satisfied. The calculated value of the **test statistic** is found using this formula:

Test Statistic for the Difference between Two Means (Independent Samples)

$$t\bigstar = \frac{(\bar{x}_1 - \bar{x}_2) - (\mu_1 - \mu_2)}{\sqrt{\left(\frac{s_1^2}{n_1}\right) + \left(\frac{s_2^2}{n_2}\right)}} \quad (10.9)$$

where df is the smaller of df_1 or df_2.

NOTE: A hypothesized difference between the two population means, $\mu_1 - \mu_2$, can be any specified value. The most common value specified is zero; however, the difference can be nonzero.

In order to examine the hypothesis procedure more closely, let's look at one- and two-tailed tests.

ONE-TAILED HYPOTHESIS TEST FOR THE DIFFERENCE BETWEEN TWO MEANS

For this test, let's suppose that we are interested in comparing the academic success of college students who belong to fraternal organizations with the academic success of those who do not belong to fraternal organizations. The reason for the comparison is the recent concern that fraternity members, on the average, are achieving at a lower academic level than nonfraternal students. (Cumulative grade point average is used to measure academic success.) Random samples of size 40 are taken from each population and are listed in Table 10.3. Using the five-step process, let's complete a hypothesis test using $\alpha = 0.05$ and assume that the grade point averages for both groups are approximately normally distributed.

Table 10.3 **Sample Information on Academic Success**

Sample	Number	Mean	Standard Deviation
Fraternity members (f)	40	2.03	0.68
Nonmembers (n)	40	2.21	0.59

STEP 1 a. **Parameter of interest:**
$\mu_n - \mu_f$, the difference between the mean grade point averages for the nonfraternity members and the fraternity members.

b. Statement of hypotheses:

$H_o: \mu_n - \mu_f = 0 \ (\leq)$ (fraternity averages are no lower)

$H_a: \mu_n - \mu_f > 0$ (fraternity averages are lower)

STEP 2 **a. Assumptions:**

Both populations are approximately normally distributed, and random samples were selected. Since the two populations are separate, the samples are independent.

b. Test statistic:

The t-distribution with df = the smaller of df_n or df_f; since both n's are 40, df = 40 − 1 = **39**; and $t\bigstar$ is calculated using formula (10.9). (When df is not in the table, use the next smaller df value.)

c. Level of significance: $\alpha = 0.05$.

Using the p-value procedure:

a. Use the right-hand tail because H_a expresses concern for values related to "greater than." $P = P(t\bigstar > 1.26$, with df = 39) as shown on the figure below.

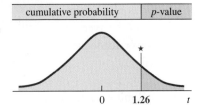

To find the p-value, use one of three methods:

1. Use Table 6 in Appendix B to place bounds on the p-value: **0.10 < P < 0.25.**
2. Use Table 7 in Appendix B to place bounds on the p-value: **0.100 < P < 0.119.**
3. Use a computer or calculator to find the p-value: **P = 0.1076.**

Specific details follow this example.

b. The p-value is not smaller than α.

STEP 3 **a. Sample information:** See Table 10.3.

b. Calculated test statistic:

$$t\bigstar = \frac{(\bar{x}_1 - \bar{x}_2) - (\mu_1 - \mu_2)}{\sqrt{\left(\frac{s_1^2}{n_1}\right) + \left(\frac{s_2^2}{n_2}\right)}};$$

$$t\bigstar = \frac{(2.21 - 2.03) - (0.00)}{\sqrt{\left(\frac{0.59^2}{40}\right) + \left(\frac{0.68^2}{40}\right)}}$$

$$= \frac{0.18}{\sqrt{0.00870 + 0.01156}} = \frac{0.18}{0.1423} = \mathbf{1.26}$$

STEP 4 **The probability distribution:**

Again, we can use either the p-value or the classical procedure:

Using the classical procedure:

a. The critical region is the right-hand tail because H_a expresses concern for values related to "greater than." The critical value is obtained from Table 6: $t(39, 0.05) = \mathbf{1.69}$.

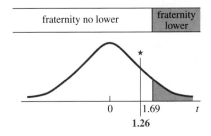

See pages 185–186 for information about critical values.

b. $t\bigstar$ is not in the critical region, as shown in red on the figure.

STEP 5 **a. Decision about H_o:** Fail to reject H_o.

b. Conclusion:

At the 0.05 level of significance, the claim that the fraternity members achieve at a lower level than nonmembers is not supported by the sample data.

To Find the p-Value Use One of Three Methods:

Method 1: Use Table 6. Find 1.26 between two entries in the df = 39 row and read the bounds for **P** from the one-tail heading at the top of the table: **0.10 < P < 0.25.**

Method 2: Use Table 7. Find $t\bigstar = 1.26$ between two rows and df=39 between two columns; read the bounds for $P(t\bigstar > 1.26|df = 39)$: **0.100 < P < 0.119.**

Method 3: If you are doing the hypothesis test with the aid of a computer or calculator, most likely it will calculate the p-value for you.

TWO-TAILED HYPOTHESIS FOR THE DIFFERENCE BETWEEN TWO MEANS

The difference between two means can also be used in a two-tailed situation. For example, many students have complained that the soft-drink vending machine A (in the student recreation room) dispenses a different amount of drink than machine B (in the faculty lounge). To test this belief, a student randomly sampled several servings from each machine and carefully measured them, with the results shown in Table 10.4.

Table 10.4 **Sample Information on Vending Machines**

Machine	Number	Mean	Standard Deviation
A	10	5.38	1.59
B	12	5.92	0.83

Does this evidence support the hypothesis that the mean amount dispensed by machine A is different from the amount dispensed by machine B? Let's assume that the amounts dispensed by both machines are normally distributed and complete the test using the five-step process, with $\alpha = 0.10$.

STEP 1 a. **Parameter of interest:**
$\mu_B - \mu_A$, the difference between the mean amount dispensed by machine B and the mean amount dispensed by machine A.

b. **Statement of hypotheses:**

$H_o: \mu_B - \mu_A = 0$ (A dispenses the same amount as B)

$H_a: \mu_B - \mu_A \neq 0$ (A dispenses a different amount than B)

STEP 2 a. **Assumptions:**
Both populations are assumed to be approximately normal, and the samples were random and independently selected.

b. **Test statistic:**
The t-distribution with df = the smaller of $n_A - 1 = 10 - 1 = 9$ or $n_B - 1 = 12 - 1 = 11$, so df = 9, and $t\bigstar$ calculated using formula (10.9).

c. **Level of significance:** $\alpha = 0.10$.

STEP 3 a. **Sample information:** See Table 10.4.

b. **Calculated test statistic:**

$$t\bigstar = \frac{(\bar{x}_B - \bar{x}_A) - (\mu_B - \mu_A)}{\sqrt{\left(\frac{s_B^2}{n_B}\right) + \left(\frac{s_A^2}{n_A}\right)}}:$$

$$t\bigstar = \frac{(5.92 - 5.38) - (0.00)}{\sqrt{\left(\frac{0.83^2}{12}\right) + \left(\frac{1.59^2}{10}\right)}}$$

$$= \frac{0.54}{\sqrt{0.0574 + 0.2528}} = \frac{0.54}{0.557} = \mathbf{0.97}$$

STEP 4 **The probability distribution:**
Again, we can use either the p-value or the classical procedure:

Using the p-value procedure:

a. Use both tails because H_a expresses concern for values related to "different than."

$\mathbf{P} = p\text{-value} = P(t\bigstar < -0.97) + P(t\bigstar > 0.97)$
$= 2 \times P(t\bigstar > 0.97 | df = 9)$
as in the figure.

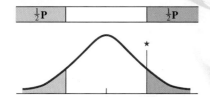

To find the p-value, you have three options:
1. Use Table 6 in Appendix B: $0.20 < \mathbf{P} < 0.50$.
2. Use Table 7 in Appendix B to place bounds on the p-value: $\mathbf{0.340 < P < 0.394}$.
3. Use a computer or calculator to find the p-value: $\mathbf{P} = 2 \times 0.1787 = \mathbf{0.3574}$.
For specific instructions, see *Methods 1, 2, and 3* on the next page.

b. The p-value is not smaller than α.

Using the classical procedure:

a. The critical region is two-tailed because H_a expresses concern for values related to "different than." The right-hand critical value is obtained from Table 6: $t(9, 0.05) = \mathbf{1.83}$. See the figure below.

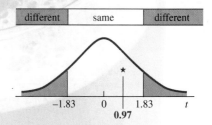

For specific instructions, see pages 185–186.

b. $t\bigstar$ is not in the critical region, as shown in **red** on the figure.

(continued on p. 221)

POLISHING A MICROCHIP

Raul has developed a new technique for polishing the reflective surface of a silicone microchip. This microchip will be used with a laser as part of his research project. As you can see in the figure below, the roughness of the surface is measured by the distance, *x*, between the surface and the place of the "highest" points on the surface, and is measured in nanometers (nms). The nanometer (one-billionth of a meter) is used to measure things that are very small, such as atoms and molecules.

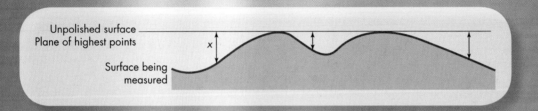

Larger values of this distance *x* (surface height), along with a large standard deviation, indicate a rougher surface. Typically *x* ranges in value from 4 to 20 nms. The table on the right lists a set of measurements taken at random locations on the microchip's unpolished surface.

Surface Heights, *x*, of an Unpolished Microchip (nms)							
8.651	11.849	7.708	8.184	7.978	4.339	9.194	9.182
5.202	6.309	10.588	8.106	9.877	7.038	9.748	12.049
8.497	7.953	5.641	4.073	7.437	14.824	11.943	8.353
14.730	9.933	7.101	18.570	4.684	8.546	8.216	8.271
10.327	9.748	12.452					

Raul's goal is to use his new technique to make the microchip's surface significantly smoother, as you can see in the illustration below. It is challenging to prove that the process makes the microchip smoother because the microchip is less than 0.25 cm-square and is thinner than a human hair. The table below is a set of measurements taken at random locations on the microchip after Raul's process was applied.

Do you think Raul's technique made a significant difference? Complete exercises 10.25 and 10.26 at the end of the chapter to complete your investigation.

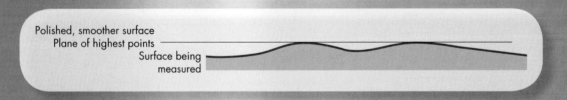

THE HUMAN EYE CANNOT
SEE 20 NMS.

Surface Heights, *x*, of a Polished Microchip (nms)							
2.077	3.096	2.110	2.264	2.039	2.437	2.181	2.510
2.354	1.732	2.120	2.545	2.054	1.562	2.231	1.480
1.775	2.230	1.465	1.548	1.979	1.993	2.263	1.913
2.177	2.201	2.861	3.241	2.183	1.639	2.342	1.428

a. Decision: Fail to reject H_o.

b. Conclusion:

The evidence is not sufficient to show that machine A dispenses a different amount of soft drink than machine B, at the 0.10 level of significance. Thus, for lack of evidence we will proceed as though the two machines dispense, on average, the same amount.

To Find the *p*-Value Use One of Three Methods:

Method 1: Use Table 6. Find 0.97 between two entries in the df = 9 row and read the bounds for **P** from the two-tail heading at the top of the table: $0.20 < P < 0.50$.

Method 2: Use Table 7. Find $t\bigstar = 0.97$ between two rows and df = 9 between two columns; read the bounds for $P(t\bigstar > 0.97|df = 9)$: $0.170 < \frac{1}{2}P < 0.197$; therefore, $0.340 < P < 0.394$.

Method 3: If you are doing the hypothesis test with the aid of a computer or calculator, most likely it will calculate the *p*-value for you (do not double it).

10.4 Inferences Concerning the Difference between Proportions Using Two Independent Samples

WE ARE OFTEN INTERESTED IN MAKING STATISTICAL COMPARISONS BETWEEN THE **PROPORTIONS, PERCENTAGES,** OR **PROBABILITIES** ASSOCIATED WITH TWO POPULATIONS.

These questions ask for such comparisons: Is the proportion of homeowners who favor a certain tax proposal different from the proportion of renters who favor it? Did a larger percentage of this semester's class than of last semester's class pass statistics? Is the probability of a Democratic candidate winning in New York greater than the probability of a Republican candidate winning in Texas? Do students' opinions about the new code

of conduct differ from those of the faculty? You have probably asked similar questions.

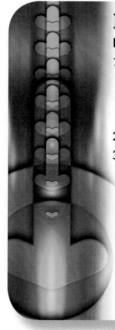

RECALL the properties of a binomial experiment:

1. The observed probability is $p' = x/n$, where x is the number of observed successes in n trials.

2. $q' = 1 - p'$.

3. p is the probability of success on an individual trial in a binomial probability experiment of n repeated independent trials.

AND the 3 "p" words (*proportion, percentage, probability*) are all the binomial parameter p, $P(\text{success})$.

In this section, we will compare two population proportions by using the difference between the observed proportions, $p'_1 - p'_2$, of two independent samples. The observed difference, $p'_1 - p'_2$, belongs to a sampling distribution with the characteristics described in the following statement:

IF independent samples of sizes n_1 and n_2 are drawn randomly from large populations with $p_1 = P_1(\text{success})$ and $p_2 = P_2(\text{success})$, respectively,

THEN the sampling distribution of $p'_1 - p'_2$ has these properties:

1. mean $\mu_{p'_1 - p'_2} = p_1 - p_2$,

2. standard error $\sigma_{p'_1 - p'_2} = \sqrt{\left(\dfrac{p_1 q_1}{n_1}\right) + \left(\dfrac{p_2 q_2}{n_2}\right)}$, (10.10) and

3. an approximately normal distribution if n_1 and n_2 are sufficiently large.

In practice, we use the following *guidelines to ensure normality*:

1. The sample sizes are both larger than 20.

2. The products $n_1 p_1$, $n_1 q_1$, $n_2 p_2$, and $n_2 q_2$ are all larger than 5.

3. The samples consist of less than 10% of their respective populations.

NOTE: p_1 and p_2 are unknown; therefore, the products mentioned in guideline 2 will be estimated by $n_1 p_1'$, $n_1 q_1'$, $n_2 p_2'$, and $n_2 q_2'$.

Inferences about the difference between two population proportions, $p_1 - p_2$, will be based on the following assumptions.

> ## Assumption for inferences about the difference between two proportions $p_1 - p_2$:
>
> The n_1 random observations and the n_2 random observations that form the two samples are selected independently from two populations that are not changing during the sampling.

Confidence Interval Procedure

When we estimate the **difference between two proportions**, $p_1 - p_2$, we will base our estimates on the **sample statistic** $p_1' - p_2'$. The point estimate, $p_1' - p_2'$, becomes the center of the confidence interval, and the confidence interval limits are found using the following formula:

Confidence Interval for the Difference between Two Proportions

$$(p_1' - p_2') - z(\alpha/2) \cdot \sqrt{\left(\frac{p_1' q_1'}{n_1}\right) + \left(\frac{p_2' q_2'}{n_2}\right)}$$

to

$$(p_1' - p_2') + z(\alpha/2) \cdot \sqrt{\left(\frac{p_1' q_1'}{n_1}\right) + \left(\frac{p_2' q_2'}{n_2}\right)} \qquad (10.11)$$

CONSTRUCTING A CONFIDENCE INTERVAL FOR THE DIFFERENCE BETWEEN TWO PROPORTIONS

In order to look more closely at how to construct this confidence interval, let's look at sample data from a campaign plan. In studying his campaign plans, Mr. Morris wishes to estimate the difference between men's and women's views regarding his appeal as a candidate. He asks his campaign manager to take two random independent samples and find the 99% confidence interval for the difference. A sample of 1,000 voters was taken from each population, with 388 men and 459 women favoring Mr. Morris. We will use the five-step process to estimate the difference between these two proportions.

STEP 1 Parameter of interest:

$p_w - p_m$, the difference between the proportion of women voters and the proportion of men voters who

plan to vote for Mr. Morris. (Note that it is customary to place the larger value first. That way, the point estimate for the difference is a positive value.)

STEP 2 a. **Assumptions:**
The samples are randomly and independently selected.

b. **Probability distribution:**
The standard normal distribution. The populations are large (all voters); the sample sizes are larger than 20; and the estimated values for $n_m p_m$, $n_m q_m$, $n_w p_w$, and $n_w q_w$ are all larger than 5. Therefore, the sampling distribution of $p_w' - p_m'$ should have an approximately normal distribution.

c. **Level of confidence:** $1 - \alpha = 0.99$.

STEP 3 Sample information:
We have $n_m = 1{,}000$, $x_m = 388$, $n_w = 1{,}000$, and $x_w = 459$.

$$p_m' = \frac{x_m}{n_m} = \frac{388}{1000} = \mathbf{0.388} \quad q_m' = 1 - 0.388 = \mathbf{0.612}$$

$$p_w' = \frac{x_w}{n_w} = \frac{459}{1000} = \mathbf{0.459} \quad q_w' = 1 - 0.459 = \mathbf{0.541}$$

STEP 4 a. **Confidence coefficient:**
This is a two-tailed situation, with $\alpha/2$ in each tail. From Table 4B, $z(\alpha/2) = z(0.005) = 2.58$. Instructions for Table 4B are on page 158.

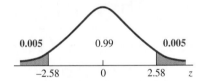

b. **Maximum error of estimate:**
Using the maximum error part of formula (10.11), we have

$$E = z(\alpha/2) \cdot \sqrt{\left(\frac{p_w' q_w'}{n_w}\right) + \left(\frac{p_m' q_m'}{n_m}\right)}$$

$$E = 2.58 \cdot \sqrt{\left(\frac{(0.459)(0.541)}{1000}\right) + \left(\frac{(0.388)(0.612)}{1000}\right)}$$

$$= 2.58\sqrt{0.000248 + 0.000237}$$

$$= (2.58)(0.022) = \mathbf{0.057}$$

c. **Lower and upper confidence limits:**

$$(p_w' - p_m') \pm E$$

$$0.071 \pm 0.057$$

$$0.071 - 0.057 = \mathbf{0.014} \text{ to } 0.071 + 0.057 = \mathbf{0.128}$$

STEP 5 **Confidence interval:**

0.014 to 0.128 is the 99% confidence interval for $p_w - p_m$. With 99% confidence, we can say that there is a difference of from 1.4% to 12.8% in Mr. Morris's voter appeal. That is, a larger proportion of women than men favor Mr. Morris, and the difference in the proportions is between 1.4% and 12.8%.

Confidence intervals and hypothesis tests can sometimes be interchanged; that is, a confidence interval can be used in place of a hypothesis test. The previous example with Mr. Morris called for a confidence interval. Now suppose that Mr. Morris asked, "Is there a difference in my voter appeal to men voters as opposed to women voters?" To answer his question, you would not need to complete a hypothesis test if you chose to test at $\alpha = 0.01$ using a two-tailed test. "No difference" would mean a difference of zero, which is not included in the interval from 0.014 to 0.128 (the interval determined in the example). Therefore, a null hypothesis of "no difference" would be rejected, thereby substantiating the conclusion that a significant difference exists in voter appeal between the two groups.

Hypothesis-Testing Procedure

When we test the null **hypothesis "There is no difference between two proportions,"** the **test statistic** will be the difference between the observed proportions divided by the **standard error**; it is found with the following formula:

Test Statistic for the Difference between Two Proportions—Population Proportion Known

$$z\bigstar = \frac{p'_1 - p'_2}{\sqrt{pq\left[\left(\frac{1}{n_1}\right) + \left(\frac{1}{n_2}\right)\right]}} \qquad (10.12)$$

NOTES:

1. The null hypothesis is $p_1 = p_2$, or $p_1 - p_2 = 0$ (the difference is zero).

2. Nonzero differences between proportions are not discussed in this section.

3. The numerator of formula (10.12) could be written as $(p'_1 - p'_2) - (p_1 - p_2)$, but since the null hypothesis is assumed to be true during the test, $p_1 - p_2 = 0$. By substitution, the numerator becomes simply $p'_1 - p'_2$.

4. Since the null hypothesis is $p_1 = p_2$, the standard error of $p'_1 - p'_2$, $\sqrt{\left|\frac{p_1 q_1}{n_1}\right| + \left|\frac{p_2 q_2}{n_2}\right|}$, can be written as $\sqrt{pq\left[\left(\frac{1}{n_1}\right) + \left(\frac{1}{n_2}\right)\right]}$, where $p = p_1 = p_2$ and $q = 1 - p$.

5. When the null hypothesis states that $p_1 = p_2$ and does not specify the value of either p_1 or p_2, the two sets of sample data will be pooled to obtain the estimate for p. This pooled probability (known as p'_p) is the total number of successes divided by the total number of observations with the two samples combined; it is found using the next formula:

$$p'_p = \frac{x_1 + x_2}{n_1 + n_2} \qquad (10.13)$$

and q'_p is its complement,

$$q'_p = 1 - p'_p \qquad (10.14)$$

When the pooled estimate, p'_p, is being used, formula (10.12) becomes formula (10.15):

Test Statistic for the Difference between Two Proportions—Population Proportion Unknown

$$z\bigstar = \frac{p'_1 - p'_2}{\sqrt{(p'_p)(q'_p)\left[\left(\frac{1}{n_1}\right) + \left(\frac{1}{n_2}\right)\right]}} \qquad (10.15)$$

ONE-TAILED HYPOTHESIS TEST FOR THE DIFFERENCE BETWEEN TWO PROPORTIONS

→ To examine the difference between two proportions more closely, think of a salesman for a new manufacturer of cellular phones. He claims not only that they cost the retailer less but also that the percentage of defective cellular phones found among his products will be no higher than the percentage of defectives found in a competitor's line. To test this statement, the retailer took random samples of each manufacturer's product. The sample summaries are given in Table 10.5. Can we reject the salesman's claim at the 0.05 level of significance? Let's use the five-step process to find out.

Table 10.5 Cellular Phone Sample Information

Product	Number Defective	Number Checked
Salesman's	15	150
Competitor's	6	150

STEP 1 a. **Population parameter of interest:**
$p_s - p_c$, the difference between the proportion of defectives in the salesman's product and the proportion of defectives in the competitor's product.

b. **Statement of hypotheses:**
The concern of the retailer is that the salesman's less expensive product may be of a

poorer quality, meaning a greater proportion of defectives. If we use the difference "suspected larger proportion − smaller proportion," then the alternative hypothesis is "The difference is positive (greater than zero)."

$H_o: p_s - p_c = 0$ (≤) (salesman's defective rate is no higher than competitor's)

$H_a: p_s - p_c > 0$ (salesman's defective rate is higher than competitor's)

STEP 2 a. **Assumptions:**
Random samples were selected from the products of two different manufacturers.

b. **Probability distribution:**
The standard normal distribution. Populations are very large (all cellular phones produced); the samples are larger than 20; and the estimated products $n_s p'_s$, $n_s q'_s$, $n_c p'_c$, and $n_c q'_c$ are all larger than 5. Therefore, the sampling distribution should have an approximately normal distribution. $z\bigstar$ will be calculated using formula (10.15).

c. **Determine the level of significance:** $\alpha = 0.05$.

STEP 3 a. **Sample information:**

$$p'_s = \frac{x_s}{n_s} = \frac{15}{150} = \mathbf{0.10}$$

$$p'_c = \frac{x_c}{n_c} = \frac{6}{150} = \mathbf{0.04}$$

$$p'_p = \frac{x_1 + x_2}{n_1 + n_2} = \frac{15 + 6}{150 + 150} = \frac{21}{300} = \mathbf{0.07}$$

$$q'_p = 1 - p'_p = 1 - 0.07 = \mathbf{0.93}$$

b. **Test statistic:**

$$z\bigstar = \frac{p'_s - p'_c}{\sqrt{(p'_p)(q'_p)\left[\left(\frac{1}{n_s}\right) + \left(\frac{1}{n_c}\right)\right]}}:$$

$$z\bigstar = \frac{0.10 - 0.04}{\sqrt{(0.07)(0.93)\left[\left(\frac{1}{150}\right) + \left(\frac{1}{150}\right)\right]}}$$

$$= \frac{0.06}{\sqrt{0.000868}} = \frac{0.06}{0.02946} = \mathbf{2.04}$$

STEP 4 The probability distribution:
Again, we can use either the *p*-value or the classical procedure:

Using the *p*-value procedure:

a. Use the right-hand tail because H_a expresses concern for values related to "higher than." $\mathbf{P} = p\text{-value} = P(z \bigstar > 2.04)$ as shown on the figure below.

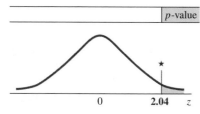

To find the *p*-value, you have three options:
1. Use Table 3 in Appendix B to calculate the *p*-value: $\mathbf{P} = 1.0000 - 0.9793 = \mathbf{0.0207}$.
2. Use Table 5 in Appendix B to place bounds on the *p*-value: $\mathbf{0.0202 < P < 0.0228}$.
3. Use a computer or calculator: $\mathbf{P} = \mathbf{0.0207}$.
For specific instructions, see pages 170–172.

b. The *p*-value is smaller than α.

Using the classical procedure:

a. The critical region is the right-hand tail because H_a expresses concern for values related to "higher than." The critical value is obtained from Table 4A: $z(0.05) = \mathbf{1.65}$.

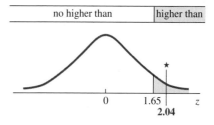

For specific instructions, see pages 175–176.

b. $z \bigstar$ is in the critical region, as shown in red on the figure.

STEP 5 The results:
a. **Decision:** Reject H_o.
b. **Conclusion:**
At the 0.05 level of significance, there is sufficient evidence to reject the salesman's claim; the proportion of his company's cellular phones that are defective is higher than the proportion of his competitor's cellular phones that are defective.

10.5 Inferences Concerning the Ratio of Variances Using Two Independent Samples

WHEN COMPARING TWO POPULATIONS, WE NATURALLY COMPARE THEIR TWO MOST FUNDAMENTAL DISTRIBUTION CHARACTERISTICS, THEIR "CENTER" AND THEIR "SPREAD," BY COMPARING THEIR MEANS AND STANDARD DEVIATIONS.

We have learned in two of the previous sections how to use the *t*-distribution to make inferences comparing two population means with either dependent or independent samples. These procedures were intended to be used with normal populations, but they work quite well even when the populations are not exactly normally distributed.

The next logical step in comparing two populations is to compare their standard deviations, the most often used measure of spread. However, sampling distributions that deal with sample standard deviations (or variances) are very sensitive to slight departures from the assumptions. Therefore, the only inference procedure to be presented here will be the **hypothesis test for**

the **equality of standard deviations (or variances)** for two normal populations.

The soft-drink bottling company discussed in Objective 9.3 (page 198) is trying to decide whether to install a modern, high-speed bottling machine. There are, of course, many concerns in making this decision, and one of them is that the increased speed may result in increased variability in the amount of fill placed in each bottle; such an increase would not be acceptable. To this concern, the manufacturer of the new system responded that the variance in fills will be no greater with the new machine than with the old. (The new system will fill several bottles in the same amount of time as the old system fills one bottle; this is the reason the change is being considered.) A test is set up to statistically test the bottling company's concern, "Standard deviation of new machine is greater than standard deviation of old," against the manufacturer's claim, "Standard deviation of new is no greater than standard deviation of old."

Writing Hypotheses for the Equality of Variances

Imagine you were given the task of stating the null and alternative hypotheses to be used for comparing the variances of the two soft-drink bottling machines. You could do this using one of several equivalent ways to express the null and alternative hypotheses, but since the test procedure uses the ratio of variances, the recommended convention is to express the null and alternative hypotheses as ratios of the population variances. Furthermore, it is recommended that the "larger" or "expected to be larger" variance be the numerator. The concern of the soft-drink company is that the new modern machine (m) will result in a larger standard deviation in the amounts of fill than its present machine (p); $\sigma_m > \sigma_p$, or equivalently $\sigma_m^2 > \sigma_p^2$, which becomes $\dfrac{\sigma_m^2}{\sigma_p^2} > 1$. We want to test the manufacturer's claim (the null hypothesis) against the company's concern (the alternative hypothesis):

$$H_o: \quad \frac{\sigma_m^2}{\sigma_p^2} = 1 \quad (\leq) \quad (m \text{ is no more variable})$$

$$H_a: \quad \frac{\sigma_m^2}{\sigma_p^2} > 1 \quad (m \text{ is more variable})$$

Using the *F*-Distribution

Inferences about the ratio of variances for two normally distributed populations use the **F-distribution**. The *F*-distribution, similar to the Student's *t*-distribution and the χ^2- distribution, is a family of probability distributions. Each *F*-distribution is identified by two numbers of degrees of freedom, one for each of the two samples involved.

Before continuing with the details of the hypothesis-testing procedure, let's learn about the *F*-distribution.

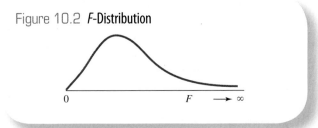

properties of the Ⓕ-distribution

1. *F* is nonnegative; it is zero or positive.

2. *F* is nonsymmetrical; it is skewed to the right.

3. *F* is distributed so as to form a family of distributions; there is a separate distribution for each pair of numbers of degrees of freedom.

For inferences discussed in this section, the number of degrees of freedom for each sample is $df_1 = n_1 - 1$ and $df_2 = n_2 - 1$. Each different combination of degrees of freedom results in a different *F*-distribution, and each *F*-distribution looks approximately like the distribution shown in Figure 10.2.

Figure 10.2 *F-Distribution*

The critical values for the *F*-distribution are identified using three values:

- df_n, the degrees of freedom associated with the sample whose variance is in the numerator of the calculated *F*,

- df_d, the degrees of freedom associated with the sample whose variance is in the denominator, and

- α, the area under the distribution curve to the right of the critical value being sought.

Therefore, the symbolic name for a critical value of *F* will be $F(df_n, df_d, \alpha)$, as shown in Figure 10.3.

Since it takes three values to identify a single critical value of *F*, making tables for *F* is not as simple as with previously studied distributions. The tables presented in this textbook are organized so as to have a different table for each different value of α, the "area

Figure 10.3 **A Critical Value of *F***

to the right." Table 9a in Appendix B shows the critical values for $F_{(df_n, df_d, \alpha)}$ when $\alpha = 0.05$; Table 9b gives the critical values when $\alpha = 0.025$; Table 9c gives the values when $\alpha = 0.01$.

If we wanted to find $F_{(5, 8, 0.05)}$, the critical *F*-value for samples of size 6 and size 9 with 5% of the area in the right-hand tail, we would need to consult Table 9a ($\alpha = 0.05$). Using the partial view of Table 9a below, notice that the intersection of column df = 5 (for the numerator) and row df = 8 (for the denominator) occurs at the value: $F_{(5, 8, 0.05)} = \mathbf{3.69}$.

Portion of Table 9a ($\alpha = 0.05$)						
		df for Numerator				
		...	5	...	8	...
df for Denom-inator	⋮					
	5				4.82	← $F_{(8, 5, 0.05)} = 4.82$
	⋮					
	8		**3.69**	←		$F_{(5, 8, 0.05)} = 3.69$
	⋮					

You can also see that if the two degrees of freedom are reversed, the resulting *F* is different: $F_{(8, 5, 0.05)}$ is 4.82. The degrees of freedom associated with the numerator and with the denominator *must* be kept in the correct order; 3.69 is different from 4.82. Check some other pairs to verify that interchanging the degrees of freedom numbers will result in different *F*-values.

Use of the *F*-distribution has a condition. That is, we make certain assumptions for inferences about the ratio of two variances: (1) the samples are randomly selected from normally distributed populations, and (2) the two samples are selected in an independent manner.

Test Statistic for Equality of Variances

$$F\bigstar = \frac{s_n^2}{s_d^2}, \quad \text{with}$$
$$df_n = n_n - 1 \text{ and } df_d = n_d - 1 \qquad (10.16)$$

The sample variances are assigned to the numerator and denominator in the order established by the null and alternative hypotheses for one-tailed tests. The calculated ratio, $F\bigstar$, will have an *F*-distribution with $df_n = n_n - 1$ (numerator) and $df_d = n_d - 1$ (denominator) when the assumptions are met and the null hypothesis is true.

ONE-TAILED HYPOTHESIS TEST FOR THE EQUALITY OF VARIANCE

Recall that our soft-drink bottling company was to make a decision about the equality of the variances of amounts of fill between its present machine and a modern high-speed outfit. Does the sample information in Table 10.6 present sufficient evidence to reject the null hypothesis (the manufacturer's claim) that the modern high-speed bottle-filling machine fills bottles with no greater variance than the company's present machine? We will assume that the amounts of fill are normally distributed for both machines, and complete the five-step process using $\alpha = 0.01$.

Table 10.6 **Sample Information on Variances of Fills**

Sample	n	s^2
Present machine (*p*)	22	0.0008
Modern high-speed machine (*m*)	25	0.0018

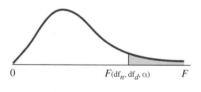 a. **Parameter of interest:**

$\dfrac{\sigma_m^2}{\sigma_p^2}$, the ratio of the variances in the amounts of fill placed in bottles for the modern machine versus the company's present machine.

b. Statement of hypotheses:
The hypotheses were established at the beginning of this section on page 226:

$$H_o: \quad \frac{\sigma_m^2}{\sigma_p^2} = 1 \quad (\leq) \quad (m \text{ is no more variable})$$

$$H_a: \quad \frac{\sigma_m^2}{\sigma_p^2} > 1 \quad (m \text{ is more variable})$$

NOTE: When the "expected to be larger" variance is in the numerator for a one-tailed test, the alternative hypothesis states "The ratio of the variances is greater than one."

STEP 2 a. Assumptions:
The sampled populations are normally distributed (given in the statement of the problem), and the samples are independently selected (drawn from two separate populations).

b. Test statistic:
The F-distribution with the ratio of the sample variances and formula (10.16).

c. Level of significance: $\alpha = 0.01$.

STEP 3 a. Sample information: See Table 10.6.

b. Calculated test statistic:
Using formula (10.16), we have

$$F\star = \frac{s_m^2}{s_p^2}: \quad F\star = \frac{0.0018}{0.0008} = 2.25$$

The number of degrees of freedom for the numerator is $df_n = 24$ (or $25 - 1$) because the sample from the modern high-speed machine is associated with the numerator, as specified by the null hypothesis. Also, $df_d = 21$ because the sample associated with the denominator has size 22.

STEP 4 The probability distribution:
Again, we can use either the p-value or the classical procedure:

Using the p-value procedure:

a. Use the right-hand tail because H_a expresses concern for values related to "more than."
$P = P(F\star > 2.25, \text{ with } df_n = 24 \text{ and } df_d = 21)$ as shown on the figure below.

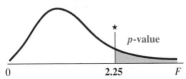

To find the p-value, you have two options:
1. Use Tables 9a and 9b in Appendix B to place bounds on the p-value: $0.025 < P < 0.05$.
2. Use a computer or calculator to find the p-value: $P = 0.0323$.
Specific instructions follow this example.

b. The p-value is not smaller than the level of significance, α (0.01).

Using the classical procedure:

a. The critical region is the right-hand tail because H_a expresses concern for values related to "more than." $df_n = 24$ and $df_d = 21$. The critical value is obtained from Table 9c: $F(24, 21, 0.01) = 2.80$.

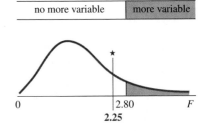

For additional instructions, see pages 226–227.

b. $F\star$ is not in the critical region, as shown in red on the figure.

STEP 5 The results:
a. **Decision:** Fail to reject H_o.
b. **Conclusion:**
At the 0.01 level of significance, the samples do not present sufficient evidence to indicate an increase in variance.

Calculating the p-Value When Using the F-Distribution

There are two methods for calculating the p-value when using the F-distribution:

Method 1: Use Table 9 in Appendix B to place bounds on the p-value. Using Tables 9a, 9b, and 9c in Appendix B to estimate the p-value is very limited. However, for the soft-drink example we just worked through comparing variances of fill amounts for the present machine and a more modern machine, the p-value can be estimated. By inspecting

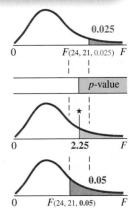

Tables 9a and 9b, you will find that $F_{(24, 21, 0.025)} = 2.37$ and $F_{(24, 21, 0.05)} = 2.05$. $F\bigstar = 2.25$ is between the values 2.37 and 2.05; therefore, the p-value is between 0.025 and 0.05: **$0.025 < P < 0.05$**. (See the figure at the bottom of the previous page.)

FYI α must still be split between the two tails for a two-tailed H_o.

Method 2: If you are doing the hypothesis test with the aid of a computer or calculator, most likely it will calculate the p-value for you.

Critical *F*-Values for One- and Two-Tailed Tests

The tables of critical values for the F-distribution give only the right-hand critical values. This will not be a problem because the right-hand critical value is the only critical value that will be needed. You can adjust the numerator–denominator order so that all the "activity" is in the right-hand tail. There are two cases: one-tailed tests and two-tailed tests.

One-tailed tests: Arrange the null and alternative hypotheses so that the alternative is always "greater than." The $F\bigstar$-value is calculated using the same order as specified in the null hypothesis (recall the soft drink bottling example).

Two-tailed tests: When the value of $F\bigstar$ is calculated, always use the sample with the larger variance for the numerator; this will make $F\bigstar$ greater than one and place it

in the right-hand tail of the distribution. Thus, you will need only the critical value for the right-hand tail.

All hypothesis tests about two variances can be formulated and completed in a way that both the critical value of F and the calculated value of $F\bigstar$ will be in the right-hand tail of the distribution. Since Tables 9a, 9b, and 9c contain only critical values for the right-hand tail, this will be convenient and you will never need critical values for the left-hand tail.

FORMAT FOR WRITING HYPOTHESES FOR THE EQUALITY OF VARIANCES

➡ To use the correct format for writing hypotheses for the equality of variances, first reorganize the alternative hypothesis so that the critical region will be the right-hand tail:

$$H_a: \sigma_1^2 < \sigma_2^2 \text{ or } \frac{\sigma_1^2}{\sigma_2^2} < 1 \text{ (population 1 is less variable)}$$

Then, reverse the direction of the inequality, and reverse the roles of the numerator and denominator.

$$H_a: \sigma_2^2 > \sigma_1^2 \text{ or } \frac{\sigma_2^2}{\sigma_1^2} > 1 \text{ (population 2 is more variable)}$$

The calculated test statistic $F\bigstar$ will be $\frac{s_2^2}{s_1^2}$.

The preceding three chapters have provided tools for using confidence intervals and hypothesis tests to answer questions about means, proportions, and standard deviations for one or two populations. In the remaining chapters, we'll add techniques to make inferences about more than two populations as well as inferences of different types.

Personality Characteristics of Police Academy Applicants

Bruce N. Carpenter and Susan M. Raza concluded that "police applicants are somewhat more like each other than are those in the normative population" when the F-test of homogeneity of variance resulted in a p-value of less than 0.005. *Homogeneity* means that the group's scores are less variable than the scores for the normative population.

COMPARISONS ACROSS SUBGROUPS AND WITH OTHER POPULATIONS

To determine whether police applicants are a more homogeneous group than the normative population, the F-test of homogeneity of variance was used.

　　With the exception of scales *F*, *K*, and 6, where the differences are nonsignificant, the results indicate that the police applicants form a somewhat more homogeneous group than the normative population [$F_{(237, 305)} = 1.36$, $p < 0.005$]. Thus, police applicants are somewhat more like each other than are individuals in the normative population.

SOURCE: Reproduced from the *Journal of Police Science and Administration,* Vol. 15, no. 1, pp. 10–17, with permission of the International Association of Chiefs of Police, PO Box 6010, 13 Firstfield Road, Gaithersburg, MD 20878.

problems

Objective 10.1

10.1 The students at a local high school were assigned to do a project for their statistics class. The project involved having sophomores take a timed test on geometric concepts. The statistics students would then use these data to determine whether there was a difference between male and female performances. Would the resulting sets of data represent dependent or independent samples? Explain.

10.2 In trying to estimate the amount of growth that took place in the trees recently planted by the County Parks Commission, 36 trees were randomly selected from the 4,000 planted. The heights of these trees were measured and recorded. One year later, another set of 42 trees was randomly selected and measured. Do the two sets of data (36 heights, 42 heights) represent dependent or independent samples? Explain.

10.3 Twenty people were selected to participate in a psychology experiment. They answered a short multiple-choice quiz about their attitudes on a particular subject and then viewed a 45-minute film. The following day the same 20 people were asked to answer a follow-up questionnaire about their attitudes. At the completion of the experiment, the experimenter will have two sets of scores. Do these two samples represent dependent or independent samples? Explain.

10.4 An insurance company is concerned that garage A charges more for repair work than garage B charges. It plans to send 25 cars to each garage and obtain separate estimates for the repairs needed for each car.
   a. How can the company do this and obtain independent samples? Explain in detail.
   b. How can the company do this and obtain dependent samples? Explain in detail.

10.5 In recent years, the sport of ballroom dancing has seen greater exposure in the media through popular television programs such as *Dancing With the Stars*. Local dance studios are seeing an increase in the number of people interested in taking ballroom dance lessons. Two samples of 15 dance students are judged before taking any lessons and then again after five lessons. The students belonging to the samples are to be randomly selected.
   a. How can data be collected if dependent samples are to be obtained? Explain in detail.
   b. How can data be collected if independent samples are to be obtained? Explain in detail.

Objective 10.2

10.6 Given this set of paired data:

Pairs	1	2	3	4	5
Sample A	3	6	1	4	7
Sample B	2	5	1	2	8

Find:
   a. The paired differences, $d = A - B$, for this set of data
   b. The mean $\bar{d}$ of the paired differences
   c. The standard deviation s_d of the paired differences

*10.7 All students who enroll in a certain memory course are given a pretest before the course begins. At the completion of the course, 10 students are selected at random and given a posttest; their scores are listed here.

Student	1	2	3	4	5	6	7	8	9	10
Before	93	86	72	54	92	65	80	81	62	73
After	98	92	80	62	91	78	89	78	71	80

MINITAB was used to find the 95% confidence interval for the mean improvement in memory resulting from taking the memory course, as measured by the difference in test scores (d = after − before). Verify the results shown on the output by calculating the values yourself. Assume normality.

Confidence Intervals					
Variable	N	Mean	StDev	SE Mean	95% C.I.
d	10	6.10	4.79	1.52	(2.67, 9.53)

*10.8 An experiment was designed to estimate the mean difference in weight gain for pigs fed ration A as compared with those fed ration B. Eight pairs of pigs were used. The pigs within each pair were littermates. The rations were assigned at random to the two animals within each pair. The gains (in pounds) after 45 days are shown in the following table.

Litter	1	2	3	4	5	6	7	8
Ration A	65	37	40	47	49	65	53	59
Ration B	58	39	31	45	47	55	59	51

Assuming weight gain is normal, find the 95% confidence interval estimate for the mean of the differences μ_d, where d = ration A − ration B.

*10.9 Two men, A and B, who usually commute to work together, decide to conduct an experiment to see whether one route is faster than the other. The men believe that their driving habits are approximately the same, and therefore they decide on the following procedure. Each morning for 2 weeks, A will drive to work on one route and B will use the other route. On the first morning, A will toss a coin. If heads appear, he will use route I; if tails appear, he will use route II. On the second morning, B will toss the coin: heads, route I; tails, route II. The times, recorded to the nearest minute, are shown in the following table. Assume commute times are normal and estimate the population mean difference with a 95% confidence interval.

Route	Day									
	M	Tu	W	Th	F	M	Tu	W	Th	F
I	29	26	25	25	25	24	26	26	30	31
II	25	26	25	25	24	23	27	25	29	30

*10.10 A single 19-year-old who has just purchased his/her own 2-year-old Honda Civic, might be asking, "Why does auto insurance cost so much?" There are many reasons according to the insurance agent, one of which is whether the driver is male or female. The insurance rates listed are for a random sample of 16 zip codes within a 50-mile radius of our 19-year-old in question. The data are for a policy whose features are: $500 deductible, $25,000/$50,000 bodily injury, $25,000 property, and $25,000/$50,000 uninsured/underinsured motorist.

Male ($)	1,215.30	996.30	1,179.30	1,254.30	1,110.30	2,086.60	856.30	1,298.30
Female ($)	1,015.30	812.30	987.30	1,045.30	916.30	1,804.60	671.30	1,132.30
Male ($)	760.30	956.30	1,304.30	1,548.30	1,760.30	1,337.30	1,037.30	1,182.30
Female ($)	606.30	771.30	1,095.30	1,278.30	1,444.30	1,095.30	812.30	940.30

a. At first glance, does there seem to be a pattern to the relationship between the insurance premiums for males and females? Describe it.

b. Describe each set of data graphically (males, females, and difference) using a histogram and one other graph.

c. Find the mean and the standard deviation for each set of data: males, females, and difference.

d. Are the assumptions for a mean of a paired difference confidence interval satisfied? Explain.

e. Using a 95% confidence interval, estimate the mean of the differences. Write a complete confidence interval statement.

f. Do your answers to the above questions suggest any evidence of a difference between the auto insurance rates for male and female 19-year-old drivers? Explain.

10.11 State the null hypothesis, H_o, and the alternative hypothesis, H_a, that would be used to test these claims:

a. There is an increase in the mean difference between posttest and pretest scores.

b. Following a special training session, it is believed that the mean of the difference in performance scores will not be zero.

c. On average, there is no difference between the readings from two inspectors on each of the selected parts.

d. The mean of the differences between pre–self-esteem and post–self-esteem scores showed improvement after involvement in a college learning community.

10.12 Determine the p-value for each hypothesis test for the mean difference.

a. $H_o: \mu_d = 0$ and $H_a: \mu_d > 0$, with $n = 20$ and $t\star = 1.86$

b. $H_o: \mu_d = 0$ and $H_a: \mu_d \neq 0$, with $n = 20$ and $t\star = -1.86$

c. $H_o: \mu_d = 0$ and $H_a: \mu_d < 0$, with $n = 29$ and $t\star = -2.63$

d. $H_o: \mu_d = 0.75$ and $H_a: \mu_d > 0.75$, with $n = 10$ and $t\star = 3.57$

10.13 Determine the test criteria that would be used with the classical approach to test the following hypotheses when t is used as the test statistic:

a. $H_o: \mu_d = 0$ and $H_a: \mu_d > 0$, with $n = 15$ and $\alpha = 0.05$

b. $H_o: \mu_d = 0$ and $H_a: \mu_d \neq 0$, with $n = 25$ and $\alpha = 0.05$

c. $H_o: \mu_d = 0$ and $H_a: \mu_d < 0$, with $n = 12$ and $\alpha = 0.10$

d. $H_o: \mu_d = 0.75$ and $H_a: \mu_d > 0.75$, with $n = 18$ and $\alpha = 0.01$

10.14 A random sample of 10 speed skaters, all of the relatively same experience level and speed, were selected to try out a new specialty blade. The difference in the short track times were measured as current blade time minus specialty blade time, resulting in a mean difference of 0.165 second with a standard deviation of 0.12 second. Does this sample provide sufficient evidence that the specialty blade is beneficial in achieving faster times? Use $\alpha = 0.05$ and assume normality.

*10.15 Ten recently diagnosed diabetics were tested to determine whether an educational program was effective in increasing their knowledge of diabetes. They were given a test, before and after the educational program, concerning self-care aspects of diabetes. The scores on the test were as follows:

Patient	1	2	3	4	5	6	7	8	9	10
Before	75	62	67	70	55	59	60	64	72	59
After	77	65	68	72	62	61	60	67	75	68

The following MINITAB output may be used to determine whether the scores improved as a result of the program. Verify the values shown on the output [mean difference (MEAN), standard deviation of the difference (STDEV), standard error of the difference (SE MEAN), $t\star$ (T-Value), and p-value] by calculating the values yourself.

Paired T for After − Before

	N	Mean	StDev	SE Mean
After	10	67.50	5.80	1.83
Before	10	64.30	6.50	2.06
Difference	10	3.200	2.741	0.867

T-Test of mean difference = 0 (vs > 0); T-Value = 3.69
p-Value = 0.002

*10.16 Ten randomly selected college students, who participated in a learning community, were given pre–self-esteem and post–self-esteem surveys. A learning community is a group of students who take two or more courses together. Typically, each learning community has a theme, and the faculty involved coordinate assignments linking the courses. Research has shown that the benefits of higher self-esteem, higher grade point averages (GPAs), and improved satisfaction in courses, as well as better retention rates, result from involvement in a learning community. The scores on the surveys are as follows:

Student	1	2	3	4	5	6	7	8	9	10
Prescore	18	14	11	23	19	21	21	21	11	22
Postscore	17	17	10	25	20	10	24	22	10	24

Does this sample of students show sufficient evidence that self-esteem scores were higher after participation in a learning community? Lower scores indicate higher self-esteem. Use the 0.05 level of significance and assume normality of scores.

*10.17 To test the effect of a physical fitness course on one's physical ability, the number of sit-ups that a person could do in 1 minute, both before and after the course, was recorded. Ten randomly selected participants scored as shown in the following table. Can you conclude that a significant amount of improvement took place? Use $\alpha = 0.01$ and assume normality.

Before	29	22	25	29	26	24	31	46	34	28
After	30	26	25	35	33	36	32	54	50	43

a. Solve using the p-value approach.
b. Solve using the classical approach.

Objective 10.3

10.18 A study comparing attitudes toward death was conducted in which organ donors (individuals who had signed organ donor cards) were compared with nondonors. The study is reported in the journal *Death Studies*. Templer's Death Anxiety Scale (DAS) was administered to both groups. On this scale, high scores indicate high anxiety concerning death. The results were reported as follows.

	n	Mean	Std. Dev.
Organ Donors	25	5.36	2.91
Nonorgan Donors	69	7.62	3.45

Construct the 95% confidence interval for the difference between the means, $\mu_{non} - \mu_{donor}$.

10.19 Women on average have 8 more pairs of shoes than do men, according to a *USA Today* Snapshot titled "Who Owns More Shoes?" (July 8, 2009). A recent study at a community college gave the following results:

	n	Mean	Std. Dev.
Males	21	8.48	4.43
Females	30	26.63	21.83

Find the 90% confidence interval for the difference between the two mean numbers of pairs of shoes for males and females.

*10.20 At a large university, a mathematics placement exam is administered to all students. Samples of 36 male and 30

female students are randomly selected from this year's student body and the following scores recorded:

Male	72	68	75	82	81	60	75	85	80	70
	71	84	68	85	82	80	54	81	86	79
	99	90	68	82	60	63	67	72	77	51
	61	71	81	74	79	76				
Female	81	76	94	89	83	78	85	91	83	83
	84	80	84	88	77	74	63	69	80	82
	89	69	74	97	73	79	55	76	78	81

Construct the 95% confidence interval for the difference between the mean scores for male and female students.

10.21 State the null and alternative hypotheses that would be used to test the following claims:
a. There is a difference between the mean age of employees at two different large companies.
b. The mean of population 1 is greater than the mean of population 2.
c. The mean yield per county of sunflower seeds in North Dakota is less than the mean yield per county in South Dakota.
d. There is no difference in the mean number of hours spent studying per week between male and female college students.

*10.22 It is a known fact that private colleges cost more than public colleges. In fact, according to the College Board, the average 2008–2009 cost (tuition, fees, room & board) for a public college is $7,020 versus $26,273 for a private college. Does this difference hold when it comes to the average cost of required textbooks per class? The following samples of size 10 were taken.

Public	Private
64.69	71.00
89.60	96.19
101.49	96.47
101.75	97.14
103.59	98.56
106.38	98.94
106.77	107.79
110.69	112.58
118.94	114.00
135.94	116.55

SOURCE: http://www.collegeboard.com/

Using the Excel output below and $\alpha = 0.05$, determine if the average cost of required textbooks per class is different between public and private colleges.
a. Solve using the p-value approach.
b. Solve using the classical approach.

t-Test: Two-Sample Assuming Unequal Variances		
	Public	Private
Mean	103.984	100.922
Variance	340.6249822	173.2995511
Observations	10	10
Hypothesized Mean Difference	0	
df	16	
t Stat	0.427125511	
P(T ≤ t) two-tail	0.674980208	
t Critical two-tail	2.119904821	

10.23 If a random sample of 18 homes south of Center Street in Provo has a mean selling price of $145,200 and a standard deviation of $4,700, and a random sample of 18 homes north of Center Street has a mean selling price of $148,600 and a standard deviation of $5,800, can you conclude that there is a significant difference between the selling price of homes in these two areas of Provo at the 0.05 level? Assume normality.
  a. Solve using the p-value approach.
  b. Solve using the classical approach.

*10.24 Are females as serious about golf as males are? If so, would the price of a driver for a male be the same as the price of a driver for a female? It was contended that drivers for females would be cheaper. Random samples of drivers were taken from the Golflink.com website. The prices (in dollars) were:

Male					
149.99	299.99	49.99	499.99	167.97	299.99
399.99	199.99	99.99	149.99		
Female					
199.99	79.99	499.99	199.97	299.99	99.99

At the 0.05 level of significance, is there sufficient evidence to support the contention that drivers for males are more expensive than drivers for females? Assume normality of golf driver prices.

*10.25 Consider the surface height data for Raul's unpolished and polished reflective surfaces of a silicone microchip given on page 220.
  a. Present and describe each set of data (unpolished and polished) using a histogram, the mean, and the standard deviation.
  b. Check each set of data for a normal distribution. 1. State what you believe to be the case based on the results found in part (a). 2. Further, find additional statistical evidence. 3. Very precisely state your conclusions regarding normality.

*10.26 Consider the surface height data for Raul's unpolished and polished reflective surfaces of a silicone microchip given on page 220 and initially investigated in Exercise 10.25.
  a. Do the two sets of data, unpolished and polished, represent independent or dependent samples? Explain your answer.

  b. Produce at least 3 graphical statistics that demonstrate the new polishing process does in fact produce a smoother reflecting surface. Explain how each graph demonstrates that the goal has been attained.
  c. Is there statistical evidence that the process has produced a surface that is significantly smoother? State the p-value.
  d. Complete the hypothesis test at the 0.01 level of significance; be sure to state your decision and conclusion.

Objective 10.4

10.27 If $n_1 = 40$, $p_1' = 0.9$, $n_2 = 50$, and $p_2' = 0.9$:
  a. Find the estimated values for both np's and both nq's.
  b. Would this situation satisfy the guidelines for approximately normal? Explain.

10.28 Calculate the estimate for the standard error of the difference between two proportions for each of the following cases:
  a. $n_1 = 40$, $p_1' = 0.8$, $n_2 = 50$, and $p_2' = 0.8$
  b. $n_1 = 33$, $p_1' = 0.6$, $n_2 = 38$, and $p_2' = 0.65$

10.29 The proportions of defective parts produced by two machines were compared, and the following data were collected:
Machine 1: $n = 150$; number of defective parts = 12
Machine 2: $n = 150$: number of defective parts = 6
Determine a 90% confidence interval for $p_1 - p_2$.

10.30 The Soap and Detergent Association issued its fifth annual Clean Hands Report Card survey for 2009. In a series of hygiene-related questions to American adults, it was found that 62% of 442 women washed their hands more than 10 times per day, while 37% of 446 men did the same. Find the 95% confidence interval for the difference in proportions of women and men that wash their hands more than 10 times a day.

10.31 In a random sample of 40 brown-haired individuals, 22 indicated that they use hair coloring. In another random sample of 40 blond individuals, 26 indicated that they use hair coloring. Use a 92% confidence interval to estimate the difference in the population proportions of brunettes and blondes who use hair coloring.

10.32 State the null hypothesis, H_o, and the alternative hypothesis, H_a, that would be used to test these claims:
  a. There is no difference between the proportions of men and women who will vote for the incumbent in next month's election.
  b. The percentage of boys who cut classes is greater than the percentage of girls who cut classes.
  c. The percentage of college students who drive old cars is higher than the percentage of noncollege students of the same age who drive old cars.

10.33 Find the value of $z\star$ that would be used to test the difference between the proportions given the following:

Sample	n	x
G	380	323
H	420	332

10.34 In a survey of families in which both parents work, 200 men and 200 women were asked the question, "Have you refused a job, promotion, or transfer because it would mean less time with your family?" 29% of the men and 24% of the women answered "Yes." Based on this information, can we conclude that there is a difference in the proportion of men and women responding "Yes" at the 0.05 level of significance?

10.35 A Harris Interactive poll found that 50% of Democrats follow professional football, whereas 59% of Republicans follow the sport. If the poll results were based on samples of 875 Democrats and 749 Republicans, determine, at the 0.05 level of significance, if the viewpoint of more Republicans following professional football is substantiated.

10.36 *Consumer Reports* conducted a survey of 1,000 adults concerning wearing bicycle helmets. One of the questions presented to 25- to 54-year-olds was whether they wore a helmet most of the time while biking or cycling. This was further broken down to whether or not they have a child at home. 87% of the age group that have a child at home reported that they wear a helmet most of the time, while 74% of those without a child at home reported they wear a helmet most of the time. If the sample size is 340 for both groups, is the proportion of helmet use significantly more when there is a child in the home, at the 0.01 level of significance?

Objective 10.5

10.37 State the null hypothesis, H_o, and the alternative hypothesis, H_a, that would be used to test the following claims:
 a. The variances of populations A and B are not equal.
 b. The standard deviation of population I is larger than the standard deviation of population II.
 c. The ratio of the variances for populations A and B is different from 1.
 d. The variability within population C is less than the variability within population D.

10.38 Using the $F(df_1, df_2, \alpha)$ notation, name each of the critical values shown on the following figures.
 a.

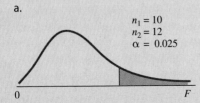

$n_1 = 10$
$n_2 = 12$
$\alpha = 0.025$

b.

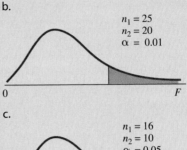

$n_1 = 25$
$n_2 = 20$
$\alpha = 0.01$

c.

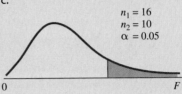

$n_1 = 16$
$n_2 = 10$
$\alpha = 0.05$

10.39 Determine the *p*-value that would be used to test the following hypotheses when F is used as the test statistic:
 a. $H_o: \sigma_1 = \sigma_2$ vs. $H_a: \sigma_1 > \sigma_2$, with $n_1 = 10$, $n_2 = 16$, and $F\star = 2.47$
 b. $H_o: \sigma_1^2 = \sigma_2^2$ vs. $H_a: \sigma_1^2 > \sigma_2^2$, with $n_1 = 25$, $n_2 = 21$, and $F\star = 2.31$
 c. $H_o: \dfrac{\sigma_1^2}{\sigma_2^2} = 1$ vs. $H_a: \dfrac{\sigma_1^2}{\sigma_2^2} \neq 1$, with $n_1 = 41$, $n_2 = 61$, and $F\star = 4.78$
 d. $H_o: \sigma_1 = \sigma_2$ vs. $H_a: \sigma_1 < \sigma_2$, with $n_1 = 10$, $n_2 = 16$, and $F\star = 2.47$

10.40 Determine the critical region and critical value(s) that would be used to test the following hypotheses using the classical approach when $F\star$ is used as the test statistic.
 a. $H_o: \sigma_1^2 = \sigma_2^2$ vs. $H_a: \sigma_1^2 > \sigma_2^2$, with $n_1 = 10$, $n_2 = 16$, and $\alpha = 0.05$
 b. $H_o: \dfrac{\sigma_1^2}{\sigma_2^2} = 1$ vs. $H_a: \dfrac{\sigma_1^2}{\sigma_2^2} \neq 1$, with $n_1 = 25$, $n_2 = 31$, and $\alpha = 0.05$
 c. $H_o: \dfrac{\sigma_1^2}{\sigma_2^2} = 1$ vs. $H_a: \dfrac{\sigma_1^2}{\sigma_2^2} > 1$, with $n_1 = 10$, $n_2 = 10$, and $\alpha = 0.01$
 d. $H_o: \sigma_1 = \sigma_2$ vs. $H_a: \sigma_1 < \sigma_2$, with $n_1 = 25$, $n_2 = 16$, and $\alpha = 0.01$

10.41 Do the lengths of boys' names have more variation than that of girls' names? With current names like Nathaniel and Christopher versus Ian and Jack, it certainly appears that boys' names cover a wide range with respect to the length. To test this theory, random samples of names of 7th grade girls and boys were taken.

Boy Names	$n = 30$	$s = 1.870$
Girl Names	$n = 30$	$s = 1.456$

At the 0.05 level of significance, do the data support the contention that the lengths of boy names have more variation than girl names?

10.42 A study in *Pediatric Emergency Care* compared the injury severity between younger and older children. One measure reported was the Injury Severity Score (ISS). The standard deviation of ISS scores for 37 children 8 years or younger was 23.9, and the standard deviation for 36 children older than 8 years was 6.8. Assume that ISS scores are normally distributed for both age groups. At the 0.01 level of significance, is there sufficient reason to conclude that the standard deviation of ISS scores for younger children is larger than the standard deviation of ISS scores for older children?

*10.43 Salaries of professional athletes are often criticized for being over the top. Many also earn even more money through endorsements. Various NBA (National Basketball Association) and MLB (Major League Baseball) players can be seen in many high-profile endorsements. Two random samples from each sport's player endorsements were selected and yielded the following amounts in millions of dollars.

NBA	16.0	15.0	21.7	12.0	9.5	0.5	15.5	0.8	2.5	5.0	16.0	15.5
MLB	6.0	8.0	2.5	0.3	3.5	0.5	0.5	0.3				

a. At the 0.05 level of significance, is there a significant difference in the variability of endorsement amounts between NBA and MLB players? Assume normality of endorsement amounts.

b. At the 0.05 level of significance, is the mean endorsement amount for NBA players significantly higher than the mean endorsement amount for MLB players?

*10.44 Americans snooze on the weekends, according to a poll of 1,506 adults for the National Sleep Foundation and reported in a *USA Today* Snapshot in April 2005.

Hours of Sleep	Weekdays	Weekends
Less than 6	0.16	0.10
6–6.9	0.24	0.15
7–7.9	0.31	0.24
8 or more	0.26	0.49

Two independent random samples were taken at a large industrial complex. The workers selected in one sample were asked, "How many hours, to the nearest 15 minutes, did you sleep on Tuesday night this week?" The workers selected for the second sample were asked, "How many hours, to nearest 15 minutes, did you sleep on Saturday night last weekend?"

	Weekday			Weekend		
5.00	7.75	7.25	9.00	7.25	8.75	7.50
9.25	7.25	8.75	6.25	5.25	9.25	9.25
7.00	7.75	6.75	7.50	8.50	8.75	6.50
9.25	7.00	7.75	8.00	8.75	9.50	8.00
9.25	9.25	6.00	8.75	7.75	8.75	7.50

a. Construct a histogram and find the mean and standard deviation for each set of data.

b. Do the distributions of "hours of sleep on weekday" and "hours of sleep on weekend" resulting from the poll appear to be similar in shape? center? spread? Discuss your responses.

c. Is it possible that both of the samples were drawn from normal populations? Justify your answer.

d. Is the mean number of hours slept on the weekend statistically greater than the mean number of hours slept on the weekday? Use $\alpha = 0.05$.

e. Is there sufficient evidence to show that the standard deviation of these two samples are statistically different? Use $\alpha = 0.05$.

f. Explain how the answers to parts (b)–(e) now affect your thoughts about your answer to part (a).

Applications of
Chi-Square

11.1 Chi-Square Statistic

THERE ARE MANY PROBLEMS FOR WHICH **ENUMERATIVE DATA** ARE CATEGORIZED AND THE RESULTS SHOWN BY WAY OF COUNTS.

To illustrate this, think about spicy foods. If you like hot foods, you probably have a preferred way to "cool" your mouth after eating a delicious spicy favorite. Some of the more common methods used by people are drinking water, milk, soda, or beer, and eating bread or other food. There are even a few people who prefer not to cool their mouth on such occasions and therefore do nothing. The "Putting Out the Fire" snapshot shown below appeared in *USA Today* and shows the top six ways adults say they cool their mouths after eating hot sauce.

Recently, 200 adults who professed to love hot, spicy food were asked to name their favorite way to cool their mouth after eating food with hot sauce. Following is a summary of the sample.

Method	Water	Milk	Soda	Beer	Bread	Other	Nothing
Number	73	35	20	19	29	11	13

Count data like these are often referred to as enumerative data.

Similarly, a set of final exam scores can be displayed as a frequency distribution. These frequency numbers are counts, the number of data that fall in each cell. A survey asks voters whether they are registered as Republican, Democrat, or Other, and whether or not they support a particular candidate. The results are usually displayed on a chart that shows the

Putting Out the Fire

| Water *43% | Bread 19% | Milk 15% | Beer 7% | Soda 7% | Don't 6% |

*Data set for this example online at cengagebrain.com.

number of voters in each possible category. Numerous examples of this way of presenting data have been given throughout the previous 10 chapters.

Data Set-Up

Suppose that we have a number of **cells** into which n observations have been sorted. (The term *cell* is synonymous with the term *class*; the terms *class* and *frequency* were defined and first used in earlier chapters. Before you continue, a brief review of Objectives 2.1, 2.2, and 3.1 might be beneficial.) The **observed frequencies** in each cell are denoted by $O_1, O_2, O_3, \ldots, O_k$ (see Table 11.1 on the next page). Note that the sum of all the observed frequencies is

$$O_1 + O_2 + \ldots + O_k = n$$

where n is the sample size.

What we would like to do is to compare the observed frequencies with some **expected**, or theoretical, **frequencies**, denoted by $E_1, E_2, E_3, \ldots, E_k$ (see Table 11.1), for each of these cells. Again, the sum of these expected frequencies must be exactly n:

$$E_1 + E_2 + \ldots + E_k = n$$

Table 11.1 **Observed Frequencies**

| | \multicolumn{6}{c}{k Categories} |
	1st	2nd	3rd	...	kth	Total
Observed frequencies	O_1	O_2	O_3	...	O_k	n
Expected frequencies	E_1	E_2	E_3	...	E_k	n

We will then decide whether the observed frequencies seem to agree or disagree with the expected frequencies. We will do this by using a **hypothesis test** with chi-square, χ^2.

Outline of Test Procedure

To conduct a hypothesis test with chi-square, you'll need to start by understanding the following test statistic:

Test Statistic for Chi-Square

$$\chi^2 \bigstar = \sum_{\text{all cells}} \frac{(O - E)^2}{E} \qquad (11.1)$$

This calculated value for chi-square is the sum of several nonnegative numbers, one from each cell (or category). The numerator of each term in the formula for $\chi^2 \bigstar$ is the square of the difference between the values of the observed and the expected frequencies. The closer together these values are, the smaller the value of $(O - E)^2$; the farther apart, the larger the value of $(O - E)^2$. The denominator for each cell puts the size of the numerator into perspective; that is, a difference $(O - E)$ of 10 resulting from frequencies of 110 (O) and 100 (E) is quite different from a difference of 10 resulting from 15 (O) and 5 (E).

These ideas suggest that small values of chi-square indicate agreement between the two sets of frequencies, whereas larger values indicate disagreement. Therefore, it is customary for these tests to be one-tailed, with the critical region on the right.

In repeated sampling, the calculated value of $\chi^2 \bigstar$ in formula (11.1) will have a sampling distribution that can be approximated by the chi-square probability distribution when n is large. This ap-

proximation is generally considered adequate when all the expected frequencies are equal to or greater than 5. Recall that the chi-square distributions, like Student's t-distributions, are a family of probability distributions, each one being identified by the parameter number of **degrees of freedom**, df. The appropriate value of df will be described with each specific test. In order to use the chi-square distribution, we must be aware of its properties, which were listed in Objective 9.3 on page 198. (Also see Figure 9.7.) The critical values for chi-square are obtained from Table 8 in Appendix B. (Specific instructions were given in Objective 9.3 on page 199.)

> **Assumption for using chi-square to make inferences based on enumerative data:**
>
> The sample information is obtained using a random sample drawn from a population in which each individual is classified according to the categorical variable(s) involved in the test.

A *categorical variable* is a variable that classifies or categorizes each individual into exactly one of several cells or classes; these cells or classes are all-inclusive and mutually exclusive. The side facing up on a rolled die is a categorical variable: The list of outcomes {1, 2, 3, 4, 5, 6} is a set of all-inclusive and mutually exclusive categories.

In this chapter we permit a certain amount of "liberalization" with respect to the null hypothesis and its testing. In previous chapters the null hypothesis was always a statement about a population parameter (μ, σ, or p). However, there are other types of hypotheses that can be tested, such as "This die is fair" or "The height and weight of individuals are independent." Notice that these hypotheses are not claims about a parameter,

Karl Pearson and the Chi-Square

Known as one of the fathers of modern statistics, Karl Pearson invented the chi-square (denoted by χ^2) in 1900. It is the oldest inference procedure still used in its original form and is often used in today's economics and business applications.

although sometimes they could be stated with parameter values specified.

Suppose that we claim "This die is fair," $p = P$(any one number) $= \frac{1}{6}$, and you want to test the claim. What would you do? Was your answer something like: Roll this die many times and record the results? Suppose that you decide to roll the die 60 times. If the die is fair, what do you expect will happen? Each number (1, 2, . . . , 6) should appear approximately $\frac{1}{6}$ of the time (that is, 10 times). If it happens that approximately 10 of each number occur, you will certainly accept the claim of fairness ($p = \frac{1}{6}$ for each value). If it happens that the die seems to favor some particular numbers, you will reject the claim. (The calculated test statistic $\chi^2\star$ will have a large value in this case, as we will soon see.)

11.2 Inferences Concerning Multinomial Experiments

THE PRECEDING DIE PROBLEM IS A GOOD EXAMPLE OF A **MULTINOMIAL EXPERIMENT**.

➡ Let's consider this problem again. Suppose that we want to test this die (at $\alpha = 0.05$) and decide whether to reject or to fail to reject the claim "This die is fair." (The probability of each number is $\frac{1}{6}$.) The die is rolled from a cup onto a smooth, flat surface 60 times, with the following observed frequencies:

Number	1	2	3	4	5	6
Observed frequency	7	12	10	12	8	11

The null hypothesis that the die is fair is assumed to be true. This allows us to calculate the expected frequencies. If the die is fair, we expect 10 occurrences of each number.

Now let's calculate an observed value of χ^2. These calculations are shown in Table 11.2. The calculated value is $\chi^2\star = 2.2$.

Now let's use our familiar five-step hypothesis-testing format.

STEP 1 a. **Parameter of interest:**
The probability with which each side faces up:
$P(1), P(2), P(3), P(4), P(5), P(6)$.

b. **Statement of hypotheses:**

H_o: The die is fair $\left(\text{each } p = \frac{1}{6}\right)$.

H_a: The die is not fair (at least one p is different from the others).

STEP 2 a. **Check the assumptions:**
The data were collected in a random manner, and each outcome is one of the six numbers.

b. **Test statistic:**
In a multinomial experiment, df $= k - 1$, where k is the number of cells.

Table 11.2 **Computations for Calculating χ^2**

Number	Observed (O)	Expected (E)	O − E	(O − E)²	$\frac{(O-E)^2}{E}$
1	7	10	−3	9	0.9
2	12	10	2	4	0.4
3	10	10	0	0	0.0
4	12	10	2	4	0.4
5	8	10	−2	4	0.4
6	11	10	1	1	0.1
Total	60	60	0 ✔		2.2

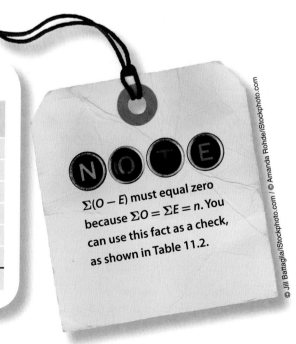

$\Sigma(O - E)$ must equal zero because $\Sigma O = \Sigma E = n$. You can use this fact as a check, as shown in Table 11.2.

The chi-square distribution and formula (11.1), with df $= k - 1 = 6 - 1 = 5$.

c. **Determine the level of significance, α:**

$\alpha = 0.05$

 The sample evidence:

a. Sample information: See Table 11.2.

b. **Calculated test statistic:**
Using formula (11.1), we have

$$\chi^2 \bigstar = \sum_{\text{all cells}} \frac{(O - E)^2}{E}:$$

$$\chi^2 \bigstar = \mathbf{2.2}$$

(calculations are shown in Table 11.2).

STEP 4 The probability distribution:
As always, we can use either the p-value or the classical procedure:

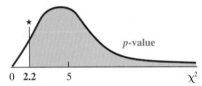

Using the p-value procedure:

a. Use the right-hand tail because "larger" values of chi-square disagree with the null hypothesis:

$$P = P(\chi^2 \bigstar > 2.2 | df = 5)$$

as shown on the figure below.

To find the p-value, you have two options:
1. Use Table 8 in Appendix B to place bounds on the p-value: $0.75 < P < 0.90$.
2. Use a computer or calculator to find the p-value: $P = 0.821$.
For specific instructions, see page 201.

b. The p-value is not smaller than the level of significance, α.

Using the classical procedure:

a. The critical region is the right-hand tail because "larger" values of chi-square disagree with the null hypothesis. The critical value is obtained from Table 8, at the intersection of row df $= 5$ and column $\alpha = 0.05$:

$$\chi^2(5, 0.05) = \mathbf{11.1}$$

For specific instructions, see page 199.

b. $\chi^2 \bigstar$ is not in the critical region, as shown in red on the figure above.

STEP 5 a. **Decision:** Fail to reject H_o.

b. **Conclusion:**
At the 0.05 level of significance, the observed frequencies are not significantly different from those expected of a fair die.

Before we look at other examples, we must define the term *multinomial experiment* and state the guidelines for completing the chi-square test for it.

The die example meets the definition of a multinomial experiment because it has all four of the characteristics described in the definition.

1. The die was rolled n (60) times in an identical fashion, and these trials were independent of each other. (The result of each trial was unaffected by the results of other trials.)

2. Each time the die was rolled, one of six numbers resulted, and each number was associated with a cell.

3. The probability associated with each cell was $\frac{1}{6}$, and this was constant from trial to trial. (Six values of $\frac{1}{6}$ sum to 1.0.)

4. When the experiment was complete, we had a list of six frequencies (7, 12, 10, 12, 8, and 11) that summed to 60, indicating that each of the outcomes was taken into account.

The testing procedure for multinomial experiments is very similar to the testing procedure described in previous chapters. The biggest change comes with the statement of the null hypothesis. It may be a verbal statement, such as in the die example: "This die is fair." Often the alternative to the null hypothesis is not stated. However, in this book the alternative hypothesis will be shown, since it aids in organizing and understanding the problem. It will not be used to determine the location of the critical region, though, as was the

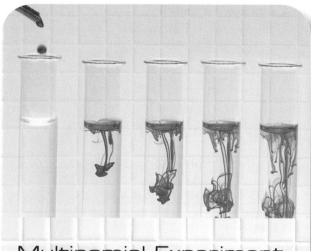

Multinomial Experiment

A multinomial experiment has the following characteristics:

1. It consists of *n* identical independent trials.
2. The outcome of each trial fits into exactly one of *k* possible cells.
3. There is a probability associated with each particular cell, and these individual probabilities remain constant during the experiment. (It must be the case that $p_1 + p_2 + \ldots + p_k = 1$.)
4. The experiment will result in a set of *k* observed frequencies, $O_1, O_2, \ldots, O_k$, where each O_i is the number of times a trial outcome falls into that particular cell. (It must be the case that $O_1 + O_2 + \ldots + O_k = n$.)

case in previous chapters. For multinomial experiments we will always use a one-tailed critical region, and it will be the right-hand tail of the χ^2-distribution because larger deviations (positive or negative) from the expected values lead to an increase in the calculated $\chi^2\bigstar$-value.

The critical value will be determined by the level of significance assigned (α) and the number of degrees of freedom (df). The number of degrees of freedom (df) will be 1 less than the number of cells (*k*) into which the data are divided:

Degrees of Freedom for Multinomial Experiments

$$df = k - 1 \tag{11.2}$$

Each expected frequency, E_i, will be determined by multiplying the total number of trials *n* by the corresponding probability (p_i) for that cell; that is,

Expected Value for Multinomial Experiments

$$E_i = n \cdot p_i \tag{11.3}$$

One guideline should be met to ensure a good approximation to the chi-square distribution: Each expected frequency should be at least 5 (that is, each $E_i \geq 5$). Sometimes it is possible to combine "smaller" cells to meet this guideline. If this guideline cannot be met, then corrective measures to ensure a good approximation should be used. These corrective measures are not covered in this book but are discussed in many other sources.

A MULTINOMIAL HYPOTHESIS TEST WITH EQUAL EXPECTED FREQUENCIES

Multinomial situations occur regularly in everyday life. Take, for instance, college course registration. College students have regularly insisted on freedom of choice when they register for courses. This semester there were seven sections of a particular mathematics course. The sections were scheduled to meet at various times with a variety of instructors. Table 11.3 shows the number of students who selected each of the seven sections. Do the data indicate that the students had a preference for certain sections, or do they indicate that each section was equally likely to be chosen?

Table 11.3 **Data on Section Enrollments**

	Section							
	1	2	3	4	5	6	7	Total
Number of students	18	12	25	23	8	19	14	119

If no preference were shown in the selection of sections, then we would expect the 119 students to be equally distributed among the seven classes: we would expect 17 students to register for each section. Using the five-step process, let's complete the hypothesis test at the 5% level of significance and see if the students are equally distributed.

STEP 1 a. **Parameter of interest:**
Preference for each section, the probability that a particular section is selected at registration.

b. Statement of hypotheses:

H_o: There was no preference shown
(equally distributed).

H_a: There was a preference shown (not
equally distributed).

 a. Assumption:

The 119 students represent a random sample of the population of all students who register for this particular course. Since no new regulations were introduced in the selection of courses and registration seemed to proceed in its usual pattern, there is no reason to believe this is other than a random sample.

b. Test statistic to be used:

The chi-square distribution and formula (11.1), with df = 6.

c. Level of significance: $\alpha = 0.05$.

 **a. Sample information:** See Table 11.3.

b. Calculated test statistic:

Using formula (11.1), we have

$$\chi^2\star = \sum_{\text{all cells}} \frac{(O - E)^2}{E}:$$

$$\chi^2\star = \frac{(18 - 17)^2}{17} + \frac{(12 - 17)^2}{17} + \frac{(25 - 17)^2}{17} +$$

$$\frac{(23 - 17)^2}{17} + \frac{(8 - 17)^2}{17} + \frac{(19 - 17)^2}{17} + \frac{(14 - 17)^2}{17}$$

$$= \frac{(1)^2}{17} + \frac{(-5)^2}{17} + \frac{(8)^2}{17} + \frac{(6)^2}{17} + \frac{(-9)^2}{17} + \frac{(2)^2}{17} + \frac{(-3)^2}{17}$$

$$= \frac{1 + 25 + 64 + 36 + 81 + 4 + 9}{17} = \frac{220}{17} = 12.9411$$

$$= \mathbf{12.94}$$

STEP 4 The probability distribution:

Again, we can use either the p-value or the classical procedure:

Using the *p*-value procedure:

a. Use the right-hand tail because "larger" values of chi-square disagree with the null hypothesis:

$$P = P(\chi^2\star > \mathbf{12.94}|df = 6)$$

as shown on the figure below:

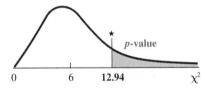

To find the p-value, you have two options:
1. Use Table 8 in Appendix B to place bounds on the p-value: $0.025 < P < 0.05$.
2. Use a computer or calculator to find the p-value: $P = 0.044$.

For specific instructions, see page 201.

b. The p-value is smaller than the level of significance, α.

Using the classical procedure:

a. The critical region is the right-hand tail because "larger" values of chi-square disagree with the null hypothesis. The critical value is obtained from Table 8, at the intersection of row df = 6 and column $\alpha = 0.05$:

$$\chi^2(6, 0.05) = \mathbf{12.6}$$

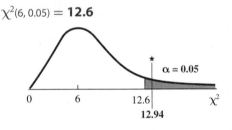

For specific instructions, see page 199.

b. $\chi^2\star$ is in the critical region, as shown in red on the figure above.

STEP 5 a. Decision: Reject H_o.

b. Conclusion:

At the 0.05 level of significance, there does seem to be a preference shown. We cannot determine from the given information

what the preference is. It could be teacher preference, time preference, or a schedule conflict.

Conclusions must be worded carefully to avoid suggesting conclusions that the data cannot support.

MULTINOMIAL HYPOTHESIS TEST WITH UNEQUAL EXPECTED FREQUENCIES

Not all multinomial experiments result in equal expected frequencies. For example, the Mendelian theory of inheritance claims that the frequencies of round and yellow, wrinkled and yellow, round and green, and wrinkled and green peas will occur in the ratio 9:3:3:1 when two specific varieties of peas are crossed. In testing this theory, Mendel obtained frequencies of 315, 101, 108, and 32, respectively. By completing our five-step hypothesis test at the 0.05 level of significance, let's find out if the sample data provide sufficient evidence to reject Mendel's theory.

STEP 1 a. **Parameter of interest:**

The proportions: P(round and yellow), P(wrinkled and yellow), P(round and green), P(wrinkled and green).

b. **Statement of hypotheses:**

H_o: 9:3:3:1 is the ratio of inheritance.

H_a: 9:3:3:1 is not the ratio of inheritance.

STEP 2 a. **Assumptions:**

We will assume that Mendel's results form a random sample.

b. **Test statistic:**

The chi-square distribution and formula (11.1), with df = 3.

c. **Level of significance:** $\alpha = 0.05$.

STEP 3 a. **Sample information:**

The observed frequencies were: 315, 101, 108, and 32.

b. **Calculated test statistic:**

The ratio 9:3:3:1 indicates probabilities of $\frac{9}{16}, \frac{3}{16}, \frac{3}{16},$ and $\frac{1}{16}$.

Therefore, the expected frequencies are $\frac{9n}{16}, \frac{3n}{16}, \frac{3n}{16},$ and $\frac{1n}{16}$. We have

$$n = \Sigma O_i = 315 + 101 + 108 + 32 = 556$$

The computations for calculating $\chi^2\star$ are shown in Table 11.4.

Table 11.4 Computations Needed to Calculate $\chi^2\star$

O	E	O−E	$\frac{(O-E)^2}{E}$	
315	312.75	2.25	0.0162	
101	104.25	−3.25	0.1013	
108	104.25	3.75	0.1349	
32	34.75	−2.75	0.2176	
556	556.00	0 ✓	**0.4700**	$\longrightarrow \chi^2\star = \sum_{\text{all cells}} \frac{(O-E)^2}{E} = 0.47$

STEP 4 The probability distribution:

Again, we can use either the p-value or the classical procedure:

Using the p-value procedure:

a. Use the right-hand tail because "larger" values of chi-square disagree with the null hypothesis:

$$P = P(\chi^2\star > 0.47 | df = 3)$$

as shown on the figure below.

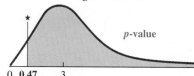

To find the p-value, you have two options:

1. Use Table 8 in Appendix B to place bounds on the p-value: $0.90 < P < 0.95$.
2. Use a computer or calculator to find the p-value: $P = 0.925$.

For specific instructions, see page 201.

b. The p-value is not smaller than the level of significance, α.

Using the classical procedure:

a. The critical region is the right-hand tail because "larger" values of chi-square disagree with the null hypothesis. The critical value is obtained from Table 8, at the intersection of row df = 3 and column $\alpha = 0.05$:

$$\chi^2(3, 0.05) = \mathbf{7.81}$$

For specific instructions, see page 199.

b. $\chi^2\star$ is not in the critical region, as shown in red on the figure above.

a. Decision: Fail to reject H_o.

b. Conclusion:

At the 0.05 level of significance, there is not sufficient evidence to reject Mendel's theory.

9:3:3:1

11.3 Inferences Concerning Contingency Tables

A **CONTINGENCY TABLE** IS AN ARRANGEMENT OF DATA IN A TWO-WAY CLASSIFICATION.

The data are sorted into cells, and the number of data in each cell is reported. The contingency table involves two factors (or variables), and a common question concerning such tables is whether the data indicate that the two variables are independent or dependent (see pages 90–93).

Two different tests use the contingency table format. The first one we will look at is the **test of independence.**

Test of Independence

To illustrate a test of independence, let's consider a random sample that shows the gender of liberal arts college students and their favorite academic area. Each person in a group of 300 students was identified as male or female

Table 11.5 **Sample Results for Gender and Subject Preference**

Gender	Favorite Subject Area			Total
	Math–Science (MS)	Social Science (SS)	Humanities (H)	
Male (M)	37	41	44	122
Female (F)	35	72	71	178
Total	72	113	115	300

and then asked whether he or she preferred taking liberal arts courses in the area of math–science, social science, or humanities. Table 11.5 is a contingency table that shows the frequencies found for these categories. Does this sample present sufficient evidence to reject the null hypothesis "Preference for math–science, social science, or humanities is independent of the gender of a college student"? To find out, we will complete the hypothesis test using the 0.05 level of significance.

a. Parameter of interest:

Determining the independence of the variables "gender" and "favorite subject area" requires us to discuss the probability of the various cases and the effect that answers about one variable have on the probability of answers about the other variable. Independence, as defined in Chapter 4, requires $P(MS|M) = P(MS|F) = P(MS)$; that is, gender has no effect on the probability of a person's choice of subject area.

b. Statement of hypothesis:

H_o: Preference for math–science, social science, or humanities is independent of the gender of a college student.

H_a: Subject area preference is not independent of the gender of the student.

a. Assumptions:

The sample information is obtained using one random sample drawn from one population, with each individual then classified according to gender and favorite subject area.

b. Test statistic to be used:

In the case of contingency tables, the number of degrees of freedom is exactly the same as the number of cells in the table that may be filled in freely when

you are given the **marginal totals.** The totals for our college students are shown in the following table:

			122
			178
72	113	115	300

Given these totals, you can fill in only two cells before the others are all determined. (The totals must, of course, remain the same.) For example, once we pick two arbitrary values (say, 50 and 60) for the first two cells of the first row, the other four cell values are fixed:

50	60	C	122
D	E	F	178
72	113	115	300

The values have to be $C = 12$, $D = 22$, $E = 53$, and $F = 103$. Otherwise, the totals will not be correct. Therefore, for this problem there are two free choices. Each free choice corresponds to 1 degree of freedom. Hence, the number of degrees of freedom for our example is 2 (df = 2).

The chi-square distribution will be used along with formula (11.1), with df = 2.

c. **Level of significance:** $\alpha = 0.05$.

STEP 3 a. **Sample information:** See Table 11.5.

b. **Calculated test statistic:**

Before we can calculate the value of chi-square, we need to determine the expected values, E, for each cell. To do this we must recall the null hypothesis, which asserts that these factors are independent. Therefore, we would expect the values to be distributed in proportion to the marginal totals. There are 122 males; we would expect them to be distributed among MS, SS, and H proportionally to the

note

We can think of the computation of the expected values in a second way. Recall that we assume the null hypothesis to be true until there is evidence to reject it. Having made this assumption in our example, we are saying in effect that the event that a student picked at random is male and the event that a student picked at random prefers math–science courses are independent. Our point estimate for the probability that a student is male is $\frac{122}{300}$, and the point estimate for the probability that the student prefers math–science courses is $\frac{72}{300}$. Therefore, the probability that both events occur is the product of the probabilities.

[Refer to formula (4.7), page 93.] Thus, $\left(\frac{122}{300}\right)\left(\frac{72}{300}\right)$ is the probability of a selected student being male and preferring math–science. The number of students out of 300 who are expected to be male and prefer math–science is found by multiplying the probability (or proportion) by the total number of students (300). Thus, the expected number of males who prefer math–science is $\left(\frac{122}{300}\right)\left(\frac{72}{300}\right)$ $(300) = \left(\frac{122}{300}\right)(72) = 29.28$. The other expected values can be determined in the same manner.

72, 113, and 115 totals. Thus, the expected cell counts for males are

$$\frac{72}{300} \cdot 122 \quad \frac{113}{300} \cdot 122 \quad \frac{115}{300} \cdot 122$$

Similarly, we would expect for the females

$$\frac{72}{300} \cdot 178 \quad \frac{113}{300} \cdot 178 \quad \frac{115}{300} \cdot 178$$

Thus, the expected values are as shown in Table 11.6. Always check the marginal totals for the expected values against the marginal totals for the observed values.

Table 11.6 **Expected Values**

	MS	SS	H	Total
Male	29.28	45.95	46.77	122.00
Female	42.72	67.05	68.23	178.00
Total	72.00	113.00	115.00	300.00

Typically the contingency table is written so that it contains all this information (see Table 11.7).

Table 11.7 **Contingency Table Showing Sample Results and Expected Values**

Gender	Favorite Subject Area			Total
	Math–Science	Social Science	Humanities	
Male	37 (29.28)	41 (45.95)	44 (46.77)	122
Female	35 (42.72)	72 (67.05)	71 (68.23)	178
Total	72	113	115	300

The calculated chi-square is

$$\chi^2 \star = \sum_{\text{all cells}} \frac{(O - E)^2}{E}:$$

$$\chi^2 \star = \frac{(37 - 29.28)^2}{29.28} + \frac{(41 - 45.95)^2}{45.95} + \frac{(44 - 46.77)^2}{46.77} +$$
$$\frac{(35 - 42.72)^2}{42.72} + \frac{(72 - 67.05)^2}{67.05} + \frac{(71 - 68.23)^2}{68.23}$$

$$= 2.035 + 0.533 + 0.164 + 1.395 + 0.365 + 0.112$$

$$= \mathbf{4.604}$$

STEP 4 The probability distribution: Again, we can use either the *p*-value or the classical procedure:

Using the *p*-value procedure:

a. Use the right-hand tail because "larger" values of chi-square disagree with the null hypothesis:

P = P($\chi^2 \star$ > 4.604|df = 2)

as shown on the figure below.

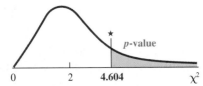

To find the *p*-value, you have two options:
1. Use Table 8 in Appendix B to place bounds on the *p*-value: **0.10 < P < 0.25.**
2. Use a computer or calculator to find the *p*-value: **P = 0.1001.**

For specific instructions, see page 201.

b. The *p*-value is not smaller than α.

Using the classical procedure:

a. The critical region is the right-hand tail because "larger" values of chi-square disagree with the null hypothesis. The critical value is obtained from Table 8, at the intersection of row df = 2 and column α = 0.05:

χ^2(2, 0.05) = **5.99**

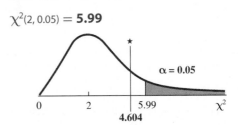

For specific instructions, see page 199.

b. $\chi^2 \star$ is not in the critical region, as shown in red on the figure above.

a. Decision: Fail to reject H_o.

b. Conclusion:

At the 0.05 level of significance, the evidence does not allow us to reject independence between the gender of a student and the student's preferred academic subject area.

In general, the **$r \times c$ contingency table** (*r* is the number of **rows**; *c* is the number of **columns**) is used to test the independence of the row factor and the column factor. The number of **degrees of freedom** is determined by:

Degrees of Freedom for Contingency Tables

$$\mathrm{df} = (r - 1) \cdot (c - 1) \tag{11.4}$$

where *r* and *c* are both greater than 1. (This value for df should agree with the number of cells counted, according to the general description on pages 244–245.)

The **expected frequencies** for an $r \times c$ contingency table are found by means of the formulas given in each cell in Table 11.8, where *n* = grand total. In general, the expected frequency at the intersection of the *i*th row and the *j*th column is given by:

Expected Frequencies for Contingency Tables

$$E_{i,j} = \frac{\text{row total} \times \text{column total}}{\text{grand total}} = \frac{R_i \times C_j}{n} \tag{11.5}$$

Table 11.8 **Expected Frequencies for an $r \times c$ Contingency Table**

Row	Column 1	2	$\cdots$	*j*th column	$\cdots$	*c*	Total
1	$\frac{R_1 \times C_1}{n}$	$\frac{R_1 \times C_2}{n}$	$\cdots$	$\frac{R_1 \times C_j}{n}$	$\cdots$	$\frac{R_1 \times C_c}{n}$	R_1
2	$\frac{R_2 \times C_1}{n}$						R_2
$\vdots$	$\vdots$						$\vdots$
*i*th row	$\frac{R_i \times C_1}{n}$		$\cdots$	$\frac{R_i \times C_j}{n}$	$\cdots$		R_i
$\vdots$	$\vdots$						$\vdots$
r	$\frac{R_r \times C_1}{n}$						$\vdots$
Total	C_1	C_2	$\cdots$	C_j	$\cdots$		*n*

We should again observe the previously mentioned guideline: Each $E_{i,j}$ should be at least 5.

Test of Homogeneity

The second type of contingency table problem is called a **test of homogeneity**. This test is used when one of the two variables is controlled by the experimenter so that the row or column totals are predetermined.

For example, suppose that we want to poll registered voters about a piece of legislation proposed by the governor. In the poll, 200 urban, 200 suburban, and 100 rural residents are randomly selected and asked whether they favor or oppose the governor's proposal. That is, a simple random sample is taken for each of these three groups. A total of 500 voters are polled. But notice that it has been predetermined (before the sample is taken) just how many are to fall within each row category, as

Notes on Notation

The notation used in Table 11.8 and formula (11.5) may be unfamiliar to you. For convenience in referring to cells or entries in a table, we use $E_{i,j}$ to denote the entry in the *i*th row and the *j*th column. That is, the first letter in the subscript corresponds to the row number and the second letter corresponds to the column number. Thus, $E_{1,2}$ is the entry in the first row, second column, and $E_{2,1}$ is the entry in the second row, first column. In Table 11.6 (page 246), $E_{1,2}$ is 45.95 and $E_{2,1}$ is 42.72. The notation used in Table 11.8 is interpreted in a similar manner; that is, R_1 corresponds to the total from row 1, and C_1 corresponds to the total from column 1.

shown in Table 11.9, and each category is sampled separately.

Table 11.9 Registered Voter Poll with Predetermined Row Totals

	Governor's Proposal		
Residence	Favor	Oppose	Total
Urban			200
Suburban			200
Rural			100
Total			500

In a test of this nature, we are actually testing the hypothesis: The distribution of proportions within the rows is the same for all rows. That is, the distribution of proportions in row 1 is the same as that in row 2, is the same as that in row 3, and so on. The alternative is: The distribution of proportions within the rows is not the same for all rows. This type of example may be thought of as a comparison of several multinomial experiments. Beyond this conceptual difference, the actual testing for independence and homogeneity with contingency tables is the same.

Let's demonstrate this hypothesis test by completing the polling example at 0.05 significance level. Does the sample evidence shown in Table 11.10 support the hypothesis: "Voters within the different residence groups have different opinions about the governor's proposal"?

Table 11.10 Sample Results for Residence and Opinion

	Governor's Proposal		
Residence	Favor	Oppose	Total
Urban	143	57	200
Suburban	98	102	200
Rural	13	87	100
Total	254	246	500

STEP 1 **a. Parameter of interest:**
The proportion of voters who favor or oppose (that is, the proportion of urban voters who favor, the proportion of suburban voters who favor, the proportion of rural voters who favor, and the proportion of all three groups, separately, who oppose).

b. Statement of hypotheses:

H_o: The proportion of voters who favor the proposed legislation is the same in all three residence groups.

H_a: The proportion of voters who favor the proposed legislation is not the same in all three groups. (That is, in at least one group the proportion is different from another.)

STEP 2 **a. Assumptions:**
The sample information is obtained using three random samples drawn from three separate populations in which each individual is classified according to his or her opinion.

b. Test statistic:
The chi-square distribution and formula (11.1), with df $= (r - 1)(c - 1) = (3 - 1)(2 - 1) = 2$

c. Level of significance: $\alpha = 0.05$.

STEP 3 **a. Sample information:** See Table 11.10.

b. Calculated test statistic:
The expected values are found by using formula (11.5) (page 247) and are given in Table 11.11.

Table 11.11 Sample Results and Expected Values

	Governor's Proposal		
Residence	Favor	Oppose	Total
Urban	143 (101.6)	57 (98.4)	200
Suburban	98 (101.6)	102 (98.4)	200
Rural	13 (50.8)	87 (49.2)	100
Total	254	246	500

NOTE: Each expected value is used twice in the calculation of $\chi^2\star$; therefore, it is a good idea to keep extra decimal places while doing the calculations.

The calculated chi-square is

$$\chi^2\star = \sum_{\text{all cells}} \frac{(O - E)^2}{E}:$$

$$\chi^2\star = \frac{(143 - 101.6)^2}{101.6} + \frac{(57 - 98.4)^2}{98.4} + \frac{(98 - 101.6)^2}{101.6}$$
$$+ \frac{(102 - 98.4)^2}{98.4} + \frac{(13 - 50.8)^2}{50.8} + \frac{(87 - 49.2)^2}{49.2}$$

$$= 16.87 + 17.42 + 0.13 + 0.13 + 28.13 + 29.04$$

$$= \mathbf{91.72}$$

STEP 4 The probability distribution:
Again, we can use either the *p*-value or the classical procedure:

Using the *p*-value procedure:

a. Use the right-hand tail because "larger" values of chi-square disagree with the null hypothesis:

$$P = P(\chi^2\star > 91.72 | df = 2)$$

as shown on the figure below.

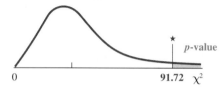

To find the *p*-value, you have two options:
1. Use Table 8 in Appendix B to place bounds on the *p*-value: **P < 0.005**.
2. Use a computer or calculator to find the *p*-value: **P = 0.000+**.

For specific instructions, see page 201.

b. The *p*-value is smaller than α.

Using the classical procedure:

a. The critical region is the right-hand tail because "larger" values of chi-square disagree with the null hypothesis. The critical value is obtained from Table 8, at the intersection of row df = 2 and column $\alpha = 0.05$:

$$\chi^2(2, 0.05) = \mathbf{5.99}$$

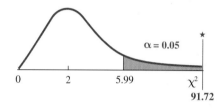

For specific instructions, see page 199.

b. $\chi^2\star$ is in the critical region, as shown in red on the figure above.

STEP 5 a. **Decision:** Reject H_o.

b. **Conclusion:** The three groups of voters do not all have the same proportions favoring the proposed legislation, at the 0.05 level of significance.

RECAP

In this chapter we have been concerned with tests of hypotheses using chi-square, with the cell probabilities associated with the multinominal experiment, and with the simple contingency table. In each case the basic assumptions are that a large number of observations have been made and that the resulting test statistic, $\Sigma\frac{(O-E)^2}{E}$, is approximately distributed as chi-square. In general, if *n* is large and the minimum allowable expected cell size is 5, then this assumption is satisfied.

A few words of caution: The correct number of degrees of freedom is critical if the test results are to be meaningful. The degrees of freedom determine, in part, the critical region, and its size is important. As in other tests of hypothesis, failure to reject H_o does not mean outright acceptance of the null hypothesis.

problems

Objective 11.1

*11.1 Referring to the sample of 200 adults surveyed in the hot foods example and "Putting Out the Fire" graph on page 236:

a. What information was collected from each adult in the sample?

b. Define the population and the variable involved in the sample.

c. Using the sample data, calculate percentages for the various methods of cooling one's mouth.

11.2 Find these critical values by using Table 8 of Appendix B.

a. $\chi^2(18, 0.01)$ b. $\chi^2(16, 0.025)$

c. $\chi^2(40, 0.10)$ d. $\chi^2(45, 0.01)$

11.3 Using the notation seen in problem 11.2, name and find the critical values of χ^2.

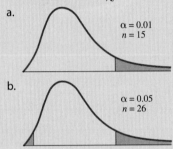

a.

     $\alpha = 0.01$
     $n = 15$

b.

     $\alpha = 0.05$
     $n = 26$

Objective 11.2

11.4 State the null hypothesis H_o and the alternative hypothesis H_a that would be used to test the following statements:

a. The four choices are all equally likely.

b. The poll showed the political party distributions of 23%, 36%, and 41% for Republicans, Democrats, and Independents, respectively.

c. Favoring responses with respect to sustainability and the four designated generation intervals were in the ratio of 11:15:8:6.

11.5 Determine the p-value for the following hypotheses tests involving the χ^2-distribution.

a. H_o: $P(1) = P(2) = P(3) = P(4) = 0.25$, with $\chi^2\bigstar = 12.25$

b. H_o: $P(I) = 0.25, P(II) = 0.40, P(III) = 0.35$, with $\chi^2\bigstar = 5.98$

11.6 Determine the critical value and critical region that would be used in the classical approach to test the null hypothesis for each of the following multinomial experiments.

a. H_o: $P(1) = P(2) = P(3) = P(4) = 0.25$, with $\alpha = 0.05$

b. H_o: $P(I) = 0.25, P(II) = 0.40, P(III) = 0.35$, with $\alpha = 0.01$

11.7 A manufacturer of floor polish conducted a consumer-preference experiment to determine which of five different floor polishes was the most appealing in appearance. A sample of 100 consumers viewed five patches of flooring that had each received one of the five polishes. Each consumer indicated the patch he or she preferred. The lighting and background were approximately the same for all patches. The results were as follows:

Polish	A	B	C	D	E	Total
Frequency	27	17	15	22	19	100

Solve the following using the p-value approach and the classical approach:

a. State the hypothesis for "no preference" in statistical terminology.

b. What test statistic will be used in testing this null hypothesis?

c. Complete the hypothesis test using $\alpha = 0.10$.

11.8 Skittles Bite Size Candies are sold with multiple colored candies in each bag, and you can "Taste the Rainbow" with their five original colors and flavors: green (lime), purple (grape), yellow (lemon), orange (orange), and red (strawberry). Unlike some of the other multicolored candies available, Skittles claims their five colors are equally likely. In an attempt to reject this claim, a 4-ounce bag of Skittles was purchased and the colors counted:

Red	Orange	Yellow	Green	Purple
18	21	23	17	27

Does this sample contradict Skittles' claim at the 0.05 level?

a. Solve using the p-value approach.

b. Solve using the classical approach.

11.9 An October 16, 2009, *USA Today* Snapshot, titled "Are Public Cell Phone Conversations Rude?" reported the following results from a Fox TV/Rasmussen Reports poll:

Poll Response	Percent
Yes	51
No	37
Not sure	12

As a member of the Civility Committee at your college, you decide to conduct a survey of students with respect to this issue. The following table shows the 300 student responses:

Poll Response	Number
Yes	126
No	118
Not sure	56

Does the distribution of responses from the college students differ significantly from the published survey results? Use a 0.01 level of significance.

11.10 National health care is currently a big issue for Americans. The October 21, 2009, *USA Today* article "Poll: Americans Skittish over Health Care Changes" reported the following percentages with respect to "Insurance company requirements you have to meet to get certain treatments covered" if the health care bill passes:

Viewpoint Sept. 11–13	Percentage
Get better	22
Not change	35
Get worse	38
Unknown	5

One month later, during October 16–19, another poll was taken of 1,521 adults. Those viewpoints are categorized in the table below.

Viewpoint Oct. 16–19	Number
Get better	380
Not change	380
Get worse	700
Unknown	61

At the 0.05 level of significance, did the distributions of viewpoints change significantly from September 2009 to October 2009?

11.11 Bird foraging behavior is being studied in a managed forest that is made up of Douglas fir (52% of canopy volume), ponderosa pine (36%), and grand fir (12%). Two hundred thirty-eight red-breasted nuthatches were observed, with 105 in Douglas fir, 92 in ponderosa pine, and 41 in grand fir. The null hypothesis being tested is "the birds forage randomly without regard to the species of tree."
   a. State the alternative hypothesis.
   b. Determine the expected values for the number of birds foraging each species of tree.
   c. Complete the hypothesis test using $\alpha = 0.05$ and carefully state the conclusion.

11.12 A certain type of flower seed will produce magenta, chartreuse, and ochre flowers in the ratio 6 : 3 : 1 (one flower per seed). A total of 100 seeds are planted and all germinate, yielding the following results:

Magenta	Chartreuse	Ochre
52	36	12

Solve the following using the *p*-value approach and the classical approach:
   a. If the null hypothesis (6 : 3 : 1) is true, what is the expected number of magenta flowers?
   b. How many degrees of freedom are associated with chi-square?
   c. Complete the hypothesis test using $\alpha = 0.10$.

*11.13 One of the major benefits of e-mail is that it makes it possible to communicate rapidly without getting a busy signal or no answer—two major criticisms of telephone calls. But does e-mail succeed in helping solve the problems people have trying to run computer software? A study polled the opinions of consumers who had tried to use e-mail to obtain help by posting a message online to their PC manufacturer or authorized representative. Results are shown in the following table.

Result of Online Query	Percent
Never got a response	14
Got a response, but it didn't help	30
Response helped, but it didn't solve problem	34
Response solved problem	22

SOURCE: "PC World's Reliability and Service Survey," *PC World*

As marketing manager for a large PC manufacturer, you decide to conduct a survey of your customers comparing your e-mail records against the published results. To ensure a fair comparison, you elect to use the same questionnaire and examine returns from 500 customers who attempted to use e-mail to get help from your technical support staff. The results follow:

Result of Online Query	Number Responding
Never got a response	35
Got a response, but it didn't help	102
Response helped, but didn't solve problem	125
Response solved problem	238
Total	500

Does the distribution of responses differ from the distribution obtained from the published survey? Test at the 0.01 level of significance.
   a. Solve using the *p*-value approach.
   b. Solve using the classical approach.

*11.14 *Nursing Magazine* reported results of a survey of more than 1,800 nurses across the country concerning job satisfaction and retention. Nurses from magnet hospitals (hospitals that successfully attract and retain nurses) describe the staffing situation in their units as follows:

Staffing Situation	Percent
1. Desperately short of help—patient care has suffered	12
2. Short, but patient care hasn't suffered	32
3. Adequate	38
4. More than adequate	12
5. Excellent	6

A survey of 500 nurses from nonmagnet hospitals gave the following responses to the staffing situation.

Staffing Situation	1	2	3	4	5
Number	165	140	125	50	20

Do the data indicate that the nurses from the nonmagnet hospitals have a different distribution of opinions? Use $\alpha = 0.05$.
   a. Solve using the *p*-value approach.
   b. Solve using the classical approach.

11.15 Does the sample of 200 adults collected in the hot foods example on page 236 show a distribution that is significantly different from the distribution shown in the "Putting Out the Fire" graph (page 236)? Use $\alpha = 0.05$.

*11.16 According to The Harris Poll, the proportion of all adults who live in households with rifles (29%), shotguns (29%), or pistols (23%) has not changed significantly since 1996. However, today more people live in households with no guns (61%). The 1,014 adults surveyed gave the following results.

	All Adults (%)	All Gun Owners (%)
Have rifle, shotgun, and pistol (3 out of 3)	16	41
Have 2 out of 3 (rifle, shotgun, or pistol)	11	27
Have 1 out of 3 (rifle, shotgun, or pistol)	11	29
Decline to answer/Not sure	1	3
Totals	39%	100%

In a survey of 2,000 adults in Memphis who said they own guns, 780 said they own all three types, 550 said they owned 2 of the 3, 560 said they owned 1 of the 3 types, and 110 declined to specify what types of guns they owned.

a. Test the null hypothesis that the distribution of number of types owned is the same in Memphis as it is nationally as reported by The Harris Poll. Use a level of significance equal to 0.05.

b. What caused the calculated value of $\chi^2\star$ to be so large? Does it seem right that one cell should have this much effect on the results? How could this test be completed differently (hopefully, more meaningfully) so that the results might not be affected as they were in part (a)? Be specific.

Objective 11.3

11.17 State the null hypothesis, H_o, and the alternative hypothesis, H_a, that would be used to test the following statements:

a. The voters expressed preferences that were not independent of their party affiliations.

b. The distribution of opinions is the same for all three communities.

c. The proportion of "yes" responses was the same for all categories surveyed.

11.18 The "test of independence" and the "test of homogeneity" are completed in identical fashion, using the contingency table to display and organize the calculations. Explain how these two hypothesis tests differ.

11.19 Find the expected value for the cell shown:

11.20 Results on seat belt usage from the 2003 Youth Risk Behavior Survey were published in a *USA Today* Snapshot on January 13, 2005. The following table outlines the results from the high school students who were surveyed in the state of Nebraska. They were asked whether they rarely or never wear seat belts when riding in someone else's car.

	Female	Male
Rarely or never use seat belt	208	324
Use seat belt	1,217	1,184

SOURCE: www.cdc.gov

Using $\alpha = 0.05$, does this sample present sufficient evidence to reject the hypothesis that gender is independent of seat belt usage?
a. Solve using the p-value approach.
b. Solve using the classical approach.

11.21 A survey of randomly selected travelers who visited the service station restrooms of a large U.S. petroleum distributor showed the following results:

	Quality of Restroom Facilities			
Gender of Respondent	Above Average	Average	Below Average	Totals
Female	7	24	28	59
Male	8	26	7	41
Totals	15	50	35	100

Using $\alpha = 0.05$, does the sample present sufficient evidence to reject the hypothesis "Quality of responses is independent of the gender of the respondent?"
a. Solve using the p-value approach.
b. Solve using the classical approach.

11.22 Tourette's syndrome is an inheritable, childhood-onset neurological disorder involving multiple motor tics and at least one vocal tic. A United States study that was published in the CDC's June 5, 2009, *Morbidity and Mortality Weekly Report* indicated that the syndrome occurs in three out of every 1,000 school-age children. Further analysis broke the data into ethnicity and race categories—see the following chart. At the 0.05 level of significance, does this sample indicate that having Tourette's is independent of ethnicity and race?

	Hispanic	Non-Hispanic White	Non-Hispanic Black
Have Tourette's	26	164	18
No Tourette's	7,321	43,602	6,427

11.23 Tourette's syndrome is an inheritable, childhood-onset neurological disorder involving multiple motor tics and at least one vocal tic. A United States study that was published in the CDC's June 5, 2009, *Morbidity and Mortality Weekly Report* indicated that the syndrome occurred in three of every 1,000 school-age children. Further analysis broke the data into household income categories with respect to the federal poverty level— see the chart below. At the 0.05 level of significance, does this sample indicate that having Tourette's is independent of household income?

	Below 200%	200% – 400%	Above 400%
Have Tourette's	65	80	80
No Tourette's	17,581	21,795	24,432

11.24 It is hypothesized that sick animals receiving a certain drug (the treated group) will survive at a more favorable rate than those that do not receive the drug (the control group). The following results from the test were recorded.

	Survived	Did Not Survive
Treated	46	18
Control	38	35

  a. Explain why the hypothesis stated in the exercise cannot be the null hypothesis.
  b. Explain why the null hypothesis is correctly stated as "Survival is independent of the drug treatment."
  c. Complete the hypothesis test, finding the p-value.
  d. If the test is completed using $\alpha = 0.02$, state the decision that must be reached.
  e. If the test is completed using $\alpha = 0.02$, carefully state the conclusion and its meaning.

11.25 The November 12, 2009, *USA Today* Snapshot "Rabies in Cats on the Rise" reported that almost 7,000 animals were reported to have rabies in 2008. Utilizing information from the *Journal of the American Veterinary Medical Association*, the following rabies cases were logged for cats and dogs:

Year	Dogs	Cats
2007	93	274
2008	75	294

At the 0.05 level of significance, is the distribution of rabies cases for dogs and cats the same for the years listed?

11.26 According to a report from the Substance Abuse and Mental Health Services Administration, food-service workers have the highest rate of cigarette smokers: 45% of food-service workers reported smoking cigarettes in the past month. Do some careers lend themselves to cigarette smokers more than others? If 100 people in each of the following occupations were asked about smoking in the last month, do the data support that some careers correspond to higher rates of smoking? Use a 0.05 level of significance.

Occupation	Construction	Production	Engineering	Politics	Education
# who smoke	43	37	17	17	12

11.27 Students use many kinds of criteria when selecting courses. "Teacher who is a very easy grader" is often one criterion. Three teachers are scheduled to teach statistics next semester. A sample of previous grade distributions for these three teachers is shown here.

	Professor		
Grades	#1	#2	#3
A	12	11	27
B	16	29	25
C	35	30	15
Other	27	40	23

At the 0.01 level of significance, is there sufficient evidence to conclude "The distribution of grades is not the same for all three professors"?
  a. Solve using the p-value approach.
  b. Solve using the classical approach.
  c. Which professor is the easiest grader? Explain, citing specific supporting evidence.

*11.28 Are younger and younger people able to obtain illegal guns? According to the October 11, 2009, Rochester, New York, *Democrat and Chronicle* article "The Gun Used to Shoot DiPonzio," which cited a 14-year-old shooting a police officer, it appears that the number of people in younger age groups found with illegal guns continues to grow. At the 0.01 level of significance, does it appear that the distribution of ages possessing illegal guns is the same for the years listed?

Year	21 and Under	22–30	31–50	51+
2005	103	93	111	33
2006	119	136	96	31
2007	155	140	130	76
2008	159	160	104	60

*remember

Problems marked with an asterisk (*) have data sets available on the CourseMate for STAT2 site. Login at cengagebrain.com.

Chapter 12

Analysis of Variance

How much time did America's workforce spend getting to their jobs this morning? How much time did your parents spend commuting to work this morning? Does everybody spend the same amount of time? What was the mean amount of time spent commuting to work this morning by people in Boston? What was the mean amount of time spent commuting to work this morning by people in Dallas? Do you think that the city will have any effect on the amount of time spent commuting to work this morning? The graphic "Longest Commute to Work" seems to suggest that some cities have longer commuting times than others. From studying previous chapters, we know that the statistics from different samples vary, even if they are drawn from the same population. The question that might be asked here is, "Is the variation between the samples greater than would be expected if the samples were all drawn from one population?"

Does there seem to be a difference between the mean one-way commute time for these six cities? Some of the techniques studied in previous chapters could help determine this, but in this chapter we will learn some new techniques designed specifically to make comparisons among multiple means.

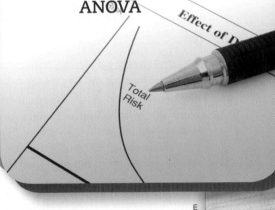

objectives

12.1 Analysis of Variance Technique—An Introduction

12.2 The Logic behind ANOVA

12.3 Applications of Single-Factor ANOVA

12.1 Analysis of Variance Technique—An Introduction

PREVIOUSLY, WE HAVE TESTED HYPOTHESES ABOUT TWO MEANS.

In this chapter we are concerned with testing a hypothesis about several means, and we begin discussing how to approach testing on page 256.

Longest Commute to Work

Here is a look at the average one-way commute times for six major U.S. Cities:

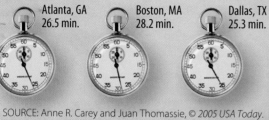

Atlanta, GA
26.5 min.

Boston, MA
28.2 min.

Dallas, TX
25.3 min.

Philadelphia, PA
30.3 min.

Seattle, WA
23.8 min.

St. Louis, MO
23.3 min.

SOURCE: Anne R. Carey and Juan Thomassie, © 2005 USA Today.

The **analysis of variance (ANOVA)** technique, which we are about to explore, will be used to test a hypothesis about several means—for example,

$$H_o: \mu_1 = \mu_2 = \mu_3 = \mu_4 = \mu_5$$

By using our former technique for testing hypotheses about two means, we could test several hypotheses if each stated a comparison of two means. For example, we could test

$H_1: \mu_1 = \mu_2$ $\qquad$ $H_2: \mu_1 = \mu_3$

$H_3: \mu_1 = \mu_4$ $\qquad$ $H_4: \mu_1 = \mu_5$

$H_5: \mu_2 = \mu_3$ $\qquad$ $H_6: \mu_2 = \mu_4$

$H_7: \mu_2 = \mu_5$ $\qquad$ $H_8: \mu_3 = \mu_4$

$H_9: \mu_3 = \mu_5$ $\qquad$ $H_{10}: \mu_4 = \mu_5$

In order to test the null hypothesis, H_o, that all five means are equal, we would have to test each of these 10 hypotheses using our former technique. Rejection of any one of the 10 hypotheses about two means would cause us to reject the null hypothesis that all five means are equal. If we failed to reject all 10 hypotheses, we would fail to reject the main null hypothesis. By testing in this manner, the overall type I error rate would become much larger than the value of α associated with a single test. The ANOVA techniques allow us to test the null hypothesis (all means are equal) against the alternative hypothesis (at least one mean value is different) with a specified value of α.

In this chapter we introduce ANOVA. ANOVA experiments can be very complex, depending on the situation. We will restrict our discussion to the most basic experimental design, the single-factor ANOVA.

Hypothesis Test for Several Means

Let's begin our discussion of the analysis of variance technique (ANOVA) by looking at an example regarding temperature and productivity. The temperature at which a manufacturing plant is maintained is believed to affect the rate of production in the plant. The data in Table 12.1 are the number, x, of units produced in one hour for randomly selected one-hour periods when the production process in the plant was operating at each of three temperature *levels*. The data values from repeated samplings are called **replicates**. Four replicates, or data values, were obtained for two of the temperatures and five were obtained for the third temperature. Do these data suggest that temperature has a significant effect on the production level at $\alpha = 0.05$?

The level of production is measured by the mean value; $\bar{x}_i$ indicates the observed production mean at level i, where $i = 1, 2,$ and 3 correspond to temperatures of 68°F, 72°F, and 76°F, respectively. There is a certain amount of variation among these means. Since sample means are not necessarily the same when repeated samples are taken from a population, some variation can be expected, even if all three population means are equal. With that in mind, is this variation among the $\bar{x}$'s due to chance, or is it due to the effect that temperature has on the production rate? We can find out using our five-step hypothesis test.

STEP 1 **a. Parameter of interest:**
The "mean" at each *level of the test factor* is of interest: the mean production rate at 68°F, μ_{68}; the mean production rate at 72°F, μ_{72}; and the mean production rate at 76°F, μ_{76}. The factor being tested, plant temperature, has three levels: 68°F, 72°F, or 76°F.

b. Statement of hypotheses:

$$H_o: \mu_{68} = \mu_{72} = \mu_{76}$$

That is, the true production mean is the same at each temperature level tested. In other words, the

> **ANOVA** Analysis of variance technique.
>
> **Replicates** The data values from repeated samplings.

Table 12.1 **Sample Results on Temperature and Production**

	Temperature Levels		
	Sample from 68°F ($i = 1$)	Sample from 72°F ($i = 2$)	Sample from 76°F ($i = 3$)
	10	7	3
	12	6	3
	10	7	5
	9	8	4
		7	
Column totals	$C_1 = 41$	$C_2 = 35$	$C_3 = 15$
	$\bar{x}_1 = 10.25$	$\bar{x}_2 = 7.0$	$\bar{x}_3 = 3.75$

temperature does not have a significant effect on the production rate. The alternative to the null hypothesis is

H_a: Not all temperature level means are equal.

Thus, we will want to reject the null hypothesis if the data show that one or more of the means are significantly different from the others.

STEP 2 **a. Assumptions:**
The data were randomly collected and are independent of each other. The effects due to chance and untested factors are assumed to be normally distributed. (See pages 262–263 for further discussion.)

© Andrea Krause/IStockphoto.com

b. Test statistic:
We will make the decision to reject H_o or fail to reject H_o by using the F-distribution and an F-test statistic.

c. Level of significance:
$\alpha = 0.05$ (given in the statement of the problem).

STEP 3 **a. Sample information:** See Table 12.1.

b. Calculate the test statistic:
Recall from Chapter 10 that the calculated value of F is the ratio of two variances. The analysis of variance procedure will separate the variation among the entire set of data into two categories. To accomplish this separation, we first work with the numerator of the fraction used to define **sample variance**, formula (2.5):

$$s^2 = \frac{\Sigma(x - \bar{x})^2}{n - 1}$$

The numerator of this fraction is called the **sum of squares:**

Total Sum of Squares

sum of squares $= \Sigma(x - \bar{x})^2$ (12.1)

We calculate the **total sum of squares, SS(total),** for the total set of data by using a formula that is equivalent to formula (12.1) but does not require the use of $\bar{x}$. This equivalent formula is

Shortcut for Total Sum of Squares

$\text{SS(total)} = \Sigma(x^2) - \frac{(\Sigma x)^2}{n}$ (12.2)

Now we can find SS(total) for our example by using formula (12.2). First,

$$\Sigma(x^2) = 10^2 + 12^2 + 10^2 + 9^2 + 7^2 +$$
$$6^2 + 7^2 + 8^2 + 7^2 + 3^2 + 3^2 +$$
$$5^2 + 4^2 = 731$$

$$\Sigma x = 10 + 12 + 10 + 9 + 7 + 6 + 7 +$$
$$8 + 7 + 3 + 3 + 5 + 4 = 91$$

Then, using formula (12.2), we have

$$\text{SS(total)} = \Sigma(x^2) - \frac{(\Sigma x)^2}{n}:$$

$$\text{SS(total)} = 731 - \frac{(91)^2}{13} = 731 - 637 = \mathbf{94}$$

Sum of squares, SS(total) The numerator of the sample variance formula.

Next, the SS(total), 94, must be separated into two parts: the sum of squares due to temperature levels, SS(temperature), and the sum of squares due to experimental error of replication, SS(error). This splitting is often called **partitioning**, since SS(temperature) + SS(error) = SS(total); that is, in our example SS(temperature) + SS(error) = 94. The sum of squares, **SS(factor)** [SS(temperature) for our example], that measures the **variation between the factor levels** (temperatures) is found by using formula (12.3):

Sum of Squares Due to Factor

$$SS(factor) = \left(\frac{C_1^2}{k_1} + \frac{C_2^2}{k_2} + \frac{C_3^2}{k_3} + \ldots \right) - \frac{(\Sigma x)^2}{n} \qquad (12.3)$$

where C_i represents the column total, k_i represents the number of replicates at each level of the factor, and n represents the total sample size ($n = \Sigma k_i$).

Now we can find SS(temperature) for our example by using formula (12.3):

$$SS(factor) = \left(\frac{C_1^2}{k_1} + \frac{C_2^2}{k_2} + \frac{C_3^2}{k_3} + \ldots \right) - \frac{(\Sigma x)^2}{n}:$$

$$SS(temperature) = \left(\frac{41^2}{4} + \frac{35^2}{5} + \frac{15^2}{4} \right) - \frac{(91)^2}{13}$$

$$= (420.25 + 245.00 + 56.25) - 637.0$$

$$= 721.5 - 637.0 = \mathbf{84.5}$$

The sum of squares, **SS(error)**, that measures the **variation within the rows** is found by using formula (12.4):

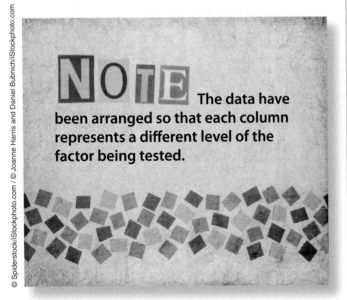

NOTE The data have been arranged so that each column represents a different level of the factor being tested.

Sum of Squares Due to Error

$$SS(error) = \Sigma(x^2) - \left(\frac{C_1^2}{k_1} + \frac{C_2^2}{k_2} + \frac{C_3^2}{k_3} + \ldots \right) \qquad (12.4)$$

The SS(error) for our example can now be found. First,

$$\Sigma(x^2) = 731 \quad \text{(found previously)}$$

$$\left(\frac{C_1^2}{k_1} + \frac{C_2^2}{k_2} + \frac{C_3^2}{k_3} + \ldots \right) = 721.5 \quad \text{(found previously)}$$

Then, using formula (12.4), we have

$$SS(error) = \Sigma(x^2) - \left(\frac{C_1^2}{k_1} + \frac{C_2^2}{k_2} + \frac{C_3^2}{k_3} + \ldots \right):$$

$$SS(error) = 731.0 - 721.5 = \mathbf{9.5}$$

NOTE: SS(total) = SS(factor) + SS(error). Inspection of formulas (12.2), (12.3), and (12.4) will verify this.

For convenience we will use an ANOVA table to record the sums of squares and to organize the rest of the calculations. The format of an ANOVA table is shown in Table 12.2.

Table 12.2 **Format for ANOVA Table**

Source	df	SS	MS
Factor		84.5	
Error		9.5	
Total		94.0	

We have calculated the three sums of squares for our example. The **degrees of freedom, df,** associated with each of the three sources are determined as follows:

1. df(factor) is 1 less than the number of levels (columns) for which the factor is tested:

Degrees of Freedom for Factor

$$df(factor) = c - 1 \qquad (12.5)$$

where c is the number of *levels for which the factor is being tested* (number of columns on the data table).

2. df(total) is 1 less than the total number of data:

Degrees of Freedom for Total

$$df(total) = n - 1 \qquad (12.6)$$

where n is the number of data in the total sample (that is, $n = k_1 + k_2 + k_3 + \ldots$, where k_i is the number of replicates at each level tested).

3. df(error) is the sum of the degrees of freedom for all the levels tested (columns in the data table). Each column has $k_i - 1$ degrees of freedom; therefore,

$$df(error) = (k_1 - 1) + (k_2 - 1) + (k_3 - 1) + \ldots$$

or

Degrees of Freedom for Error

$$df(error) = n - c \qquad (12.7)$$

The degrees of freedom for our example are

$$df(temperature) = c - 1 = 3 - 1 = \textbf{2}$$
$$df(total) = n - 1 = 13 - 1 = \textbf{12}$$
$$df(error) = n - c = 13 - 3 = \textbf{10}$$

The sums of squares and the degrees of freedom must check; that is,

$$SS(factor) + SS(error) = SS(total) \qquad (12.8)$$

and

$$df(factor) + df(error) = df(total) \qquad (12.9)$$

The **mean square** for the factor being tested, **MS(factor)**, and for error, **MS(error)**, are obtained by dividing the sum-of-squares value by the corresponding number of degrees of freedom:

Mean Square for Factor

$$MS(factor) = \frac{SS(factor)}{df(factor)} \qquad (12.10)$$

Mean Square for Error

$$MS(error) = \frac{SS(error)}{df(error)} \qquad (12.11)$$

The mean squares for our example are

$$MS(temperature) = \frac{SS(temperature)}{df(temperature)} = \frac{84.5}{2} = \textbf{42.25}$$

$$MS(error) = \frac{SS(error)}{df(error)} = \frac{9.5}{10} = \textbf{0.95}$$

The complete ANOVA table appears in Table 12.3.

Table 12.3 ANOVA Table for Temperature Example p. 256

Source	df	SS	MS
Temperature	2	84.5	42.25
Error	10	9.5	0.95
Total	12	94.0	

The hypothesis test is now completed using the two mean squares as the measures of variance. The calculated value of the test statistic, $F\star$, is found by dividing the MS(factor) by the MS(error):

Test Statistic for ANOVA

$$F\star = \frac{MS(factor)}{MS(error)} \qquad (12.12)$$

The calculated value of F for our example is found by using formula (12.12):

$$F\star = \frac{MS(factor)}{MS(error)}:$$

$$F\star = \frac{MS(temperature)}{MS(error)} = \frac{42.25}{0.95} = \textbf{44.47}$$

NOTE

Since the calculated value of F, $F\star$, is found by dividing MS(temperature) by MS(error), the number of degrees of freedom for the numerator is df(temperature) = 2 and the number of degrees of freedom for the denominator is df(error) = 10.

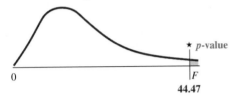

STEP 4 The probability distribution:

As always, we can use either the p-value or the classical procedure:

Using the p-value procedure:

a. Use the right-hand tail because larger values of $F\star$ indicate "not all equal" as expressed by H_a. $P = P(F\star > 44.47 | df_n = 2, df_d = 10)$, as shown on the figure below.

To find the p-value, you have two options:

1. Use Table 9c in Appendix B to place bounds on the p-value: **P < 0.01.**
2. Use a computer or calculator to find the p-value: **P = 0.00001.**

For additional instructions, see pages 228–229.

b. The p-value is smaller than the level of significance, α (0.05).

Using the classical procedure:

a. The critical region is the right-hand tail because larger values of $F\star$ indicate "not all equal" as expressed by H_a. $df_n = 2$ and $df_d = 10$. The critical value is obtained from Table 9a:

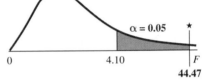

$F_{(2, 10, 0.05)} = 4.10$

For additional instructions, see page 227.

b. $F\star$ is in the critical region, as shown in red on the figure above.

STEP 5 a. **Decision:** Reject H_o.

b. **Conclusion:**

At least one of the room temperatures does have a significant effect on the production rate at the 0.05 level of significance. The differences in the mean production rates at the tested temperature levels were found to be significant.

In this section we have seen how the ANOVA technique separated the variance among the sample data into two measures of variance: (1) MS(factor), the measure of variance between the levels being tested, and (2) MS(error), the measure of variance within the levels being tested. Then these measures of variance can be compared. For our example, the between-level variance was found to be significantly greater than the within-level variance (experimental error). This led us to the conclusion that temperature did have a significant effect on the variable x, the number of units of production completed per hour.

In the next section we will demonstrate the logic of the analysis of variance technique.

12.2 The Logic behind ANOVA

ANOVA TECHNIQUES HELP US MAKE DECISIONS CONCERNING THE EFFECT THAT LEVELS OF THE TEST FACTORS HAVE ON THE **RESPONSE (OBSERVED) VARIABLE.**

The logic of the analysis of variance technique proceeds like this: In order to compare the means of the levels of the test factor, a measure of the **variation between the levels** (between the columns on the data table), the **MS(factor)**, will be compared to a measure of the **variation within the levels** (within the columns on the data table), the **MS(error)**.

FORMULAS

Total sum of squares

$$\text{sum of squares} = \Sigma(x - \bar{x})^2 \tag{12.1}$$

Shortcut for total sum of squares

$$SS(\text{total}) = \Sigma(x^2) - \frac{(\Sigma x)^2}{n} \tag{12.2}$$

Sum of squares due to factor

$$SS(\text{factor}) = \left| \frac{C_1^2}{k_1} + \frac{C_2^2}{k_2} + \frac{C_3^2}{k_3} + \dots \right| - \frac{(\Sigma x)^2}{n} \tag{12.3}$$

Sum of squares due to error

$$SS(\text{error}) = \Sigma(x^2) - \left| \frac{C_1^2}{k_1} + \frac{C_2^2}{k_2} + \frac{C_3^2}{k_3} + \dots \right| \tag{12.4}$$

Degrees of freedom for factor

$$df(\text{factor}) = c - 1 \tag{12.5}$$

Degrees of freedom for total

$$df(\text{total}) = n - 1 \tag{12.6}$$

Degrees of freedom for error

$$df(\text{error}) = n - c \tag{12.7}$$

Total sum of squares

$$SS(\text{factor}) + SS(\text{error}) = SS(\text{total}) \tag{12.8}$$

Total sum of degrees of freedom

$$df(\text{factor}) + df(\text{error}) = df(\text{total}) \tag{12.9}$$

Mean square for factor

$$MS(\text{factor}) = \frac{SS(\text{factor})}{df(\text{factor})} \tag{12.10}$$

Mean square for error

$$MS(\text{error}) = \frac{SS(\text{error})}{df(\text{error})} \tag{12.11}$$

Test statistic $F\star$

$$F\star = \frac{MS(\text{factor})}{MS(\text{error})} \tag{12.12}$$

If MS(factor) is significantly larger than MS(error), then we will conclude that the means for the factor levels being tested are not all the same. This implies that the factor being tested does have a significant effect on the response variable. If, however, MS(factor) is not significantly larger than MS(error), then we will not be able to reject the null hypothesis that all means are equal.

For example, look at Table 12.4. Do the data provide sufficient evidence to conclude that there is a difference in the three population means μ_F, μ_G, and μ_H?

Table 12.4 **Sample Results**

| | Factor Levels | |
Sample from Level F	Sample from Level G	Sample from Level H
3	5	8
2	6	7
3	5	7
4	5	8
$C_F = 12$	$C_G = 21$	$C_H = 30$
$\bar{x}_F = 3.00$	$\bar{x}_G = 5.25$	$\bar{x}_H = 7.50$

We can visualize the relative relationship among the three samples in Figure 12.1. A quick look at the figure suggests that the three sample means are different from each other, implying that the sampled populations have different mean values. These three samples demonstrate relatively little within-sample variation, although there is a relatively large amount of between-sample variation.

Figure 12.1 **Data from Table 12.4**

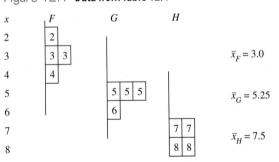

In addition to visualizing differences among several means, we can use a similar process to visualize

equality among several means. Look at Table 12.5. Do the data provide sufficient evidence to conclude that there is a difference in the three population means μ_J, μ_K, and μ_L?

Table 12.5 **Sample Results**

| | Factor Levels | |
Sample from Level J	Sample from Level K	Sample from Level L
3	5	6
8	4	2
6	3	7
4	7	5
$C_J = 21$	$C_K = 19$	$C_L = 20$
$\bar{x}_J = 5.25$	$\bar{x}_K = 4.75$	$\bar{x}_L = 5.00$

Again, we can see the relative relationship in Figure 12.2. A quick look at the figure does not suggest that the three sample means are different from each other. There is little **between-sample variation** for these three samples (that is, the sample means are relatively close in value), whereas the **within-sample variation** is relatively large (that is, the data values within each sample cover a relatively wide range of values).

Figure 12.2 **Data from Table 12.5**

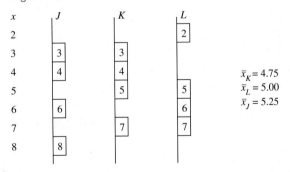

Before we complete a hypothesis test for analysis of variance, we must agree on some ground rules, or **assumptions**. In this chapter we will use the following three basic assumptions:

1. Our goal is to investigate the effect that various levels of the factor being tested have on the response variable. Typically, we want to find the level that yields the most advantageous values of the response (observed) variable. This, of course, means that we probably will want to reject the null hypothesis in favor of the alternative hypothesis. Then a follow-up study could determine the "best" level of the factor.

2. We must assume that the effects due to chance and due to untested factors are normally distributed and that the variance caused by these effects is constant throughout the experiment.

3. We must assume independence among all observations of the experiment. (Recall that independence means that the results of one observation of the experiment do not affect the results of any other observation.) We will usually conduct the tests in a **randomized** order to ensure independence. This technique also helps to avoid data contamination.

12.3 Applications of Single-Factor ANOVA

BEFORE CONTINUING OUR ANOVA DISCUSSION, LET'S IDENTIFY THE NOTATION, PARTICULARLY THE SUBSCRIPTS THAT ARE USED.

Notice in Table 12.6 that each piece of data has two subscripts; the first subscript indicates the column number (test factor level), and the second subscript identifies the row (replicate) number. The column totals, C_i, are listed across the bottom of the table. The grand total, T, is equal to the sum of all x's and is found by adding the column totals. Row totals can be used as a cross-check but serve no other purpose.

Table 12.6 **Notation Used in ANOVA**

	Factor Levels				
Replicates	Sample from Level 1	Sample from Level 2	Sample from Level 3	$\cdots$	Sample from Level c
$k = 1$	$x_{1,1}$	$x_{2,1}$	$x_{3,1}$		$x_{c,1}$
$k = 2$	$x_{1,2}$	$x_{2,2}$	$x_{3,2}$		$x_{c,2}$
$k = 3$	$x_{1,3}$	$x_{2,3}$	$x_{3,3}$		$x_{c,3}$
$\vdots$					
Column Totals	C_1	C_2	C_3	$\cdots$	C_c	T

$$T = \text{grand total} = \text{sum of all } x\text{'s} = \Sigma x = \Sigma C_i$$

A **mathematical model** (equation) is often used to express a particular situation. In Chapter 3 we used a mathematical model to help explain the relationship between the values of bivariate data. The equation $\hat{y} = b_0 + b_1 x$ served as the model when we believed that a straight-line relationship existed. The probability functions studied in Chapter 5 are also examples of mathematical models. For the single-factor ANOVA, the mathematical model, formula (12.13), is an expression of the composition of each piece of data entered in our data table:

Mathematical Model for Single-Factor ANOVA

$$x_{c,k} = \mu + F_c + \epsilon_{k(c)} \tag{12.13}$$

We interpret each term of this model as follows:

$x_{c,k}$ is the value of the variable at the kth replicate of level c.

μ is the mean value for all the data without respect to the test factor.

F_c is the effect that the factor being tested has on the response variable at each different level c.

$\epsilon_{k(c)}$ (ϵ is the lowercase Greek letter epsilon) is the **experimental error** that occurs among the k replicates in each of the c columns.

Hypothesis Test for the Equality of Several Means

Let's look at another hypothesis test using an analysis of variance (ANOVA). A gun club performed an experiment with a randomly selected group of first-time shooters. The purpose of the experiment was to determine whether shooting accuracy is affected by the method of sighting used: only the right eye open, only the left eye open, or both eyes open. Fifteen first-time shooters were selected and divided into three groups. Each group experienced the same training and practicing procedures with one exception: the method of sighting used. After completing training, each shooter was given the same number of rounds and asked to shoot at a target. Their scores are listed in Table 12.7.

*Table 12.7 **Sample Results on Target Shooting**

Method of Sighting		
Right Eye	Left Eye	Both Eyes
12	10	16
10	17	14
18	16	16
12	13	11
14		20
		21

At the 0.05 level of significance, is there sufficient evidence to reject the claim that the three methods of sighting are equally effective?

In this experiment the factor is method of sighting and the levels are the three different methods of sighting (right eye, left eye, and both eyes open). The replicates are the scores received by the shooters in each group. The null hypothesis to be tested is that the three methods of sighting are equally effective, or the mean scores attained using each of the three methods are the same. Using this hypothesis, we can begin the five-step process to determine whether or not there is evidence to reject it.

STEP 1 a. **Parameter of interest:**

The "mean" at each level of the test factor is of interest: the mean score using the right eye, μ_R; the mean score using the left eye, μ_L; and the mean score using both eyes, μ_B. The factor being tested, "method of sighting," has three levels: right, left, and both.

b. **Statement of hypotheses:**

$$H_o: \mu_R = \mu_L = \mu_B$$

$$H_a: \text{The means are not all equal (that is, at least one mean is different).}$$

STEP 2 a. **Assumptions:**

The shooters were randomly assigned to the method, and their scores are independent of each other. The effects due to chance and untested factors are assumed to be normally distributed.

b. **Test statistic to be used:**

The F-distribution and formula (12.12) will be used with df(numerator) = df(method) = 2 and df(denominator) = df(error) = 12.

c. **Level of significance:** $\alpha = 0.05$.

STEP 3 a. **Sample information:** See Table 12.8.

b. **Calculated test statistic:**

The test statistic is $F\star$: Table 12.8 is used to find the column totals.

First, the summations Σx and Σx^2 need to be calculated:

$$\Sigma x = 12 + 10 + 18 + 12 + 14 + 10 + 17 + \ldots + 21$$
$$= 220$$

$$(\text{Or } 66 + 56 + 98 = 220 \checkmark)$$

$$\Sigma x^2 = 12^2 + 10^2 + 18^2 + 12^2 + 14^2 + 10^2 + \ldots + 21^2$$
$$= 3{,}392$$

© Yunus Arakoni/iStockphoto.com

Using formula (12.2), we find

$$SS(total) = \Sigma(x^2) - \frac{(\Sigma x)^2}{n}:$$

$$SS(total) = 3{,}392 - \frac{(220)^2}{15}$$

$$= 3{,}392 - 3{,}226.67 = \mathbf{165.33}$$

Using formula (12.3), we find

$$SS(method) = \left(\frac{C_1^2}{k_1} + \frac{C_2^2}{k_2} + \frac{C_3^2}{k_3} + \ldots \right) - \frac{(\Sigma x)^2}{n}:$$

$$SS(method) = \left(\frac{66^2}{5} + \frac{56^2}{4} + \frac{98^2}{6} \right) - \frac{(220)^2}{15}$$

$$= (871.2 + 784 + 1{,}600.67) - 3{,}226.67$$

$$= 3{,}255.87 - 3{,}226.67 = \mathbf{29.20}$$

To find SS(error) we need first:

$$\Sigma(x^2) = 3{,}392 \quad \text{(found previously)}$$

$$\left(\frac{C_1^2}{k_1} + \frac{C_2^2}{k_2} + \frac{C_3^2}{k_3} + \ldots \right) = 3{,}255.87 \quad \text{(found previously)}$$

Then using formula (12.4), we have

$$SS(error) = \Sigma(x^2) - \left(\frac{C_1^2}{k_1} + \frac{C_2^2}{k_2} + \frac{C_3^2}{k_3} + \ldots \right):$$

$$SS(error) = 3{,}392 - 3{,}255.87 = \mathbf{136.13}$$

We use formula (12.8) to check the sum of squares:

$$SS(method) + SS(error) = SS(total):$$

$$29.20 + 136.13 = 165.33$$

The degrees of freedom are found using formulas (12.5), (12.6), and (12.7):

$$df(method) = c - 1 = 3 - 1 = \mathbf{2}$$

$$df(total) = n - 1 = 15 - 1 = \mathbf{14}$$

$$df(error) = n - c = 15 - 3 = \mathbf{12}$$

Using formulas (12.10) and (12.11), we find

$$MS(method) = \frac{SS(method)}{df(method)}:$$

$$MS(method) = \frac{29.20}{2} = \mathbf{14.60}$$

$$MS(error) = \frac{SS(error)}{df(error)}:$$

$$MS(error) = \frac{136.13}{12} = \mathbf{11.34}$$

The results of these computations are recorded in the ANOVA table in Table 12.9.

The calculated value of the test statistic is then found using formula (12.12):

$$F\bigstar = \frac{MS(factor)}{MS(error)}:$$

$$F\bigstar = \frac{MS(method)}{MS(error)} = \frac{14.60}{11.34} = \mathbf{1.287}$$

Table 12.8 **Sample Results for Target Shooting**

| Replicates | Factor Levels: Method of Sighting | | |
	Right Eye	Left Eye	Both Eyes
$k=1$	12	10	16
$k=2$	10	17	14
$k=3$	18	16	16
$k=4$	12	13	11
$k=5$	14		20
$k=6$			21
Totals	$C_R = 66$	$C_L = 56$	$C_B = 98$

Table 12.9 **ANOVA Table for Target Shooting Example**

Source	df	SS	MS
Method	2	29.20	14.60
Error	12	136.13	11.34
Total	14	165.33	

The probability distribution:

Again, we can use either the p-value or the classical procedure:

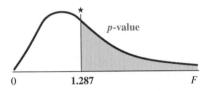

Using the p-value procedure:

a. Use the right-hand tail: $\mathbf{P} = P(F\bigstar > 1.287$, with $df_n = 2$ and $df_d = 12$) as shown on the figure below.

```
        ★
       /|
      / |\        p-value
     /  | \
    /   |  \___
   /    |      \____
  /     |           _____
 0    1.287              F
```

To find the p-value, you have two options:
1. Use Table 9a in Appendix B to place bounds on the p-value: **P > 0.05.**
2. Use a computer or calculator to find the p-value: **P = 0.312.**

For additional instructions, see pages 228–229.

b. The p-value is not smaller than the level of significance, α (0.05).

Using the classical procedure:

a. The critical region is the right-hand tail; the critical value is obtained from Table 9a:

$$F(2, 12, 0.05) = 3.89$$

```
        ★
       /|
      / |
     /  |
    /   |       α = 0.05
   /    |    ____
  /     |   /    \____
 0    1.287  3.89     F
```

For additional instructions, see page 227.

b. $F\bigstar$ is not in the critical region, as shown in red on the figure above.

STEP 5   a. **Decision:** Fail to reject H_o.

b. **Conclusion:**

At the 0.05 level of significance, the data show no evidence to reject the null hypothesis that the three methods are equally effective.

Recall the null hypothesis: "There is no difference between the levels of the factor being tested." A "fail to reject H_o" decision must be interpreted as the conclusion that there is no evidence of a difference due to the levels of the tested factor, whereas the rejection of H_o implies that there is a difference between the levels. That is, at least one level is different from the others. If there is a difference, the next task is to locate the level or levels that are different. Locating this difference may be the main objective of the analysis. In order to find the difference, the only method that is appropriate at this stage is to inspect the data. It may be obvious which level(s) caused the rejection of H_o. In the earlier example on the effect of temperature on production, it seems quite obvious that at least one of the levels [level 1 (68°F) or level 3 (76°F) because they have the largest and the smallest sample means] is different from the other two. If the higher values are more desirable for finding the "best" level to use, we would choose that corresponding level of the factor.

> Note that we restricted our development to one-factor experiments. This one-factor technique represents only a beginning to the study of analysis of variance techniques.

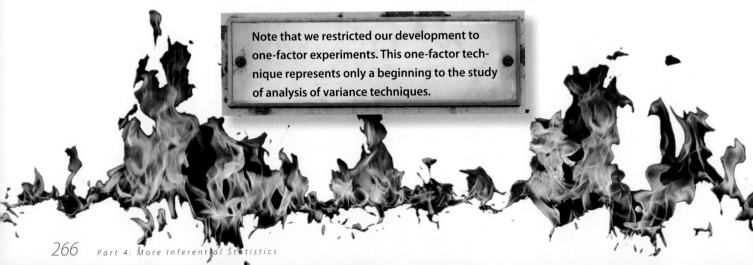

problems

Objective 12.1

*12.1 To compare commuting times in various locations, independent random samples were obtained from the six cities presented in the "Longest Commute to Work" graphic on page 255. The samples were from workers who commute to work during the 8:00 A.M. rush hour.

One-Way Travel to Work in Minutes

Atlanta	Boston	Dallas	Philadelphia	Seattle	St. Louis
29	18	42	29	30	15
21	37	25	20	23	24
20	37	36	33	31	42
15	25	32	37	39	23
37	32	20	42	14	33
26	34	26			18
	48	35			

a. Construct a graphic representation of the data using six side-by-side dotplots.

b. Visually estimate the mean commute time for each city and locate it with an X.

c. Does it appear that different cities have different effects on the average amount of time spent by workers who commute to work during the 8:00 A.M. rush hour? Explain.

d. Does it visually appear that different cities have different effects on the variation in the amount of time spent by workers who commute to work during the 8:00 A.M. rush hour? Explain.

12.2 a. Calculate the mean commute time for each city depicted in problem 12.1.

b. Does there seem to be a difference among the mean one-way commute times for these six cities?

c. Calculate the standard deviation for each city's commute time.

d. Does there seem to be a difference among the standard deviations between the one-way commute times for these six cities?

12.3 a. Construct the 95% confidence interval for the mean commute time for Atlanta and Boston using the data in problem 12.1.

b. Based on the confidence intervals found in (a), does it appear that the mean commute time is the same or different for these two cities? Explain.

c. Construct the 95% confidence interval for the mean commute time for Dallas.

d. Based on the confidence intervals found in (a) and (c), does it appear that the mean commute time is the same or different for Boston and Dallas? Explain.

e. Based on the confidence intervals found in (a) and (c), does it appear that the mean commute time is the same or different for the set of three cities, Atlanta, Boston, and Dallas? Explain.

f. How do your confidence intervals compare to the intervals given for Atlanta, Boston, and Dallas in "Longest Commute to Work" on page 255?

12.4 Each department at a large industrial plant is rated weekly. State the hypotheses used to test that "The mean weekly ratings are the same in three departments."

12.5 Refer to the following ANOVA table.

Source	df	SS	MS
Factor	3		
Error		40.4	
Total	20	164.2	

a. Find the four missing values.
b. Find the calculated value for F, $F\star$.

Objective 12.2

12.6 Do the data shown in the boxplot have a greater amount of variability within levels A, B, C, and D or between the four levels? Explain.

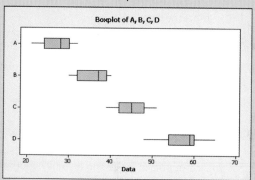

12.7 Do the data shown in the boxplot have a greater amount of variability within levels A, B, C, and D or between the four levels? Explain.

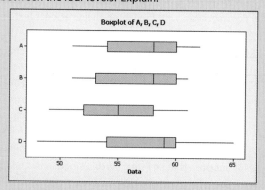

Objective 12.3

12.8 Consider the following table for a single-factor ANOVA. Find the following:

a. $x_{1,2}$ b. $x_{2,1}$ c. C_1 d. $\sum x$ e. $\sum (C_i)^2$

continued on p. 268

Replicates	Level of Factor		
	1	2	3
1	3	2	7
2	0	5	4
3	1	4	5

12.9 The following table of data is to be used for single-factor ANOVA. Find each of the following:

a. $x_{3,2}$ b. $x_{4,3}$ c. C_3 d. $\sum x$ e. $\sum (C_i)^2$

Replicates	Level of Factor			
	1	2	3	4
1	13	12	16	14
2	17	8	18	11
3	9	15	10	19

12.10 State the null hypothesis, H_o, and the alternative hypothesis, H_a, that would be used to test the following statements:

a. The mean value of x is the same at all five levels of the experiment.

b. The scores are the same at all four locations.

c. The four levels of the test factor do not significantly affect the data.

d. The three different methods of treatment do affect the variable.

12.11 Find the p-value for each of the following situations:

a. $F\star = 3.852$, df(factor) = 3, df(error) = 12

b. $F\star = 4.152$, df(factor) = 5, df(error) = 18

c. $F\star = 4.572$, df(factor) = 5, df(error) = 22

12.12 For the following ANOVA experiments, determine the critical region(s) and critical value(s) that are used in the classical approach for testing the null hypothesis.

a. $H_o: \mu_1 = \mu_2 = \mu_3 = \mu_4$, with $n = 18$ and $\alpha = 0.05$

b. $H_o: \mu_1 = \mu_2 = \mu_3 = \mu_4 = \mu_5$, with $n = 15$ and $\alpha = 0.01$

c. $H_o: \mu_1 = \mu_2 = \mu_3$, with $n = 25$ and $\alpha = 0.05$

12.13 Why does df(factor), the number of degrees of freedom associated with the factor, always appear first in the critical value notation F [df(factor), df(error), α]?

12.14 Suppose that an F-test (as described in this chapter using the p-value approach) has a p-value of 0.04.

a. What is the interpretation of p-value = 0.04?

b. What is the interpretation of the situation if you had previously decided on a 0.05 level of significance?

c. What is the interpretation of the situation if you had previously decided on a 0.02 level of significance?

12.15 Suppose that an F-test (as described in this chapter using the classical approach) has a critical value of 2.2, as shown in this figure:

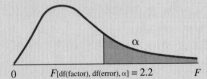

$F[\text{df(factor), df(error)}, \alpha] = 2.2$

a. What is the interpretation of a calculated value of F larger than 2.2?

b. What is the interpretation of a calculated value of F smaller than 2.2?

c. What is the interpretation if the calculated F were 0.1? 0.01?

12.16 An article titled "The Effectiveness of Biofeedback and Home Relaxation Training on Reduction of Borderline Hypertension" (SOURCE: *Health Education*) compared different methods of reducing blood pressure. Biofeedback ($n = 13$ subjects), biofeedback/relaxation ($n = 15$), and relaxation ($n = 14$) were the three methods compared. There were no differences among the three groups on pretest diastolic or systolic blood pressure readings. There was a significant post-test difference between groups on the systolic measure, $F(2, 39) = 4.14$, $p < 0.025$, and diastolic measure, $F(2, 39) = 5.56$, $p < 0.008$.

a. Verify that df(method) = 2 and df(error) = 39.

b. Use Tables 9A, 9B, and 9C in Appendix B to verify that for systolic, $p < 0.025$, and that for diastolic, $p < 0.008$.

*12.17 An employment agency wants to see which of three types of ads in the help-wanted section of local newspapers is the most effective. Three types of ads (big headline, straightforward, and bold print) were randomly alternated over a period of weeks, and the number of people responding to the ads was noted each week. Do these data support the null hypothesis that there is no difference in the effectiveness of the ads, as measured by the mean number responding, at the 0.01 level of significance?

Number of responses (replicates)	Type of Advertisement		
	Big Headline	Straightforward	Bold Print
1	23	19	28
2	42	31	33
3	36	18	46
4	48	24	29
5	33	26	34
6	26		34

a. Solve using the p-value approach.

b. Solve using the classical approach.

*12.18 A new operator was recently assigned to a crew of workers who perform a certain job. From the records of the number of units of work completed by each worker each day last month, a sample of size 5 was randomly selected for each of the two experienced workers and the new worker. At the 0.05 level of significance, does the evidence provide sufficient reason to reject the claim that there is no difference in the amount of work done by the three workers?

	Workers		
	New	A	B
Units of work (replicates)	8	11	10
	10	12	13
	9	10	9
	11	12	12
	8	13	13

a. Solve using the *p*-value approach.
b. Solve using the classical approach.

*12.19 Random samples of 2009 pickup trucks with 4-cylinder, 6-cylinder, and 8-cylinder engines were obtained. Each pickup truck was tested for miles per gallons in city driving.

4 Cyl (C)	6 Cyl (C)	8 Cyl (C)
21	19	19
18	18	19
19	20	15
17	21	20
18	20	19
18	19	21
19	19	18
18	20	19
	20	20
	19	16

Is there significant evidence to reject the hypothesis that the MPG for pickup trucks is the same for all three engine sizes? Use $\alpha = 0.05$.

*12.20 Random samples of 2009 pickup trucks with 4-cylinder, 5-cylinder, 6-cylinder, and 8-cylinder engines were obtained. Each pickup truck was tested for miles per gallons in highway driving.

4 Cyl (H)	5 Cyl (H)	6 Cyl (H)	8 Cyl (H)
24	21	19	20
23	21	19	19
22	23	19	19
24	21	18	20
24	18	21	16
23	22	20	18
23	23	19	15
24	18	20	21
24	20	19	
23	20	19	

Is there significant evidence to show that the MPG for pickup trucks is not the same for all four engine sizes? Use $\alpha = 0.01$.

*12.21 A number of sports enthusiasts have argued that major league baseball teams in the Central Division have an unfair advantage over teams in the Western and Eastern Divisions. This is because of the impact on Western and Eastern Division players due to the differences in time is likely to be greater when playing on the road (i.e., games away from home). Players from teams on the coasts could gain (going west) or lose (going east) up to three hours, whereas Central Division players would seldom gain or lose more than one hour. The following table shows the win/loss percentages for games played on the road by teams in all three divisions of Major League Baseball for the 2009 season:

Major League Baseball		
East	Central	West
56.8	46.9	59.3
48.1	42.7	48.1
39.5	44.4	45.7
38.3	37.0	43.2
30.9	39.5	55.6
59.3	55.6	50.6
54.3	45.7	44.4
56.8	49.4	40.7
35.8	46.9	42.0
32.1	37.0	
	27.5	

SOURCE: MLB.com

Complete an ANOVA table for win/loss percentages by teams representing each division. Test the null hypothesis that when teams play on the road, the mean win/loss percentage is the same for each of the three divisions. Use the 0.05 level of significance.

a. Solve using the *p*-value approach.
b. Solve using the classical approach.

*12.22 Does highest attained level of education influence the number of hours of television that people watch per day? Random samples were identified from each level of education and the hours of television watched per day by each person obtained.

Less than HS	High School	Associate	Bachelor	Graduate
2.1	3.7	3.9	4.6	1.9
6.3	4.4	3.0	4.1	2.5
4.5	4.4	2.0	0.1	0.7
5.9	3.3	2.2	4.9	1.7
3.5	3.3	0.6	4.5	1.2
4.0	3.3	0.6	4.0	3.5
1.7	4.4	2.7	6.3	2.5
5.2	4.9	3.0	5.0	3.3
4.5	2.4	3.8		0.5
2.2	2.7	4.1		3.0
4.4	2.3	2.3		2.4
		0.6		

a. Does the sample data show significant evidence to conclude that the level of education does influence the amount of television watched? Use $\alpha = 0.01$.
b. Provide explanations why the discrepancies among the categories may exist.

*12.23 The NBA is a big man's game. Average height for the league is about 6 feet 7 inches, as reported on the NBA website for the 2007–2008 season. Generally guards average 6 feet 4 inches, forwards average 6 feet 9 inches and centers average 7 feet.

A random sample of 2008 NBA players was selected and each player's height recorded to the nearest inch.

Guards	Forwards	Centers
78	81	84
74	84	90
78	80	83
74	84	83
77	82	85
73	81	83
72	82	87
80	80	84
	80	

SOURCE: NBA.com

a. Do you expect to find the mean heights for the three positions to be different from each other? Do you expect to find more variation between the positions or within the positions? Explain.

b. Construct a side-by-side graph (dotplot, boxplot, other) of your choice.

c. Does the graph in (b) show a relative large amount of variability between the positions? Explain in detail what you can determine from the graph.

d. Is there a significant difference in the heights of NBA players by position? Use $\alpha = 0.05$.

e. Do the results found in part (d) confirm your answer to part (c)? Explain.

f. Are the results what you anticipated they would be? Explain why or why not.

*12.24 To compare the commuting times in various locations, independent and random samples were obtained in each of six different U.S. cities displayed in the "Longest Commute to Work" graphic in Section 12.1. Using the commute time data located in problem 12.1 on page 267, complete the following:

a. Construct a boxplot showing the six cities side by side.

b. Does your graph show visual evidence suggesting that the city has an effect on the average morning commute time? Justify your answer.

c. Using the ANOVA technique, does this data show sufficient evidence to claim that the city has an effect on the average morning commute time? Use $\alpha = 0.05$.

d. Does the statistical answer found in part (c) agree with your graphical display in part (a) and your response in part (b)? Explain why your answers agree or disagree, citing statistical information learned in this chapter.

e. "Does the sample show that the city has an effect on the *amount of time* spent commuting to work?" "Does the sample show that the city has an effect on the *average amount of time* spent commuting to work?" Are these different questions? Explain.

*12.25 As a student at a community college, Stacey figured that the majority of her fellow students probably work a job in addition to their studies. Looking at only daytime classes, Stacey considered both gender and type of course may have an effect on the number of hours the students may work. The following data was randomly collected from three of Stacey's courses during the fall 2009 semester.

	Male	Female
Geography	40	40
	38	25
	47	30
Accounting	25	42
	30	35
	30	28
Music	26	16
	30	15
	33	18

a. At the 0.05 level of significance, does the data provide sufficient reason to support that type of course has an effect on the number of hours worked by a student?

b. At the 0.05 level of significance, does the data provide sufficient reason to support that gender has an effect on the number of hours worked by a student?

*12.26 The March 14, 2009, *Boston Globe* article "Bargains on the Menu—and a Side of Jitters" reported concerns that Restaurant Week would not be well attended due to the economy this year. Local restaurants were coming up with special promotions to draw past and new customers. Keeping this in mind, a waitress wondered if the "reduced" prices would also "reduce" tip percentages—especially on the lower-cost promotions and midweek shifts. To test her hypotheses, she collected the following data:

	Percent Tip		
Amount of Bill	Tuesday	Thursday	Saturday
$0–$29	21	15	12
	19	17	18
	15	18	19
	19	14	13
$30–$59	17	1	10
	18	16	16
	14	17	22
	18	12	17

(data continued)

Amount of Bill	Percent Tip		
	Tuesday	Thursday	Saturday
$60–$89	20	21	31
	14	19	25
	15	16	24
	24	15	30

a. At the 0.05 level of significance, does the data provide sufficient reason to support that the day of the week had an effect on the percent tip received?

b. At the 0.05 level of significance, does the data provide sufficient reason to support that the amount of the bill had an effect on the percent tip received?

*12.27 A local packaging plant implements various production lines based on the product to be packaged. Each line is for a different product, some being more complicated than others. With several lines running on a daily basis, concern over production rates surfaced due to the variation in the rates. Management decided to keep records to see if certain days of the week produced better production rates than others. The results follow:

Monday	Tuesday	Wednesday	Thursday	Friday
128	114	115	113	81
118	109	77	101	98
87	114	117	115	80
88	62	110	78	75
95	71	78	72	75
92	69	77	76	90
92	102	113	112	104
103	106	92	133	114
132	127	93	79	81

a. Using a one-way ANOVA, test the claim that the mean production rate is not the same for all five days of the week. Use a level of significance of 0.05.

b. Explain the meaning of the conclusion in part (a). Does the conclusion tell you which days are different? Which days have the larger means?

c. Construct a side-by-side boxplot of the data. Explain how the multiple boxplot coupled with the hypothesis test in part (a) help identify the difference between days.

d. How might the packaging company use this information?

e. Could there be other factors affecting the company's production rate problems? If so, name a few.

*12.28 Thirty-nine counties from the six-state Upper Midwest region of the United States were randomly selected from the USDA's National Agricultural Statistics Service website, and the following data on oat-production yield per acre were obtained.

County	IA	MN	ND	NE	SD	WI
1	76.2	53.0	71.4	60.0	76.5	52.0
2	65.3	70.0	64.3	37.0	50.0	53.0
3	86.0	71.0	66.7	53.0	42.0	72.0
4	73.6	54.0	61.4	50.0	62.5	81.0
5	61.3	64.0	66.0	56.0	55.7	57.0
6	74.3	40.0		58.0	59.1	64.0
7	58.3				59.3	
8	56.0					
9	61.4					

SOURCE: http://www.usda.gov/nass/graphics

a. Do these data show a significant difference in the mean yield rates for the six states? Use $\alpha = 0.05$.

b. Draw a graph that demonstrates the results found in part (a).

c. Explain the meaning of the results, including an explanation of how the graph in part (b) portrays the results.

*12.29 A study was conducted to assess the effectiveness of treating vertigo (motion sickness) with the transdermal therapeutic system (TTS—a patch worn on the skin). Two other treatments, both oral (one pill containing a drug and one a placebo), were used. The age and the gender of the patients for each treatment are listed here.

TTS		Antivert		Placebo	
47-f	53-m	51-f	43-f	67-f	38-m
41-f	58-f	53-f	56-f	52-m	59-m
63-m	62-f	27-m	48-m	47-m	33-f
59-f	34-f	29-f	52-f	35-f	32-f
62-f	47-f	31-f	19-f	37-f	26-f
24-m	35-f	25-f	31-f	40-f	37-m
43-m	34-f	52-f	48-m	31-f	49-f
20-m	63-m	55-f	53-m	45-f	49-m
55-f	46-f	32-f	63-m	41-f	38-f
		51-f	54-m	49-m	
		21-f			

Is there a significant difference between the mean age of the three test groups? Use $\alpha = 0.05$. Use a computer or calculator to complete this problem.

a. Solve using the p-value approach.

b. Solve using the classical approach.

*remember

P roblems marked with an asterisk (*) have data sets available on the CourseMate for STAT2 site. Login at cengagebrain.com.

Linear Correlation
and Regression Analysis

When you look at married couples, generally speaking the husband is 2 to 6 inches taller than the wife. It is unclear and debatable whether such preferences are innate or the function of height discrimination in a particular society. Certainly, newspapers and magazines make a big deal out of height differences in celebrity couples—particularly if the husband is shorter than his wife.

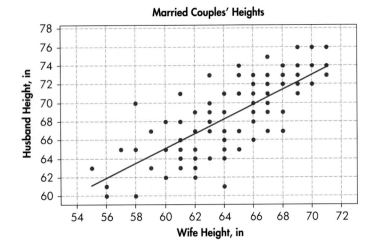

Married Couples' Heights

Based on the scatter diagram, "Married Couples' Heights," there appears to be a linear relationship between husbands' and wives' heights. As the heights of the wives increase, so do the heights of the husbands.

The basic ideas of regression and linear correlation analysis were introduced in Chapter 3. (If these concepts are not fresh in your mind, review Chapter 3 before beginning this chapter.) Chapter 3 was only a first look: a presentation of the basic graphic (the scatter diagram) and descriptive statistical aspects of **linear correlation** and regression analysis. In this chapter we take a second, more detailed look at linear correlation and regression analysis.

objectives

13.1 Linear Correlation Analysis

13.2 Inferences about the Linear Correlation Coefficient

13.3 Linear Regression Analysis

13.4 Inferences Concerning the Slope of the Regression Line

13.5 Confidence Intervals for Regression

13.6 Understanding the Relationship between Correlation and Regression

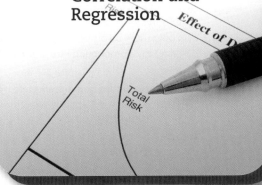

© Big Cheese Photo/Jupiterimages

13.1 Linear Correlation Analysis

NOW LET'S TAKE A SECOND LOOK AT THIS CONCEPT AND SEE HOW r, THE COEFFICIENT OF LINEAR CORRELATION, WORKS.

Intuitively, we want to think about how to measure the mathematical linear dependency of one variable on another. As x increases, does y tend to increase or decrease? How strong (consistent) is this tendency? We are going to use two measures of dependence—covariance and the coefficient of linear correlation—to measure the relationship between

two variables. We'll begin our discussion by examining a sample of **bivariate data** (ordered pairs) and identifying some related facts as we prepare to define covariance. Consider a sample of six bivariate data: (2, 1), (3, 5), (6, 3), (8, 2), (11, 6), (12, 1).

Figure 13.1 plots each ordered pair, providing the groundwork on which to measure covariance. We must also calculate the mean values of the six x's and y's. The mean of the six x values (2, 3, 6, 8, 11, 12) is $\bar{x} = 7$. The mean of the six y values (1, 5, 3, 2, 6, 1) is $\bar{y} = 3$.

The point $(\bar{x}, \bar{y})$, which is (7, 3) (generated from the means of the six x and y values), is located as shown on the graph of the sample points in Figure 13.2. The point $(\bar{x}, \bar{y})$ is called the **centroid** of the data. A vertical and a horizontal line drawn through the centroid divide the graph into four sections, as shown in Figure 13.2. Each point (x, y) lies a certain distance from each of these two lines: $(x - \bar{x})$ is the horizontal distance from (x, y) to the vertical line that passes through the centroid, and $(y - \bar{y})$ is the vertical distance from (x, y) to the horizontal line that passes through the centroid. Both the horizontal and vertical distances of each data point from the centroid can be measured, as shown in Figure 13.3. The distances may be positive, negative, or zero, depending on the position of the point (x, y) in relation to $(\bar{x}, \bar{y})$. [Figure 13.3 shows $(x - \bar{x})$ and $(y - \bar{y})$ represented by braces, with positive or negative signs.]

One measure of linear dependency is covariance. The **covariance of x and y** is defined as the sum of the products of the distances of all values of x and y from the centroid, $\Sigma[(x - \bar{x})(y - \bar{y})]$, divided by $n - 1$:

Covariance of x and y

$$\text{covar}(x, y) = \frac{\sum_{i=1}^{n}(x_i - \bar{x})(y_i - \bar{y})}{n - 1} \tag{13.1}$$

Calculations for the covariance for the data in this example are given in Table 13.1. The covariance, written as covar(x, y), of the data is $\frac{3}{5} = 0.6$.

The covariance is positive if the graph is dominated by points to the upper right and to the lower left of the

Figure 13.2 **The Point (7, 3) Is the Centroid**

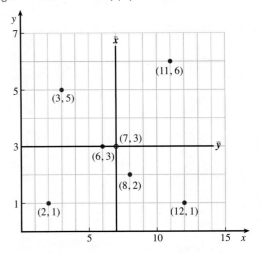

Figure 13.3 **Measuring the Distance of Each Data Point from the Centroid**

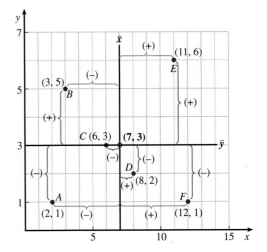

Figure 13.1 **Graph of Bivariate Data**

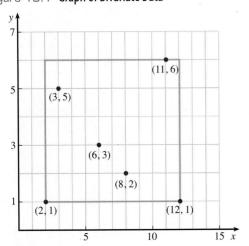

Table 13.1 Calculations for Finding covar(x, y) for the Sample Data

Points	$x - \bar{x}$	$y - \bar{y}$	$(x - \bar{x})(y - \bar{y})$
(2, 1)	−5	−2	10
(3, 5)	−4	2	−8
(6, 3)	−1	0	0
(8, 2)	1	−1	−1
(11, 6)	4	3	12
(12, 1)	5	−2	−10
Total	0 (ck)	0 (ck)	3

Notes:

1. $\Sigma(x - \bar{x}) = 0$ and $\Sigma(y - \bar{y}) = 0$. This will always happen. Why? (See page 40.)

2. Even though the variance of a single set of data is always positive, the covariance of bivariate data can be negative.

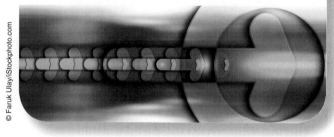

© Faruk Ulay/iStockphoto.com

centroid. The products of $(x - \bar{x})$ and $(y - \bar{y})$ are positive in these two sections. If the majority of the points are to the upper left and the lower right of the centroid, then the sum of the products is negative. Figure 13.4 shows data that represent (a) a positive dependency, (b) a negative dependency, and (c) little or no dependency. The covariances for these three situations would definitely be positive in part (a), negative in (b), and near zero in (c). (The sign of the covariance is always the same as the sign of the slope of the regression line.)

The biggest disadvantage of covariance

as a measure of linear dependency is that it is not a standardized unit of measure. One reason for this is that the spread of the data is a strong factor in the size of the covariance. For example, if we multiply each data point in our example by 10, we have (20, 10), (30, 50), (60, 30), (80, 20), (110, 60), and (120, 10). The relationship of the points to each other is changed only in that they are much more spread out. However, the covariance for this new set of data is 60. Does this mean that the dependency between the x and y variables is stronger than in the original case? No, it does not; the relationship is the same, even though each data value has been multiplied by 10. This is the trouble with covariance as a measure. We must find some way to eliminate the effect of the spread of the data when we measure dependency.

If we standardize x and y by dividing the distance of each from the respective mean by the respective standard deviation:

$$x' = \frac{x - \bar{x}}{s_x} \quad \text{and} \quad y' = \frac{y - \bar{y}}{s_y}$$

and then compute the covariance of x' and y', we will have a covariance that is not affected by the spread of the data. This is exactly what is accomplished by the linear correlation coefficient. It divides the covariance of x and y by a measure of the spread of x and by a measure of the spread of y (the standard deviations of x and of y are used as measures of spread). Therefore, by definition, the **coefficient of linear correlation** is:

Coefficient of Linear Correlation

$$r = \text{covar}(x', y') = \frac{\text{covar}(x, y)}{s_x \cdot s_y} \qquad (13.2)$$

The coefficient of linear correlation standardizes the measure of dependency and allows us to compare

Figure 13.4 Data and Covariance

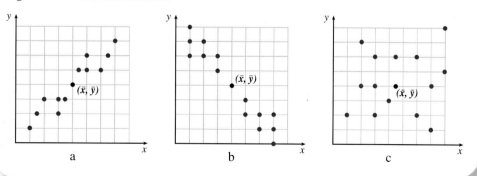

the relative strengths of dependency of different sets of data. [Formula (13.2) for linear correlation is also commonly referred to as **Pearson's product moment, r.**]

We can find the value of r, the coefficient of linear correlation, for the data in the previous example by calculating the two standard deviations and then dividing:

$$s_x = 4.099 \quad \text{and} \quad s_y = 2.098$$

$$r = \frac{\text{covar}(x, y)}{s_x \cdot s_y}: \quad r = \frac{0.6}{(4.099)(2.098)} = \mathbf{0.07}$$

Finding the correlation coefficient using formula (13.2) can be a very tedious arithmetic process. We can write the formula in a more workable form, however, as it was in Chapter 3:

Shortcut for Coefficient of Linear Correlation

$$r = \frac{\text{covar}(x, y)}{s_x \cdot s_y} = \frac{\dfrac{\Sigma[(x - \bar{x})(y - \bar{y})]}{n - 1}}{s_x \cdot s_y}$$

$$= \frac{\text{SS}(xy)}{\sqrt{\text{SS}(x) \cdot \text{SS}(y)}} \tag{13.3}$$

Formula (13.3) avoids the separate calculations of $\bar{x}$, $\bar{y}$, s_x, and s_y as well as the calculations of the deviations from the means. Therefore, formula (13.3) is much easier to use, and more important, it is more accurate when decimals are involved because it minimizes round-off error.

13.2 Inferences about the Linear Correlation Coefficient

IN OBJECTIVE 13.1 WE LEARNED THAT COVARIANCE IS A MEASURE OF LINEAR DEPENDENCY.

Also noted was the fact that its value is affected by the spread of the data; therefore, we standardize the covariance by dividing it by the standard deviations of both x and y. This standardized form is known as r, the coefficient of linear correlation. Standardizing enables us to compare different sets of data, thereby allowing r to play a role much like z or t does for $\bar{x}$. The calculated r value becomes $r\bigstar$, the test statistic for inferences about ρ, the population correlation coefficient. (ρ is the lowercase Greek letter rho.)

Assumptions for inferences about the linear correlation coefficient:

The set of (x, y) ordered pairs forms a random sample, and the y values at each x have a normal distribution. Inferences use the t-distribution with $n - 2$ degrees of freedom.

Francis Galton and the Correlation Coefficient

The complete name of the correlation coeffiecient deceives many into believing that Karl Pearson developed this statistical measure himself. Although Pearson did develop a rigorous treatment of the mathematics of the Pearson Product Moment Correlation (PPMC), it was the imagination of Sir Francis Galton that originally conceived modern notions of correlation and regression. Galton's fascination with genetics and heredity provided the initial inspiration that led to regression and the PPMC.

ANTHROPOMETRIC LABORATORY

For the measurement in various ways of Human Form and Faculty.

Entered from the Science Collection of the S. Kensington Museum.

This laboratory is established by Mr. Francis Galton for the following purposes:—

1. For the use of those who desire to be accurately measured in many ways, either to obtain timely warning of remediable faults in development, or to learn their powers.

2. For keeping a methodical register of the principal measurements of each person, of which he may at any future time obtain a copy under reasonable restrictions. His initials and date of birth will be entered in the register, but not his name. The names are indexed in a separate book.

3. For supplying information on the methods, practice, and uses of human measurement.

4. For anthropometric experiment and research, and for obtaining data for statistical discussion.

Charges for making the principal measurements:
THREEPENCE each, to those who are already on the Register.
FOURPENCE each, to those who are not:—one page of the Register will thenceforward be assigned to them, and a few extra measurements will be made, chiefly for future identification.

The Superintendent is charged with the control of the laboratory and with determining in each case, which, if any, of the extra measurements may be made, and under what conditions.

H. & W. Brown, Printers, 20 Fulton Road, S.W.

Confidence Interval Procedure

As with other parameters, a **confidence interval** may be used to estimate the value of ρ, the linear correlation coefficient of the population. Usually this is accomplished by using a table that shows **confidence belts**. Table 10 in Appendix B gives confidence belts for 95% confidence intervals. This table is a bit tricky to read and utilizes n, the sample size, so be extra careful when you use it.

Let's look at an example that demonstrates the procedure for estimating ρ. A random sample of 15 ordered pairs of data has a calculated r value of 0.35. Let's find the 95% confidence interval for ρ, the population linear correlation coefficient, using the five-step process.

STEP 1 **Parameter of interest:** The linear correlation coefficient for the population, ρ.

STEP 2 **a. Assumptions:** The ordered pairs form a random sample, and we will assume that the y values at each x have a normal distribution.

b. Formula: The calculated linear correlation coefficient, r.

c. Level of confidence: $1 - \alpha = 0.95$.

STEP 3 **Sample information:** $n = 15$ and $r = 0.35$.

STEP 4 **Confidence interval:**
The confidence interval is read from Table 10 in Appendix B. Find $r = 0.35$ at the bottom of Table 10. (See the arrow on Figure 13.5.) Visualize a vertical line drawn through that point. Find the two points where the belts marked for the correct sample size cross the vertical line. The sample size is 15. These two points are circled in Figure 13.5. Now look horizontally from the two circled points to the vertical scale on the left and read the confidence interval. The values are −0.20 and 0.72.

Figure 13.5 **Using Table 10 of Appendix B, Confidence Belts for the Correlation Coefficient**

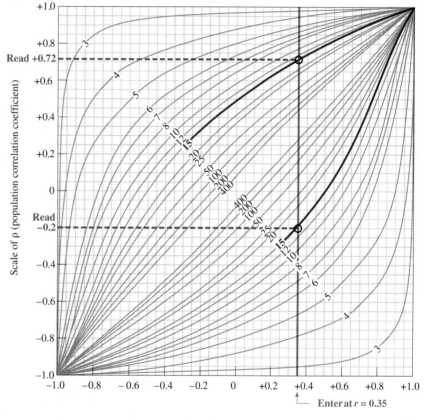

Scale of r (sample correlation)

STEP 5 **Confidence interval:**
The 95% confidence interval for ρ, the population coefficient of linear correlation, is −0.20 to 0.72.

Hypothesis-Testing Procedure

After the linear correlation coefficient, r, has been calculated for the sample data, it seems necessary to ask this question: Does the value of r indicate that there is a linear dependency between the two variables in the population from which the sample was drawn? To answer this question we can perform a **hypothesis test**.

CAUTION Inferences about the linear correlation coefficient are about the pattern of behavior of the two variables involved and the usefulness of one variable in predicting the other. *Significance of the linear correlation coefficient does not mean that you have established a cause-and-effect relationship.* Cause and effect is a separate issue. (See the causation discussion on pages 63–64.)

The null hypothesis is: The two variables are linearly unrelated ($\rho = 0$), where ρ is the linear correlation coefficient for the population. The alternative hypothesis may be either one-tailed or two-tailed. Most frequently it is two-tailed, $\rho \neq 0$. However, when we suspect that there is only a positive or only a negative correlation, we should use a one-tailed test. The alternative hypothesis of a one-tailed test is $\rho > 0$ or $\rho < 0$.

The area that represents the *p*-value or the critical region for the test is on the right when a positive correlation is expected and on the left when a negative correlation is expected. The test statistic used to test the null hypothesis is the calculated value of *r* from the sample data. Probability bounds for the *p*-value or critical values for *r* are found in Table 11 of Appendix B (page 346). The number of degrees of freedom for the *r* statistic is 2 less than the sample size, df $= n - 2$. Specific details for using Table 11 are discussed shortly.

Rejection of the null hypothesis means that there is evidence of a linear relationship between the two variables in the population. Failure to reject the null hypothesis is interpreted as meaning that a linear relationship between the two variables in the population has not been shown.

Two-Tailed Hypothesis Test

Now let's look at an example of a hypothesis test studying 15 randomly selected ordered pairs where $r = 0.548$.

Using the usual five-step process, we can determine if this linear correlation coefficient is significantly different from zero at the 0.02 level of significance.

STEP 1 **a. Parameter of interest:**
The linear correlation coefficient for the population, ρ.

b.   Statement of hypotheses:

$$H_o: \rho = 0$$
$$H_a: \rho \neq 0$$

STEP 2 **a. Assumptions:**
The ordered pairs form a random sample, and we will assume that the *y* values at each *x* have a normal distribution.

b.   Test statistic:
$r\star$, formula (13.3), with df $= n - 2 = 15 - 2 = 13$.

c.   Level of significance:
$\alpha = 0.02$ (given in the statement of the problem).

STEP 3 **a. Sample information:**
$n = 15$ and $r = 0.548$.

b.   Value of the test statistic:
The calculated sample linear correlation coefficient is the test statistic: $r\star = 0.548$.

STEP 4 **Probability distribution:**
As always, we can use either the *p*-value or the classical procedure:

Using the *p*-value procedure:

a.   Use both tails because H_a expresses concern for values related to "different from."

$$\mathbf{P} = P(r < -0.548) + P(r > 0.548)$$
$$= 2 \cdot P(r > 0.548),$$

with df $= 13$ as shown in the figure below.

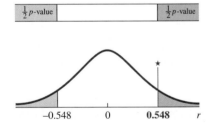

Use Table 11 (Appendix B) to place bounds on the *p*-value: $0.02 < \mathbf{P} < 0.05$.
Specific details follow this illustration.

b.   The *p*-value is not smaller than the level of significance, α.

Using the classical procedure:

a.   The critical region is both tails because H_a expresses concern for values related to "different from."
The critical value is found at the intersection of the df $= 13$ row and the two-tailed 0.02 column of Table 11: **0.592**.

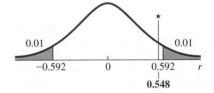

Specific details follow this illustration.

b. $r\star$ is not in the critical region, as shown in red on the figure above.

STEP 5 a. **Decision:** Fail to reject H_o.

b. **Conclusion:**
At the 0.02 level of significance, we have failed to show that x and y are correlated.

CALCULATING THE *p*-VALUE

Use Table 11 in Appendix B to "place bounds" on the p-*value*. By inspecting the df = 13 row of Table 11, you can determine an interval within which the p-value lies. Locate $r\star$ along the row labeled df = 13. If $r\star$ is not listed, locate the two table values it falls between and read the bounds for the p-value from the top of the table. In this case, $r\star = 0.548$ is between 0.514 and 0.592; therefore, **P** is between 0.02 and 0.05. Table 11 shows only two-tailed values. When the alternative hypothesis is two-tailed, the bounds for the p-value are read directly from the table.

Portion of Table 11				
	Amount of α in Two Tails			
df	... 0.05	**P**	0.02 ⟶	0.02 < **P** < 0.05
⋮	↑	⋮	↑	
13	0.514	**0.548**	0.592	

NOTE: When H_a is one-tailed, divide the column headings by 2 to place bounds on the p-value.

Use Table 11 in Appendix B to find the critical values. The critical value is at the intersection of the df = 13 and the two-tailed α = 0.02 column. Table 11 shows only two-tailed values. Since the alternative hypothesis is two-tailed, the critical values are read directly from the table.

Portion of Table 11		
	Amount of α in Two Tails	
df	... 0.02	...
⋮	⋮	
13	0.592 ⟶	Critical values: ± 0.592

NOTE: When H_a is one-tailed, divide the column headings by 2.

Correlation can have very practical applications, as you can see in the article on page 280.

13.3 Linear Regression Analysis

RECALL THAT THE **LINE OF BEST FIT** RESULTS FROM AN ANALYSIS OF TWO (OR MORE) RELATED QUANTITATIVE VARIABLES. (WE WILL RESTRICT OUR WORK TO TWO VARIABLES.)

When two variables are studied jointly, we often would like to control one variable by controlling the other. Or we might want to predict the value of a variable based on knowledge about another variable. In both cases we want to find the line of best fit, if one exists, that will best predict the value of the dependent, or output, variable from a value of the independent, or input, variable. Recall that the variable we know or can control is called the *independent*, or input, variable; the variable that results from using the equation of the line of best fit is called the *dependent*, or predicted, variable.

In Chapter 3 we developed the method of least squares. From this concept, formulas (3.7) and (3.6) were obtained and used to calculate b_0 (the **y-intercept**) and b_1 (the **slope** of the line of best fit):

$$b_0 = \frac{\Sigma y - (b_1 \cdot \Sigma x)}{n} \tag{3.7}$$

$$b_1 = \frac{SS(xy)}{SS(x)} \tag{3.6}$$

Then these two coefficients are used to write the equation of the line of best fit in the form

$$\hat{y} = b_0 + b_1 x$$

When the line of best fit is plotted, it does more than just show us a pictorial representation of the line. It also tells us two things: (1) whether or not there really is a linear relationship between the two variables and (2) the quantitative (equation) relationship between the two variables. When there is no relationship between the variables, a horizontal line of best fit will result. A horizontal line has a slope of zero, which implies that the value of the input variable has no effect on the output variable. (This idea will be amplified later in this chapter.)

The result of regression analysis is the mathematical equation of the line of best fit. We will, as mentioned

Use of Correlation in a Medical Study

Correlation of Activated Clotting Time and Activated Partial Thromboplastin Time to Plasma Heparin Concentration

Study Objective: Determine the correlation between activated clotting time (ACT) or activated partial thromboplastin time (aPTT) and plasma heparin concentration

Design: Two-phase prospective study

Patients: Thirty patients receiving continuous-infusion intravenous heparin

Interventions: Measurement of ACT, aPTT, and plasma heparin concentrations

Heparin has been administered for more than 50 years as an anticoagulant and is known to have a narrow therapeutic range. Underdosing of heparin is associated with recurrent thromboembolism, whereas excessive dosing may increase the risk of hemorrhagic complications. Several clotting time tests are available to monitor heparin, including whole blood clotting time, activated partial thromboplastin time (aPTT), and activated clotting time (ACT).

The study was conducted in two phases. In phase 1 (intraperson phase), sequential blood draws from five patients were evaluated. The goal was to determine if there was a significant relationship between plasma heparin concentrations and clotting time tests within an individual. In phase 2 (interperson phase), single random blood draws from 25 additional patients were evaluated with the same collection technique and analysis as in phase 1. Blood draws were performed within 48 hours after the start of heparin therapy. The goal of phase 2 was to determine the quantitative relationship between ACT or aPTT and plasma heparin concentration between individuals.

For both phases, correlations between ACT or aPTT results and plasma heparin concentrations were performed using the Pearson moment R correlation test. Phase 1: Linear correlation coefficients (r) for the five patients were 0.93 ($p = 0.02$), 0.99 ($p = 0.009$), 0.89

($p = 0.12$), 0.96 ($p = 0.04$), and 0.90 ($p = 0.10$). Phase 2: Correlation coefficient for these data was 0.58 (linear, $p = 0.008$). The linear regression line formula is $137 + (52.9)$(plasma heparin concentration), which, for a therapeutic heparin range of 0.3–0.7 U/ml (by antifactor Xa), equates to an ACT range of 153–174 seconds. Linear regression lines for aPTT versus plasma heparin concentration are shown in Figure A. Correlation coefficient for these data was 0.89 (linear, $p = 0.0001$). The linear regression line formula was $14.4 + (135.4)$(plasma heparin concentration), which, for the same therapeutic heparin range, equates to an aPTT range of 55–109 seconds.

The decision analysis results indicate that a standard clotting time test therapeutic range (not derived from heparin concentration) often results in incorrect patient management decisions. The ACT based on a standard therapeutic range may result in dosage adjustment decisions that may increase the risk of bleeding (in 43% of patients). The aPTT based on a standard therapeutic range may result in dosage adjustment decisions that may increase the risk of thrombosis (in 37% of patients). A larger study in 200 patients is under way to confirm these results using heparin concentration-derived therapeutic ranges for both aPTT and ACT

SOURCE: John M. Koerber, B.S.; Maureen A. Smythe, Pharm.D.; Robert L. Begle, M.D.; Joan C. Mattson, M.D.; Beverly P. Kershaw, M.S.; and Susan J. Westley, M.T. (ASCP). *Pharmacotherapy*, 19(8):922–931. © Pharmacotherapy Publications, http://www.medscape.com/viewarticle/418017_3. Reprinted with permission.

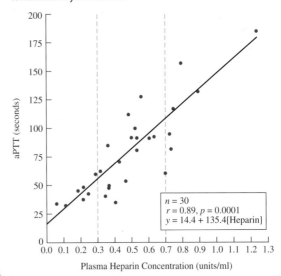

Figure A Linear aPTT versus Plasma Heparin Concentration for Phase 2 (Interperson Correlation and Regression).

Vertical dashed lines indicate the therapeutic range for plasma heparin concentration by antifactor Xa.

$n = 30$
$r = 0.89, p = 0.0001$
$y = 14.4 + 135.4$[Heparin]

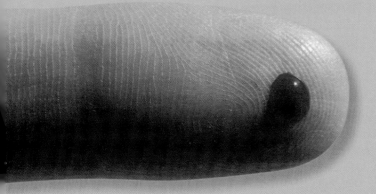

before, restrict our work to the *simple linear* case—that is, one input variable and one output variable, where the line of best fit is straight. However, you should be aware that not all relationships are of this nature. If the **scatter diagram** suggests something other than a straight line, the relationship may be **curvilinear regression.** In cases of this type we must introduce terms of higher powers, x^2, x^3, and so on, or other functions, e^x, $\log x$, and so on; or we must introduce other input variables. Maybe two or three input variables would improve the usefulness of our regression equation. These possibilities are examples of curvilinear regression and **multiple regression.**

The linear model used to explain the behavior of linear **bivariate data** in the population is:

Linear Model

$$y = \beta_0 + \beta_1 x + \epsilon \qquad (13.4)$$

This equation represents the linear relationship between the two variables in a population. β_0 is the y-intercept, and β_1 is the slope. ϵ (also shown as ε) is the random **experimental error** in the observed value of y at a given value of x.

The **regression line** from the sample data gives us b_0, which is our estimate of β_0, and b_1, our estimate of β_1. The error ϵ is approximated by $e = y - \hat{y}$, the difference between the observed value of y and the **predicted value of y, $\hat{y}$,** at a given value of x:

Estimate of the Experimental Error

$$e = y - \hat{y} \qquad (13.5)$$

The random variable e (also known as the "residual") is positive when the observed value of y is larger than the predicted value, $\hat{y}$; e is negative when y is less than $\hat{y}$. The sum of the errors (residuals) for all values of y for a given value of x is exactly zero. (This is part of the least squares criteria.) Thus, the mean value of the experimental error is zero; its variance is σ_ϵ^2.

Variance of the Experimental Error

Our next goal is to estimate this **variance of the experimental error, σ_ϵ^2.** But before we estimate the variance of ϵ, let's try to understand exactly what the error represents: ϵ is the amount of error in our observed value of y. That is, it is the difference between the observed value of y and the mean value of y at that particular

value of x. Since we do not know the mean value of y, we will use the regression equation and estimate it with $\hat{y}$, the **predicted value of y** at this same value of x. Thus, the best estimate that we have for ϵ is $e = y - \hat{y}$, as shown in Figure 13.6.

Figure 13.6 **The Error**

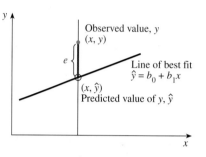

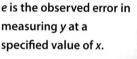

Note:

e is the observed error in measuring *y* at a specified value of *x*.

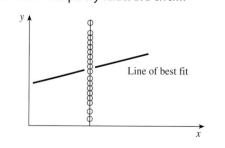

© Yenwen Lu/iStockphoto.com

If we were to observe several values of y at a given value of x, we could plot a distribution of y values about the line of best fit (about $\hat{y}$, in particular). Figure 13.7 shows a sample of bivariate values that share a common x value. Figure 13.8 on the next page shows the theoretical distribution of all possible y values at a given x value. A similar distribution occurs at each different value of x. The mean of the observed y's at a given value of x varies, but it can be estimated by $\hat{y}$.

Before we can make any inferences about a regression line, we must assume that the distribution of y's

Figure 13.7 **Sample of y Values at a Given x**

Figure 13.8 Theoretical Distribution of y Values for a Given x

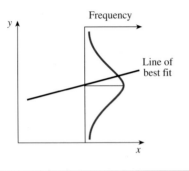

Figure 13.8 **Theoretical Distribution of y Values for a Given x**

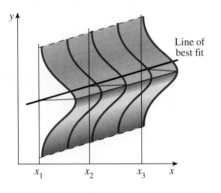

Figure 13.9 **Standard Deviation of the Distribution of y Values Is the Same for All x**

is approximately normal and that the variances of the distributions of y at all values of x are the same. That is, the standard deviation of the distribution of y about $\hat{y}$ is the same for all values of x, as shown in Figure 13.9.

Before we look at the variance of e, let's review the definition of sample variance. The sample variance, s^2, is defined as $\frac{\Sigma(x - \bar{x})^2}{n - 1}$, the sum of the squares of each deviation divided by the number of degrees of freedom, $n - 1$, associated with a sample of size n. The variance of y involves an additional complication: there is a different mean for y at each value of x. (Notice the many distributions in Figure 13.9.) However, each of these "means" is actually the predicted value, $\hat{y}$, that corresponds to the x that fixes the distribution. So the variance of the estimated error e is given by the formula:

Variance of the Estimated Error, *e*

$$s_e^2 = \frac{\Sigma(y - \hat{y})^2}{n - 2} \qquad (13.6)$$

where $n - 2$ is the number of degrees of freedom.

Formula (13.6) can be rewritten by substituting $b_0 + b_1 x$ for $\hat{y}$. Since $\hat{y} = b_0 + b_1 x$, we have

$$s_e^2 = \frac{\Sigma(y - b_0 - b_1 x)^2}{n - 2} \qquad (13.7)$$

With some algebra and some patience, this formula can be rewritten once again into a more workable form. The form we will use is

Variance of the Estimated Error, *e*

$$s_e^2 = \frac{(\Sigma y^2) - (b_0)(\Sigma y) - (b_1)(\Sigma xy)}{n - 2} \qquad (13.8)$$

For ease of discussion, let's agree to call the numerator of formulas (13.6), (13.7), and (13.8) the **sum of squares for error** (SSE).

We need one more formula before we can put all this information to use:

Standard Deviation of the Estimated Error, *e* (Standard Error of the Estimate)

$$s_e = \sqrt{s_e^2} \qquad (13.9)$$

Now let's see how we can use all of this information in determining the variance of y about the regression line. Suppose you move to a new city and find a job. You will, of course, be concerned about the problems you will face commuting to and from work. For example, you would like to know how long it will take you to drive to work each morning. Let's use "one-way distance to work" as a measure of where you live. You live x miles away from work and want to know how long it will take you to commute each day. Fifteen of your co-workers were asked to give their one-way travel times and distances to work. The resulting data are shown in Table 13.2. Let's find the line of best fit and the variance of y about the line of best fit, s_e^2.

The extensions and summations needed for this problem are shown in Table 13.2. The line of best fit can now be calculated using formulas (2.8), (3.4), (3.6), and (3.7). From formula (2.8):

$$SS(x) = \Sigma x^2 - \frac{(\Sigma x)^2}{n}:$$

$$SS(x) = 2{,}616 - \frac{(184)^2}{15} = 358.9333$$

From formula (3.4):

$$SS(xy) = \Sigma xy - \frac{\Sigma x \Sigma y}{n}:$$

$$SS(xy) = 5{,}623 - \frac{(184)(403)}{15} = 679.5333$$

We use formula (3.6) for the slope:

$$b_1 = \frac{SS(xy)}{SS(x)}:$$

$$b_1 = \frac{679.5333}{358.9333} = 1.893202 = \mathbf{1.89}$$

*Table 13.2 **Data on Commute Distances and Times**

Co-Worker	Miles (x)	Minutes (y)	x^2	xy	y^2
1	3	7	9	21	49
2	5	20	25	100	400
3	7	20	49	140	400
4	8	15	64	120	225
5	10	25	100	250	625
6	11	17	121	187	289
7	12	20	144	240	400
8	12	35	144	420	1,225
9	13	26	169	338	676
10	15	25	225	375	625
11	15	35	225	525	1,225
12	16	32	256	512	1,024
13	18	44	324	792	1,936
14	19	37	361	703	1,369
15	20	45	400	900	2,025
Total	184	403	2,616	5,623	12,493

We use formula (3.7) for the y-intercept:

$$b_0 = \frac{\Sigma y - (b_1 \cdot \Sigma x)}{n}:$$

$$b_0 = \frac{403 - (1.893202)(184)}{15} = 3.643387 = \mathbf{3.64}$$

Auto Crashes and Insurance Premiums

The graphic to the right reports the effect each at-fault accident had on the average annual auto insurance premium in 2009. Does it appear that the variable "number of at-fault accidents" did have an effect on the average annual premiums? How much of an effect? The graphic reports only one value for premiums for each number of accidents, but each dollar amount reported summarizes the amount of many premiums. How does this relate to the underlying assumptions for regression analysis?

SOURCE: 2008 insurance.com study, reprinted in *USA Today*, August 12, 2009

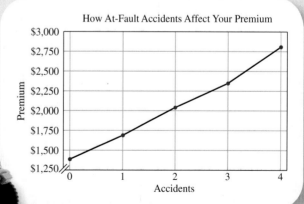

Therefore, the equation of the line of best fit is

$$\hat{y} = 3.64 + 1.89x$$

The variance of y about the regression line is calculated by using formula (13.8):

$$s_e^2 = \frac{(\Sigma y^2) - (b_0)(\Sigma y) - (b_1)(\Sigma xy)}{n - 2}:$$

$$s_e^2 = \frac{(12{,}493) - (3.643387)(403) - (1.893202)(5{,}623)}{15 - 2}$$

$$= \frac{379.2402}{13} = \textbf{29.17}$$

$$s_e = \sqrt{29.17} = \textbf{5.40}$$

$s_e^2 = 29.17$ is the variance of the 15 e's and $s_e = 5.40$ is the standard deviation of the 15 e's. In Figure 13.10 the 15 e's are shown as vertical line segments.

Figure 13.10 **The 15 Random Errors as Line Segments**

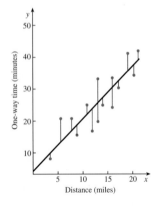

NOTE: Extra decimal places are often needed for this type of calculation. Notice that b_1 (1.893202) was multiplied by 5,623. If 1.89 had been used instead, that one product would have changed the numerator by approximately 18. That, in turn, would have changed the final answer by almost 1.4—a sizable round-off error.

In the sections that follow, we will use the variance of e in much the same way as the variance of x (as calculated in Chapter 2) was used in Chapters 8, 9, and 10 to complete the statistical inferences studied there.

13.4 Inferences Concerning the Slope of the Regression Line

NOW THAT THE EQUATION OF THE LINE OF BEST FIT HAS BEEN FOUND AND THE LINEAR MODEL HAS BEEN VERIFIED (BY INSPECTION OF THE SCATTER DIAGRAM), WE ARE READY TO DETERMINE WHETHER WE CAN USE THE EQUATION TO PREDICT y.

We will test the null hypothesis: The equation of the line of best fit is of no value in predicting y given x. That is, the null hypothesis to be tested is: β_1 (the slope of the relationship in the population) is zero. If $\beta_1 = 0$, then the linear equation will be of no real use in predicting y. Before we look at the confidence interval or the hypothesis test, let's discuss the **sampling distribution** of the slope. If random samples of size n are repeatedly taken from a bivariate population, then the calculated slopes, the b_1's, will form a sampling distribution that is normally distributed with a mean of β_1, the population value of the slope, and with a variance of $\sigma_{b_1}^2$, where

$$\sigma_{b_1}^2 = \frac{\sigma_\epsilon^2}{\Sigma(x - \bar{x})^2} \tag{13.10}$$

Remember: Computer and calculator instructions to find the regression line for bivariate data can be found on your Tech card for Chapter 3.

provided there is no lack of fit. An appropriate estimator for $\sigma_{b_1}^2$ is obtained by replacing σ_ϵ^2 by s_e^2, the estimate of the variance of the error about the regression line:

$$s_{b_1}^2 = \frac{s_e^2}{\Sigma(x - \bar{x})^2} \qquad (13.11)$$

This formula may be rewritten in the following, more manageable form:

Estimate for Variance of Slope

$$s_{b_1}^2 = \frac{s_e^2}{\Sigma x^2 - \frac{(\Sigma x)^2}{n}} \qquad (13.12)$$

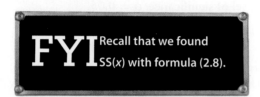

FYI Recall that we found SS(x) with formula (2.8).

NOTE: The "**standard error** of __" is the standard deviation of the sampling distribution of __. Therefore, the *standard error of regression* (slope) is σ_{b_1} and is estimated by s_{b_1}.

Estimate for Standard Error of Regression (slope)

$$s_{b_1} = \sqrt{s_{b_1}^2} \qquad (13.13)$$

In our example of commute times and distances, the variance and the standard error among the b_1's is estimated by using formulas (13.12) and (13.13):

$$s_{b_1}^2 = \frac{s_e^2}{\Sigma x^2 - \frac{(\Sigma x)^2}{n}}:$$

$$s_{b_1}^2 = \frac{29.1723}{358.9333} = 0.081275 = \mathbf{0.0813}$$

$$s_{b_1} = \sqrt{0.0813} = \mathbf{0.285}$$

Assumptions for inferences about the linear regression:

The set of (x, y) ordered pairs forms a random sample, and the y values at each x have a normal distribution. Since the population standard deviation is unknown and replaced with the sample standard deviation, the t-distribution will be used with $n - 2$ degrees of freedom.

Confidence Interval Procedure

The slope β_1 of the regression line of the population can be estimated by means of a confidence interval.

Confidence Interval for Slope

$$b_1 \pm t(n - 2, \alpha/2) \cdot s_{b_1} \qquad (13.14)$$

Let's look at how to construct a confidence interval for β_1, the population slope of the line of best fit, by finding the 95% confidence interval for the population's slope, β_1. Using the data from Table 13.2 (page 283), we'll once again follow the five-step process.

STEP 1 Parameter of interest:
The slope, β_1, of the line of best fit for the population.

STEP 2 a. Assumptions:
The ordered pairs form a random sample, and we will assume that the y values (minutes) at each x (miles) have a normal distribution.
b. **Probability distribution and formula:**
Student's t-distribution and formula (13.14).
c. **Level of confidence:** $1 - \alpha = 0.95$.

STEP 3 Sample information:
$n = 15$, $b_1 = 1.89$, and $s_{b_1}^2 = 0.0813$.

STEP 4 a. **Confidence coefficients:**
From Table 6 in Appendix B, we find $t(\text{df}, \alpha/2) = t(13, 0.025) = 2.16$.
b. **Maximum error of estimate:**
We use formula (13.14) to find $E = t(n - 2, \alpha/2) \cdot s_{b_1}$:
$$E = (2.16) \cdot \sqrt{0.0813} = 0.6159 = 0.62$$
c. **Lower and upper confidence limits:**
$$b_1 - E \quad \text{to} \quad b_1 + E$$
$$1.89 - 0.62 \quad \text{to} \quad 1.89 + 0.62$$
Thus, 1.27 to 2.51 is the 95% confidence interval for β_1.

STEP 5 Confidence interval:
We can say that the slope of the line of best fit of the population from which the sample was drawn is between 1.27 and 2.51 with 95% confidence. That is, we are 95% confident that, on average, every extra mile will take between 1.27 minutes (1 min, 16 sec) and 2.51 minutes (2 min, 31 sec) of time to make the commute.

Hypothesis-Testing Procedure

We are now ready to test the hypothesis $\beta_1 = 0$. That is, we want to determine whether the equation of the line of best fit is of any real value in predicting y. For this hypothesis test, the null hypothesis is always $H_o: \beta_1 = 0$. It will be tested using Student's t-distribution with df $= n - 2$ and the test statistic $t\bigstar$ found using formula (13.15):

Test Statistic for Slope

$$t\bigstar = \frac{b_1 - \beta_1}{s_{b_1}} \qquad (13.15)$$

With that information we can now determine if the slope of the line of best fit is significant enough to show that one-way distance is useful in predicting one-way travel time in our continuing example on commute time. We will use our five-step procedure and $\alpha = 0.05$ significance level.

STEP 1 **a. Parameter of interest:**
β_1, the slope of the line of best fit for the population.

b. Statement of hypotheses:
$H_o: \beta_1 = 0$ (This implies that x is of no use in predicting y; that is, $\hat{y} = \bar{y}$ would be as effective.)

The alternative hypothesis can be either one-tailed or two-tailed. If we suspect that the slope is positive, as the previous example on the one-way distance to work, a one-tailed test is appropriate.

$H_a: \beta_1 > 0$ (We expect travel time y to increase as the distance x increases.)

STEP 2 **a. Assumptions:**
The ordered pairs form a random sample, and we will assume that the y values (minutes) at each x (miles) have a normal distribution.

b. Probability distribution and test statistic:
The t-distribution with df $= n - 2 = 13$, and the test statistic $t\bigstar$ from formula (13.15).

c. Level of significance: $\alpha = 0.05$.

STEP 3 **a. Sample information:**
$n = 15$, $b_1 = 1.89$, and $s_{b_1}^2 = 0.0813$.

b. Test statistic:
Using formula (13.15), we find the observed value of t:

$$t\bigstar = \frac{b_1 - \beta_1}{s_{b_1}}: \quad t\bigstar = \frac{1.89 - 0.0}{\sqrt{0.0813}} = 6.629 = 6.63$$

STEP 4 **Probability distribution:**
Again, we can use either the p-value or the classical procedure:

Using the p-value procedure:

a. Use the right-hand tail because H_a expresses concern for values related to "positive."

$\mathbf{P} = P(t\bigstar > 6.63$, with df $= 13)$

as shown in the figure below.

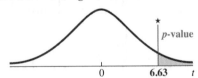

To find the p-value, use one of three methods:
1. Use Table 6 (Appendix B) to place bounds on the p-value: $\mathbf{P} < 0.005$.
2. Use Table 7 (Appendix B) to place bounds on the p-value: $\mathbf{P} < 0.001$.
3. Use a computer or calculator to find the p-value: $\mathbf{P} = 0.0000082$.
Specific details are on page 189.

b. The p-value is smaller than the level of significance, α.

Using the classical procedure:

a. The critical region is the right-hand tail because H_a expresses concern for values related to "positive." The critical value is found in Table 6:

$t(13, 0.05) = \mathbf{1.77}$

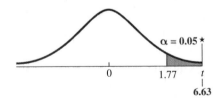

Specific instructions are in Objective 9.1 on pages 185–186.

b. $t\bigstar$ is in the critical region, as shown in red in the figure above.

STEP 5 **a. Decision:** Reject H_o.

b. Conclusion:

At the 0.05 level of significance, we conclude that the slope of the line of best fit in the population is greater than zero. The evidence indicates that there is a linear relationship and that the one-way distance (x) is useful in predicting the travel time to work (y).

13.5 Confidence Intervals for Regression

ONCE THE EQUATION OF THE LINE OF BEST FIT HAS BEEN OBTAINED AND DETERMINED TO BE USABLE, WE ARE READY TO USE THE EQUATION TO MAKE ESTIMATES AND PREDICTIONS.

We can estimate the mean of the population y values at a given value of x, written $\mu_{y|x_0}$. We can also predict the individual y value selected at random that will occur at a given value of x, written y_{x_0}. The best point estimate, or prediction, for both $\mu_{y|x_0}$ and y_{x_0} is $\hat{y}$. This is the y value obtained when an x value is substituted into the equation of the line of best fit. Like other point estimates, it is seldom correct. The calculated value of $\hat{y}$ will vary above and below the actual values for both $\mu_{y|x_0}$ and y_{x_0}.

Before we develop interval estimates of $\mu_{y|x_0}$ and y_{x_0}, recall the development of confidence intervals for the population mean μ in Chapter 8, when the variance was known, and in Chapter 9, when the variance was estimated. The sample mean, $\bar{x}$, was the best point estimate of μ. We used the fact that $\bar{x}$ is normally distributed, or approximately normally distributed, with a standard deviation of $\frac{\sigma}{\sqrt{n}}$ to construct formula (8.1) for the confidence interval for μ. When σ has to be estimated, we use formula (9.1) for the confidence interval.

The **confidence interval for** $\mu_{y|x_0}$ and the **prediction interval for** y_{x_0} are constructed in a similar fashion, with $\hat{y}$ replacing $\bar{x}$ as our point estimate. If we were to

randomly select several samples from the population, construct the line of best fit for each sample, calculate $\hat{y}$ for a given x using each regression line, and plot the various $\hat{y}$ values (they would vary because each sample would yield a slightly different regression line), we would find that the $\hat{y}$ values form a normal distribution. That is, the **sampling distribution of** $\hat{y}$ is normal, just as the sampling distribution of $\bar{x}$ is normal. What about the appropriate standard deviation of $\hat{y}$? The standard deviation in both cases ($\mu_{y|x_0}$ and y_{x_0}) is calculated by multiplying the square root of the variance of the error by an appropriate correction factor. Recall that the variance of the error, s_e^2, is calculated by means of formula (13.8).

Before we look at the correction factors for the two cases, let's see why they are necessary. Recall that the line of best fit passes through the point $(\bar{x}, \bar{y})$, the centroid. In Objective 13.4 we formed a confidence interval for the slope β_1 by using formula (13.14). If we draw lines with slopes equal to the extremes of that confidence interval, 1.27 to 2.51, through the point $(\bar{x}, \bar{y})$ [which is (12.3, 26.9)] on the scatter diagram, we will see that the value for $\hat{y}$ fluctuates considerably for different values of x (Figure 13.11). Therefore, we should suspect a need for a wider confidence interval as we select values of x that are farther away from $\bar{x}$. Hence we need a correction factor to adjust for the distance between x_0 and $\bar{x}$. This factor must also adjust for the variation of the y values about $\hat{y}$.

First, let's estimate the mean value of y at a given value of x, $\mu_{y|x_0}$. The confidence interval formula is:

$$\hat{y} \pm t(n-2, \alpha/2) \cdot s_e \cdot \sqrt{\frac{1}{n} + \frac{(x_0 - \bar{x})^2}{\Sigma(x - \bar{x})^2}} \qquad (13.16)$$

Figure 13.11 **Lines Representing the Confidence Interval for Slope**

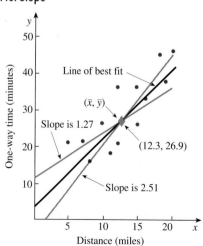

NOTE: The numerator of the second term under the radical sign is the square of the distance of x_0 from $\bar{x}$. The denominator is closely related to the variance of x and has a "standardizing effect" on this term.

Formula (13.16) can be modified for greater ease of calculation. Here is the new form:

Confidence Interval for $\mu_{y|x_0}$

$$\hat{y} \pm t(n-2, \alpha/2) \cdot s_e \cdot \sqrt{\frac{1}{n} + \frac{(x_0 - \bar{x})^2}{SS(x)}} \qquad (13.17)$$

Let's continue to examine our commuter example and construct a 95% confidence interval for the mean travel time for the co-workers who travel 7 miles to work. The set-up data are given in Objective 13.3.

STEP 1 Parameter of interest:
$\mu_{y|x=7}$, the mean travel time for co-workers who travel 7 miles to work.

STEP 2 a. Assumptions:
The ordered pairs form a random sample, and we will assume that the y values (minutes) at each x (miles) have a normal distribution.

b. Probability distribution and formula:
Student's t-distribution and formula (13.17).

c. Level of confidence: $1 - \alpha = 0.95$.

STEP 3 Sample information:

$s_e^2 = 29.17$

$s_e = \sqrt{29.17} = 5.40$ (found earlier on page 284)

$\hat{y} = 3.64 + 1.89x = 3.64 + 1.89(7) = 16.87$

STEP 4 a. Confidence coefficient:
$t(13, 0.025) = 2.16$ (from Table 6 in Appendix B).

b. Maximum error of estimate:
Using formula (13.17), we have

$$E = t(n-2, \alpha/2) \cdot s_e \cdot \sqrt{\frac{1}{n} + \frac{(x_0 - \bar{x})^2}{SS(x)}}:$$

$$E = (2.16)(5.40)\sqrt{\frac{1}{15} + \frac{(7 - 12.27)^2}{358.933}}$$

$$= (2.16)(5.40)\sqrt{0.06667 + 0.07738}$$

$$= (2.16)(5.40)(0.38) = 4.43$$

c. Lower and upper confidence limits:

$$\hat{y} - E \quad \text{to} \quad \hat{y} + E$$

$$16.87 - 4.43 \quad \text{to} \quad 16.87 + 4.43$$

Thus, **12.44 to 21.30** is the 95% confidence interval for $\mu_{y|x=7}$. That is, with 95% confidence,

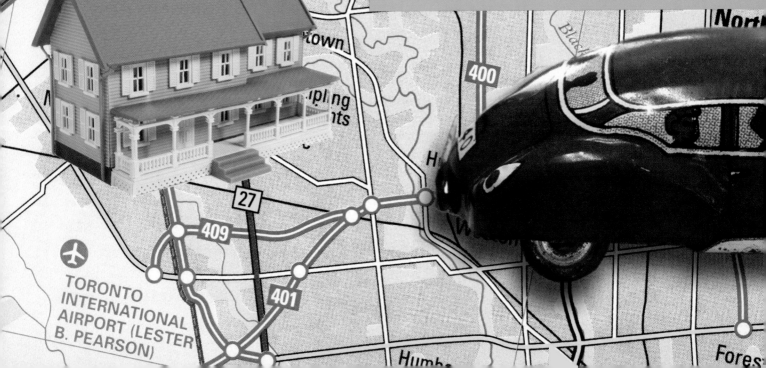

7 miles

the mean travel time for commuters who travel 7 miles is between 12.44 minutes (12 min, 26 sec) and 21.30 minutes (21 min, 18 sec).

This confidence interval is shown in Figure 13.12 by the dark red vertical line. The confidence belt showing the upper and lower boundaries of all intervals at 95% confidence is also shown in red. Notice that the boundary lines for the x values far away from $\bar{x}$ become close to the two lines that represent the equations with slopes equal to the extreme values of the 95% confidence interval for the slope (see Figure 13.12).

Prediction Intervals

Often we want to predict the value of an individual y. For example, you live 7 miles from your place of business and you are interested in an estimate of how long it will take you to get to work. You are somewhat less interested in the average time for all of those who live 7 miles away. The formula for the prediction interval of the value of a single randomly selected y is

Prediction Interval for $y_{x = x_0}$

$$\hat{y} \pm t_{(n-2, \alpha/2)} \cdot s_e \cdot \sqrt{1 + \frac{1}{n} + \frac{(x_0 - \bar{x})^2}{SS(x)}} \qquad (13.18)$$

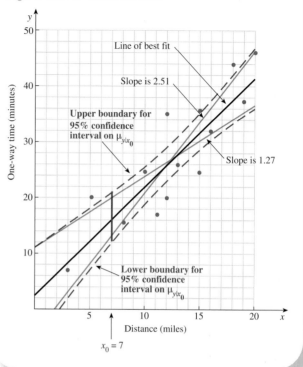

Figure 13.12 **Confidence Belts**

So, what is the 95% prediction interval for the time it will take you to commute to work if you live 7 miles away?

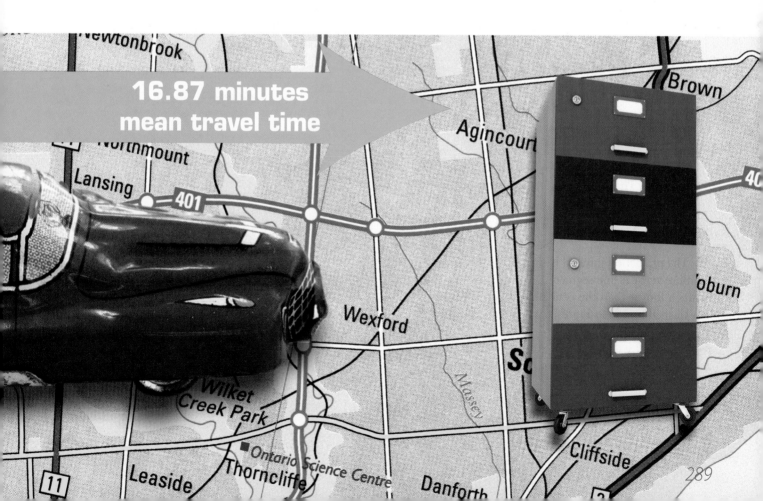

16.87 minutes mean travel time

STEP 1 Parameter of interest:

$y_{x=7}$, the travel time for one co-worker, or you, who travels 7 miles to work.

STEP 2 a. Assumptions:

The ordered pairs form a random sample, and we will assume that the y values (minutes) at each x (miles) have a normal distribution.

b. **Probability distribution and formula:**

Student's t-distribution and formula (13.18).

c. **Level of confidence:** $1 - \alpha = 0.95$.

STEP 3 Sample information:

$s_e = 5.40$ and $\hat{y}_{x=7} = 16.87$ (from page 288).

STEP 4 a. Confidence coefficient:

$t(13, 0.025) = 2.16$ (from Table 6 in Appendix B).

b. **Maximum error of estimate:**

Using formula (13.18), we have

$$E = t(n - 2, \alpha/2) \cdot s_e \cdot \sqrt{1 + \frac{1}{n} + \frac{(x_0 - \bar{x})^2}{SS(x)}}:$$

$$E = (2.16)(5.40)\sqrt{1 + \frac{1}{15} + \frac{(7 - 12.27)^2}{358.933}}$$

$$= (2.16)(5.40)\sqrt{1 + 0.06667 + 0.07738}$$

$$= (2.16)(5.40)\sqrt{1.14405}$$

$$= (2.16)(5.40)(1.0696) = 12.48$$

c. **Lower and upper confidence limits:**

$$\hat{y} - E \quad \text{to} \quad \hat{y} + E$$

$$16.87 - 12.48 \quad \text{to} \quad 16.87 + 12.48$$

Thus, **4.39 to 29.35** is the 95% prediction interval for $y_{x=7}$. That is, with 95% confidence, the individual travel times for commuters who travel 7 miles is between 4.39 minutes (4 min, 23 sec) and 29.35 minutes (29 min, 21 sec).

The prediction interval is shown in Figure 13.13 as the blue vertical line segment at $x_0 = 7$. Notice that it is much longer than the confidence interval for $\mu_{y|x=7}$. The dashed blue lines represent the prediction belts, the upper and lower boundaries of the prediction intervals for individual y values for all given x values.

Can you justify the fact that the prediction interval for individual values of y is wider than the confidence interval for the mean values? Think about "individual values" and "mean values" and study Figure 13.14.

Figure 13.13 **Prediction Belts**

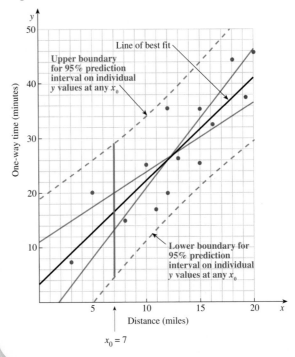

Figure 13.14 **Confidence Belts for the Mean Value of y and Prediction Belts for Individual y's**

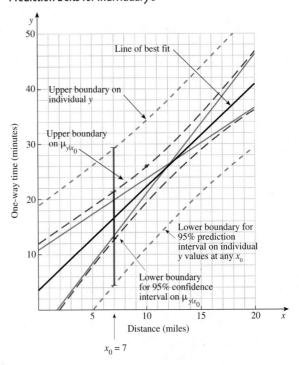

There are three basic precautions that you need to be aware of as you work with regression analysis:

1. Remember that the regression equation is meaningful only in the domain of the x variable studied. Estimation outside this domain is extremely dangerous; it requires that we know or assume that the relationship between x and y remains the same outside the domain of the sample data. For example, Joe says that he lives 75 miles from work, and he wants to know how long it will take him to commute. We certainly can use $x = 75$ in all the formulas, but we do not expect the answers to have the confidence or validity of the values of x between 3 and 20, which were in the sample. The 75 miles may represent a distance to the heart of a nearby major city. Do you think the estimated times, which were based on local distances of 3 to 20 miles, would be good predictors in this situation? Also, at $x = 0$, the equation has no real meaning. However, although projections outside the interval may be somewhat dangerous, they may be the best predictors available.

2. Don't get caught by the common fallacy of applying the regression results inappropriately. For example, this fallacy would include applying the results of our one-way commuting example to another company. Suppose that the second company has a city location, whereas the first company has a rural location, or vice versa. Do you think the results for a rural location would be valid for a city location? Basically, the results of one sample should not be used to make inferences about a population other than the one from which the sample was drawn.

3. Don't jump to the conclusion that the results of the regression prove that x *causes* y to change. (This is perhaps the most common fallacy.) Regressions only measure the movement between x and y; they never prove causation. (See Objective 3.2 for a discussion of causation.) A judgment of causation can be made only when it is based on theory or knowledge of the relationship separate from the regression results. The most common

Remember that the regression equation is meaningful only in the domain of the *x* variable studied. Don't get caught by the common fallacy of applying the regression results inappropriately. Don't jump to the conclusion that the results of the regression prove that *x* *causes* *y* to change.

difficulty in this regard occurs because of what is called the *missing variable,* or *third-variable, effect.* That is, we observe a relationship between x and y because a third variable, one that is not in the regression, affects both x and y.

USING REGRESSION CONFIDENCE INTERVALS IN AN ENVIRONMENTAL STUDY

→ We can use regression confidence intervals to help study real-world problems, such as those surrounding our environment. Much time, money, and effort are spent studying our environmental problems so that effective and appropriate management practices might be implemented. Here are excerpts from a study in South Florida in which linear regression analysis was an important tool.

A major concern in many coastal areas across the nation is the ecological health of bays and estuaries. One common problem in many of these areas is nutrient enrichment as a result of agricultural and urban activities. Nutrients are essential compounds for the growth and maintenance of all organisms and especially for the productivity of aquatic environments. Nitrogen and phosphorus compounds are especially important to seagrass, macroalgae, and phytoplankton. However, heavy nutrient loads transported to bays and estuaries can result in conditions conducive to eutrophication and the attendant problems of algal blooms and high phytoplankton productivity. Additionally, reduced light penetration in the water column because of phytoplankton blooms can adversely affect seagrasses, which many commercial and sport fish rely on for their habitat.

The purpose of the environmental report was to present methodology that could be used to estimate nutrient loads discharged from the east coast canals into Biscayne Bay in southeastern Florida. Water samples were collected from the gated control structures at the east coast canal sites in Miami-Dade County for the purpose of developing models that could be used to estimate nitrogen and phosphorus loads.

An ordinary least-squares regression technique was used to develop predictive equations for the purpose of estimating total nitrogen and total phosphorus loads discharged from the east coast canals to Biscayne Bay. The predictive equations can be used to estimate the value of a dependent variable from observations on a related or independent variable.

In this study, load was used as the dependent or response variable and discharge as the independent or explanatory variable. All of the total nitrogen load models had p-values less than 0.05, indicating that they were statistically significant at an alpha level of 0.05. Plots showing total nitrogen load as a function of discharge at the east coast canal sites are shown below. (Sites S25 and S27 are shown here.)

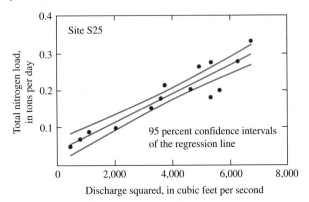

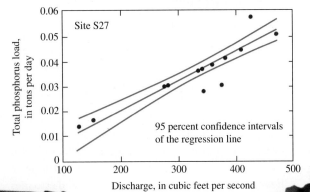

SOURCE: U.S. Geological Survey, Water-Resources Investigations Report 99-4094, by A. C. Lietz

13.6 Understanding the Relationship between Correlation and Regression

NOW THAT WE HAVE TAKEN A CLOSER LOOK AT BOTH CORRELATION AND REGRESSION ANALYSIS, IT IS NECESSARY TO DECIDE WHEN TO USE THEM. DO YOU SEE ANY DUPLICATION OF WORK?

The primary use of the linear correlation coefficient is in answering the question: Are these two variables linearly related? Other words may be used to ask this basic question—for example: Is there a linear correlation between the annual consumption of alcoholic beverages and the salary paid to firefighters?

The linear correlation coefficient can be used to indicate the usefulness of x as a predictor of y in the case where the linear model is appropriate. The test concerning the slope of the regression line ($H_o: \beta_1 = 0$) tests this same basic concept. Either one of the two is sufficient to determine the answer to this query.

Regression analysis should be used to answer questions about the relationship between two variables. Questions such as "What is the relationship?" and "How are two variables related?" require this regression analysis.

RECAP

This chapter basically applied many of the topics from earlier chapters. You should now be aware of the basic concepts of regression analysis and be able to collect the data for and do a complete analysis on any two-variable linear relationship.

problems

Objective 13.1

13.1 Consider the "Married Couples' Heights" scatter diagram presented on page 272.

  a. What is the independent variable for this set of data? How does it show on the scatter diagram?

  b. What is the dependent variable for this set of data? How does it show on the scatter diagram?

  c. Does there appear to be a relationship between a wife's and husband's heights? How does this show on the scatter diagram?

*13.2 a. Construct a scatter diagram of the following bivariate data.

Point	A	B	C	D	E	F	G	H	I	J
x	1	1	3	3	5	5	7	7	9	9
y	1	2	2	3	3	4	4	5	5	6

  b. Calculate the covariance.

  c. Calculate s_x and s_y.

  d. Calculate r using formula (13.2).

  e. Calculate r using formula (13.3).

*13.3 a. Draw a scatter diagram of the following bivariate data.

Point	A	B	C	D	E	F	G	H	I	J
x	0	1	1	2	3	4	5	6	6	7
y	6	6	7	4	5	2	3	0	1	1

  b. Calculate the covariance.

  c. Calculate s_x and s_y.

  d. Calculate r using formula (13.2).

  e. Calculate r using formula (13.3).

*13.4 NFL football enthusiasts often look at a team's total points scored for (Pts F) and total points scored against (Pts A) as a way of comparing the relative strength of teams. The season totals for the 32 teams in the NFL 2009 season appear in the following table.

Pts F	Pts A	Pts F	Pts A	Pts F	Pts A	Pts F	Pts A
427	285	305	291	416	307	454	320
348	236	391	261	388	333	326	324
360	390	368	324	354	402	197	379
258	326	245	375	290	380	294	424
361	250	470	312	510	341	375	325
429	337	461	297	363	325	330	281
402	427	327	375	315	308	280	390
266	336	262	494	244	400	175	436

SOURCE: http://www.cbssports.com/nfl/standings

  a. Calculate the linear correlation coefficient (Pearson's product moment r) for the points scored for and against.

  b. What conclusion can you draw from the answer in part (a)?

  c. Construct the scatter diagram and comment on how it supports or disagrees with your comments in part (b).

13.5 A formula that is sometimes given for computing the correlation coefficient is

$$r = \frac{n(\Sigma xy) - (\Sigma x)(\Sigma y)}{\sqrt{n(\Sigma x^2) - (\Sigma x)^2}\sqrt{n(\Sigma y^2) - (\Sigma y)^2}}$$

Use this expression as well as the formula

$$r = \frac{SS(xy)}{\sqrt{SS(x) \cdot SS(y)}}$$

to compute r for the data in the following table.

x	2	4	3	4	0
y	6	7	5	6	3

Objective 13.2

13.6 Using Figure 13.5 on page 277, find the 95% interval when a sample of $n = 25$ results in $r = 0.35$.

13.7 a. Using Figure 13.5 on page 277, find the 95% interval when a sample of $n = 100$ results in $r = 0.35$.

  b. Compare your answer in part (a) with the confidence interval formed in problem 13.7. Describe what occurred when you increased the sample size.

13.8 Use Table 10 of Appendix B to determine a 95% confidence interval for the true population linear correlation coefficient based on the following sample statistics:

  a. $n = 50, r = 0.60$ b. $n = 12, r = -0.45$

  c. $n = 6, r = +0.80$ d. $n = 200, r = -0.56$

*13.9 The test–retest method is one way of establishing the reliability of a test. The test is administered, and then, at a later date, the same test is readministered to the same individuals. The correlation coefficient is computed between the two sets or scores. The following test scores were obtained in a test–retest situation.

First Score	75	87	60	75	98	80	68	84	47	72
Second Score	72	90	52	75	94	78	72	80	53	70

Find r and set a 95% confidence interval for ρ.

*13.10 California is noted for its dry Chardonnay wines. Listed in the table below are five varieties, along with their *Wine Spectator* score and price per bottle. *Wine Spectator* rates wines on a 100-point scale, and all wines are blind-tasted.

Name	Score	Price
Ridge Chardonnay Monte Bello 2006	95	$57.99
Rodney Strong Chardonnay Reserve 2006	94	$33.99
Chalone Chardonnay 2007	92	$22.99
Lincourt Chardonnay Santa Rita Hills 2007	91	$19.99
Rombauer Vineyards Chardonnay Carneros 2007	91	$17.00

  a. Calculate r.

  b. Set a 95% confidence interval of ρ.

  c. Describe the meaning of your answer in part (b).

  d. Explain the meaning of the width of the interval answer in part (b).

13.11 State the null hypothesis, H_o, and the alternative hypothesis, H_a, that would be used to test the following statements:

a. The linear correlation coefficient is positive.

b. There is no linear correlation.

c. There is evidence of negative correlation.

d. There is a positive linear relationship.

13.12 Determine the bounds on the p-value that would be used in testing each of the following null hypotheses using the p-value approach:

a. $H_o: \rho = 0$ vs. $H_a: \rho \neq 0$, with $n = 32$ and $r = 0.41$

b. $H_o: \rho = 0$ vs. $H_a: \rho > 0$, with $n = 9$ and $r = 0.75$

c. $H_o: \rho = 0$ vs. $H_a: \rho < 0$, with $n = 15$ and $r = -0.83$

13.13 Determine the critical values that would be used in testing each of the following null hypotheses using the classical approach.

a. $H_o: r = 0$ vs. $H_a: r \neq 0$, with $n = 18$ and $\alpha = 0.05$

b. $H_o: r = 0$ vs. $H_a: r > 0$, with $n = 32$ and $\alpha = 0.01$

c. $H_o: r = 0$ vs. $H_a: r < 0$, with $n = 16$ and $\alpha = 0.05$

13.14 If a sample of size 18 has a linear correlation coefficient of -0.50, is there significant reason to conclude that the linear correlation coefficient of the population is negative? Use $\alpha = 0.01$.

*13.15 Two indicators of the level of economic activity in a given geographical area are its median household income and the percent of population in poverty. The table below lists the data for seven states for the year 2008:

State	Median Household Income	Percent in Poverty
Colorado	$57,184	11.2
Kansas	$50,174	11.3
Missouri	$46,847	13.5
Nebraska	$49,731	10.8
New Mexico	$43,719	17.0
Oklahoma	$42,836	15.7
Wyoming	$54,735	9.5

SOURCE: http://www.census.gov/

a. Calculate the correlation coefficient between the two variables.

b. Test for a significant correlation at the 0.05 level of significance and draw your conclusion.

*13.16 Consider the "Married Couples' Heights" scatter diagram on page 272.

a. Calculate r.

b. Set a 95% confidence interval for ρ.

c. Test for a significant positive correlation at the 0.05 level of significance.

d. Explain the meaning of the results found in parts (b) and (c).

Objective 13.3

13.17 Hailstones nationwide cause about $1 billion in property and crop damage each year. According to "Thunderstorm Hazards—Hail" on the National Weather Service website, the speed of a thunderstorm's updraft is one of the factors affecting the size of a hailstone. The following data was given in the article:

x–wind updraft speed (mph)	35	40	64	84
y–hailstone size (in.)	0.5	0.75	1.75	3.0

SOURCE: http://www.srh.noaa.gov/jetstream//tstorms/hail.htm

Refer to the following computer output and verify that the equation of the line of best fit is $\hat{y} = -1.279 + 0.0499x$ and that $s_e = 0.1357$ by calculating these values yourself.

The regression equation is size $= -1.279 + 0.0499$ speed

Predictor	Coef	SE Coef	T	P
Constant	−1.2789	0.2041	−6.27	0.025
Speed	0.049846	0.003453	14.44	0.005

$S = 0.135718$ R-Sq $= 99.0\%$ R-Sq(adj) $= 98.6\%$

*13.18 The National Basketball Association calculates many statistics, as do other professional sports organizations. Using the data below, investigate the relationship between the average number of points per game and the number of All-Star appearances for six of the NBA's best Big Men. Include a scatter diagram, the linear correlation coefficient and line of best fit, and a statement about their meaning.

Player	Points	All-Star
George Mikan	22.6	4
Bill Russell	15.1	12
Wilt Chamberlain	30.1	13
Kareem Abdul-Jabbar	24.6	19
Hakeem Olajuwon	21.8	12
Shaquille O'Neal	16.8	12

*13.19 A diamond is often thought of as a cherished item with a personal value well in excess of its monetary value. The monetary value of a diamond is determined by its exact quality as defined by the 4 Cs: cut, color, clarity, and carat weight. The price (dollars) and the carat weight of a diamond are its two most known characteristics. In order to understand the role carat weight has in determining the price of a diamond, the carat weight and price of 20 loose round diamonds, all of color D and clarity VS1, were obtained January 7, 2010, on the Internet.

Carat Weight	Price	Carat Weight	Price	Carat Weight	Price
0.56	2,055	0.40	1,242	0.62	2,384
0.90	5,433	0.80	4,182	0.54	1,746
0.50	1,735	0.57	2,085	0.30	894
0.53	1,962	0.71	3,117	0.50	1,871
0.92	5,554	0.40	1,176	0.54	1,746
0.51	1,900	0.30	855	0.70	3,074
0.41	1,264	0.40	1,153		

SOURCE: http://www.overnightdiamonds.com/

a. Draw a scatter diagram of the data: carat weight (x) and price (y).

b. Do the data suggest a linear relationship for the domain 0.30 to 0.92 carat? Discuss your findings in part (a).

c. Diamonds smaller than 0.30 carat and diamonds larger than 0.92 carat may not fit the linear pattern demonstrated by this data. Explain.

d. Find the equation for the line of best fit.

e. According to this information, what would be a typical price for a 0.75 carat loose diamond of this quality?

f. On average, by how much does the price increase for each extra 0.01 carat in weight? Within what interval of x-values would you expect this to be true?

g. Find the variance of y about the regression line. What characteristics in the scatter diagram support this large value?

13.20 The following data show the number of hours studied for an exam, x, and the grade received on the exam, y (y is measured in tens, that is, $y = 8$ means that the grade, rounded to the nearest 10 points, is 80.)

x	2	3	3	4	4	5	5	6	6	6	7	7	7	8	8
y	5	5	7	5	7	7	8	6	9	8	7	9	10	8	9

a. Draw a scatter diagram of the data.

b. Find the equation of the line of best fit and graph it on the scatter diagram.

c. Find the ordinates $\hat{y}$ that correspond to $x = 2, 3, 4, 5, 6, 7,$ and 8.

d. Find the five values of e that are associated with the points where $x = 3$ and $x = 6$.

e. Find the variance s_e^2 of all the points about the line of best fit.

Objective 13.4

13.21 Calculate the estimated standard error of regression, s_{b_1}, for the number of hours studied–exam grade relationship in problem 13.20.

13.22 Using the estimated standard error of regression, s_{b_1}, found in problem 13.21 for the number of hours studied–exam grade relationship, find the 95% confidence interval for the population slope β_1. The equation for the line of best fit was: $\hat{y} = 3.96 + 0.625x$.

13.23 Is the time spent watching television taking over a young person's reading time? A quick random survey of 7th-grade girls provided the following results.

Television Time (minutes)	Number of Books Read in Last Year
75	10
45	9
120	4
60	7
30	22

Let y be the number of books read in the last year and x be the time spent watching television each week night.

a. Find the equation of the line of best fit.

b. Find a 95% confidence interval for β_1.

c. Explain the meaning of the interval found in part (b).

*13.24 Interstate 90 is the longest of the east-west U.S. interstate highways with its 3,112 miles stretching from Boston, MA at I-93 on the eastern end to Seattle, WA at the Kingdome on the western end. It travels across 13 northern states; the number of miles and number of intersections in each of those states is listed below.

State	WA	ID	MT	WY	SD	MN	WI	IL	IN	OH	PA	NY	MA
Inter-sections	57	15	83	23	61	52	40	19	21	40	14	48	18
Miles	298	73	558	207	412	275	188	103	157	244	47	391	159

SOURCES: Rand McNally and http://www.ihoz.com/I90.html

a. Construct a scatter diagram of the data.

b. Find the equation for the line of best fit using x = miles and y = intersections.

c. Using the equation found in part (b), estimate the average number of intersections per mile along I-90.

d. Find a 95% confidence interval for β_1.

e. Explain the meaning of the interval found in part (d).

13.25 State the null hypothesis, H_o, and the alternative hypothesis, H_a, that would be used to test the following statements:

a. The slope for the line of best fit is positive.

b. The slope of regression line is not significant.

c. The negative slope for the regression is significant.

*13.26 Each student in a sample of 10 was asked for the distance and the time required to commute to college yesterday. The data collected are shown in the table.

Distance	1	3	5	5	7	7	8	10	10	12
Time	5	10	15	20	15	25	20	25	35	35

a. Draw a scatter diagram of these data.

b. Find the equation that describes the regression line for these data.

c. Does the value of b_1 show sufficient strength to conclude that β_1 is greater than zero at the $\alpha = 0.05$ level?

d. Find the 98% confidence interval for the estimation of β_1. (Retain these answers for use in problem 13.31.)

*13.27 "The fast-food hamburger remains to be the single largest seller in fast-food restaurants in America" according to http://www.loseweightgroup.com/. All fast-food restaurants are required to provide nutritional facts about their various burgers. Do the calories due to fat determine the cholesterol mg in a hamburger? The following data were obtained from the website.

Fast Food	Fat Calories	Cholesterol (mg)
Big Mac	270	80
Quarter-Pounder with Cheese	220	95
Double Cheeseburger	210	80
Whopper with Cheese	420	150
Double Whopper with Cheese	580	195
Classic Triple with Everything	700	260
1/2 lb Bacon Cheddar Double Melt	380	150

SOURCE: http://www.loseweightgroup.com/

a. Determine the equation for the line of best fit.
b. Determine if number of fat calories is an effective predictor of cholesterol, at the 0.05 level of significance.
c. Find the 95% confidence interval for β_1.

13.28 Consider the "Married Couples' Heights" scatter diagram presented on page 272.
a. Find the equation for the line of best fit.
b. Is the value of β_1 significantly greater than zero at the 0.05 level?
c. Find the 95% confidence interval for β_1.

Objective 13.5

13.29 A study in *Physical Therapy* reports on seven different methods to determine crutch length plus two new techniques using linear regression. One of the regression techniques uses the patient's reported height. The study included 107 individuals. The mean of the self-reported heights was 68.84 inches. The regression equation determined was $y = 0.68x - 4.8$, where y = crutch length and x = self-reported height. The MSE (s_e^2) was reported to be 0.50. In addition, the standard deviation of the self-reported heights was 7.35 inches. Use this information to determine a 95% confidence interval estimate for the mean crutch length for individuals who say they are 70 inches tall.

*13.30 Cicadas are flying, plant-eating insects. One particular species, the 13-year cicadas (*Magicicada*), spends five juvenile stages in underground burrows. During the 13 years underground, the cicadas grow from approximately the size of a small ant to nearly the size of an adult cicada. The adult body weights (BW) in grams and wing lengths (WL) in millimeters are given for three different species of these 13-year cicadas in the following table.

BW	WL	Species	BW	WL	Species
0.15	28	tredecula	0.18	29	tredecula
0.29	32	tredecim	0.21	27	tredecassini
0.17	27	tredecim	0.15	30	tredecula
0.18	30	tredecula	0.17	27	tredecula
0.39	35	tredecim	0.13	27	tredecassini
0.26	31	tredecim	0.17	29	tredecassini
0.17	29	tredecassini	0.23	30	tredecassini
0.16	28	tredecassini	0.12	22	tredecim

BW	WL	Species	BW	WL	Species
0.14	25	tredecassini	0.26	30	tredecula
0.14	28	tredecassini	0.19	30	tredecula
0.28	25	tredecassini	0.20	30	tredecassini
0.12	28	tredecim	0.14	23	tredecula

a. Draw a scatter diagram with body weight as the independent variable and wing length as the dependent variable. Find the equation of the line of best fit.
b. Is body weight an effective predictor of wing length for a 13-year cicada? Use a 0.05 level of significance.
c. Give a 90% confidence interval for the mean wing length for all 0.20-gram cicada body weights.

13.31 Use the data and the answers found in problem 13.26 to make the following estimates.
a. Give a point estimate for the mean time required to commute 4 miles.
b. Give a 90% confidence interval for the mean travel time required to commute 4 miles.
c. Give a 90% prediction interval for the travel time required for one person to commute the 4 miles.
d. Answer parts (a)–(c) for $x = 9$.

*13.32 An experiment was conducted to study the effect of a new drug in lowering the heart rate in adults. The data collected are shown in the following table.

x, Drug Dose (mg)	0.50	0.75	1.00	1.25	1.50	1.75	2.00	2.25	2.50	2.75
y, Heart Rate Reduction	10	7	15	12	15	14	20	20	18	21

a. Find the 95% confidence interval for the mean heart rate reduction for a dose of 2.00 mg.
b. Find the 95% prediction interval for the heart rate reduction expected for an individual receiving a dose of 2.00 mg.

*13.33 Use the "Married Couples' Heights" data presented on page 272 and the answers found in problem 13.28 to make the following estimates.
a. Give a point estimate for the mean husband height for a wife height of 59 inches.
b. Give a 95% confidence interval for the mean husband height for a wife height of 59 inches.
c. Give a 95% prediction interval for the husband height expected for a wife height of 59 inches.
d. Answer parts (a), (b), and (c) for $x = 68$.

*13.34 Mr. B, the manager at a large retail store, is investigating several variables while measuring the level of his business. His store is open every day during the year except for New Year's Day, Christmas, and all Sundays. From his records, which cover several years prior, Mr. B has randomly identified 62 days and collected data for the daily total for three variables: number of paying customers, number of items purchased, and total cost of items purchased.

Day	Month	Customers	Items	Sales
2	1	425	1,311	$12,707.00
1	1	412	1,123	$11,467.50

***The remainder of the data can be found online at cengagebrain.com.

Data are actual values; store name withheld for privacy reasons.
Day Code: 1 = M, 2 = Tu, 3 = W, 4 = Th, 5 = F, 6 = Sa
Month Code: 1 = Jan, 2 = Feb, 3 = Mar, ..., 12 = Dec

Is there evidence to claim a linear relationship between the two variables, number of customers and number of items purchased?

The computer output that follows resulted from analysis of the data.

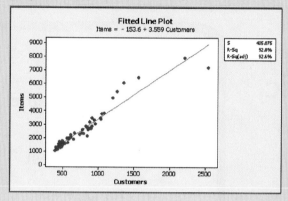

Fitted Line Plot
Items = - 153.6 + 3.559 Customers

S   405.075
R-Sq   92.8%
R-Sq(adj) 92.6%

```
Regression Analysis: Items versus Customers
The regression equation is
Items = -154 + 3.56 Customers
Predictor    Coef     SE Coef      T        P
Constant    -153.6    108.2     -1.42    0.161
Customers    3.5591   0.1284    27.71    0.000
S = 405.075    R-Sq = 92.8%    R-Sq(adj) = 92.6%
```

Inspect the preceding scatter diagram and the regression analysis output for number of customers versus number of items purchased. Look for evidence that either supports or contradicts the claim, "There is a linear relationship between the two variables."

a. Describe the graphical evidence found and discuss how it shows lack of linearity for the entire range of values. Which ordered pairs appear to be different from the others?

b. Describe how the numerical evidence shown indicates that the linear model does fit this data. Explain.

c. Some of the evidence seems to indicate the linear model is the correct model, and some evidence indicates the opposite. What months provided the points that are separate from the rest of the pattern? What is going on in those months that might cause this?

*13.35 Do you think your height and shoe size are related? Probably so. There is a known "quick" relationship that says your height (in inches) can be approximated by doubling your shoe size and adding 50 ($y = 2x + 50$). A random sample of 30 community college students' heights and shoe sizes was taken to test this relationship.

Height	Shoe Size
74	13.0
71	10.0

***The remainder of the data can be found online at cengagebrain.com.

a. Construct a scatter diagram of the data with shoe size as the independent variable (x) and height as the dependent variable (y). Comment on the visual linear relationship.

b. Calculate the correlation coefficient, r. Is it significant at the 0.05 level of significance?

c. Calculate the line of best fit.

d. Compare the slope and intercept from part (c) to the slope and intercept of $y = 2x + 50$. List similarities and differences.

e. Estimate the height for a student with a size 10 shoe, first using the line of best fit found in part (c) and then using the relationship $y = 2x + 50$. Compare your results.

f. Construct the 95% confidence interval for the mean height of all community college students with a size 10 shoe using the equation formed in part (c). Is your estimate using $y = 2x + 50$ for a size 10 included in this interval?

g. Construct the 95% prediction interval for the individual heights of all community college students with a size 10 shoe using the equation formed in part (c).

h. Comment on the widths of the two intervals formed in parts (f) and (g). Explain.

Chapter 14

Elements of
Nonparametric Statistics

14.1 Nonparametric Statistics

MOST OF THE STATISTICAL PROCEDURES WE HAVE STUDIED IN THIS BOOK ARE KNOWN AS **PARAMETRIC METHODS.**

For a statistical procedure to be parametric, either we assume that the parent population is at least approximately normally distributed or we rely on the central limit theorem to give us a normal approximation. This is particularly true of the statistical methods studied in Chapters 8, 9, and 10.

The **nonparametric methods**, or **distribution-free methods** as they are also known, do not depend on the distribution of the population being sampled. The nonparametric statistics are usually subject to much less confining restrictions than are their parametric counterparts. Some, for example, require only that the parent population be continuous. The recent popularity of nonparametric statistics can be attributed to the following characteristics:

1. Nonparametric methods require few assumptions about the parent population.

2. Nonparametric methods are generally easier to apply than their parametric counterparts.

3. Nonparametric methods are relatively easy to understand.

4. Nonparametric methods can be used in situations in which the normality assumptions cannot be made.

5. Nonparametric methods are generally only slightly less efficient than their parametric counterparts.

objectives

14.1 Nonparametric Statistics

14.2 The Sign Test

14.3 The Mann–Whitney U Test

14.4 The Runs Test

14.5 Rank Correlation

Parametric methods Statistical procedures for which either we assume that the parent population is at least approximately normally distributed, or we rely on the central limit theorem to give us a normal approximation.

Nonparametric methods (distribution-free methods) Statistical procedures that do not depend on the distribution of the population being sampled.

© Ali Johnson Photography/Flickr/Getty Images

Comparing Statistical Tests

This chapter presents only a very small sampling of the many different nonparametric tests that exist. The selections presented demonstrate their ease of application and variety of technique. Many of the nonparametric tests can be used in place of certain parametric tests. The question is, then: Which statistical test do we use, the parametric or the nonparametric? Sometimes there is also more than one nonparametric test to choose from.

The decision about which test to use must be based on the answer to the question: Which test will do the job best? First, let's agree that when we compare two or more tests, they must be equally qualified for use. That is, each test has a set of assumptions that must be satisfied before it can be applied. From this starting point we will attempt to define "best" to mean the test that is best able to control the risks of error and at the same time keep the size of the sample to a number that is reasonable to work with. (Sample size means cost—cost to you or your employer.)

Table 14.1 Comparison of Parametric and Nonparametric Tests

Test Situation	Parametric Test	Nonparametric Test	Efficiency of Nonparametric Test
One mean	*t*-test (p. 187)	Sign test (p. 301)	0.63
Two independent means	*t*-test (p. 217)	*U* test (p. 307)	0.95
Two dependent means	*t*-test (p. 212)	Sign test (p. 303)	0.63
Correlation	Pearson's (pp. 275–276)	Spearman test (p. 315)	0.91
Randomness		Runs test (p. 312)	Not meaningful; there is no parametric test for comparison

POWER AND EFFICIENCY CRITERIA

Let's look first at the ability to control the risk of error. The risk associated with a type I error is controlled directly by the level of significance α. Recall that P(type I error) = α and P(type II error) = β. Therefore, it is β that we must control. Statisticians like to talk about *power* (as do others), and the **power of a statistical test** is defined to be $1 - \beta$. Thus, the power of a test, $1 - \beta$, is the probability that we reject the null hypothesis when we should have rejected it. If two tests with the same α are equal candidates for use, then the one with the greater power is the one you would want to choose.

The other factor is the sample size required to do a job. Suppose that you set the levels of risk you can tolerate, α and β, and then you are able to determine the sample size it would take to meet your specified challenge. The test that required the smaller sample size would seem to have the edge. Statisticians usually use the term **efficiency** to talk about this concept. *Efficiency* is the ratio of the sample size of the best parametric test to the sample size of the best nonparametric test when compared under a fixed set of risk values. For example, the efficiency rating for the sign test is approximately 0.63. This means that a sample of size 63 with a parametric test will do the same job as a sample of size 100 will do with the sign test.

> **Power of a statistical test** $1 - \beta$, or the probability that we reject the null hypothesis when we should have rejected it.

The power and the efficiency of a test cannot be used alone to determine the choice of test. Sometimes you will be forced to use a certain test because of the data you are given. When there is a decision to be made, the final decision rests in a trade-off of three factors: (1) the power of the test, (2) the efficiency of the test, and (3) the data (and the sample size) available. Table 14.1 shows how the nonparametric tests discussed in this chapter compare with the parametric tests covered in previous chapters.

14.2 The Sign Test

THE **SIGN TEST** IS A VERSATILE AND EXCEPTIONALLY EASY-TO-APPLY NONPARAMETRIC METHOD THAT USES ONLY PLUS AND MINUS SIGNS.

Three sign test applications are presented here: (1) a confidence interval for the median of one population, (2) a hypothesis test concerning the value of the median for one population, and (3) a hypothesis test concerning the median difference (paired difference) for two **dependent samples.** These sign tests are carried out using the same basic confidence interval and hypothesis test procedures as described in earlier chapters. They are the nonparametric alternatives to the *t*-tests used for one mean (see Objective 9.1) and for the difference between two dependent means (see Objective 10.2).

Assumptions for inferences about the population single-sample median using the sign test:

The *n* random observations that form the sample are selected independently, and the population is continuous in the vicinity of the median *M*.

Single-Sample Confidence Interval Procedure

The sign test can be applied to obtain a confidence interval for the unknown **population median, M.** To accomplish this we will need to arrange the sample data in ascending order (smallest to largest). The data are identified as x_1 (smallest), $x_2, x_3, \ldots, x_n$ (largest). The critical value, k (known as the "maximum allowable number of signs"), is obtained from Table 12 in Appendix B, and it tells us the number of positions to be dropped from each end of the ordered data. The remaining extreme values become the bounds of the $1 - \alpha$ confidence interval. That is, the lower boundary for the confidence interval is x_{k+1}, the $(k + 1)$th data value; the upper boundary is x_{n-k}, the $(n - k)$th data value.

In general, the two data values that bound the confidence interval occupy positions $k + 1$ and $n - k$, where k is the critical value read from Table 12. Thus,

$$x_{k+1} \text{ to } x_{n-k}, 1 - \alpha \text{ confidence interval for } M$$

To clarify this procedure, let's suppose that we have a random sample of 12 daily high temperature readings in ascending order, [50, 62, 64, 76, 76, 77, 77, 77, 80, 86, 92, 94], and we wish to form a 95% confidence interval for the population median. Table 12 shows a critical value of 2 ($k = 2$) for $n = 12$ and $\alpha = 0.05$ for a two-tailed hypothesis test. This means that we drop the last two values on each end (50 and 62 on left; 92 and 94 on the right). The confidence interval is bounded inclusively by the remaining end values, 64 and 86. That is, the 95% confidence interval is 64 to 86 and is expressed as:

64° to 86°, the 95% confidence interval for the median daily high temperature

Single-Sample Hypothesis-Testing Procedure

The sign test can be used when the null hypothesis to be tested concerns the value of the population median M. The test may be either one- or two-tailed.

In order to demonstrate this test procedure, consider a random sample of 75 students. They were selected and each student was asked to carefully measure the amount of time it takes to commute from his or her front door to the college parking lot. The data collected were used to test the hypothesis, "The median time required for students to commute is 15 minutes," against the alternative that the median is unequal to 15 minutes. The 75 pieces of data were summarized as follows:

Under 15:	18
15:	12
Over 15:	45

We will use the sign test to test the null hypothesis against the alternative hypothesis.

The data are converted to + and − signs according to whether each data value is more or less than 15. A plus sign will be assigned to each value larger than 15, a minus sign to each value smaller than 15, and a zero to those equal to 15. The sign test uses only the plus and minus signs; therefore, the zeros are discarded and the usable sample size becomes 63. That is, $n(+) = 45$, $n(-) = 18$, and $n = n(+) + n(-) = 45 + 18 = 63$.

STEP 1 a. **Parameter of interest:**
M, the population median time to commute.

b. **Statement of hypotheses:**

$$H_o: M = 15$$

$$H_a: M \neq 15$$

STEP 2 a. **Assumptions:**
The 75 observations were randomly selected, and the variable, commute time, is continuous.

b. **Test statistic:**
The test statistic that will be used is the number of the less frequent sign: the smaller of $n(+)$ and $n(-)$, which is $n(-)$ for our example. We will want to reject the null hypothesis whenever the number of

the less frequent sign is extremely small. Table 12 in Appendix B gives the maximum allowable number of the less frequent sign, k, that will allow us to reject the null hypothesis. That is, if the number of the less frequent sign is less than or equal to the critical value in the table, we will reject H_o. If the observed value of the less frequent sign is larger than the table value, we will fail to reject H_o. In the table, n is the total number of signs, not including zeros. The test statistic $= x\bigstar = n$(least frequent sign).

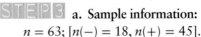

c. **Level of significance:**
$\alpha = 0.05$ for a two-tailed test.

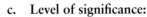 **a. Sample information:**
$n = 63$; $[n(-) = 18, n(+) = 45]$.

b. **Test statistic:**
The observed value of the test statistic is

$$x\bigstar = n(-) = \mathbf{18}.$$

STEP 4 Probability distribution:
As always, we can use either the p-value or the classical procedure:

Using the *p*-value procedure:

a. Since the concern is for values "not equal to," the p-value is the area of both tails. We will find the left tail and double it: $\mathbf{P} = 2 \times P(x \le 18$, for $n = 63)$.

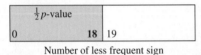

Number of less frequent sign

To find the p-value, you have two options:
1. Use Table 12 (Appendix B) to place bounds on the p-value. Table 12 lists only two-tailed values (do not double): $\mathbf{P} < 0.01$
2. Use a computer or calculator to find the p-value: $\mathbf{P} = 0.0011$.
Specific instructions follow this example.

b. The p-value is smaller than α.

Using the classical procedure:

a. The critical region is split into two equal parts because H_a expresses concern for values related to "not equal to." Since the table is for two-tailed tests, the critical value is located at the intersection of the $\alpha = 0.05$ column and the $n = 63$ row of Table 12: **23**.

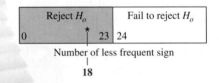

Number of less frequent sign
18

b. $x\bigstar$ is in the critical region, as shown in the figure above.

The sample shows sufficient evidence at the 0.05 level to conclude that the median commute time is not equal to 15 minutes.

© RTimages/iStockphoto.com

STEP 5 a. Decision: Reject H_o.

b. **Conclusion:**
The sample shows sufficient evidence at the 0.05 level to conclude that the median commute time is not equal to 15 minutes.

Calculating the *p*-Value When Using the Sign Test

Method 1: Use Table 12 in Appendix B to place bounds on the p-value. By inspecting the $n = 63$ row of Table 12, you can determine an interval within which the p-value lies. Locate the value of x along the $n = 63$ row and read the bounds from the top of the table. Table 12 lists only two-tailed values (therefore, do not double): **P < 0.01.**

Method 2: If you are doing the hypothesis test with the aid of a computer or graphing calculator, most likely it will calculate the *p*-value for you.

Two-Sample Hypothesis-Testing Procedure

The sign test may also be applied to a hypothesis test dealing with the median difference between **paired data** that result from **two dependent samples.** A familiar application is the use of before-and-after testing to determine the effectiveness of some activity. In a test of this nature, the signs of the differences are used to carry out the test. Again, zeros are disregarded.

> **Assumptions for inferences about the median of paired differences using the sign test:**
> The paired data are selected independently, and the variables are ordinal or numerical.

Let's put the sign test to work looking at a new no-exercise, no-starve weight-reducing plan that has been developed and advertised. To test the claim that "you will lose weight within 2 weeks or . . . ," a local statistician obtained the before-and-after weights of 18 people who had used this plan. Table 14.2 lists the people, their weights, and a minus (−) for those who lost weight during the 2 weeks, a 0 for those whose weight remained the same, and a plus (+) for those who actually gained weight.

The claim being tested is that people lose weight. The null hypothesis that will be tested is, "There is no weight loss (or the median weight loss is zero)," meaning that only a rejection of the null hypothesis will allow us to conclude in favor of the advertised claim. Actually we will be testing to see whether there are significantly more minus signs than plus signs. If the weight-reducing plan is of absolutely no value, we would expect to find an equal number of plus and minus signs. If it works, there should be significantly more minus signs than plus signs. Thus, the test performed here will be a one-tailed test. (We want to reject the null hypothesis in favor of the advertised claim if there are "many" minus signs.)

*Table 14.2 Sample Results on Weight-Reducing Plan

Person	Weight Before	Weight After	Sign of Difference, After − Before
Mrs. Smith	146	142	−
Mr. Brown	175	178	+
Mrs. White	150	147	−
Mr. Collins	190	187	−
Mr. Gray	220	212	−
Ms. Collins	157	160	+
Mrs. Allen	136	135	−
Mrs. Noss	146	138	−
Ms. Wagner	128	132	+
Mr. Carroll	187	187	0
Mrs. Black	172	171	−
Mrs. McDonald	138	135	−
Ms. Henry	150	151	+
Ms. Greene	124	126	+
Mr. Tyler	210	208	−
Mrs. Williams	148	148	0
Mrs. Moore	141	138	−
Mrs. Sweeney	164	159	−

STEP 1 **a. Parameter of interest:**
M, the median weight loss.
b. Statement of hypotheses:

$$H_o: M = 0 \text{ (no weight loss)}$$
$$H_a: M < 0 \text{ (weight loss)}$$

STEP 2 **a. Assumptions:**
The 18 observations were randomly selected, and the variables, weight before and weight after, are both continuous.
b. Test statistic:
The number of the less frequent sign: the test statistic $= x\bigstar = n(\text{least frequent sign})$.
c. Level of significance:
$\alpha = 0.05$ for a one-tailed test.

STEP 3 **a. Sample information:**
$n = 16$ [$n(+) = 5, n(−) = 11$].
b. Test statistic:
The observed value of the test statistic is
$$x\bigstar = n(+) = 5.$$

STEP 4 **Probability Distribution:**
Again, we can use either the p-value or the classical procedure:

Using the *p*-value procedure:

a. Since the concern is for values "less than," the p-value is the area to the left: $\mathbf{P} = P(x \leq 5$, for $n = 16)$.

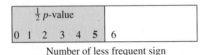

$\frac{1}{2}p$-value	
0 1 2 3 4 **5**	6

Number of less frequent sign

To find the p-value, you have two options:
1. Use Table 12 in Appendix B to estimate the p-value. Table 12 lists only two-tailed α (this is one-tailed, so divide α by two): $\mathbf{P} \approx \mathbf{0.125}$.
2. Use a computer or calculator to find the p-value: $\mathbf{P} = \mathbf{0.1051}$.
For specific instructions, see page 302.

b. The p-value is not smaller than α.

Using the classical procedure:

a. The critical region is one-tailed because H_a expresses concern for values related to "less than." Since the table is for two-tailed tests, the critical value is located at the intersection of the $\alpha = 0.10$ column ($\alpha = 0.05$ in each tail) and the $n = 16$ row of Table 12:

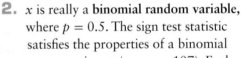

Reject H_o	Fail to reject H_o
0 1 2 3 4	★ 5

Number of less frequent sign
5

b. x★ is not in the critical region, as shown in the figure above.

STEP 5 **a.** **Decision:** Fail to reject H_o.
b. **Conclusion:**
The evidence observed is not sufficient to allow us to reject the no-weight-loss null hypothesis at the 0.05 level of significance.

Normal Approximation

The sign test may be carried out by means of a normal approximation using the standard normal variable z. The normal approximation may be used if Table 12 does not show the particular levels of significance desired or if n is large.

When using a normal approximation, keep in mind the following:

1. x may be the number of the less frequent sign or the more frequent sign. You will have to determine this in such a way that the direction is consistent with the interpretation of the situation.

2. x is really a **binomial random variable,** where $p = 0.5$. The sign test statistic satisfies the properties of a binomial experiment (see page 107). Each sign is the result of an independent trial. There are n trials, and each trial has two possible outcomes ($+$ or $-$). Since the median is used, the probabilities for each outcome are both 0.5. Therefore, the mean, μ_x, is equal to

$$\mu_x = \frac{n}{2} \quad \left[\mu = np = n \cdot \frac{1}{2} = \frac{n}{2}\right]$$

and the standard deviation, σ_x, is equal to

$$\sigma_x = \frac{1}{2}\sqrt{n} \quad \left[\sigma = \sqrt{npq} = \sqrt{n \cdot \frac{1}{2} \cdot \frac{1}{2}} = \frac{1}{2}\sqrt{n}\right]$$

3. x is a discrete variable. But recall that the normal distribution must be used only with continuous variables. However, although the binomial random variable is discrete, it does become approximately normally distributed for large n. Nevertheless, when using the normal distribution for testing, we should make an adjustment in the variable so that the approximation is more accurate. (See Objective 6.5, page 129, on the normal approximation.) This adjustment is illustrated in Figure 14.1 and is called a **continuity correction**. For this discrete variable, the area that represents the probability is a rectangular bar. Its width is 1 unit wide, from $\frac{1}{2}$ unit below to $\frac{1}{2}$ unit above the value of interest. Therefore, when z is to be used, we will need to make a $\frac{1}{2}$-unit adjustment before calculating the observed value of z. So x' will be the adjusted value for x. If x is larger than $\frac{n}{2}$, then $x' = x - \frac{1}{2}$. If x is smaller than $\frac{n}{2}$, then $x' = x + \frac{1}{2}$. The test is then completed by the usual procedure, using x'.

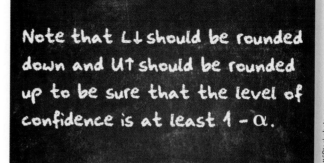

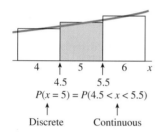

Figure 14.1 Continuity Correction

$$4 \quad 4.5 \quad 5 \quad 5.5 \quad 6 \quad x$$
$$P(x = 5) = P(4.5 < x < 5.5)$$

Discrete   Continuous

CONFIDENCE INTERVAL PROCEDURE

If the normal approximation is to be used (including the continuity correction), the position numbers for a $1 - \alpha$ confidence interval for M are found using the formula:

$$\frac{1}{2}(n) \pm \left(\frac{1}{2} + \frac{1}{2} \cdot z(\alpha/2) \cdot \sqrt{n} \right) \qquad (14.1)$$

The interval is

$\quad x_L$ to x_U, $1 - \alpha$ confidence interval for M (median)

where

$$L = \frac{n}{2} - \frac{1}{2} - \frac{1}{2} \cdot z(\alpha/2) \cdot \sqrt{n} \quad \text{and}$$

$$U = \frac{n}{2} + \frac{1}{2} + \frac{1}{2} \cdot z(\alpha/2) \cdot \sqrt{n}$$

➤ One way that we can use normal approximation is to estimate the population median daily high temperature. We will run the test with a 95% confidence interval based on the following random sample of 60 daily high temperature readings. (Note: temperatures have been arranged in ascending order.)

43(x_1)	55(x_2)	59	60	67	73	73	73	73	73
73	75	75	76	78	78	78	79	79	80
80	80	80	80	80	80	82	82	82	82
83	83	83	83	83	84	84	84	85	85
86	86	87	87	88	88	88	88	88	88
88	89	89	89	89	90	92	93	94	98(x_{60})

When we use formula (14.1), the position numbers L and U are

$$\frac{1}{2}(n) \pm \left(\frac{1}{2} + \frac{1}{2} \cdot z(\alpha/2) \cdot \sqrt{n} \right):$$

$$\frac{1}{2}(60) \pm \left(\frac{1}{2} + \frac{1}{2} \cdot 1.96 \cdot \sqrt{60} \right)$$

$$30 \pm (0.50 + 7.59)$$

$$30 \pm 8.09$$

Thus,

$L = 30 - 8.09 = 21.91$, rounded down becomes 21 (21st data value)

$U = 30 + 8.09 = 38.09$, rounded up becomes 39 (39th data value)

Therefore,

80° to 85°, the 95% confidence interval for the median high daily temperature

HYPOTHESIS TESTING-PROCEDURE

When a hypothesis test is to be completed using the standard normal distribution, z will be calculated with the formula.

$$z\star = \frac{x' - \frac{n}{2}}{\frac{1}{2} \cdot \sqrt{n}} \qquad (14.2)$$

(See Note 3 on page 304 with regard to x'.)

Let's look at a one-tailed situation and use the sign test to test the hypothesis that the median number of hours, M, worked by students at a certain college is at least 15 hours per week. A survey of 120 students was taken; a plus sign was recorded if the number of hours the student worked last week was equal to or greater than 15, and a minus sign was recorded if the number of hours was less than 15. Totals showed 80 minus signs and 40 plus signs.

STEP 1 **a. Parameter of interest:**
M, the median number of hours worked by students.

b. Statement of hypotheses:

$H_o: M = 15\ (\geq)$ (at least as many plus signs as minus signs)

$H_a: M < 15$ (fewer plus signs than minus signs)

STEP 2 **a. Assumptions:**
The random sample of 120 students was independently surveyed, and the variable, hours worked, is continuous.

b. **Probability distribution and test statistic:**
The standard normal z and formula (14.2).

c. **Level of significance:** $\alpha = 0.05$.

STEP 3 **a.** **Sample information:**
$n(+) = 40$ and $n(-) = 80$; therefore, $n = 120$ and x is the number of plus signs, $x = 40$.

b. **Test statistic:** Using formula (14.2), we have

$$z\star = \frac{x' - \frac{n}{2}}{\frac{1}{2} \cdot \sqrt{n}}:$$

$$z\star = \frac{40.5 - \frac{120}{2}}{\frac{1}{2} \cdot \sqrt{120}} = \frac{40.5 - 60}{\frac{1}{2} \cdot (10.95)} = \frac{-19.5}{5.475}$$

$$= -3.562 = \mathbf{-3.56}$$

STEP 4 **Probability distribution:**
Again, we can use either the p-value or the classical procedure:

Using the *p*-value procedure:

a. Use the left-hand tail because H_a expresses concern for values related to "fewer than."
$\mathbf{P} = P(z < -3.56)$ as shown in the figure below.

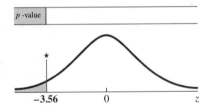

To find the p-value, you have three options:
1. Use Table 3 (Appendix B) to calculate the p-value: $\mathbf{P = 0.0002}$.
2. Use Table 5 (Appendix B) to place bounds on the p-value: $\mathbf{P = 0.0002}$.
3. Use a computer or calculator to find the p-value: $\mathbf{P = 0.0002}$.
For specific instructions, see page 171.

b. The p-value is smaller than α.

Using the classical procedure:

a. The critical region is the left-hand tail because H_a expresses concern for values related to "fewer than." The critical value is obtained from Table 4A:

$-z(0.05) = -1.65$

Specific instructions for finding critical values are on page 175.

b. $z\star$ is in the critical region, as shown in red in the figure above.

STEP 5 **a.** **Decision:** Reject H_o.

b. **Conclusion:**
At the 0.05 level, there are significantly more minus signs than plus signs, thereby implying that the median is less than the claimed 15 hours.

14.3 The Mann–Whitney U Test

THE **MANN–WHITNEY *U* TEST** IS A NONPARAMETRIC ALTERNATIVE TO THE *t*-TEST FOR THE DIFFERENCE BETWEEN TWO INDEPENDENT MEANS.

The usual two-sample situation occurs when the experimenter wants to see whether the difference between the two samples is sufficient to reject the null hypothesis that the two sampled populations are identical.

Hypothesis-Testing Procedure

Assumptions for inferences about two populations using the Mann–Whitney *U* test:

The two **independent random samples** are independent within each sample as well as between samples, and the random variables are ordinal or numerical.

This test is often used in situations in which the two samples are drawn from the same population of subjects but different "treatments" are used on each set. For example, in a large lecture class, when a 1-hour exam is given, the instructor gives two "equivalent" examinations. It is reasonable to ask: Are these two different exams equivalent? Students in even-numbered seats take exam A, and those in the odd-numbered seats take exam B. To test this "equivalent" hypothesis, two random samples were taken. Table 14.3 lists the exam scores of the two samples.

*Table 14.3 **Data on Exam Scores**

Exam A	52	78	56	90	65	86	64	90	49	78
Exam B	72	62	91	88	90	74	98	80	81	71

If we assume that the odd- or even-numbered seats had no effect, does the sample present sufficient evidence to reject the hypothesis "The exam forms yielded scores that had identical distributions"? We will perform the test using $\alpha = 0.05$.

STEP 1 a. **Parameter of interest:**
The distribution of scores for each version of the exam.

b. **Statement of hypotheses:**

> H_o: Exam A and exam B have test scores with identical distributions.

> H_a: The two distributions are not the same.

STEP 2 a. **Assumptions:**
The two samples are independent, and the random variable, exam score, is numerical.

b. **Test statistic:** The Mann–Whitney U statistic.

c. **Level of significance:** $\alpha = 0.05$.

STEP 3 a. **Sample information:**
The sample data are listed in Table 14.3.

b. **Test statistic:**
The size of the individual samples will be called n_a and n_b; actually, it makes no difference which way these are assigned. In our example they both have the value 10. The two samples are combined into one sample (all $n_a + n_b$) and ordered from smallest to largest:

49	52	56	62	64	65	71	72	74	78
78	80	81	86	88	90	90	90	91	98

Each is then assigned a **rank** number. The smallest (49) is assigned rank 1, the next smallest (52) is assigned rank 2, and so on, up to the largest, which is assigned rank $n_a + n_b$ (20). Ties are handled by assigning to each of the tied observations the mean rank of those rank positions that they occupy. For example, in our example there are two 78s; they are the 10th and 11th. The mean rank

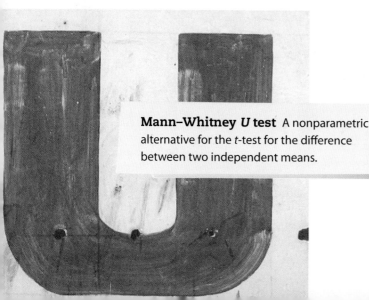

Mann–Whitney *U* test A nonparametric alternative for the *t*-test for the difference between two independent means.

© an_pop/iStockphoto.com

for each is then $\dfrac{10 + 11}{2} = 10.5$. In the case of the three 90s—the 16th, 17th, and 18th data values—each is assigned 17 because $\dfrac{16 + 17 + 18}{3} = 17$. The rankings and source are shown in Table 14.4.

Figure 14.2 shows the relationship between the two sets of data, first by using the data values and second by comparing the rank numbers for the data.

The calculation of the **test statistic U** is a two-step procedure. We first determine the sum of the ranks for each of the two samples. Then, using the two sums of ranks, we calculate a U score for each sample. The smaller U score is the test statistic.

The sum of ranks R_a for sample A is computed as

$$R_a = 1 + 2 + 3 + 5 + 6 + 10.5 + 10.5 + 14 + 17 + 17 = \textbf{86}$$

Table 14.4 **Ranked Exam Score Data**

Ranked Data	Rank	Source	Ranked Data	Rank	Source
49	1	A	78	10.5	A
52	2	A	80	12	B
56	3	A	81	13	B
62	4	B	86	14	A
64	5	A	88	15	B
65	6	A	90	17	A
71	7	B	90	17	A
72	8	B	90	17	B
74	9	B	91	19	B
78	10.5	A	98	20	B

The sum of ranks R_b for sample B is

$$R_b = 4 + 7 + 8 + 9 + 12 + 13 + 15 + 17 + 19 + 20 = \textbf{124}$$

The U score for each sample is obtained by using the following pair of formulas:

Mann–Whitney U Test Statistics

$$U_a = n_a \cdot n_b + \dfrac{(n_b)(n_b + 1)}{2} - R_b \qquad (14.3)$$

$$U_b = n_a \cdot n_b + \dfrac{(n_a)(n_a + 1)}{2} - R_a \qquad (14.4)$$

$U\bigstar$, the test statistic, is the smaller of U_a and U_b.

For our example, we obtain

$$U_a = (10)(10) + \dfrac{(10)(10 + 1)}{2} - 124 = 31$$

$$U_b = (10)(10) + \dfrac{(10)(10 + 1)}{2} - 86 = 69$$

Therefore, $U\bigstar = \textbf{31}$.

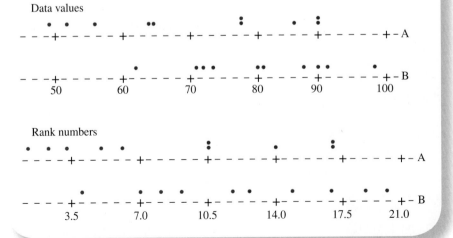

Figure 14.2 **Comparing the Data of Two Samples**

Before we carry out the test for this example, let's try to understand some of the underlying possibilities. Recall that the null hypothesis is that the distributions are the same and that we will most likely want to conclude from this that the averages are approximately equal. Suppose for a moment that the distributions are indeed quite different; say, all of one sample comes before the smallest data value in the second sample when they are ranked together. This would certainly mean that we would want to reject the null hypothesis. What kind of a value can we expect for U in this case? Suppose that the 10 A values had ranks 1 through 10 and that the 10 B values had ranks 11 through 20. Then we would obtain

$$R_a = 55 \quad \text{and} \quad R_b = 155$$

$$U_a = (10)(10) + \frac{(10)(10 + 1)}{2} - 155 = 0$$

$$U_b = (10)(10) + \frac{(10)(10 + 1)}{2} - 55 = 100$$

Therefore, $U\bigstar = 0$.

If this were the case, we certainly would want to reach the decision: Reject the null hypothesis.

Suppose, on the other hand, that both samples were perfectly matched; that is, a score in each set is identical to one in the other.

54	54	62	62	71	71	72	72	...
A	B	A	B	A	B	A	B	...
1.5	1.5	3.5	3.5	5.5	5.5	7.5	7.5	...

Now what would happen?

$$R_a = R_b = 105$$

$$U_a = U_b = (10)(10) + \frac{(10)(10 + 1)}{2} - 105 = 50$$

Therefore, $U\bigstar = 50$. If this were the case, we certainly would want to reach the decision: Fail to reject the null hypothesis.

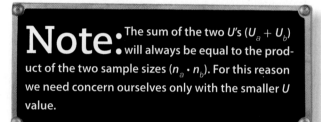

Note: The sum of the two U's $(U_a + U_b)$ will always be equal to the product of the two sample sizes $(n_a \cdot n_b)$. For this reason we need concern ourselves only with the smaller U value.

Now, let's return to our hypothesis test.

STEP 4 **Probability distribution:**
Again, we can use either the p-value or the classical procedure:

Using the *p*-value procedure:

a. Since the concern is for values related to "not the same," the p-value is the probability of both tails. It will be found by finding the probability of the left tail and doubling:

P $= 2 \times P(U \leq 31$ for $n_1 = 10$ and $n_2 = 10)$

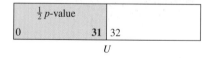

To find the p-value, you have two options:
1. Use Table 13 in Appendix B to place bounds on the p-value: **P > 0.10**.
2. Use a computer or calculator to find the p-value: **P = 0.1612**.
Specific instructions follow this example.
b. The p-value is not smaller than α.

Using the classical procedure:

a. The critical region is two-tailed because H_a expresses concern for values related to "not the same." Use Table 13A for two-tailed $\alpha = 0.05$. The critical value is at the intersection of column $n_1 = 10$ and row $n_2 = 10$: **23**. The critical region is $U \leq 23$.

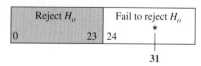

b. $U\bigstar$ is not in the critical region, as shown in the figure above.

STEP 5 a. **Decision:** Fail to reject H_o.
b. **Conclusion:**
At the 0.05 level of significance, we do not have sufficient evidence to reject the "equivalent" hypothesis.

Calculating the *p*-Value When Using the Mann–Whitney Test

Method 1: Use Table 13 in Appendix B to place bounds on the *p*-value. By inspecting Table 13A and B at the intersection of column $n_1 = 10$ and row $n_2 = 10$, you can determine that the *p*-value is greater than 0.10; the larger two-tailed value of α is 0.10 in Table 13B.

Method 2: If you are doing the hypothesis test with the aid of a computer or graphing calculator, most likely it will calculate the *p*-value for you.

Normal Approximation

If the samples are larger than size 20, we may make the test decision with the aid of the standard normal variable, *z*. This is possible because the distribution of *U* is approximately normal with a mean

$$\mu_U = \frac{n_a \cdot n_b}{2} \tag{14.5}$$

and a standard deviation

$$\sigma_U = \sqrt{\frac{n_a \cdot n_b \cdot (n_a + n_b + 1)}{12}} \tag{14.6}$$

The hypothesis test is then completed using the **test statistic $z\star$**:

$$z\star = \frac{U\star - \mu_U}{\sigma_U} \tag{14.7}$$

The standard normal distribution may be used whenever n_a and n_b are both greater than 10.

→ We can demonstrate the normal approximation procedure for the Mann–Whitney *U* test by examining a dog-obedience trainer who is training 27 dogs to obey a certain command. The trainer is using two different training techniques: (I) the reward-and-encouragement method and (II) the no-reward method. Table 14.5 shows the numbers of obedience sessions that were necessary before the dogs would obey the command. Does the trainer have sufficient evidence to

Will sit for treats

*Table 14.5 **Data on Dog Training**

Method I	29	27	32	25	27	28	23	31
Method II	40	44	33	26	31	29	34	31
Method I	37	28	22	24	28	31	34	
Method II	38	33	42	35				

Table 14.6 **Rankings for Training Methods**

Number of Sessions	Group	Rank	
22	I	1	
23	I	2	
24	I	3	
25	I	4	
26	II	5	
27	I	6	6.5
27	I	7	6.5
28	I	8	9
28	I	9	9
28	I	10	9
29	I	11	11.5
29	II	12	11.5
31	I	13	14.5
31	I	14	14.5
31	II	15	14.5
31	II	16	14.5
32	I	17	
33	II	18	18.5
33	II	19	18.5
34	I	20	20.5
34	II	21	20.5
35	II	22	
37	I	23	
38	II	24	
40	II	25	
42	II	26	
44	II	27	

claim that the reward method will, on average, require fewer obedience sessions ($\alpha = 0.05$)? To find out, we can perform the five-step hypothesis test.

STEP 1 a. Parameter of interest:
The distribution of needed obedience sessions for each technique.

b. Statement of hypotheses:

H_o: The distributions of the needed obedience sessions are the same for both methods.

H_a: The reward method, on average, requires fewer sessions.

STEP 2 a. Assumptions:
The two samples are independent, and the random variable, training time, is numerical.

b. Test statistic: The Mann–Whitney U statistic.

c. Level of significance: $\alpha = 0.05$.

STEP 3 a. Sample information:
The sample data are listed in Table 14.5.

b. Test statistic:
The two sets of data are ranked jointly, and ranks are assigned as shown in Table 14.6.
The sums are:

$$R_I = 1 + 2 + 3 + 4 + 6.5 + \ldots + 20.5 + 23 = 151.0$$

$$R_{II} = 5 + 11.5 + 14.5 + \ldots + 26 + 27 = 227.0$$

The U scores are found using formulas (14.3) and (14.4):

$$U_I = (15)(12) + \frac{(12)(12+1)}{2} - 227$$
$$= 180 + 78 - 227 = 31$$

$$U_{II} = (15)(12) + \frac{(15)(15+1)}{2} - 151$$
$$= 180 + 120 - 151 = 149$$

Therefore, $U\bigstar = 31$. Now we use formulas (14.5), (14.6), and (14.7) to determine the z-statistic:

$$\mu_U = \frac{n_a \cdot n_b}{2}: \quad \mu_U = \frac{12 \cdot 15}{2} = 90$$

$$\sigma_U = \sqrt{\frac{n_a \cdot n_b \cdot (n_a + n_b + 1)}{12}}:$$

$$\sigma_U = \sqrt{\frac{12 \cdot 15 \cdot (12 + 15 + 1)}{12}}$$

$$= \sqrt{\frac{(180)(28)}{12}} = \sqrt{420} = 20.49$$

$$z\bigstar = \frac{U\bigstar - \mu_U}{\sigma_U}: \quad z\bigstar = \frac{31 - 90}{20.49}$$

$$= \frac{-59}{20.49} = -2.879 = -2.88$$

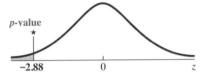

STEP 4 **Probability distribution:**

Again, we can use either the *p*-value or the classical procedure:

Using the *p*-value procedure:

a. Use the left-hand tail because H_a expresses concern for values related to "fewer than." $P = P(z < -2.88)$ as shown in the figure below.

p-value
★

−2.88　　0　　　　z

To find the *p*-value, you have three options:
1. Use Table 3 (Appendix B) to calculate the *p*-value: **P = 0.0020.**
2. Use Table 5 (Appendix B) to place bounds on the *p*-value: **0.0019 < P < 0.0022.**
3. Use a computer or calculator to find the *p*-value: **P = 0.0020.**
For specific instructions, see page 171.

b. The *p*-value is smaller than α.

Using the classical procedure:

a. The critical region is the left-hand tail because H_a expresses concern for values related to "fewer than." The critical value is obtained from Table 4A:

$$-z(0.05) = -\mathbf{1.65}$$

0.05
★

　−1.65　　0　　　　z
−2.88

Specific instructions for finding critical values are on page 175.

b. $z\bigstar$ is in the critical region, as shown in red in the figure above.

STEP 5 a. **Decision:** Reject H_o.

b. **Conclusion:**

At the 0.05 level of significance, the data show sufficient evidence to conclude that the reward method does, on average, require fewer training sessions.

14.4 The Runs Test

THE **RUNS TEST** IS USED MOST FREQUENTLY TO TEST THE **RANDOMNESS** (OR LACK OF RANDOMNESS) OF DATA.

A **run** is a sequence of data that possess a common property. One run ends and another starts when an observation does not display the property in question. The **test statistic** in this test is **V**, the number of runs observed.

➡ To illustrate the idea of runs, let's draw a sample of 10 single-digit numbers from the telephone book, listing the next-to-last digit from each of the selected telephone numbers:

Sample: 2 3 1 1 4 2 6 6 6 7

Let's consider the property of "odd" (o) or "even" (e). The sample, as it was drawn, becomes e ooo eeeee o, which displays four runs:

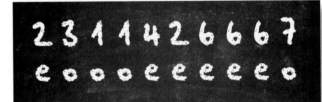

Thus, $V\bigstar = 4$.

If the sample contained no randomness, there would be only two runs—all the evens, then all the odds, or the other way around. We would also not expect to see them alternate—odd, even, odd, even. The maximum number of possible runs would be $n_1 + n_2$ or fewer (provided n_1 and n_2 are not equal), where n_1 and n_2 are the numbers of data that have each of the two properties being identified.

Assumption for inferences about randomness using the runs test:

Each piece of sample data can be classified into one of two categories.

The runs test is generally a two-tailed test. We will reject the hypothesis when there are too few runs because this indicates that the data are "separated" according to the two properties. We will also reject the hypothesis when there are too many runs because that indicates that the data alternate between the two properties too often to be random. For example, if the data alternated all the way down the line, we might suspect that the data had been tampered with. There are many aspects to the concept of randomness. The occurrence of odd and even demonstrated by the telephone book example is one aspect. Another aspect of randomness that we might wish to check is the ordering of fluctuations of the data above or below the mean or median of the sample.

Let's consider the following sample and determine whether the data points form a random sequence with regard to being above or below the median value.

| 2 | 5 | 3 | 8 | 4 | 2 | 9 | 3 | 2 | 3 | 7 | 1 | 7 | 3 | 3 |
| 6 | 3 | 4 | 1 | 9 | 5 | 2 | 5 | 5 | 2 | 4 | 3 | 4 | 0 | 4 |

We will test the null hypothesis that this sequence is random using $\alpha = 0.05$.

STEP 1   **a.   Parameter of interest:**

Randomness of the values above or below the median.

b.   Statement of hypotheses:

H_o: The numbers in the sample form a random sequence with respect to the two properties "above" and "below" the median value.

H_a: The sequence is not random.

STEP 2   **a.   Assumptions:**

Each sample data value can be classified as "above" or "below" the median.

b.   Test statistic:

V, the number of runs in the sample data.

c.   Level of significance: $\alpha = 0.05$.

STEP 3   **a.   Sample information:**

The sample data are listed at the beginning of the example.

b.   Test statistic:

First we must rank the data and find the median. The ranked data are

| 0 | 1 | 1 | 2 | 2 | 2 | 2 | 2 | 3 | 3 | 3 | 3 | 3 | 3 | 3 |
| 4 | 4 | 4 | 4 | 4 | 5 | 5 | 5 | 5 | 6 | 7 | 7 | 8 | 9 | 9 |

Since there are 30 data values, the depth of the median is at the $d(\tilde{x}) = 15.5$ position. Thus, $\tilde{x} = \dfrac{3 + 4}{2} = 3.5$. By comparing each number in the original sample to the value of the median, we obtain the following sequence of a's (above) and b's (below):

b a b a a b a b b b a b a b b b a b a b a b a a b a a b a b a b a

We observe $n_a = 15$, $n_b = 15$, and 24 runs. So $V\bigstar = 24$. If n_1 and n_2 are both less than or equal to 20 and a two-tailed test at $\alpha = 0.05$ is desired, then Table 14 in Appendix B is used to complete the hypothesis test.

STEP 4   **Probability distribution:**

Again, we can use either the p-value or the classical procedure:

Using the p-value procedure:

a. Since the concern is for values related to "not random," the test is two-tailed. The p-value is found by finding the probability of the right tail and doubling:

$\mathbf{P} = 2 \times P(V \geq 24$ for $n_a = 15$ and $n_b = 15)$

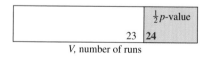

V, number of runs

To find the p-value, you have two options:
1. Use Table 14 (Appendix B) to place bounds on the p-value: $P < 0.05$.
2. Use a computer or calculator to find the p-value: $\mathbf{P} = 0.003$.

Specific instructions follow this example.

b. The p-value is smaller than α.

Using the classical procedure:

a. Since the concern is for values related to "not random," the test is two-tailed. Use Table 14 for two-tailed $\alpha = 0.05$. The critical values are at the intersection of column $n_1 = 15$ and row $n_2 = 15$: 10 and 22. The critical region is $V \leq 10$ or $V \geq 22$.

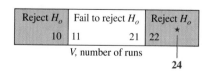

V, number of runs

b. $V\bigstar$ is in the critical region, as shown in the figure above.

STEP 5 a. **Decision:** Reject H_o.

b. **Conclusion:**

We are able to reject the hypothesis of randomness at the 0.05 level of significance and conclude that the sequence is not random with regard to above and below the median.

Calculating the *p*-Value When Using the Runs Test

Method 1: Use Table 14 in Appendix B to place bounds on the *p*-value. By inspecting Table 14 at the intersection of column $n_1 = 15$ and row $n_2 = 15$, you can determine that the *p*-value is less than 0.05; the observed value of $V\bigstar = 24$ is larger than the larger critical value listed.

Method 2: If you are doing the hypothesis test with the aid of a computer or graphing calculator, most likely it will calculate the *p*-value for you. Specific instructions are given on the Chapter 14 Tech Card.

Normal Approximation

To complete the hypothesis test about randomness when n_1 and n_2 are larger than 20 or when α is other than 0.05, we will use *z*, the standard normal random variable. *V* is approximately normally distributed, with a mean of μ_V and standard deviation of σ_V. The formulas for the mean and standard deviation of the *V* statistic and the test statistic $z\bigstar$ follow:

$$\mu_V = \frac{2n_1 \cdot n_2}{n_1 + n_2} + 1 \tag{14.8}$$

$$\sigma_V = \sqrt{\frac{(2n_1 \cdot n_2) \cdot (2n_1 \cdot n_2 - n_1 - n_2)}{(n_1 + n_2)^2(n_1 + n_2 - 1)}} \tag{14.9}$$

$$z\bigstar = \frac{V\bigstar - \mu_V}{\sigma_V} \tag{14.10}$$

Let's test the null hypothesis that the sequence of sample data in Table 14.7 is a random sequence with regard to each data value being odd or even using $\alpha = 0.10$.

*Table 14.7 **Sample Data for Sequence Test***

1	2	3	0	2	4	3	4	8	1
2	1	2	4	3	9	6	2	4	1
5	6	3	3	2	2	1	2	4	2
3	6	3	5	1	7	3	3	0	1
4	4	1	2	7	2	1	7	5	3

STEP 1 a. **Parameter of interest:**

Randomness of odd and even numbers.

b. **Statement of hypotheses:**

H_o: The sequence of odd and even numbers is random.

H_a: The sequence is not random.

STEP 2 a. **Assumptions:**

Each sample value can be classified as either odd or even.

b. **Test statistic:**

V, the number of runs in the sample data.

c. **Level of significance:** $\alpha = 0.10$.

STEP 3 a. **Sample information:**

The data are given in Table 14.7.

b. **Test statistic:**

The sample data, when converted to "o" for odd and "e" for even, become

o e o e e e o e e o e o e e o o e e e o o e o o e

e o e e e o e o o o o o o e o e e o e o e o e o o o o

and reveal $n_o = 26$, $n_e = 24$, and 29 runs, so $V\bigstar = 29$. Now use formulas (14.8), (14.9), and (14.10) to determine the *z*-statistic:

$$\mu_V = \frac{2n_1 \cdot n_2}{n_1 + n_2} + 1:$$

$$\mu_V = \frac{2 \cdot 26 \cdot 24}{26 + 24} + 1$$

$$= 24.96 + 1 = 25.96$$

$$\sigma_V = \sqrt{\frac{(2n_1 \cdot n_2) \cdot (2n_1 \cdot n_2 - n_1 - n_2)}{(n_1 + n_2)^2(n_1 + n_2 - 1)}}$$

$$\sigma_V = \sqrt{\frac{(2 \cdot 26 \cdot 24) \cdot (2 \cdot 26 \cdot 24 - 26 - 24)}{(26 + 24)^2(26 + 24 - 1)}}$$

$$= \sqrt{\frac{(1,248)(1,198)}{(50)^2 \cdot (49)}} = \sqrt{12.20493} = 3.49$$

$$z\bigstar = \frac{V\bigstar - \mu_V}{\sigma_V}:$$

$$z\bigstar = \frac{29 - 25.96}{3.49} = \frac{3.04}{3.49} = \mathbf{0.87}$$

STEP 4 **Probability distribution:**
Again, we can use either the *p*-value or the classical procedure:

Using the *p*-value procedure:

a. A two-tailed test is used:

$$\mathbf{P} = 2 \times P(z > 0.87)$$

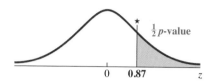

$\frac{1}{2}p\text{-value}$

$0 \quad 0.87 \quad z$

To find the *p*-value, you have three options:
1. Use Table 3 (Appendix B) to calculate the *p*-value: $\mathbf{P} = 2(1.0000 - 0.8079) = 0.3842$.
2. Use Table 5 (Appendix B) to place bounds on the *p*-value: $0.3682 < \mathbf{P} < 0.3954$.
3. Use a computer or calculator to find the *p*-value: $\mathbf{P} = 0.3843$.
For specific instructions, see page 171.

b. The *p*-value is not smaller than α.

Using the classical procedure:

a. A two-tailed test is used. The critical values are obtained from Table 4A:

$$-z(0.05) = -\mathbf{1.65} \text{ and } z(0.05) = \mathbf{1.65}$$

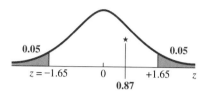

$0.05 \qquad 0.05$

$z = -1.65 \quad 0 \quad +1.65 \quad z$

0.87

Specific instructions for finding critical values are on page 175.

b. $z\bigstar$ is not in the critical region, as shown in red in the figure above.

STEP 5 a. **Decision:** Fail to reject H_o.
b. **Conclusion:**
At the 0.10 level of significance, we are unable to reject the hypothesis of randomness and conclude that these data are a random sequence.

14.5 Rank Correlation

CHARLES SPEARMAN DEVELOPED THE RANK CORRELATION COEFFICIENT IN THE EARLY 1900s.

It is a nonparametric alternative to the linear correlation coefficient (Pearson's product moment, *r*) that was discussed in Chapters 3 and 13.

The **Spearman rank correlation coefficient**, r_s, is found by using this formula:

Spearman Rank Correlation Coefficient

$$r_s = 1 - \frac{6\Sigma(d_i)^2}{n(n^2 - 1)} \tag{14.11}$$

where d_i is the difference in the **paired rankings** and *n* is the number of pairs of data. The value of r_s will range from $+1$ to -1 and will be used in much the same manner as Pearson's linear correlation coefficient, *r*, was used.

The Spearman rank coefficient is defined by using formula (3.1), with data rankings substituted for quantitative *x* and *y* values. The original data may be rankings, or if the data are quantitative, each variable must be ranked separately; then the rankings are used as pairs. If there are no ties in the rankings, formula (14.11) is equivalent to formula (3.1). Formula (14.11) provides us with an easier procedure to use for calculating the r_s statistic.

Assumptions for inferences about rank correlation:

The *n* ordered pairs of data form a random sample, and the variables are ordinal or numerical.

The null hypothesis that we will be testing is: "There is no correlation between the two rankings." The alternative hypothesis may be either two-tailed if there is correlation, or one-tailed if we anticipate either positive or negative correlation. The critical region will be on the side(s) corresponding to the specific alternative that is expected. For example, if we suspect negative correlation, then the critical region will be in the left-hand tail.

Let's consider a hypothetical situation in which four judges rank five contestants in a contest. Let's identify the judges as A, B, C, and D and the contestants as a, b, c, d, and e. Table 14.8 lists the awarded rankings.

Table 14.8 **Rankings for Five Contestants**

Contestant	Judge			
	A	B	C	D
a	1	5	1	5
b	2	4	2	2
c	3	3	3	1
d	4	2	4	4
e	5	1	5	3

When we compare judges A and B, we see that they ranked the contestants in exactly the opposite order:

perfect disagreement (see Table 14.9). From our previous work with correlation, we expect the calculated value for r_s to be exactly −1 for these data. We have:

Table 14.9 **Rankings of A and B**

Contestant	A	B	$d_i = A - B$	$(d_i)^2$
a	1	5	−4	16
b	2	4	−2	4
c	3	3	0	0
d	4	2	2	4
e	5	1	4	16
				40

$$r_s = 1 - \frac{6\Sigma(d_i)^2}{n(n^2 - 1)}:$$

$$r_s = 1 - \frac{(6)(40)}{5(5^2 - 1)} = 1 - \frac{240}{120} = 1 - 2 = -1$$

When judges A and C are compared, we see that their rankings of the contestants are identical (see Table

14.10). We would expect to find a calculated correlation coefficient of $+1$ for these data:

Table 14.10 Rankings of A and C

Contestant	A	C	$d_i = A - C$	$(d_i)^2$
a	1	1	0	0
b	2	2	0	0
c	3	3	0	0
d	4	4	0	0
e	5	5	0	0
				0

$$r_s = 1 - \frac{6\Sigma(d_i)^2}{n(n^2 - 1)}:$$

$$r_s = 1 - \frac{(6)(0)}{5(5^2 - 1)} = 1 - \frac{0}{120} = 1 - 0 = 1$$

By comparing the rankings of judge A with those of judge B and then with those of judge C, we have seen the extremes: total agreement and total disagreement. Now let's compare the rankings of judge A with those of judge D (see Table 14.11). There seems to be no real agreement or disagreement here. Let's compute r_s:

Table 14.11 Rankings of A and D

Contestant	A	D	$d_i = A - D$	$(d_i)^2$
a	1	5	-4	16
b	2	2	0	0
c	3	1	2	4
d	4	4	0	0
e	5	3	2	4
				24

$$r_s = 1 - \frac{6\Sigma(d_i)^2}{n(n^2 - 1)}:$$

$$r_s = 1 - \frac{(6)(24)}{5(5^2 - 1)} = 1 - \frac{144}{120} = 1 - 1.2 = -0.2$$

The result is fairly close to zero, which is what we should have suspected, since there was no real agreement or disagreement.

The test of significance will result in a failure to reject the null hypothesis when r_s is close to zero; the test will result in a rejection of the null hypothesis when r_s is found to be close to $+1$ or -1. The critical values in Table 15 in Appendix B are the positive critical values only. Since the null hypothesis is, "The population correlation coefficient is zero (i.e., $\rho_s = 0$)," we have a symmetrical test statistic. Hence we need only add a plus or minus sign to the value found in the table, as appropriate. The sign is determined by the specific alternative that we have in mind.

When there are only a few ties, it is common practice to use formula (14.11). Even though the resulting value of r_s is not exactly equal to the value that would occur if formula (3.1) were used, it is generally considered to be an acceptable estimate. The example below on how quickly students complete exams shows the procedure for handling ties and uses formula (14.11) for the calculation of r_s.

When ties occur in either set of the ordered pairs of rankings, assign each tied observation the mean of the ranks that would have been assigned had there been no ties, as was done for the Mann–Whitney U test (see Objective 14.4).

Let's look at an example. Students who finish exams more quickly than the rest of the class are often thought to be smarter. Table 14.12 presents the scores and order of finish for 12 students on a recent 1-hour exam. At the 0.01 level, do these data support the alternative hypothesis that the first students to complete an exam have higher grades?

*Table 14.12 Data on Exam Scores

Order of Finish	1	2	3	4	5	6
Exam Score	90	78	76	60	92	86
Order of Finish	7	8	9	10	11	12
Exam Score	74	60	78	70	68	64

STEP 1 a. **Parameter of interest:**
The rank correlation coefficient between score and order of finish, ρ_s.

b. **Statement of hypotheses:**

H_o: Order of finish has no relationship to exam score.

H_a: The first to finish tend to have higher grades.

a. Assumptions:

The 12 ordered pairs of data form a random sample; order of finish is an ordinal variable and test score is numerical.

b. **Test statistic:**

The Spearman rank correlation coefficient, r_s.

c. **Level of significance:** $\alpha = 0.01$ for a one-tailed test.

STEP3 a. **Sample information:**

The data are given in Table 14.12.

b. **Test statistic:**

Rank the scores from highest to lowest, assigning the highest score the rank number 1, as shown. (Order of finish is already ranked.)

92	90	86	78	78	76	74	70	68	64	60	60
1	2	3	4	5	6	7	8	9	10	11	12
			4.5	4.5						11.5	11.5

The rankings and preliminary calculations are shown in Table 14.13.

Table 14.13 **Rankings of Test Scores and Differences**

Order of Finish	Test Score Rank	Difference (d_i)	$(d_i)^2$
1	2	−1	1.00
2	4.5	−2.5	6.25
3	6	−3	9.00
4	11.5	−7.5	56.25
5	1	4	16.00
6	3	3	9.00
7	7	0	0.00
8	11.5	−3.5	12.25
9	4.5	4.5	20.25
10	8	2	4.00
11	9	2	4.00
12	10	2	4.00
			142.00

Using formula (14.11), we obtain

$$r_s = 1 - \frac{6\Sigma(d_i)^2}{n(n^2 - 1)}:$$

$$r_s = 1 - \frac{(6)(142.0)}{12(12^2 - 1)}$$

$$= 1 - \frac{852}{1{,}716} = 1 - 0.497 = 0.503$$

Thus, $r_s\bigstar = 0.503$.

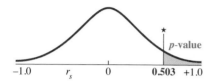

STEP 4 Probability distribution:

Again, we can use either the *p*-value or the classical procedure:

Using the *p*-value procedure:

a. Since the concern is for positive values, the *p*-value is the area to the right:

$$\mathbf{P} = P(r_s \geq 0.503 \text{ for } n = 12)$$

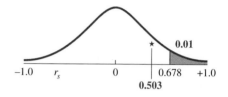

To find the *p*-value, you have two options:

1. Use Table 15 (Appendix B) to place bounds on the *p*-value. Table 15 lists only two-tailed α (this test is one-tailed, so divide the column heading by 2): **0.025 < P < 0.05**.
2. Use a computer or calculator to find the *p*-value: **P = 0.048**.

Specific instructions follow this example.

b. The *p*-value is not smaller than α.

Using the classical procedure:

a. The critical region is one-tailed because H_a expresses concern for values related to "positive." Since the table is for two-tailed, the critical value is located at the intersection of the α = 0.02 column (α = 0.01 in each tail) and the *n* = 12 row of Table 15: **0.678**.

b. $r_s\bigstar$ is not in the critical region, as shown in the figure above.

STEP 5 a. **Decision:** Fail to reject H_o.

b. **Conclusion:**

These sample data do not provide sufficient evidence to enable us to conclude that the first students to finish have higher grades, at the 0.01 level of significance.

Calculating the *p*-Value for the Spearman Rank Correlation Test

Method 1: Use Table 15 in Appendix B to place bounds on the *p*-value. By inspecting the *n* = 12 row of Table 15, you can determine an interval within which the *p*-value lies. Locate the value of r_s along the *n* = 12 row and read the bounds from the top of the table. Table 15 lists only two-tailed values (therefore, you must divide by 2 for a one-tailed test). We find **0.025 < P < 0.05**.

Method 2: If you are doing the hypothesis test with the aid of a computer or graphing calculator, most likely it will calculate the *p*-value for you.

RECAP

After reading this chapter, you should now understand some of the basic assumptions that are needed when the parametric techniques of the earlier chapters are encountered. You now have seen a variety of tests, many of which somewhat duplicate the job done by others. Keep in mind that you should use the best test for your particular needs.

problems

Objective 14.2

14.1 a. Describe why the sign test may be the easiest test procedure of all to use.

b. What population parameter can be tested using the sign test? What property of that parameter allows the sign test to be used? Explain.

*14.2 Ten randomly selected shut-ins were each asked how many hours of television they watched last week. The results are as follows:

82	66	90	84	75	88	80	94	110	91

Determine the 90% confidence interval estimate for the median number of hours of television watched per week by shut-ins.

*14.3 The following daily high temperatures (°F) were recorded in the city of Rochester, New York, on 20 randomly selected days in December.

47	46	40	40	46	35	34	59	54	33
65	39	48	47	48	46	42	36	45	38

Use the sign test to determine the 95% confidence interval for the median daily high temperature in Rochester, New York, during December.

*14.4 A sample of the daily rental-car rates for a compact car was collected in order to estimate the average daily cost of renting a compact car.

39.93	41.00	42.99	38.99	42.93	35.00	40.95	29.99	49.93	50.95
34.95	28.99	43.93	43.00	41.99	42.99	36.93	34.95	35.99	31.99
45.93	46.50	34.90	29.80	32.93	29.70	32.99	27.94	53.93	46.00
35.94	34.99	29.93	28.70	34.99	31.48	37.93	37.90	37.92	35.99

Find the 99% confidence interval for the median daily rental cost.

14.5 State the null hypothesis, H_o, and the alternative hypothesis, H_a, that would be used to test the following statements:

a. The median length of vacation time is less than 18 days.

b. The median value is at least 32.

c. The median tax rate is 4.5%.

14.6 In "The Annual Report on the Economic Status of the Profession" done by American Association of University Professors, the mean salary of a full professor was reported to be $83,282. The following table lists the average salary (in dollars for a random sample of institutions in Colorado.

54,500	63,000	83,600	67,000	49,700	60,800	47,700	82,200	86,800	73,900
57,700	58,200	62,200	82,000	78,500	70,000	96,100	89,700	57,200	55,400

Using the computer output that follows, test the claim that the median salary of full professors in Colorado is lower than the mean for the whole country by writing the hypotheses and verifying the number below and the number above the median. Complete the test by using the p-value given and $\alpha = 0.05$. Verify the given p-value using Table 12 in Appendix B.

Sign Test for Median: C1					
Sign test of median $= 83282$ versus < 83292					
	N	Below	Equal	Above	P
C1	20	16	0	4	0.0059

14.7 Determine the p-value for the following hypothesis tests involving the sign test:

a. $H_o: P(+) = 0.5$ vs. $H_a: P(+) \neq 0.5$, with $n = 18$ and $x\bigstar = n(-) = 3$

b. $H_o: P(+) = 0.5$ vs. $H_a: P(+) > 0.5$, with $n = 78$ and $x\bigstar = n(-) = 30$

c. $H_o: P(+) = 0.5$ vs. $H_a: P(+) < 0.5$, with $n = 38$ and $x\bigstar = n(+) = 10$

d. $H_o: P(+) = 0.5$ vs. $H_a: P(+) \neq 0.5$, with $n = 148$ and $z\bigstar = -2.56$

14.8 Determine the critical value that would be used to test the null hypothesis for the following situations using the classical approach and the sign test:

a. $H_o: P(+) = 0.5$ vs. $H_a: P(+) \neq 0.5$, with $n = 18$ and $\alpha = 0.05$

b. $H_o: P(+) = 0.5$ vs. $H_a: P(+) > 0.5$, with $n = 78$ and $\alpha = 0.05$

c. $H_o: P(+) = 0.5$ vs. $H_a: P(+) < 0.5$, with $n = 38$ and $\alpha = 0.05$

d. $H_o: P(+) = 0.5$ vs. $H_a: P(+) \neq 0.5$, with $n = 148$ and $\alpha = 0.05$

14.9 An article titled "Graft-versus-Host Disease" (http://www.nature.com/bmt/journal/v35/n10/abs/1704957a.html) gives the median age of 42 years for the 87 patients with acute myeloid leukemia (AML) who received hematopoietic stem cell transplants from unrelated donors after standard conditioning. Clinical outcome after the use of two different antithymocyte globulins (ATG) for the prevention of graft-versus-host disease (GvHD) were then analyzed. Suppose that a sample of 100 patients with AML was recently selected for a study and it found that 40 of the patients were older than 42 and 60 were younger than 42 years of age. Test the null hypothesis that the median age of the population from which the 100 patients were selected equals 42 years versus the alternative that the median does not equal 42 years. Use $\alpha = 0.05$.

*14.10 Every year sixth-grade students in Ohio schools take proficiency tests. The list of sixth-grade reading score changes from the prior year are given below. Negative values indicate a decrease in score, positive values show an increase, and a zero shows no change from the prior year. Use the sign test to test the hypothesis that, "On average, reading scores have decreased from the prior year." Use $\alpha = 0.05$.

−4	−4	−10	−9	−30	6	18	−3	2	−5	−6
−12	−9	1	−1	−2	19	6	−1	−14	−13	5
12	−8	6	−3	−8	−14	−16	−6	2	0	16
−7	6	−11	6	−8	−4	13	9	−12	12	−10

*14.11 Part of the results from the Third International Mathematics and Science Study was a comparison of eighth-grade science achievement by nation from 1999, 2003, and 2007. The table below gives the average scores for the nations with all three scores.

Nation	1999	2003	2007
Bulgaria	518	479	470
Cyprus	460	441	452
Hong Kong	530	556	530
Hungary	552	543	539
Iran, Islamic Republic of	448	453	459
Japan	550	552	554
Korea, Republic of	549	558	553
Lithuania	488	519	519
Romania	472	470	462
Russian Federation	529	514	530
Singapore	568	578	567
United States	515	527	520

a. Construct a table showing the sign of the difference between the years 1999 and 2003 for each country.

b. Using $\alpha = 0.05$, has there been a significant improvement in science scores?

c. Repeat parts (a) and (b) for the years 2003 and 2007.

14.12 A blind taste test was used to determine people's preference for the taste of the "classic" cola and "new" cola. The results were as follows:

645 preferred the new

583 preferred the old

272 had no preference

Is the preference for the taste of the new cola significantly greater than one-half? Use $\alpha = 0.01$.

14.13 According to an American Optometric Association survey, 57% of adults wear eyeglasses as their type of corrective lenses. Suppose we wish to test the null hypothesis "one-half of college students wear 'eyeglasses' as their type of corrective lenses" against the alternative that the proportion is greater than one-half. Let + represent "wear eyeglasses" and − represent "some other corrective lenses or none." If a random sample of 1,500 students is tested, what value of x, the number of the least frequent sign, will be the critical value at the 0.05 level of significance?

Objective 14.3

14.14 State the null hypothesis, H_o, and the alternative hypothesis, H_a, that would be used to test the following statements:

a. The students in the new reading program scored higher on the comprehension test than the students in the traditional reading programs.

b. Men on the grapefruit diet lose more weight than men not on the grapefruit diet.

c. There is no difference in growth between the use of the two fertilizers.

14.15 Determine the p-value that will result when testing the following hypotheses for experiments involving two independent samples:

a. H_o: Average(A) = Average(B)
H_a: Average(A) > Average(B)
with $n_A = 18$, $n_B = 15$, and $U = 95$.

b. H_o: Average(I) = Average(II)
H_a: Average(I) $\neq$ Average(II)
with $n_I = 8$, $n_{II} = 10$, and $U = 13$.

c. H_o: The average height is the same for both groups.
H_a: Group I average heights are less than those for group II
with $n_I = 50$, $n_{II} = 45$, and $z = -2.37$.

14.16 Determine the critical value that would be used to test the following hypotheses for experiments involving two independent samples, using the classical method.

a. H_o: Average(A) = Average(B)
H_a: Average(A) > Average(B)
with $n_A = 78$, $n_B = 45$, and $\alpha = 0.05$

b. H_o: The average score is the same for both groups.
H_a: Group I average scores are less than those for group II
with $n_I = 18$, $n_{II} = 15$, and $\alpha = 0.05$

*14.17 Given the following data from Group 1 and Group 2:

Group 1	30	35	40	42	45	36
Group 2	25	32	27	39	30	

a. Combine the two datasets in ranked order. Compute the sum of the ranks for Group 1, R_1. Compute the sum of the ranks for Group 2, R_2.

b. Compute the U score for each group, U_1 and U_2, and determine the $U\bigstar$ statistic.

c. Use the Mann–Whitney U statistic to test the null hypothesis that the average is the same for both groups versus the alternative that the Group 1 average is greater than the Group 2 average.

*14.18 An article in the *International Journal of Sports Medicine* discusses the use of the Mann–Whitney U test to compare the total cholesterol (mg/dL) of 35 adipose (obese) boys with that of 27 adipose girls. No significant difference was found between the two groups with respect to total cholesterol. A similar study involving six adipose boys and eight adipose girls gave the following total cholesterol values.

Adipose Boys	175	185	160	200	170	150		
Adipose Girls	160	190	175	190	185	150	140	195

Use the Mann–Whitney U test to test the research hypothesis that the total cholesterol values differ for the two groups, using the 0.05 level of significance.

*14.19 Ten North Carolina and 13 Texas peanut-producing counties were randomly identified and the 2008 peanut yield rate, in pounds of peanuts harvested per acre, was recorded.

NC County	NC Yield	TX County	TX Yield
Edgecombe	3,360	Donley	3,640
Hertford	3,560	Terry	3,335
Northhampton	3,700	Collingsworth	2,555
Greene	3,815	Cochran	3,120
Pitt	3,530	Frio	3,685
Bladen	4,265	Yoakum	3,530
Robeson	3,750	Bailey	3,120
Chowan	4,000	Wheeler	2,880
Halifax	3,310	Hall	3,700
Nash	3,435	Hockley	3,280
		Andrews	3,665
		Gaines	3,845
		Dawson	3,565

SOURCE: http://www.nass.usda.gov/

Use the Mann–Whitney U statistic to test the hypothesis that the average yield is different for the two states. Use $\alpha = 0.05$.

14.20 The Oregon Health & Science University's news website (http://www.ohsu.edu/) gives information on a study that found that some commercial cigarette brands contain 10 to 20 times higher percentages of nicotine in the "free-base" form, that is, the form thought to be the most addictive. Consider another study designed to compare the nicotine content of two different brands of cigarettes. The nicotine content was determined for 25 cigarettes of brand A and 25 cigarettes of brand B. The sum of ranks for brand A equals 688, and the sum of ranks for brand B equals 587. Use the Mann–Whitney U statistic to test the null hypothesis that the average nicotine content is the same for the two brands versus the alternative that the average nicotine content differs. Use $\alpha = 0.01$.

*14.21 As part of a study to determine whether cloud seeding increased rainfall, clouds were randomly seeded with silver nitrate, or were not seeded. The amounts of rainfall that followed are listed here.

Unseeded	4.9	41.1	21.7	372.4	26.3	17.3	36.6	26.1
	47.3	95.0	147.8	321.2	11.5	68.5	29.0	24.4
	1,202.6	87.0	28.6	830.1	81.2	4.9	163.0	345.5
	244.3							
Seeded	129.6	334.1	274.7	198.6	430.0	274.7	31.4	115.3
	1,656.0	118.3	489.1	302.8	255.0	32.7	119.0	17.5
	242.5	2,745.6	7.7	40.6	978.0	200.7	703.4	92.4
	1,697.8							

Do these data show that cloud seeding significantly increases the average amount of rainfall? Use $\alpha = 0.05$.

*14.22 Are more hours spent watching sporting events on television or watching reality shows? A community college student hypothesized that males more likely watch sports and female students are more likely to watch reality shows. With this premise, the student randomly gathered data from 30 males and 30 females at his community college on the hours of television watched in a week.

Hours Males Spent Watching Sports on Television														
4	10	15	26	10	20	13	4	5	3	1	20	60	35	3
6	10	26	3	0	15	5	8	8	6	14	15	3	2	4

Hours Females Spent Watching Reality Shows on Television										
2	10	5	8	10	3	4	3	3	2	3
3	1	14	2	4	5	32.5	6	5	20	1
3	10	6	7	15	2	20	12			

a. Do these data show that more time is spent by males watching sporting events than females watching reality shows in a week? Use a 0.05 level of significance.

b. Comment on the meaning of the relationship of α and the p-value.

Objective 14.4

14.23 State the null hypothesis, H_o, and the alternative hypothesis, H_a, that would be used to test the following statements:

a. The data did not occur in a random order about the median.

b. The sequence of odd and even is not random.

c. The gender of customers entering a grocery store was recorded; the entry is not random in order.

14.24 Determine the p-value that would be used to complete the following runs tests:

a. H_o: The sequence of gender of customers coming into the gym was random.
   H_a: The sequence was not random.
   with $n(A) = 10$, $n(B) = 12$, and $V = 5$.

b. H_o: The home prices collected occurred in random order above and below the median.
   H_a: The home prices did not occur in random order.
   with $z = 1.31$.

14.25 Determine the critical values that would be used to complete the following runs tests using the classical approach:

a. H_o: The results collected occurred in random order above and below the median.
   H_a: The results were not random.
   with $n(A) = 14$, $n(B) = 15$, and $\alpha = 0.05$.

b. H_o: The two properties alternated randomly.
   H_a: The two properties did not occur in random fashion.
   with $n(I) = 78$, $n(II) = 45$, and $\alpha = 0.05$.

14.26 Jessica did not believe she was playing a game with a fair die. She thought that if the die was fair, the tossing of the die should result in a random order of even and odd output. She performed her experiment 14 times. After each toss, Jessica recorded the results. The following data were reported (E = 2, 4, 6; O = 1, 3, 5).

O	E	O	O	O	O	O	E	E	O	O	O	E	E	O

Use the runs test at a 5% level of significance to test the claim that the results reported are random.

14.27 A manufacturing firm hires both men and women. The following shows the gender of the last 20 individuals hired (M = male, F = female).

M	M	F	M	F	F	F	M	M	M	M	M	M	F	M	M	F	M	M	M

At the $\alpha = 0.05$ level of significance, are we correct in concluding that this sequence is not random?

14.28 A student was asked to perform an experiment that involved tossing a coin 25 times. After each toss, the student recorded the results. The following data were reported (H = heads, T = tails).

H	T	H	T	H	T	H	T	H	H	H	T	T	H	H	T	T	H	T	H	T	H	T	H	T	H

Use the runs test at a 5% level of significance to test the student's claim that the results reported are random.

*14.29 The following data were collected in an attempt to show that the number of minutes the city bus is late is steadily growing larger. The data are in order of occurrence.

Minutes:	6	1	3	9	10	10	2	5	5	6	12	3	7	8	9	4	5	8	11	14

At $\alpha = 0.05$, do these data show sufficient lack of randomness to support the claim?

*14.30 According to a new survey from the Boston Indicators Project website, Boston city schools had an average of 3.6 students per computer for the 2007–2008 school year. The average was the same for the state of Massachusetts but higher than the average for any of the suburban districts. A sample of Boston city schools reported their average number of students per computer in the sequence that follows:

3.5	2.6	3.8	5.7	2.6	3.4	2.7	4.6	3.4	3.6	4.2	3.7	4.6	2.9

a. Determine the median and the number of runs above and below the median.
b. Use the runs test to test these data for randomness about the median. Use $\alpha = 0.05$.

*14.31 On June 24, 2009, the 2008 American Time Use Survey was released by the Bureau of Labor Statistics. Within the many statistics provided was information on leisure activities, including the average amount of time spent in various categories. For 15- to 19-year-olds, the average time spent using the computer for leisure or playing games was 42 minutes per day. Suppose 20 15- to 19-year-olds were randomly selected and monitored for a day and the number of minutes they spent on these leisure activities was recorded. The resulting sequence of times was given:

50	59	16	34	43	47	46	27	43	12
45	50	51	89	63	42	23	39	43	28

a. Determine the median and the number of runs above and below the median.
b. Use the runs test to test these data for randomness about the median.
c. State your conclusion.

14.32 According to an August 26, 2008, press release from the U.S. Census Bureau, the median household income for 2007 was $50,023. A random sample of 250 incomes shows a median value different from any of the 250 incomes in the sample. The data contains 105 runs above and below the median. Use the above information to test the null hypothesis that the incomes in the sample form a random sequence with respect to the two properties above and below the median value versus the alternative that the sequence is not random at $\alpha = 0.05$.

14.33 The number of absences recorded at a lecture that met at 8 A.M. on Mondays and Thursdays last semester were (in order of occurrence):

					n(absences)							
5	16	6	9	18	11	16	21	14	17	12	14	10
6	8	12	13	4	5	5	6	1	7	18	26	6

Do these data show a randomness about the median value at $\alpha = 0.05$? Complete this test by using (a) critical values in Table 14 in Appendix B and (b) the standard normal distribution.

Objective 14.5

14.34 State the null hypothesis, H_o, and the alternative hypothesis, H_a, that would be used to test the following statements:
a. There is no relationship between the two rankings.
b. The two variables are unrelated.
c. There is a positive correlation between the two variables.
d. Refrigerator age has a decreasing effect on monetary value.

14.35 Determine the p-value that would be used to test the null hypothesis for the following Spearman rank correlation experiments:
a. H_o: No relationship between the two variables.
 H_a: There is a positive relationship.
 with $n = 21$ and $r_s = 0.55$.
b. H_o: No correlation.
 H_a: There is a relationship.
 with $n = 27$ and $r_s = 0.71$.
c. H_o: Variable A has no effect on variable B.
 H_a: Variable B decreases as variable A increases.
 with $n = 10$ and $r_s = -0.62$.

14.36 Determine the test criteria that would be used to test the null hypothesis for the following Spearman rank correlation experiments:

a. H_o: No relationship between the two variables.
H_a: There is a relationship.
with $n = 14$ and $\alpha = 0.05$.

b. H_o: No correlation.
H_a: Positively correlated.
with $n = 27$ and $\alpha = 0.05$.

c. H_o: Variable A has no effect on variable B.
H_a: Variable B decreases as variable A increases.
with $n = 18$ and $\alpha = 0.01$.

*14.37 When it comes to getting workers to produce, money is not everything; feeling appreciated is more important. Do the rankings assigned by workers and the boss show a significant difference in what each person thinks is important? (Ratings: 1 = most important; 10 = least important.) Test using $\alpha = 0.05$.

Component of Job Satisfaction	Worker Ranking	Boss Ranking
Full appreciation of work done	1	8
Feeling of being in on things	2	10
Sympathetic help on personal problems	3	9
Job security	4	2
Good wages	5	1
Interesting work	6	5
Promotion and growth in the organization	7	3
Personal loyalty to employees	8	6
Good working conditions	9	4
Tactful disciplining	10	7

SOURCE: *Philadelphia Inquirer*

*14.38 Consumer product testing groups commonly supply ratings of all sorts of products to consumers in an effort to assist them in their purchase decisions. Different manufacturers' products are usually tested for their performance and then given an overall rating. *PC World* ranked the top ten 17-inch computer monitors and also supplied the street price (dollars). The ranks of each are shown in the table that follows, with the highest-priced monitor given a rank of 1 and the lowest a rank of 10.

Overall Rating	Street Price Rank	Overall Rating	Street Price Rank
1	3	6	2
2	4	7	8.5
3	6.5	8	6.5
4	8.5	9	10
5	5	10	1

SOURCE: *PC World*

a. Compute the Spearman rank correlation coefficient for the overall rating and the street price of the 17-inch monitors.

b. Does a higher price yield a higher rating? Test the null hypothesis that there is no relationship between the overall ratings of the monitors and their street prices versus the alternative that there is a positive relationship between them. Use $\alpha = 0.05$.

*14.39 What can I do with a two-year degree from a community college? Consider the information taken

from the September 2, 2009, *USA Today* article titled "Health Care, Energy among 'Hot' Jobs."

Top Jobs	% Growth	Median Income
Physical therapist assistants	32.4	41,360
Dental hygienists	30.1	62,800
Environmental technicians	28.0	38,090
Cardiovascular technicians	25.5	42,300
Occupational therapist asst.	25.4	42,060
Radiation therapists	24.8	66,170
Environmental eng. techn.	24.8	40,560
Court reporters	24.5	45,610
Registered nurses	23.5	57,280
Computer specialists	15.1	68,570

a. Rank the percentage growth and median incomes in ascending order (high to low).

b. Compute the Spearman rank correlation coefficient for the two rankings.

c. At the 0.05 level of significance, determine whether there is a significant relationship between a job's growth and its corresponding median income.

d. Discuss the relationship between a job's growth and its corresponding income.

*14.40 The following data represent the ages of 12 subjects and the mineral concentration (in parts per million) in their tissue samples.

Age, x	82	83	64	53	47	50	70	62	34	27	75	28
Mineral Concentration, y	170	40	64	5	15	5	48	34	3	7	50	10

Refer to the following MINITAB output and verify that the Spearman rank correlation coefficient equals 0.753 by calculating it yourself.

Correlations: xRank, yRank
Correlation of xRank and yRank = 0.753,
p-Value = 0.005

*14.41 Many people are concerned about eating foods that have high sodium content. They are also advised of the benefits of obtaining sufficient fiber in their diets. Do foods high in fiber tend to have more sodium? The following table was obtained by selecting 11 soups from a list published in *Nutrition Action Healthletter*. The soups were measured on the basis of both sodium content and fiber:

Soup	Sodium	Fiber	Soup	Sodium	Fiber
A	480	12	G	420	2
B	830	0	H	290	4
C	510	1	I	450	10
D	460	5	J	430	6
E	490	3	K	390	9
F	580	7			

SOURCE: *Nutrition Action Healthletter*

a. Rank the soups in ascending order on the basis of their sodium content and on their fiber content. Show your results in a table.

b. Compute the Spearman rank order correlation coefficient for the two sets of rankings.

c. Does higher sodium content accompany foods that are higher in fiber? Test the null hypothesis that there is no relationship between the fiber and sodium content of the soups—versus the alternative that there is a relationship between them. Use $\alpha = 0.05$.

*14.42 The *Journal of Professional Nursing* article titled "The Graduate Record Examination as an Admission Requirement for the Graduate Nursing Program" reported a significant correlation between undergraduate grade point average (GPA) and GPA at graduation from a graduate nursing program. The following data were collected on 10 nursing students who graduated from a graduate nursing program.

Undergraduate GPA	3.5	3.1	2.7	3.7	2.5	3.3	3.0	2.9	3.8	3.2
GPA at Graduation	3.4	3.2	3.0	3.6	3.1	3.4	3.0	3.4	3.7	3.8

Compute the Spearman rank coefficient and test the null hypothesis of no relationship versus a positive relationship. Use a level of significance equal to 0.05.

*14.43 The November 24, 2009, *USA Today* article "Flier Satisfaction with Airlines Goes up a Bit" reported the results from a Zagat Survey of 5,900 fliers who averaged 17 flights a year. The survey rated airlines on a 30-point scale. The airline ratings for coach passengers on flights within the United States are given in the following table.

Airline	Comfort	Service	Web
Midwest	23	22	18
Virgin America	23	24	23
JetBlue	23	22	22
Alaska	17	20	21
Hawaiian	16	19	19
Continental	15	17	22
Southwest	16	21	23
Frontier	16	18	16
AirTran	14	15	18
Delta	13	13	19
American	12	13	20
United	12	12	19
US Airways	11	10	15
Spirit	11	10	14

a. Construct a new table ranking the percentages for comfort, service, and Web separately.

b. Using Spearman's rank correlation and a 0.05 level of significance, determine if there is a relationship between comfort and service.

c. Using Spearman's rank correlation and a 0.05 level of significance, determine if there is a relationship between comfort and Web.

d. Using Spearman's rank correlation and a 0.05 level of significance, determine if there is a relationship between service and Web.

e. Review the results of parts (b), (c), and (d), and comment on your combined findings.

*14.44 "Survey of Home Buyer Preferences" was conducted by the National Association of Home Builders to determine the features that home buyers really want. Respondents were to rate each feature desirable (as well as if essential). The following table shows the results.

Feature	Desirable	Essential
Laundry room	40	52
Linen closet	56	32
Exhaust fan	44	42
Dining room	43	36
Walk-in pantry	59	19
Island work area	55	16
Separate shower enclosure	49	20
Temperature control faucets	49	18
Whirlpool tub	46	12
White bathroom fixtures	40	16
Ceramic wall tiles	43	12
Solid-surface countertops	48	7
Den/library	43	11
Wood burning fireplace	39	15
Special use storage	47	6

SOURCE: National Association of Home Builders

It is not surprising that the ratings in the "desireable" column of the table are considerably higher than the ratings in the "essentials" column. There is no question about there being a difference in the ratings; however, an appropriate question is, "Do the items on the list appear in the same order of preference in both columns?"

a. Use the Mann–Whitney U test to test the hypothesis that the items follow essentially the same distribution using $\alpha = 0.05$.

b. Use the Spearman rank correlation coefficient to test the hypothesis that the rankings of the items are not correlated using $\alpha = 0.05$.

c. State your conclusion.

*remember

P roblems marked with an asterisk (*) have data sets available on the CourseMate for STAT2 site. Login at cengagebrain.com.

Introductory
Concepts and Review Lessons

Appendix A is available online at <u>cengagebrain.com</u>.

Tables

Table 1 **Random Numbers**

10 09 73 25 33	76 52 01 35 86	34 67 35 48 76	80 95 90 91 17	39 29 27 49 45
37 54 20 48 05	64 89 47 42 96	24 80 52 40 37	20 63 61 04 02	00 82 29 16 65
08 42 26 89 53	19 64 50 93 03	23 20 90 25 60	15 95 33 43 64	35 08 03 36 06
99 01 90 25 29	09 37 67 07 15	38 31 13 11 65	88 67 67 43 97	04 43 62 76 59
12 80 79 99 70	80 15 73 61 47	64 03 23 66 53	98 95 11 68 77	12 17 17 68 33
66 06 57 47 17	34 07 27 68 50	36 69 73 61 70	65 81 33 98 85	11 19 92 91 70
31 06 01 08 05	45 57 18 24 06	35 30 34 26 14	86 79 90 74 39	23 40 30 97 32
85 26 97 76 02	02 05 16 56 92	68 66 57 48 18	73 05 38 52 47	18 62 38 85 79
63 57 33 21 35	05 32 54 70 48	90 55 35 75 48	28 46 82 87 09	83 49 12 56 24
73 79 64 57 53	03 52 96 47 78	35 80 83 42 82	60 93 52 03 44	35 27 38 84 35
98 52 01 77 67	14 90 56 86 07	22 10 94 05 58	60 97 09 34 33	50 50 07 39 98
11 80 50 54 31	39 80 82 77 32	50 72 56 82 48	29 40 52 42 01	52 77 56 78 51
83 45 29 96 34	06 28 89 80 83	13 74 67 00 78	18 47 54 06 10	68 71 17 78 17
88 68 54 02 00	86 50 75 84 01	36 76 66 79 51	90 36 47 64 93	29 60 91 10 62
99 59 46 73 48	87 51 76 49 69	91 82 60 89 28	93 78 56 13 68	23 47 83 41 13
65 48 11 76 74	17 46 85 09 50	58 04 77 69 74	73 03 95 71 86	40 21 81 65 44
80 12 43 56 35	17 72 70 80 15	45 31 82 23 74	21 11 57 82 53	14 38 55 37 63
74 35 09 98 17	77 40 27 72 14	43 23 60 02 10	45 52 16 42 37	96 28 60 26 55
69 91 62 68 03	66 25 22 91 48	36 93 68 72 03	76 62 11 39 90	94 40 05 64 18
09 89 32 05 05	14 22 56 85 14	46 42 75 67 88	96 29 77 88 22	54 38 21 45 98
91 49 91 45 23	68 47 92 76 86	46 16 28 35 54	94 75 08 99 23	37 08 92 00 48
80 33 69 45 98	26 94 03 68 58	70 29 73 41 35	54 14 03 33 40	42 05 08 23 41
44 10 48 19 49	85 15 74 79 54	32 97 92 65 75	57 60 04 08 81	22 22 20 64 13
12 55 07 37 42	11 10 00 20 40	12 86 07 46 97	96 64 48 94 39	28 70 72 58 15
63 60 64 93 29	16 50 53 44 84	40 21 95 25 63	43 65 17 70 82	07 20 73 17 90
61 19 69 04 46	26 45 74 77 74	51 92 43 37 29	65 39 45 95 93	42 58 26 05 27
15 47 44 52 66	95 27 07 99 53	59 36 78 38 48	82 39 61 01 18	33 21 15 94 66
94 55 72 85 73	67 89 75 43 87	54 62 24 44 31	91 19 04 25 92	92 92 74 59 73
42 48 11 62 13	97 34 40 87 21	16 86 84 87 67	03 07 11 20 59	25 70 14 66 70
23 52 37 83 17	73 20 88 98 37	68 93 59 14 16	26 25 22 96 63	05 52 28 25 62
04 49 35 24 94	75 24 63 38 24	45 86 25 10 25	61 96 27 93 35	65 33 71 24 72
00 54 99 76 54	64 05 18 81 59	96 11 96 38 96	54 69 28 23 91	23 28 72 95 29
35 96 31 53 07	26 89 80 93 54	33 35 13 54 62	77 97 45 00 24	90 10 33 93 33
59 80 80 83 91	45 42 72 68 42	83 60 94 97 00	13 02 12 48 92	78 56 52 01 06
46 05 88 52 36	01 39 09 22 86	77 28 14 40 77	93 91 08 36 47	70 61 74 29 41
32 17 90 05 97	87 37 92 52 41	05 56 70 70 07	86 74 31 71 57	85 39 41 18 38
69 23 46 14 06	20 11 74 52 04	15 95 66 00 00	18 74 39 24 23	97 11 89 63 38
19 56 54 14 30	01 75 87 53 79	40 41 92 15 85	66 67 43 68 06	84 96 28 52 07
45 15 51 49 38	19 47 60 72 46	43 66 79 45 43	59 04 79 00 33	20 82 66 95 41
94 86 43 19 94	36 16 81 08 51	34 88 88 15 53	01 54 03 54 56	05 01 45 11 76
98 08 62 48 26	45 24 02 84 04	44 99 90 88 96	39 09 47 34 07	35 44 13 18 80
33 18 51 62 32	41 94 15 09 49	89 43 54 85 81	88 69 54 19 94	37 54 87 30 43
80 95 10 04 06	96 38 27 07 74	20 15 12 33 87	25 01 62 52 98	94 62 46 11 71
79 75 24 91 40	71 96 12 82 96	69 86 10 25 91	74 85 22 05 39	00 38 75 95 79
18 63 33 25 37	98 14 50 65 71	31 01 02 46 74	05 45 56 14 27	77 93 89 19 36

Specific details about using this table can be found on page 14; Appendix A on page 326

Table 1 (Continued)

54 17 84 56 11	80 99 33 71 43	05 33 51 29 69	56 12 71 92 55	36 04 09 03 24
11 66 44 98 83	52 07 98 48 27	59 38 17 15 39	09 97 33 34 40	88 46 12 33 56
48 32 47 79 28	31 24 96 47 10	02 29 53 68 70	32 30 75 75 46	15 02 00 99 94
69 07 49 41 38	87 63 79 19 76	35 58 40 44 01	10 51 82 16 15	01 84 87 69 38
09 18 82 00 97	32 82 53 95 27	04 22 08 63 04	83 38 98 73 74	64 27 85 80 44
90 04 58 54 97	51 98 15 06 54	94 93 88 19 97	91 87 07 61 50	68 47 66 46 59
73 18 95 02 07	47 67 72 62 69	62 29 06 44 64	27 12 46 70 18	41 36 18 27 60
75 76 87 64 90	20 97 18 17 49	90 42 91 22 72	95 37 50 58 71	93 82 34 31 78
54 01 64 40 56	66 28 13 10 03	00 68 22 73 98	20 71 45 32 95	07 70 61 78 13
08 35 86 99 10	78 54 24 27 85	13 66 15 88 73	04 61 89 75 53	31 22 30 84 20
28 30 60 32 64	81 33 31 05 91	40 51 00 78 93	32 60 46 04 75	94 11 90 18 40
53 84 08 62 33	81 59 41 36 28	51 21 59 02 90	28 46 66 87 95	77 76 22 07 91
91 75 75 37 41	61 61 36 22 69	50 26 39 02 12	55 78 17 65 14	83 48 34 70 55
89 41 59 26 94	00 39 75 83 91	12 60 71 76 46	48 94 97 23 06	94 54 13 74 08
77 51 30 38 20	86 83 42 99 01	68 41 48 27 74	51 90 81 39 80	72 89 35 55 07
19 50 23 71 74	69 97 92 02 88	55 21 02 97 73	74 28 77 52 51	65 34 46 74 15
21 81 85 93 13	93 27 88 17 57	05 68 67 31 56	07 08 28 50 46	31 85 33 84 52
51 47 46 64 99	68 10 72 36 21	94 04 99 13 45	42 83 60 91 91	08 00 74 54 49
99 55 96 83 31	62 53 52 41 70	69 77 71 28 30	74 81 97 81 42	43 86 07 28 34
33 71 34 80 07	93 58 47 28 69	51 92 66 47 21	58 30 32 98 22	93 17 49 39 72
85 27 48 68 93	11 30 32 92 70	28 83 43 41 37	73 51 59 04 00	71 14 84 36 43
84 13 38 96 40	44 03 55 21 66	73 85 27 00 91	61 22 26 05 61	62 32 71 84 23
56 73 21 62 34	17 39 59 61 31	10 12 39 16 22	85 49 65 75 60	81 60 41 88 80
65 13 85 68 06	87 60 88 52 61	34 31 36 58 61	45 87 52 10 69	85 64 44 72 77
38 00 10 21 76	81 71 91 17 11	71 60 29 29 37	74 21 96 40 49	65 58 44 96 98
37 40 29 63 97	01 30 47 75 86	56 27 11 00 86	47 32 46 26 05	40 03 03 74 38
97 12 54 03 48	87 08 33 14 17	21 81 53 92 50	75 23 76 20 47	15 50 12 95 78
21 82 64 11 34	47 14 33 40 72	64 63 88 59 02	49 13 90 64 41	03 85 65 45 52
73 13 54 27 42	95 71 90 90 35	85 79 47 42 96	08 78 98 81 56	64 69 11 92 02
07 63 87 79 29	03 06 11 80 72	96 20 74 41 56	23 82 19 95 38	04 71 36 69 94
60 52 88 34 41	07 95 41 98 14	59 17 52 06 95	05 53 35 21 39	61 21 20 64 55
83 59 63 56 55	06 95 89 29 83	05 12 80 97 19	77 43 35 37 83	92 30 15 04 98
10 85 06 27 46	99 59 91 05 07	13 49 90 63 19	53 07 57 18 39	06 41 01 93 62
39 82 09 89 52	43 62 26 31 47	64 42 18 08 14	43 80 00 93 51	31 02 47 31 67
59 58 00 64 78	75 56 97 88 00	88 83 55 44 86	23 76 80 61 56	04 11 10 84 08
38 50 80 73 41	23 79 34 87 63	90 82 29 70 22	17 71 90 42 07	95 95 44 99 53
30 69 27 06 68	94 68 81 61 27	56 19 68 00 91	82 06 76 34 00	05 46 26 92 00
65 44 39 56 59	18 28 82 74 37	49 63 22 40 41	08 33 76 56 76	96 29 99 08 36
27 26 75 02 64	13 19 27 22 94	07 47 74 46 06	17 98 54 89 11	97 34 13 03 58
91 30 70 69 91	19 07 22 42 10	36 69 95 37 28	28 82 53 57 93	28 97 66 62 52
68 43 49 46 88	84 47 31 36 22	62 12 69 84 08	12 84 38 25 90	09 81 59 31 46
48 90 81 58 77	54 74 52 45 91	35 70 00 47 54	83 82 45 26 92	54 13 05 51 60
06 91 34 51 97	42 67 27 86 01	11 88 30 95 28	63 01 19 89 01	14 97 44 03 44
10 45 51 60 19	14 21 03 37 12	91 34 23 78 21	88 32 58 08 51	43 66 77 08 83
12 88 39 73 43	65 02 76 11 84	04 28 50 13 92	17 97 41 50 77	90 71 22 67 69
21 77 83 09 76	38 80 73 69 61	31 64 94 20 96	63 28 10 20 23	08 81 64 74 49
19 52 35 95 15	65 12 25 96 59	86 28 36 82 58	69 57 21 37 98	16 43 59 15 29
67 24 55 26 70	35 58 31 65 63	79 24 68 66 86	76 46 33 42 22	26 65 59 08 02
60 58 44 73 77	07 50 03 79 92	45 13 42 65 29	26 76 08 36 37	41 32 64 43 44
53 85 34 13 77	36 06 69 48 50	58 83 87 38 59	49 36 47 33 31	96 24 04 36 42
24 63 73 97 36	74 38 48 93 42	52 62 30 79 92	12 36 91 86 01	03 74 28 38 73
83 08 01 24 51	38 99 22 28 15	07 75 95 17 77	97 37 72 75 85	51 97 23 78 67
16 44 42 43 34	36 15 19 90 73	27 49 37 09 39	85 13 03 25 52	54 84 65 47 59
60 79 01 81 57	57 17 86 57 62	11 16 17 85 76	45 81 95 29 79	65 13 00 48 60

From tables of the RAND Corporation. Reprinted from Wilfred J. Dixon and Frank J. Massey, Jr., *Introduction to Statistical Analysis.* 3rd ed. (New York: McGraw-Hill, 1969), pp. 446–447. Reprinted by permission of the RAND Corporation.

Table 2 **Binomial Probabilities** $\left[\binom{n}{x} \cdot p^x \cdot q^{n-x}\right]$

n	x							P							x
		0.01	0.05	0.10	0.20	0.30	0.40	0.50	0.60	0.70	0.80	0.90	0.95	0.99	
2	0	.980	.902	.810	.640	.490	.360	.250	.160	.090	.040	.010	.002	0+	0
	1	.020	.095	.180	.320	.420	.480	.500	.480	.420	.320	.180	.095	.020	1
	2	0+	.002	.010	.040	.090	.160	.250	.360	.490	.640	.810	.902	.980	2
3	0	.970	.857	.729	.512	.343	.216	.125	.064	.027	.008	.001	0+	0+	0
	1	.029	.135	.243	.384	.441	.432	.375	.288	.189	.096	.027	.007	0+	1
	2	0+	.007	.027	.096	.189	.288	.375	.432	.441	.384	.243	.135	.029	2
	3	0+	0+	.001	.008	.027	.064	.125	.216	.343	.512	.729	.857	.970	3
4	0	.961	.815	.656	.410	.240	.130	.062	.026	.008	.002	0+	0+	0+	0
	1	.039	.171	.292	.410	.412	.346	.250	.154	.076	.026	.004	0+	0+	1
	2	.001	.014	.049	.154	.265	.346	.375	.346	.265	.154	.049	.014	.001	2
	3	0+	0+	.004	.026	.076	.154	.250	.346	.412	.410	.292	.171	.039	3
	4	0+	0+	0+	.002	.008	.026	.062	.130	.240	.410	.656	.815	.961	4
5	0	.951	.774	.590	.328	.168	.078	.031	.010	.002	0+	0+	0+	0+	0
	1	.048	.204	.328	.410	.360	.259	.156	.077	.028	.006	0+	0+	0+	1
	2	.001	.021	.073	.205	.309	.346	.312	.230	.132	.051	.008	.001	0+	2
	3	0+	.001	.008	.051	.132	.230	.312	.346	.309	.205	.073	.021	.001	3
	4	0+	0+	0+	.006	.028	.077	.156	.259	.360	.410	.328	.204	.048	4
	5	0+	0+	0+	0+	.002	.010	.031	.078	.168	.328	.590	.774	.951	5
6	0	.941	.735	.531	.262	.118	.047	.016	.004	.001	0+	0+	0+	0+	0
	1	.057	.232	.354	.393	.303	.187	.094	.037	.010	.002	0+	0+	0+	1
	2	.001	.031	.098	.246	.324	.311	.234	.138	.060	.015	.001	0+	0+	2
	3	0+	.002	.015	.082	.185	.276	.312	.276	.185	.082	.015	.002	0+	3
	4	0+	0+	.001	.015	.060	.138	.234	.311	.324	.246	.098	.031	.001	4
	5	0+	0+	0+	.002	.010	.037	.094	.187	.303	.393	.354	.232	.057	5
	6	0+	0+	0+	0+	.001	.004	.016	.047	.118	.262	.531	.735	.941	6
7	0	.932	.698	.478	.210	.082	.028	.008	.002	0+	0+	0+	0+	0+	0
	1	.066	.257	.372	.367	.247	.131	.055	.017	.004	0+	0+	0+	0+	1
	2	.002	.041	.124	.275	.318	.261	.164	.077	.025	.004	0+	0+	0+	2
	3	0+	.004	.023	.115	.227	.290	.273	.194	.097	.029	.003	0+	0+	3
	4	0+	0+	.003	.029	.097	.194	.273	.290	.227	.115	.023	.004	0+	4
	5	0+	0+	0+	.004	.025	.077	.164	.261	.318	.275	.124	.041	.002	5
	6	0+	0+	0+	0+	.004	.017	.055	.131	.247	.367	.372	.257	.066	6
	7	0+	0+	0+	0+	0+	.002	.008	.028	.082	.210	.478	.698	.932	7
8	0	.923	.663	.430	.168	.058	.017	.004	.001	0+	0+	0+	0+	0+	0
	1	.075	.279	.383	.336	.198	.090	.031	.008	.001	0+	0+	0+	0+	1
	2	.003	.051	.149	.294	.296	.209	.109	.041	.010	.001	0+	0+	0+	2
	3	0+	.005	.033	.147	.254	.279	.219	.124	.047	.009	0+	0+	0+	3
	4	0+	0+	.005	.046	.136	.232	.273	.232	.136	.046	.005	0+	0+	4
	5	0+	0+	0+	.009	.047	.124	.219	.279	.254	.147	.033	.005	0+	5
	6	0+	0+	0+	.001	.010	.041	.109	.209	.296	.294	.149	.051	.003	6
	7	0+	0+	0+	0+	.001	.008	.031	.090	.198	.336	.383	.279	.075	7
	8	0+	0+	0+	0+	0+	.001	.004	.017	.058	.168	.430	.663	.923	8

For specific details about using this table, see pages 109–110.

Table 2 was generated using Excel.

Table 2 (Continued)

n	x	0.01	0.05	0.10	0.20	0.30	0.40	0.50	0.60	0.70	0.80	0.90	0.95	0.99	x
								P							
9	0	.914	.630	.387	.134	.040	.010	.002	0+	0+	0+	0+	0+	0+	0
	1	.083	.299	.387	.302	.156	.060	.018	.004	0+	0+	0+	0+	0+	1
	2	.003	.063	.172	.302	.267	.161	.070	.021	.004	0+	0+	0+	0+	2
	3	0+	.008	.045	.176	.267	.251	.164	.074	.021	.003	0+	0+	0+	3
	4	0+	.001	.007	.066	.172	.251	.246	.167	.074	.017	.001	0+	0+	4
	5	0+	0+	.001	.017	.074	.167	.246	.251	.172	.066	.007	.001	0+	5
	6	0+	0+	0+	.003	.021	.074	.164	.251	.267	.176	.045	.008	0+	6
	7	0+	0+	0+	0+	.004	.021	.070	.161	.267	.302	.172	.063	.003	7
	8	0+	0+	0+	0+	0+	.004	.018	.060	.156	.302	.387	.299	.083	8
	9	0+	0+	0+	0+	0+	0+	.002	.010	.040	.134	.387	.630	.914	9
10	0	.904	.599	.349	.107	.028	.006	.001	0+	0+	0+	0+	0+	0+	0
	1	.091	.315	.387	.268	.121	.040	.010	.002	0+	0+	0+	0+	0+	1
	2	.004	.075	.194	.302	.233	.121	.044	.011	.001	0+	0+	0+	0+	2
	3	0+	.010	.057	.201	.267	.215	.117	.042	.009	.001	0+	0+	0+	3
	4	0+	.001	.011	.088	.200	.251	.205	.111	.037	.006	0+	0+	0+	4
	5	0+	0+	.001	.026	.103	.201	.246	.201	.103	.026	.001	0+	0+	5
	6	0+	0+	0+	.006	.037	.111	.205	.251	.200	.088	.011	.001	0+	6
	7	0+	0+	0+	.001	.009	.042	.117	.215	.267	.201	.057	.010	0+	7
	8	0+	0+	0+	0+	.001	.011	.044	.121	.233	.302	.194	.075	.004	8
	9	0+	0+	0+	0+	0+	.002	.010	.040	.121	.268	.387	.315	.091	9
	10	0+	0+	0+	0+	0+	0+	.001	.006	.028	.107	.349	.599	.904	10
11	0	.895	.569	.314	.086	.020	.004	0+	0+	0+	0+	0+	0+	0+	0
	1	.099	.329	.384	.236	.093	.027	.005	.001	0+	0+	0+	0+	0+	1
	2	.005	.087	.213	.295	.200	.089	.027	.005	.001	0+	0+	0+	0+	1
	3	0+	.014	.071	.221	.257	.177	.081	.023	.004	0+	0+	0+	0+	3
	4	0+	.001	.016	.111	.220	.236	.161	.070	.017	.002	0+	0+	0+	4
	5	0+	0+	.002	.039	.132	.221	.226	.147	.057	.010	0+	0+	0+	5
	6	0+	0+	0+	.010	.057	.147	.226	.221	.132	.039	.002	0+	0+	6
	7	0+	0+	0+	.002	.017	.070	.161	.236	.220	.111	.016	.001	0+	7
	8	0+	0+	0+	0+	.004	.023	.081	.177	.257	.221	.071	.014	0+	8
	9	0+	0+	0+	0+	.001	.005	.027	.089	.200	.295	.213	.087	.005	9
	10	0+	0+	0+	0+	0+	.001	.005	.027	.093	.236	.384	.329	.099	10
	11	0+	0+	0+	0+	0+	0+	0+	.004	.020	.086	.314	.569	.895	11
12	0	.886	.540	.282	.069	.014	.002	0+	0+	0+	0+	0+	0+	0+	0
	1	.107	.341	.377	.206	.071	.017	.003	0+	0+	0+	0+	0+	0+	1
	2	.006	.099	.230	.283	.168	.064	.016	.002	0+	0+	0+	0+	0+	2
	3	0+	.017	.085	.236	.240	.142	.054	.012	.001	0+	0+	0+	0+	3
	4	0+	.002	.021	.133	.231	.213	.121	.042	.008	.001	0+	0+	0+	4
	5	0+	0+	.004	.053	.158	.227	.193	.101	.029	.003	0+	0+	0+	5
	6	0+	0+	0+	.016	.079	.177	.226	.177	.079	.016	0+	0+	0+	6
	7	0+	0+	0+	.003	.029	.101	.193	.227	.158	.053	.004	0+	0+	7
	8	0+	0+	0+	.001	.008	.042	.121	.213	.231	.133	.021	.002	0+	8
	9	0+	0+	0+	0+	.001	.012	.054	.142	.240	.236	.085	.017	0+	9
	10	0+	0+	0+	0+	0+	.002	.016	.064	.168	.283	.230	.099	.006	10
	11	0+	0+	0+	0+	0+	0+	.003	.017	.071	.206	.377	.341	.107	11
	12	0+	0+	0+	0+	0+	0+	0+	.002	.014	.069	.282	.540	.886	12

Table 2 was generated using Excel.

Table 2 (Continued)

n	x	0.01	0.05	0.10	0.20	0.30	0.40	0.50	0.60	0.70	0.80	0.90	0.95	0.99	x
13	0	.878	.513	.254	.055	.010	.001	0+	0+	0+	0+	0+	0+	0+	0
	1	.115	.351	.367	.179	.054	.011	.002	0+	0+	0+	0+	0+	0+	1
	2	.007	.111	.245	.268	.139	.045	.010	.001	0+	0+	0+	0+	0+	2
	3	0+	.021	.100	.246	.218	.111	.035	.006	.001	0+	0+	0+	0+	3
	4	0+	.003	.028	.154	.234	.184	.087	.024	.003	0+	0+	0+	0+	4
	5	0+	0+	.006	.069	.180	.221	.157	.066	.014	.001	0+	0+	0+	5
	6	0+	0+	.001	.023	.103	.197	.209	.131	.044	.006	0+	0+	0+	6
	7	0+	0+	0+	.006	.044	.131	.209	.197	.103	.023	.001	0+	0+	7
	8	0+	0+	0+	.001	.014	.066	.157	.221	.180	.069	.006	0+	0+	8
	9	0+	0+	0+	0+	.003	.024	.087	.184	.234	.154	.028	.003	0+	9
	10	0+	0+	0+	0+	.001	.006	.035	.111	.218	.246	.100	.021	0+	10
	11	0+	0+	0+	0+	0+	.001	.010	.045	.139	.268	.245	.111	.007	11
	12	0+	0+	0+	0+	0+	0+	.002	.011	.054	.179	.367	.351	.115	12
	13	0+	0+	0+	0+	0+	0+	0+	.001	.010	.055	.254	.513	.878	13
14	0	.869	.488	.229	.044	.007	.001	0+	0+	0+	0+	0+	0+	0+	0
	1	.123	.359	.356	.154	.041	.007	.001	0+	0+	0+	0+	0+	0+	1
	2	.008	.123	.257	.250	.113	.032	.006	.001	0+	0+	0+	0+	0+	2
	3	0+	.026	.114	.250	.194	.085	.022	.003	0+	0+	0+	0+	0+	3
	4	0+	.004	.035	.172	.229	.155	.061	.014	.001	0+	0+	0+	0+	4
	5	0+	0+	.008	.086	.196	.207	.122	.041	.007	0+	0+	0+	0+	5
	6	0+	0+	.001	.032	.126	.207	.183	.092	.023	.002	0+	0+	0+	6
	7	0+	0+	0+	.009	.062	.157	.209	.157	.062	.009	0+	0+	0+	7
	8	0+	0+	0+	.002	.023	.092	.183	.207	.126	.032	.001	0+	0+	8
	9	0+	0+	0+	0+	.007	.041	.122	.207	.196	.086	.008	0+	0+	9
	10	0+	0+	0+	0+	.001	.014	.061	.155	.229	.172	.035	.004	0+	10
	11	0+	0+	0+	0+	0+	.003	.022	.085	.194	.250	.114	.026	.0+	11
	12	0+	0+	0+	0+	0+	.001	.006	.032	.113	.250	.257	.123	.008	12
	13	0+	0+	0+	0+	0+	0+	.001	.007	.041	.154	.356	.359	.123	13
	14	0+	0+	0+	0+	0+	0+	0+	.001	.007	.044	.229	.488	.869	14
15	0	.860	.463	.206	.035	.005	0+	0+	0+	0+	0+	0+	0+	0+	0
	1	.130	.366	.343	.132	.031	.005	0+	0+	0+	0+	0+	0+	0+	1
	2	.009	.135	.267	.231	.092	.022	.003	0+	0+	0+	0+	0+	0+	2
	3	0+	.031	.129	.250	.170	.063	.014	.002	0+	0+	0+	0+	0+	3
	4	0+	.005	.043	.188	.219	.127	.042	.007	.001	0+	0+	0+	0+	4
	5	0+	.001	.010	.103	.206	.186	.092	.024	.003	0+	0+	0+	0+	5
	6	0+	0+	.002	.043	.147	.207	.153	.061	.012	.001	0+	0+	0+	6
	7	0+	0+	0+	.014	.081	.177	.196	.118	.035	.003	0+	0+	0+	7
	8	0+	0+	0+	.003	.035	.118	.196	.177	.081	.014	0+	0+	0+	8
	9	0+	0+	0+	.001	.012	.061	.153	.207	.147	.043	.002	0+	0+	9
	10	0+	0+	0+	0+	.003	.024	.092	.186	.206	.103	.010	.001	0+	10
	11	0+	0+	0+	0+	.001	.007	.042	.127	.219	.188	.043	0.05	0+	11
	12	0+	0+	0+	0+	0+	.002	.014	.063	.170	.250	.129	.031	0+	12
	13	0+	0+	0+	0+	0+	0+	.003	.022	.092	.231	.267	.135	.009	13
	14	0+	0+	0+	0+	0+	0+	0+	.005	.031	.132	.343	.366	.130	14
	15	0+	0+	0+	0+	0+	0+	0+	0+	.005	.035	.206	.463	.860	15

Table 2 was generated using Excel.

Table 3 Cumulative Areas of the Standard Normal Distribution

The entries in this table are the cumulative probabilities for the standard normal distribution z (that is, the normal distribution with mean 0 and standard deviation 1). The shaded area under the curve of the standard normal distribution represents the cumulative probability to the left of a z-value in the **left-hand tail**.

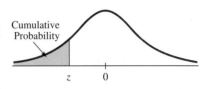

z	0.00	0.01	0.02	0.03	0.04	0.05	0.06	0.07	0.08	0.09
−5.0	0.0000003									
−4.5	0.000003									
−4.0	0.00003	0.00003	0.00003	0.00003	0.00003	0.00003	0.00002	0.00002	0.00002	0.00002
−3.9	0.00005	0.00005	0.00004	0.00004	0.00004	0.00004	0.00004	0.00004	0.00003	0.00003
−3.8	0.00007	0.00007	0.00007	0.00006	0.00006	0.00006	0.00006	0.00005	0.00005	0.00005
−3.7	0.00011	0.00010	0.00010	0.00010	0.00009	0.00009	0.00008	0.00008	0.00008	0.00008
−3.6	0.0002	0.0002	0.0002	0.00014	0.00014	0.00013	0.00013	0.00012	0.00012	0.00011
−3.5	0.0002	0.0002	0.0002	0.0002	0.0002	0.0002	0.0002	0.0002	0.0002	0.0002
−3.4	0.0003	0.0003	0.0003	0.0003	0.0003	0.0003	0.0003	0.0003	0.0003	0.0002
−3.3	0.0005	0.0005	0.0005	0.0004	0.0004	0.0004	0.0004	0.0004	0.0004	0.0004
−3.2	0.0007	0.0007	0.0006	0.0006	0.0006	0.0006	0.0006	0.0005	0.0005	0.0005
−3.1	0.0010	0.0009	0.0009	0.0009	0.0008	0.0008	0.0008	0.0008	0.0007	0.0007
−3.0	0.0014	0.0013	0.0013	0.0012	0.0012	0.0011	0.0011	0.0011	0.0010	0.0010
−2.9	0.0019	0.0018	0.0018	0.0017	0.0016	0.0016	0.0015	0.0015	0.0014	0.0014
−2.8	0.0026	0.0025	0.0024	0.0023	0.0023	0.0022	0.0021	0.0021	0.0020	0.0019
−2.7	0.0035	0.0034	0.0033	0.0032	0.0031	0.0030	0.0029	0.0028	0.0027	0.0026
−2.6	0.0047	0.0045	0.0044	0.0043	0.0042	0.0040	0.0039	0.0038	0.0037	0.0036
−2.5	0.0062	0.0060	0.0059	0.0057	0.0055	0.0054	0.0052	0.0051	0.0049	0.0048
−2.4	0.0082	0.0080	0.0078	0.0076	0.0073	0.0071	0.0070	0.0068	0.0066	0.0064
−2.3	0.0107	0.0104	0.0102	0.0099	0.0096	0.0094	0.0091	0.0089	0.0087	0.0084
−2.2	0.0139	0.0136	0.0132	0.0129	0.0126	0.0122	0.0119	0.0116	0.0113	0.0110
−2.1	0.0179	0.0174	0.0170	0.0166	0.0162	0.0158	0.0154	0.0150	0.0146	0.0143
−2.0	0.0228	0.0222	0.0217	0.0212	0.0207	0.0202	0.0197	0.0192	0.0188	0.0183
−1.9	0.0287	0.0281	0.0274	0.0268	0.0262	0.0256	0.0250	0.0244	0.0239	0.0233
−1.8	0.0359	0.0352	0.0344	0.0336	0.0329	0.0322	0.0314	0.0307	0.0301	0.0294
−1.7	0.0446	0.0436	0.0427	0.0418	0.0409	0.0401	0.0392	0.0384	0.0375	0.0367
−1.6	0.0548	0.0537	0.0526	0.0516	0.0505	0.0495	0.0485	0.0475	0.0465	0.0455
−1.5	0.0668	0.0655	0.0643	0.0630	0.0618	0.0606	0.0594	0.0582	0.0571	0.0559
−1.4	0.0808	0.0793	0.0778	0.0764	0.0749	0.0735	0.0721	0.0708	0.0694	0.0681
−1.3	0.0968	0.0951	0.0934	0.0918	0.0901	0.0885	0.0869	0.0853	0.0838	0.0823
−1.2	0.1151	0.1131	0.1112	0.1094	0.1075	0.1057	0.1038	0.1020	0.1003	0.0985
−1.1	0.1357	0.1335	0.1314	0.1292	0.1271	0.1251	0.1230	0.1210	0.1190	0.1170
−1.0	0.1587	0.1563	0.1539	0.1515	0.1492	0.1469	0.1446	0.1423	0.1401	0.1379
−0.9	0.1841	0.1814	0.1788	0.1762	0.1736	0.1711	0.1685	0.1660	0.1635	0.1611
−0.8	0.2119	0.2090	0.2061	0.2033	0.2005	0.1977	0.1949	0.1922	0.1894	0.1867
−0.7	0.2420	0.2389	0.2358	0.2327	0.2297	0.2266	0.2236	0.2207	0.2177	0.2148
−0.6	0.2743	0.2709	0.2676	0.2643	0.2611	0.2578	0.2546	0.2514	0.2483	0.2451
−0.5	0.3085	0.3050	0.3015	0.2981	0.2946	0.2912	0.2877	0.2843	0.2810	0.2776
−0.4	0.3446	0.3409	0.3372	0.3336	0.3300	0.3264	0.3228	0.3192	0.3156	0.3121
−0.3	0.3821	0.3783	0.3745	0.3707	0.3669	0.3632	0.3594	0.3557	0.3520	0.3483
−0.2	0.4207	0.4168	0.4129	0.4090	0.4052	0.4013	0.3974	0.3936	0.3897	0.3859
−0.1	0.4602	0.4562	0.4522	0.4483	0.4443	0.4404	0.4364	0.4325	0.4286	0.4247
0.0	0.5000	0.4960	0.4920	0.4880	0.4840	0.4801	0.4761	0.4721	0.4681	0.4641

For specific details about using this table to find probabilities, see pages 121–123, 127–128; p-values, page 170–171.

Table 3 was generated using Minitab.

Table 3 (Continued)

The entries in this table are the cumulative probabilities for the standard normal distribution z (that is, the normal distribution with mean 0 and standard deviation 1). The shaded area under the curve of the standard normal distribution represents the cumulative probability to the left of a z-value in the **right-hand tail**.

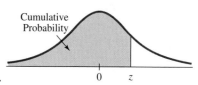

z	0.00	0.01	0.02	0.03	0.04	0.05	0.06	0.07	0.08	0.09
0.0	0.5000	0.5040	0.5080	0.5120	0.5160	0.5199	0.5239	0.5279	0.5319	0.5359
0.1	0.5398	0.5438	0.5478	0.5517	0.5557	0.5596	0.5636	0.5675	0.5714	0.5754
0.2	0.5793	0.5832	0.5871	0.5910	0.5948	0.5987	0.6026	0.6064	0.6103	0.6141
0.3	0.6179	0.6217	0.6255	0.6293	0.6331	0.6368	0.6406	0.6443	0.6480	0.6517
0.4	0.6554	0.6591	0.6628	0.6664	0.6700	0.6736	0.6772	0.6808	0.6844	0.6879
0.5	0.6915	0.6950	0.6985	0.7019	0.7054	0.7088	0.7123	0.7157	0.7190	0.7224
0.6	0.7258	0.7291	0.7324	0.7357	0.7389	0.7422	0.7454	0.7486	0.7518	0.7549
0.7	0.7580	0.7612	0.7642	0.7673	0.7704	0.7734	0.7764	0.7794	0.7823	0.7852
0.8	0.7881	0.7910	0.7939	0.7967	0.7996	0.8023	0.8051	0.8079	0.8106	0.8133
0.9	0.8159	0.8186	0.8212	0.8238	0.8264	0.8289	0.8315	0.8340	0.8365	0.8389
1.0	0.8413	0.8438	0.8461	0.8485	0.8508	0.8531	0.8554	0.8577	0.8599	0.8621
1.1	0.8643	0.8665	0.8686	0.8708	0.8729	0.8749	0.8770	0.8790	0.8810	0.8830
1.2	0.8849	0.8869	0.8888	0.8907	0.8925	0.8944	0.8962	0.8980	0.8997	0.9015
1.3	0.9032	0.9049	0.9066	0.9082	0.9099	0.9115	0.9131	0.9147	0.9162	0.9177
1.4	0.9192	0.9207	0.9222	0.9236	0.9251	0.9265	0.9279	0.9292	0.9306	0.9319
1.5	0.9332	0.9345	0.9357	0.9370	0.9382	0.9394	0.9406	0.9418	0.9430	0.9441
1.6	0.9452	0.9463	0.9474	0.9485	0.9495	0.9505	0.9515	0.9525	0.9535	0.9545
1.7	0.9554	0.9564	0.9573	0.9582	0.9591	0.9599	0.9608	0.9616	0.9625	0.9633
1.8	0.9641	0.9649	0.9656	0.9664	0.9671	0.9678	0.9686	0.9693	0.9700	0.9706
1.9	0.9713	0.9719	0.9726	0.9732	0.9738	0.9744	0.9750	0.9756	0.9762	0.9767
2.0	0.9773	0.9778	0.9783	0.9788	0.9793	0.9798	0.9803	0.9808	0.9812	0.9817
2.1	0.9821	0.9826	0.9830	0.9834	0.9838	0.9842	0.9846	0.9850	0.9854	0.9857
2.2	0.9861	0.9865	0.9868	0.9871	0.9875	0.9878	0.9881	0.9884	0.9887	0.9890
2.3	0.9893	0.9896	0.9898	0.9901	0.9904	0.9906	0.9909	0.9911	0.9913	0.9916
2.4	0.9918	0.9920	0.9922	0.9925	0.9927	0.9929	0.9931	0.9932	0.9934	0.9936
2.5	0.9938	0.9940	0.9941	0.9943	0.9945	0.9946	0.9948	0.9949	0.9951	0.9952
2.6	0.9953	0.9955	0.9956	0.9957	0.9959	0.9960	0.9961	0.9962	0.9963	0.9964
2.7	0.9965	0.9966	0.9967	0.9968	0.9969	0.9970	0.9971	0.9972	0.9973	0.9974
2.8	0.9974	0.9975	0.9976	0.9977	0.9977	0.9978	0.9979	0.9980	0.9980	0.9981
2.9	0.9981	0.9982	0.9983	0.9983	0.9984	0.9984	0.9985	0.9985	0.9986	0.9986
3.0	0.9987	0.9987	0.9987	0.9988	0.9988	0.9989	0.9989	0.9989	0.9990	0.9990
3.1	0.9990	0.9991	0.9991	0.9991	0.9992	0.9992	0.9992	0.9992	0.9993	0.9993
3.2	0.9993	0.9993	0.9994	0.9994	0.9994	0.9994	0.9994	0.9995	0.9995	0.9995
3.3	0.9995	0.9995	0.9996	0.9996	0.9996	0.9996	0.9996	0.9996	0.9996	0.9997
3.4	0.9997	0.9997	0.9997	0.9997	0.9997	0.9997	0.9997	0.9997	0.9998	0.9998
3.5	0.9998	0.9998	0.9998	0.9998	0.9998	0.9998	0.9998	0.9998	0.9998	0.9998
3.6	0.99984	0.99985	0.99985	0.99986	0.99986	0.99987	0.99987	0.99988	0.99988	0.99989
3.7	0.99989	0.99990	0.99990	0.99990	0.99991	0.99991	0.99992	0.99992	0.99992	0.99992
3.8	0.99993	0.99993	0.99993	0.99994	0.99994	0.99994	0.99994	0.99995	0.99995	0.99995
3.9	0.99995	0.99995	0.99996	0.99996	0.99996	0.99996	0.99996	0.99996	0.99997	0.99997
4.0	0.99997	0.99997	0.99997	0.99997	0.99997	0.99997	0.99998	0.99998	0.99998	0.99998
4.5	0.999997									
5.0	0.9999997									

Table 3 was generated using Minitab.

Table 4 Critical Values of Standard Normal Distribution

A   One-Tailed Situations

The entries in this table are the critical values for *z* for which the area under the curve representing α is in the right-hand tail. Critical values for the left-hand tail are found by symmetry.

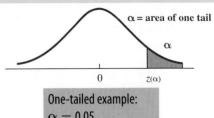

α = area of one tail

	Amount α of in one tail						
α	0.25	0.10	0.05	0.025	0.02	0.01	0.005
$z(\alpha)$	0.67	1.28	1.65	1.96	2.05	2.33	2.58

One-tailed example:
$\alpha = 0.05$
$z(\alpha) = z(0.05) = 1.65$

B   Two-Tailed Situations

The entries in this table are the critical values for *z* for which the area under the curve representing α is split equally between the two tails.

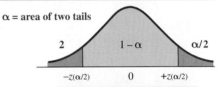

α = area of two tails

	Amount of α in two tails					
α	0.25	0.20	0.10	0.05	0.02	0.01
$z(\alpha/2)$	1.15	1.28	1.65	1.96	2.33	2.58
$1-\alpha$	0.75	0.80	0.90	0.95	0.98	0.99

Area in the "center"

Two-tailed example:
$\alpha = 0.05$ or $1 - \alpha = 0.95$
$\alpha/2 = 0.025$
$z(\alpha/2) = z(0.025) = 1.96$

For specific details about using: Table A to find critical values, see page 175. Table B to find confidence coefficients, see pages 157–161; critical values, pages 175, 177.

Table 5 *p*-Values for Standard Normal Distribution

The entries in this table are the *p*-values related to the right-hand tail for the calculated *z*★ for the standard normal distribution.

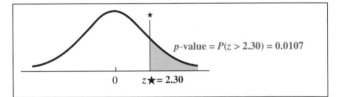

$p\text{-value} = P(z > 2.30) = 0.0107$

z★ = 2.30

z★	*p*-value	z★	*p*-value	z★	*p*-value	z★	*p*-value	z★	*p*-value
0.00	0.5000	0.80	0.2119	1.60	0.0548	2.40	0.0082	3.20	0.0007
0.05	0.4801	0.85	0.1977	1.65	0.0495	2.45	0.0071	3.25	0.0006
0.10	0.4602	0.90	0.1841	1.70	0.0446	2.50	0.0062	3.30	0.0005
0.15	0.4404	0.95	0.1711	1.75	0.0401	2.55	0.0054	3.35	0.0004
0.20	0.4207	1.00	0.1587	1.80	0.0359	2.60	0.0047	3.40	0.0003
0.25	0.4013	1.05	0.1469	1.85	0.0322	2.65	0.0040	3.45	0.0003
0.30	0.3821	1.10	0.1357	1.90	0.0287	2.70	0.0035	3.50	0.0002
0.35	0.3632	1.15	0.1251	1.95	0.0256	2.75	0.0030	3.55	0.0002
0.40	0.3446	1.20	0.1151	2.00	0.0228	2.80	0.0026	3.60	0.0002
0.45	0.3264	1.25	0.1056	2.05	0.0202	2.85	0.0022	3.65	0.0001
0.50	0.3085	1.30	0.0968	2.10	0.0179	2.90	0.0019	3.70	0.0001
0.55	0.2912	1.35	0.0885	2.15	0.0158	2.95	0.0016	3.75	0.0001
0.60	0.2743	1.40	0.0808	2.20	0.0139	3.00	0.0013	3.80	0.0001
0.65	0.2578	1.45	0.0735	2.25	0.0122	3.05	0.0011	3.85	0.0001
0.70	0.2420	1.50	0.0668	2.30	0.0107	3.10	0.0010	3.90	0+
0.75	0.2266	1.55	0.0606	2.35	0.0094	3.15	0.0008	3.95	0+

For specific details about using this table to find *p*-values, see pages 170, 172.

Table 5 was generated using Minitab.

Table 6 Critical Values of Student's *t*-Distribution

The entries in this table are the critical values of the Student's *t*-distribution, for which the area under the curve is: a) in the right-hand tail, or b) in two tails. See the illustrations at the bottom of the page.

a) Area in One Tail

	0.25	0.10	0.05	0.025	0.01	0.005
b) Area in Two Tails						
df	0.50	0.20	0.10	0.05	0.02	0.01
3	0.765	1.64	2.35	3.18	4.54	5.84
4	0.741	1.53	2.13	2.78	3.75	4.60
5	0.727	1.48	2.02	2.57	3.36	4.03
6	0.718	1.44	1.94	2.45	3.14	3.71
7	0.711	1.41	1.89	2.36	3.00	3.50
8	0.706	1.40	1.86	2.31	2.90	3.36
9	0.703	1.38	1.83	2.26	2.82	3.25
10	0.700	1.37	1.81	2.23	2.76	3.17
11	0.697	1.36	1.80	2.20	2.72	3.11
12	0.695	1.36	1.78	2.18	2.68	3.05
13	0.694	1.35	1.77	2.16	2.65	3.01
14	0.692	1.35	1.76	2.14	2.62	2.98
15	0.691	1.34	1.75	2.13	2.60	2.95
16	0.690	1.34	1.75	2.12	2.58	2.92
17	0.689	1.33	1.74	2.11	2.57	2.90
18	0.688	1.33	1.73	2.10	2.55	2.88
19	0.688	1.33	1.73	2.09	2.54	2.86
20	0.687	1.33	1.72	2.09	2.53	2.85
21	0.686	1.32	1.72	2.08	2.52	2.83
22	0.686	1.32	1.72	2.07	2.51	2.82
23	0.685	1.32	1.71	2.07	2.50	2.81
24	0.685	1.32	1.71	2.06	2.49	2.80
25	0.684	1.32	1.71	2.06	2.49	2.79
26	0.684	1.31	1.70	2.05	2.47	2.77
27	0.684	1.31	1.70	2.05	2.47	2.77
28	0.683	1.31	1.70	2.05	2.47	2.76
29	0.683	1.31	1.70	2.05	2.46	2.76
30	0.683	1.31	1.70	2.04	2.46	2.75
35	0.682	1.31	1.69	2.03	2.44	2.72
40	0.681	1.30	1.68	2.02	2.42	2.70
50	0.679	1.30	1.68	2.01	2.40	2.68
70	0.678	1.29	1.67	1.99	2.38	2.65
100	0.677	1.29	1.66	1.98	2.36	2.63
df > 100	0.675	1.28	1.65	1.96	2.33	2.58

One-tailed example:
df = 9 and α = 0.10
$t(df, \alpha) = t(9, 0.10) = 1.38$

Two-tailed example:
df = 14, α = 0.02, 1 − α = 0.98
$t(df, \alpha/2) = t(14, 0.01) = 2.62$

For specific details about using this table to find confidence coefficients, see pages 185–187; *p*-values, pages 189–191; critical values, pages 185, 188.
Table 6 was generated using Minitab.

Table 7 Probability-Values for Student's *t*-distribution

The entries in this table are the *p*-values related to the right-hand tail for the calculated *t*★ value for the *t*-distribution of df degrees of freedom.

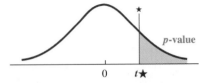

p-value

t★	3	4	5	6	7	8	10	12	15	18	21	25	29	35	df≥45
0.0	0.500	0.500	0.500	0.500	0.500	0.500	0.500	0.500	0.500	0.500	0.500	0.500	0.500	0.500	0.500
0.1	0.463	0.463	0.462	0.462	0.462	0.461	0.461	0.461	0.461	0.461	0.461	0.461	0.461	0.460	0.460
0.2	0.427	0.426	0.425	0.424	0.424	0.423	0.423	0.422	0.422	0.422	0.422	0.422	0.421	0.421	0.421
0.3	0.392	0.390	0.388	0.387	0.386	0.386	0.385	0.385	0.384	0.384	0.384	0.383	0.383	0.383	0.383
0.4	0.358	0.355	0.353	0.352	0.351	0.350	0.349	0.348	0.347	0.347	0.347	0.346	0.346	0.346	0.346
0.5	0.326	0.322	0.319	0.317	0.316	0.315	0.314	0.313	0.312	0.312	0.311	0.311	0.310	0.310	0.310
0.6	0.295	0.290	0.287	0.285	0.284	0.283	0.281	0.280	0.279	0.278	0.277	0.277	0.277	0.276	0.276
0.7	0.267	0.261	0.258	0.255	0.253	0.252	0.250	0.249	0.247	0.246	0.246	0.245	0.245	0.244	0.244
0.8	0.241	0.234	0.230	0.227	0.225	0.223	0.221	0.220	0.218	0.217	0.216	0.216	0.215	0.215	0.214
0.9	0.217	0.210	0.205	0.201	0.199	0.197	0.195	0.193	0.191	0.190	0.189	0.188	0.188	0.187	0.186
1.0	0.196	0.187	0.182	0.178	0.175	0.173	0.170	0.169	0.167	0.165	0.164	0.163	0.163	0.162	0.161
1.1	0.176	0.167	0.161	0.157	0.154	0.152	0.149	0.146	0.144	0.143	0.142	0.141	0.140	0.139	0.139
1.2	0.158	0.148	0.142	0.138	0.135	0.132	0.129	0.127	0.124	0.123	0.122	0.121	0.120	0.119	0.118
1.3	0.142	0.132	0.125	0.121	0.117	0.115	0.111	0.109	0.107	0.105	0.104	0.103	0.102	0.101	0.100
1.4	0.128	0.117	0.110	0.106	0.102	0.100	0.096	0.093	0.091	0.089	0.088	0.087	0.086	0.085	0.084
1.5	0.115	0.104	0.097	0.092	0.089	0.086	0.082	0.080	0.077	0.075	0.074	0.073	0.072	0.071	0.070
1.6	0.104	0.092	0.085	0.080	0.077	0.074	0.070	0.068	0.065	0.064	0.062	0.061	0.060	0.059	0.058
1.7	0.094	0.082	0.075	0.070	0.066	0.064	0.060	0.057	0.055	0.053	0.052	0.051	0.050	0.049	0.048
1.8	0.085	0.073	0.066	0.061	0.057	0.055	0.051	0.049	0.046	0.044	0.043	0.042	0.041	0.040	0.039
1.9	0.077	0.065	0.058	0.053	0.050	0.047	0.043	0.041	0.038	0.037	0.036	0.035	0.034	0.033	0.032
2.0	0.070	0.058	0.051	0.046	0.043	0.040	0.037	0.034	0.032	0.030	0.029	0.028	0.027	0.027	0.026
2.1	0.063	0.052	0.045	0.040	0.037	0.034	0.031	0.029	0.027	0.025	0.024	0.023	0.022	0.022	0.021
2.2	0.058	0.046	0.040	0.035	0.032	0.029	0.026	0.024	0.022	0.021	0.020	0.019	0.018	0.017	0.016
2.3	0.052	0.041	0.035	0.031	0.027	0.025	0.022	0.020	0.018	0.017	0.016	0.015	0.014	0.014	0.013
2.4	0.048	0.037	0.031	0.027	0.024	0.022	0.019	0.017	0.015	0.014	0.013	0.012	0.012	0.011	0.010
2.5	0.044	0.033	0.027	0.023	0.020	0.018	0.016	0.014	0.012	0.011	0.010	0.010	0.009	0.009	0.008
2.6	0.040	0.030	0.024	0.020	0.018	0.016	0.013	0.012	0.010	0.009	0.008	0.008	0.007	0.007	0.006
2.7	0.037	0.027	0.021	0.018	0.015	0.014	0.011	0.010	0.008	0.007	0.007	0.006	0.006	0.005	0.005
2.8	0.034	0.024	0.019	0.016	0.013	0.012	0.009	0.008	0.007	0.006	0.005	0.005	0.005	0.004	0.004
2.9	0.031	0.022	0.017	0.014	0.011	0.010	0.008	0.007	0.005	0.005	0.004	0.004	0.004	0.003	0.003
3.0	0.029	0.020	0.015	0.012	0.010	0.009	0.007	0.006	0.004	0.004	0.003	0.003	0.003	0.002	0.002
3.1	0.027	0.018	0.013	0.011	0.009	0.007	0.006	0.005	0.004	0.003	0.003	0.002	0.002	0.002	0.002
3.2	0.025	0.016	0.012	0.009	0.008	0.006	0.005	0.004	0.003	0.002	0.002	0.002	0.002	0.001	0.001
3.3	0.023	0.015	0.011	0.008	0.007	0.005	0.004	0.003	0.002	0.002	0.002	0.001	0.001	0.001	0.001
3.4	0.021	0.014	0.010	0.007	0.006	0.005	0.003	0.003	0.002	0.002	0.001	0.001	0.001	0.001	0.001
3.5	0.020	0.012	0.009	0.006	0.005	0.004	0.003	0.002	0.002	0.001	0.001	0.001	0.001	0.001	0.001
3.6	0.018	0.011	0.008	0.006	0.004	0.004	0.002	0.002	0.001	0.001	0.001	0.001	0.001	0+	0+
3.7	0.017	0.010	0.007	0.005	0.004	0.003	0.002	0.002	0.001	0.001	0.001	0.001	0+	0+	0+
3.8	0.016	0.010	0.006	0.004	0.003	0.003	0.002	0.001	0.001	0.001	0.001	0+	0+	0+	0+
3.9	0.015	0.009	0.006	0.004	0.003	0.002	0.001	0.001	0.001	0.001	0+	0+	0+	0+	0+
4.0	0.014	0.008	0.005	0.004	0.003	0.002	0.001	0.001	0.001	0+	0+	0+	0+	0+	0+

Degrees of Freedom

For specific details about using this table to find *p*-values, see pages 189–191.

Table 7 was generated using Minitab.

Table 8 Critical Values of χ^2 ("Chi-Square") Distribution

The entries in this table are the critical values for the χ^2-distribution for which the area under the curve is: a) to the right-hand tail, or b) to the left-hand tail (the cumulative area). See the illustrations at the bottom of the page.

a) Area to the Right						Median						
0.995	0.99	0.975	0.95	0.90	0.75	0.50	0.25	0.10	0.05	0.025	0.01	0.005

b) Area to the Left (the Cumulative Area)						Median						
df 0.005	0.01	0.025	0.05	0.10	0.25	0.50	0.75	0.90	0.95	0.975	0.99	0.995
1 0.0000393	0.000157	0.000982	0.00393	0.0158	0.102	0.455	1.32	2.71	3.84	5.02	6.63	7.88
2 0.0100	0.0201	0.0506	0.103	0.211	0.575	1.39	2.77	4.61	5.99	7.38	9.21	10.6
3 0.0717	0.115	0.216	0.352	0.584	1.21	2.37	4.11	6.25	7.81	9.35	11.3	12.8
4 0.207	0.297	0.484	0.711	1.06	1.92	3.36	5.39	7.78	9.49	11.1	13.3	14.9
5 0.412	0.554	0.831	1.15	1.61	2.67	4.35	6.63	9.24	11.1	12.8	15.1	16.7
6 0.676	0.872	1.24	1.64	2.20	3.45	5.35	7.84	10.6	12.6	14.4	16.8	18.5
7 0.989	1.24	1.69	2.17	2.83	4.25	6.35	9.04	12.0	14.1	16.0	18.5	20.3
8 1.34	1.65	2.18	2.73	3.49	5.07	7.34	10.2	13.4	15.5	17.5	20.1	22.0
9 1.73	2.09	2.70	3.33	4.17	5.90	8.34	11.4	14.7	16.9	19.0	21.7	23.6
10 2.16	2.56	3.25	3.94	4.87	6.74	9.34	12.5	16.0	18.3	20.5	23.2	25.2
11 2.60	3.05	3.82	4.57	5.58	7.58	10.34	13.7	17.3	19.7	21.9	24.7	26.8
12 3.07	3.57	4.40	5.23	6.30	8.44	11.34	14.8	18.5	21.0	23.3	26.2	28.3
13 3.57	4.11	5.01	5.89	7.04	9.30	12.34	16.0	19.8	22.4	24.7	27.7	29.8
14 4.07	4.66	5.63	6.57	7.79	10.2	13.34	17.1	21.1	23.7	26.1	29.1	31.3
15 4.60	5.23	6.26	7.26	8.55	11.0	14.34	18.2	22.3	25.0	27.5	30.6	32.8
16 5.14	5.81	6.91	7.96	9.31	11.9	15.34	19.4	23.5	26.3	28.8	32.0	34.3
17 5.70	6.41	7.56	8.67	10.1	12.8	16.34	20.5	24.8	27.6	30.2	33.4	35.7
18 6.26	7.01	8.23	9.39	10.9	13.7	17.34	21.6	26.0	28.9	31.5	34.8	37.2
19 6.84	7.63	8.91	10.1	11.7	14.6	18.34	22.7	27.2	30.1	32.9	36.2	38.6
20 7.43	8.26	9.59	10.9	12.4	15.5	19.34	23.8	28.4	31.4	34.2	37.6	40.0
21 8.03	8.90	10.3	11.6	13.2	16.3	20.34	24.9	29.6	32.7	35.5	38.9	41.4
22 8.64	9.54	11.0	12.3	14.0	17.2	21.34	26.0	30.8	33.9	36.8	40.3	42.8
23 9.26	10.2	11.7	13.1	14.8	18.1	22.34	27.1	32.0	35.2	38.1	41.6	44.2
24 9.89	10.9	12.4	13.8	15.7	19.0	23.34	28.2	33.2	36.4	39.4	43.0	45.6
25 10.5	11.5	13.1	14.6	16.5	19.9	24.34	29.3	34.4	37.7	40.6	44.3	46.9
26 11.2	12.2	13.8	15.4	17.3	20.8	25.34	30.4	35.6	38.9	41.9	45.6	48.3
27 11.8	12.9	14.6	16.2	18.1	21.7	26.34	31.5	36.7	40.1	43.2	47.0	49.6
28 12.5	13.6	15.3	16.9	18.9	22.7	27.34	32.6	37.9	41.3	44.5	48.3	51.0
29 13.1	14.3	16.0	17.7	19.8	23.6	28.34	33.7	39.1	42.6	45.7	49.6	52.3
30 13.8	15.0	16.8	18.5	20.6	24.5	29.34	34.8	40.3	43.8	47.0	50.9	53.7
40 20.7	22.2	24.4	26.5	29.1	33.7	39.34	45.6	51.8	55.8	59.3	63.7	66.8
50 28.0	29.7	32.4	34.8	37.7	42.9	49.33	56.3	63.2	67.5	71.4	76.2	79.5
60 35.5	37.5	40.5	43.2	46.5	52.3	59.33	67.0	74.4	79.1	83.3	88.4	92.0
70 43.3	45.4	48.8	51.7	55.3	61.7	69.33	77.6	85.5	90.5	95.0	100.4	104.2
80 51.2	53.5	57.2	60.4	64.3	71.1	79.33	88.1	96.6	101.9	106.6	112.3	116.3
90 59.2	61.8	65.6	69.1	73.3	80.6	89.33	98.6	107.6	113.1	118.1	124.1	128.3
100 67.3	70.1	74.2	77.9	82.4	90.1	99.33	109.1	118.5	124.3	129.6	135.8	140.2

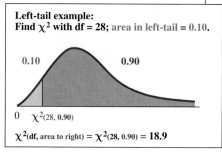

Left-tail example:
Find χ^2 with df = 28; area in left-tail = 0.10.

0.10   0.90

$0 \quad \chi^2_{(28, 0.90)}$

$\chi^2(\text{df, area to right}) = \chi^2_{(28, 0.90)} = \mathbf{18.9}$

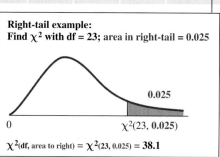

Right-tail example:
Find χ^2 with df = 23; area in right-tail = 0.025

0.025

$0 \quad \chi^2_{(23, 0.025)}$

$\chi^2(\text{df, area to right}) = \chi^2_{(23, 0.025)} = \mathbf{38.1}$

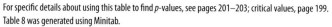

For specific details about using this table to find *p*-values, see pages 201–203; critical values, page 199.
Table 8 was generated using Minitab.

Table 9a Critical Values of the F Distribution ($\alpha = 0.05$)

The entries in this table are critical values of F for which the area under the curve to the right is equal to 0.05.

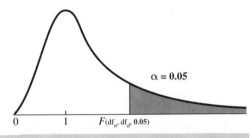

		Degrees of Freedom for Numerator									
		1	2	3	4	5	6	7	8	9	10
Degrees of Freedom for Denominator	1	161.	200.	216.	225.	230.	234.	237.	239.	241.	242.
	2	18.5	19.0	19.2	19.2	19.3	19.3	19.4	19.4	19.4	19.4
	3	10.1	9.55	9.28	9.12	9.01	8.94	8.89	8.85	8.81	8.79
	4	7.71	6.94	6.59	6.39	6.26	6.16	6.09	6.04	6.00	5.96
	5	6.61	5.79	5.41	5.19	5.05	4.95	4.88	4.82	4.77	4.74
	6	5.99	5.14	4.76	4.53	4.39	4.28	4.21	4.15	4.10	4.06
	7	5.59	4.74	4.35	4.12	3.97	3.87	3.79	3.73	3.68	3.64
	8	5.32	4.46	4.07	3.84	3.69	3.58	3.50	3.44	3.39	3.35
	9	5.12	4.26	3.86	3.63	3.48	3.37	3.29	3.23	3.18	3.14
	10	4.96	4.10	3.71	3.48	3.33	3.22	3.14	3.07	3.02	2.98
	11	4.84	3.98	3.59	3.36	3.20	3.09	3.01	2.95	2.90	2.85
	12	4.75	3.89	3.49	3.26	3.11	3.00	2.91	2.85	2.80	2.75
	13	4.67	3.81	3.41	3.18	3.03	2.92	2.83	2.77	2.71	2.67
	14	4.60	3.74	3.34	3.11	2.96	2.85	2.76	2.70	2.65	2.60
	15	4.54	3.68	3.29	3.06	2.90	2.79	2.71	2.64	2.59	2.54
	16	4.49	3.63	3.24	3.01	2.85	2.74	2.66	2.59	2.54	2.49
	17	4.45	3.59	3.20	2.96	2.81	2.70	2.61	2.55	2.49	2.45
	18	4.41	3.55	3.16	2.93	2.77	2.66	2.58	2.51	2.46	2.41
	19	4.38	3.52	3.13	2.90	2.74	2.63	2.54	2.48	2.42	2.38
	20	4.35	3.49	3.10	2.87	2.71	2.60	2.51	2.45	2.39	2.35
	21	4.32	3.47	3.07	2.84	2.68	2.57	2.49	2.42	2.37	2.32
	22	4.30	3.44	3.05	2.82	2.66	2.55	2.46	2.40	2.34	2.30
	23	4.28	3.42	3.03	2.80	2.64	2.53	2.44	2.37	2.32	2.27
	24	4.26	3.40	3.01	2.78	2.62	2.51	2.42	2.36	2.30	2.25
	25	4.24	3.39	2.99	2.76	2.60	2.49	2.40	2.34	2.28	2.24
	30	4.17	3.32	2.92	2.69	2.53	2.42	2.33	2.27	2.21	2.16
	40	4.08	3.23	2.84	2.61	2.45	2.34	2.25	2.18	2.12	2.08
	60	4.00	3.15	2.76	2.53	2.37	2.25	2.17	2.10	2.04	1.99
	120	3.92	3.07	2.68	2.45	2.29	2.18	2.09	2.02	1.96	1.91
	10,000	3.84	3.00	2.61	2.37	2.21	2.10	2.01	1.94	1.88	1.83

For specific details about using this table to find p-values, see page 228; critical values, page 227.

Table 9a was generated using Minitab.

		Degrees of Freedom for Numerator								
		12	15	20	24	30	40	60	120	10,000
Degrees of Freedom for Denominator	1	244.	246.	248.	249.	250.	251.	252.	253.	254.
	2	19.4	19.4	19.4	19.5	19.5	19.5	19.5	19.5	19.5
	3	8.74	8.70	8.66	8.64	8.62	8.59	8.57	8.55	8.53
	4	5.91	5.86	5.80	5.77	5.75	5.72	5.69	5.66	5.63
	5	4.68	4.62	4.56	4.53	4.50	4.46	4.43	4.40	4.37
	6	4.00	3.94	3.87	3.84	3.81	3.77	3.74	3.70	3.67
	7	3.57	3.51	3.44	3.41	3.38	3.34	3.30	3.27	3.23
	8	3.28	3.22	3.15	3.12	3.08	3.04	3.01	2.97	2.93
	9	3.07	3.01	2.94	2.90	2.86	2.83	2.79	2.75	2.71
	10	2.91	2.85	2.77	2.74	2.70	2.66	2.62	2.58	2.54
	11	2.79	2.72	2.65	2.61	2.57	2.53	2.49	2.45	2.41
	12	2.69	2.62	2.54	2.51	2.47	2.43	2.38	2.34	2.30
	13	2.60	2.53	2.46	2.42	2.38	2.34	2.30	2.25	2.21
	14	2.53	2.46	2.39	2.35	2.31	2.27	2.22	2.18	2.13
	15	2.48	2.40	2.33	2.29	2.25	2.20	2.16	2.11	2.07
	16	2.42	2.35	2.28	2.24	2.19	2.15	2.11	2.06	2.01
	17	2.38	2.31	2.23	2.19	2.15	2.10	2.06	2.01	1.96
	18	2.34	2.27	2.19	2.15	2.11	2.06	2.02	1.97	1.92
	19	2.31	2.23	2.16	2.11	2.07	2.03	1.98	1.93	1.88
	20	2.28	2.20	2.12	2.08	2.04	1.99	1.95	1.90	1.84
	21	2.25	2.18	2.10	2.05	2.01	1.96	1.92	1.87	1.81
	22	2.23	2.15	2.07	2.03	1.98	1.94	1.89	1.84	1.78
	23	2.20	2.13	2.05	2.01	1.96	1.91	1.86	1.81	1.76
	24	2.18	2.11	2.03	1.98	1.94	1.89	1.84	1.79	1.73
	25	2.16	2.09	2.01	1.96	1.92	1.87	1.82	1.77	1.71
	30	2.09	2.01	1.93	1.89	1.84	1.79	1.74	1.68	1.62
	40	2.00	1.92	1.84	1.79	1.74	1.69	1.64	1.58	1.51
	60	1.92	1.84	1.75	1.70	1.65	1.59	1.53	1.47	1.39
	120	1.83	1.75	1.66	1.61	1.55	1.50	1.43	1.35	1.26
	10,000	1.75	1.67	1.57	1.52	1.46	1.40	1.32	1.22	1.03

Table 9a was generated using Minitab.

Table 9b Critical Values of the F Distribution ($\alpha = 0.025$)

The entries in this table are critical values of F for which the area under the curve to the right is equal to 0.025.

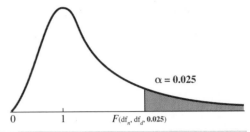

$\alpha = 0.025$

$F(\text{df}_n, \text{df}_d, 0.025)$

		Degrees of Freedom for Numerator									
		1	2	3	4	5	6	7	8	9	10
Degrees of Freedom for Denominator	1	648.	800.	864.	900.	922.	937.	948.	957.	963.	969.
	2	38.5	39.0	39.2	39.2	39.3	39.3	39.4	39.4	39.4	39.4
	3	17.4	16.0	15.4	15.1	14.9	14.7	14.6	14.5	14.5	14.4
	4	12.2	10.6	9.98	9.60	9.36	9.20	9.07	8.98	8.90	8.84
	5	10.0	8.43	7.76	7.39	7.15	6.98	6.85	6.76	6.68	6.62
	6	8.81	7.26	6.60	6.23	5.99	5.82	5.70	5.60	5.52	5.46
	7	8.07	6.54	5.89	5.52	5.29	5.12	4.99	4.90	4.82	4.76
	8	7.57	6.06	5.42	5.05	4.82	4.65	4.53	4.43	4.36	4.30
	9	7.21	5.71	5.08	4.72	4.48	4.32	4.20	4.10	4.03	3.96
	10	6.94	5.46	4.83	4.47	4.24	4.07	3.95	3.85	3.78	3.72
	11	6.72	5.26	4.63	4.28	4.04	3.88	3.76	3.66	3.59	3.53
	12	6.55	5.10	4.47	4.12	3.89	3.73	3.61	3.51	3.44	3.37
	13	6.41	4.97	4.35	4.00	3.77	3.60	3.48	3.39	3.31	3.25
	14	6.30	4.86	4.24	3.89	3.66	3.50	3.38	3.28	3.21	3.15
	15	6.20	4.77	4.15	3.80	3.58	3.41	3.29	3.20	3.12	3.06
	16	6.12	4.69	4.08	3.73	3.50	3.34	3.22	3.12	3.05	2.99
	17	6.04	4.62	4.01	3.66	3.44	3.28	3.16	3.06	2.98	2.92
	18	5.98	4.56	3.95	3.61	3.38	3.22	3.10	3.01	2.93	2.87
	19	5.92	4.51	3.90	3.56	3.33	3.17	3.05	2.96	2.88	2.82
	20	5.87	4.46	3.86	3.51	3.29	3.13	3.01	2.91	2.84	2.77
	21	5.83	4.42	3.82	3.48	3.25	3.09	2.97	2.87	2.80	2.73
	22	5.79	4.38	3.78	3.44	3.22	3.05	2.93	2.84	2.76	2.70
	23	5.75	4.35	3.75	3.41	3.18	3.02	2.90	2.81	2.73	2.67
	24	5.72	4.32	3.72	3.38	3.15	2.99	2.87	2.78	2.70	2.64
	25	5.69	4.29	3.69	3.35	3.13	2.97	2.85	2.75	2.68	2.61
	30	5.57	4.18	3.59	3.25	3.03	2.87	2.75	2.65	2.57	2.51
	40	5.42	4.05	3.46	3.13	2.90	2.74	2.62	2.53	2.45	2.39
	60	5.29	3.93	3.34	3.01	2.79	2.63	2.51	2.41	2.33	2.27
	120	5.15	3.80	3.23	2.89	2.67	2.52	2.39	2.30	2.22	2.16
	10,000	5.03	3.69	3.12	2.79	2.57	2.41	2.29	2.19	2.11	2.05

For specific details about using this table to find p-values, see page 227; critical values, page 228.

Table 9b was generated using Minitab.

		Degrees of Freedom for Numerator								
		12	15	20	24	30	40	60	120	10,000
	1	977.	985.	993.	997.	1,001.	1,006.	1,010.	1,014.	1,018.
	2	39.4	39.4	39.4	39.5	39.5	39.5	39.5	39.5	39.5
	3	14.3	14.3	14.2	14.1	14.1	14.0	14.0	13.9	13.9
	4	8.75	8.66	8.56	8.51	8.46	8.41	8.36	8.31	8.26
	5	6.52	6.43	6.33	6.28	6.23	6.18	6.12	6.07	6.02
	6	5.37	5.27	5.17	5.12	5.07	5.01	4.96	4.90	4.85
	7	4.67	4.57	4.47	4.42	4.36	4.31	4.25	4.20	4.14
	8	4.20	4.10	4.00	3.95	3.89	3.84	3.78	3.73	3.67
	9	3.87	3.77	3.67	3.61	3.56	3.51	3.45	3.39	3.33
	10	3.62	3.52	3.42	3.37	3.31	3.26	3.20	3.14	3.08
	11	3.43	3.33	3.23	3.17	3.12	3.06	3.00	2.94	2.88
Degrees of Freedom for Denominator	12	3.28	3.18	3.07	3.02	2.96	2.91	2.85	2.79	2.73
	13	3.15	3.05	2.95	2.89	2.84	2.78	2.72	2.66	2.60
	14	3.05	2.95	2.84	2.79	2.73	2.67	2.61	2.55	2.49
	15	2.96	2.86	2.76	2.70	2.64	2.59	2.52	2.46	2.40
	16	2.89	2.79	2.68	2.63	2.57	2.51	2.45	2.38	2.32
	17	2.82	2.72	2.62	2.56	2.50	2.44	2.38	2.32	2.25
	18	2.77	2.67	2.56	2.50	2.44	2.38	2.32	2.26	2.19
	19	2.72	2.62	2.51	2.45	2.39	2.33	2.27	2.20	2.13
	20	2.68	2.57	2.46	2.41	2.35	2.29	2.22	2.16	2.09
	21	2.64	2.53	2.42	2.37	2.31	2.25	2.18	2.11	2.04
	22	2.60	2.50	2.39	2.33	2.27	2.21	2.14	2.08	2.00
	23	2.57	2.47	2.36	2.30	2.24	2.18	2.11	2.04	1.97
	24	2.54	2.44	2.33	2.27	2.21	2.15	2.08	2.01	1.94
	25	2.51	2.41	2.30	2.24	2.18	2.12	2.05	1.98	1.91
	30	2.41	2.31	2.20	2.14	2.07	2.01	1.94	1.87	1.79
	40	2.29	2.18	2.07	2.01	1.94	1.88	1.80	1.72	1.64
	60	2.17	2.06	1.94	1.88	1.82	1.74	1.67	1.58	1.48
	120	2.05	1.95	1.82	1.76	1.69	1.61	1.53	1.43	1.31
	10,000	1.95	1.83	1.71	1.64	1.57	1.49	1.39	1.27	1.04

Table 9b was generated using Minitab.

Table 9c Critical Values of the *F* Distribution ($\alpha = 0.01$)

The entries in the table are critical values of *F* for which the area under the curve to the right is equal to 0.01.

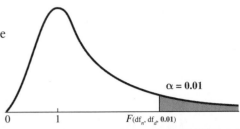

$\alpha = 0.01$

$F_{(df_n,\, df_d,\, 0.01)}$

		Degrees of Freedom for Numerator									
		1	**2**	**3**	**4**	**5**	**6**	**7**	**8**	**9**	**10**
Degrees of Freedom for Denominator	**1**	4,052.	5,000.	5,403.	5,625.	5,764.	5,859.	5,928.	5,981.	6,022.	6,056.
	2	98.5	99.0	99.2	99.2	99.3	99.3	99.4	99.4	99.4	99.4
	3	34.1	30.8	29.5	28.7	28.2	27.9	27.7	27.5	27.3	27.2
	4	21.2	18.0	16.7	16.0	15.5	15.2	15.0	14.8	14.7	14.5
	5	16.3	13.3	12.1	11.4	11.0	10.7	10.5	10.3	10.2	10.1
	6	13.7	10.9	9.78	9.15	8.75	8.47	8.26	8.10	7.98	7.87
	7	12.2	9.55	8.45	7.85	7.46	7.19	6.99	6.84	6.72	6.62
	8	11.3	8.65	7.59	7.01	6.63	6.37	6.18	6.03	5.91	5.81
	9	10.6	8.02	6.99	6.42	6.06	5.80	5.61	5.47	5.35	5.26
	10	10.0	7.56	6.55	5.99	5.64	5.39	5.20	5.06	4.94	4.85
	11	9.65	7.21	6.22	5.67	5.32	5.07	4.89	4.74	4.63	4.54
	12	9.33	6.93	5.95	5.41	5.06	4.82	4.64	4.50	4.39	4.30
	13	9.07	6.70	5.74	5.21	4.86	4.62	4.44	4.30	4.19	4.10
	14	8.86	6.51	5.56	5.04	4.70	4.46	4.28	4.14	4.03	3.94
	15	8.68	6.36	5.42	4.89	4.56	4.32	4.14	4.00	3.89	3.80
	16	8.53	6.23	5.29	4.77	4.44	4.20	4.03	3.89	3.78	3.69
	17	8.40	6.11	5.19	4.67	4.34	4.10	3.93	3.79	3.68	3.59
	18	8.29	6.01	5.09	4.58	4.25	4.01	3.84	3.71	3.60	3.51
	19	8.18	5.93	5.01	4.50	4.17	3.94	3.77	3.63	3.52	3.43
	20	8.10	5.85	4.94	4.43	4.10	3.87	3.70	3.56	3.46	3.37
	21	8.02	5.78	4.87	4.37	4.04	3.81	3.64	3.51	3.40	3.31
	22	7.95	5.72	4.82	4.31	3.99	3.76	3.59	3.45	3.35	3.26
	23	7.88	5.66	4.76	4.26	3.94	3.71	3.54	3.41	3.30	3.21
	24	7.82	5.61	4.72	4.22	3.90	3.67	3.50	3.36	3.26	3.17
	25	7.77	5.57	4.68	4.18	3.86	3.63	3.46	3.32	3.22	3.13
	30	7.56	5.39	4.51	4.02	3.70	3.47	3.30	3.17	3.07	2.98
	40	7.31	5.18	4.31	3.83	3.51	3.29	3.12	2.99	2.89	2.80
	60	7.08	4.98	4.13	3.65	3.34	3.12	2.95	2.82	2.72	2.63
	120	6.85	4.79	3.95	3.48	3.17	2.96	2.79	2.66	2.56	2.47
	10,000	6.64	4.61	3.78	3.32	3.02	2.80	2.64	2.51	2.41	2.32

For specific details about using this table to find *p*-values, see page 228; critical values, page 227.

Table 9c was generated using Minitab.

		Degrees of Freedom for Numerator								
		12	15	20	24	30	40	60	120	10,000
Degrees of Freedom for Denominator	1	6,106.	6,157.	6,209.	6,235.	6,261.	6,287.	6,313.	6,339.	6,366.
	2	99.4	99.4	99.4	99.5	99.5	99.5	99.5	99.5	99.5
	3	27.1	26.9	26.7	26.6	26.5	26.4	26.3	26.2	26.1
	4	14.4	14.2	14.0	13.9	13.8	13.7	13.7	13.6	13.5
	5	9.89	9.72	9.55	9.47	9.38	9.29	9.20	9.11	9.02
	6	7.72	7.56	7.40	7.31	7.23	7.14	7.06	6.97	6.88
	7	6.47	6.31	6.16	6.07	5.99	5.91	5.82	5.74	5.65
	8	5.67	5.52	5.36	5.28	5.20	5.12	5.03	4.95	4.86
	9	5.11	4.96	4.81	4.73	4.65	4.57	4.48	4.40	4.31
	10	4.71	4.56	4.41	4.33	4.25	4.17	4.08	4.00	3.91
	11	4.40	4.25	4.10	4.02	3.94	3.86	3.78	3.69	3.60
	12	4.16	4.01	3.86	3.78	3.70	3.62	3.54	3.45	3.36
	13	3.96	3.82	3.66	3.59	3.51	3.43	3.34	3.25	3.17
	14	3.80	3.66	3.51	3.43	3.35	3.27	3.18	3.09	3.01
	15	3.67	3.52	3.37	3.29	3.21	3.13	3.05	2.96	2.87
	16	3.55	3.41	3.26	3.18	3.10	3.02	2.93	2.84	2.75
	17	3.46	3.31	3.16	3.08	3.00	2.92	2.83	2.75	2.65
	18	3.37	3.23	3.08	3.00	2.92	2.84	2.75	2.66	2.57
	19	3.30	3.15	3.00	2.92	2.84	2.76	2.67	2.58	2.49
	20	3.23	3.09	2.94	2.86	2.78	2.69	2.61	2.52	2.42
	21	3.17	3.03	2.88	2.80	2.72	2.64	2.55	2.46	2.36
	22	3.12	2.98	2.83	2.75	2.67	2.58	2.50	2.40	2.31
	23	3.07	2.93	2.78	2.70	2.62	2.54	2.45	2.35	2.26
	24	3.03	2.89	2.74	2.66	2.58	2.49	2.40	2.31	2.21
	25	2.99	2.85	2.70	2.62	2.54	2.45	2.36	2.27	2.17
	30	2.84	2.70	2.55	2.47	2.39	2.30	2.21	2.11	2.01
	40	2.66	2.52	2.37	2.29	2.20	2.11	2.02	1.92	1.81
	60	2.50	2.35	2.20	2.12	2.03	1.94	1.84	1.73	1.60
	120	2.34	2.19	2.03	1.95	1.86	1.76	1.66	1.53	1.38
	10,000	2.19	2.04	1.88	1.79	1.70	1.59	1.48	1.33	1.05

Table 9c was generated using Minitab.

Table 10 Confidence Belts for the Correlation Coefficient $(1 - \alpha) = 0.95$

The numbers on the curves are sample sizes.

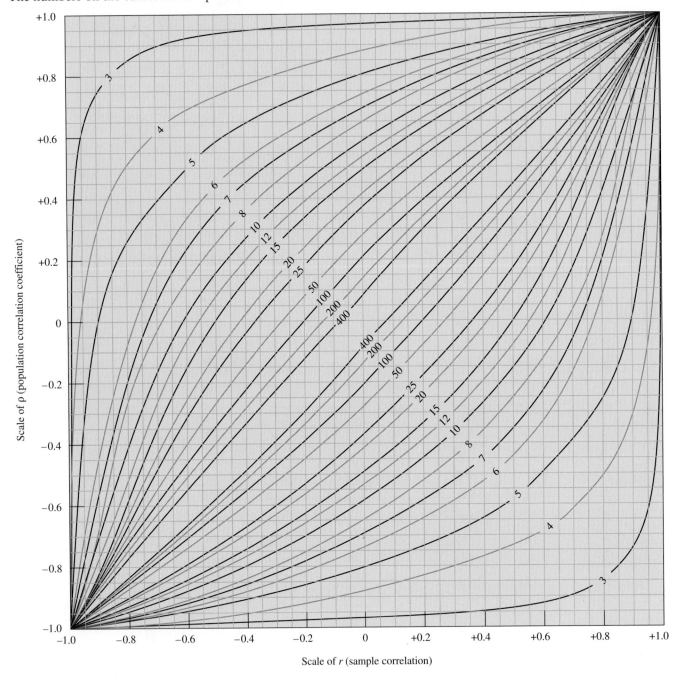

Scale of r (sample correlation)

For specific details about using this table to find confidence intervals, see page 277.

Table 11 Critical Values of r When $\rho = 0$

The entries in this table are the critical values of r for a two-tailed test at α. For simple correlation, df $= n - 2$, where n is the number of pairs of data in the sample. For a one-tailed test, the value of α shown at the top of the table is double the value of α being used in the hypothesis test.

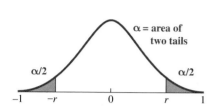

df	α			
	0.10	0.05	0.02	0.01
1	0.988	0.997	1.000	1.000
2	0.900	0.950	0.980	0.990
3	0.805	0.878	0.934	0.959
4	0.729	0.811	0.882	0.917
5	0.669	0.754	0.833	0.875
6	0.621	0.707	0.789	0.834
7	0.582	0.666	0.750	0.798
8	0.549	0.632	0.715	0.765
9	0.521	0.602	0.685	0.735
10	0.497	0.576	0.658	0.708
11	0.476	0.553	0.634	0.684
12	0.458	0.532	0.612	0.661
13	0.441	0.514	0.592	0.641
14	0.426	0.497	0.574	0.623
15	0.412	0.482	0.558	0.606
16	0.400	0.468	0.543	0.590
17	0.389	0.456	0.529	0.575
18	0.378	0.444	0.516	0.561
19	0.369	0.433	0.503	0.549
20	0.360	0.423	0.492	0.537
25	0.323	0.381	0.445	0.487
30	0.296	0.349	0.409	0.449
35	0.275	0.325	0.381	0.418
40	0.257	0.304	0.358	0.393
45	0.243	0.288	0.338	0.372
50	0.231	0.273	0.322	0.354
60	0.211	0.250	0.295	0.325
70	0.195	0.232	0.274	0.302
80	0.183	0.217	0.256	0.283
90	0.173	0.205	0.242	0.267
100	0.164	0.195	0.230	0.254

For specific details about using this table to find p-values and critical values, pages 278–279.

Table 12 Critical Values of the Sign Test

The entries in this table are the critical values for the number of the least frequent sign for a two-tailed test at α for the binomial $p = 0.5$. For a one-tailed test, the value of α shown at the top of the table is double the value of α being used in the hypothesis test.

	α					α			
n	0.01	0.05	0.10	0.25	n	0.01	0.05	0.10	0.25
1					51	15	18	19	20
2					52	16	18	19	21
3				0	53	16	18	20	21
4				0	54	17	19	20	22
5			0	0	55	17	19	20	22
6		0	0	1	56	17	20	21	23
7		0	0	1	57	18	20	21	23
8	0	0	1	1	58	18	21	22	24
9	0	1	1	2	59	19	21	22	24
10	0	1	1	2	60	19	21	23	25
11	0	1	2	3	61	20	22	23	25
12	1	2	2	3	62	20	22	24	25
13	1	2	3	3	63	20	23	24	26
14	1	2	3	4	64	21	23	24	26
15	2	3	3	4	65	21	24	25	27
16	2	3	4	5	66	22	24	25	27
17	2	4	4	5	67	22	25	26	28
18	3	4	5	6	68	22	25	26	28
19	3	4	5	6	69	23	25	27	29
20	3	5	5	6	70	23	26	27	29
21	4	5	6	7	71	24	26	28	30
22	4	5	6	7	72	24	27	28	30
23	4	6	7	8	73	25	27	28	31
24	5	6	7	8	74	25	28	29	31
25	5	7	7	9	75	25	28	29	32
26	6	7	8	9	76	26	28	30	32
27	6	7	8	10	77	26	29	30	32
28	6	8	9	10	78	27	29	31	33
29	7	8	9	10	79	27	30	31	33
30	7	9	10	11	80	28	30	32	34
31	7	9	10	11	81	28	31	32	34
32	8	9	10	12	82	28	31	33	35
33	8	10	11	12	83	29	32	33	35
34	9	10	11	13	84	29	32	33	36
35	9	11	12	13	85	30	32	34	36
36	9	11	12	14	86	30	33	34	37
37	10	12	13	14	87	31	33	35	37
38	10	12	13	14	88	31	34	35	38
39	11	12	13	15	89	31	34	36	38
40	11	13	14	15	90	32	35	36	39
41	11	13	14	16	91	32	35	37	39
42	12	14	15	16	92	33	36	37	39
43	12	14	15	17	93	33	36	38	40
44	13	15	16	17	94	34	37	38	40
45	13	15	16	18	95	34	37	38	41
46	13	15	16	18	96	34	37	39	41
47	14	16	17	19	97	35	38	39	42
48	14	16	17	19	98	35	38	40	42
49	15	17	18	19	99	36	39	40	43
50	15	17	18	20	100	36	39	41	44

From Wilfred J. Dixon and Frank J. Massey, Jr., *Introduction to Statistical Analysis,* 3d ed. (New York: McGraw-Hill, 1969), p. 509. Reprinted by permission.

For specific details about using this table: confidence interval, page 301; *p*-values, pages 302–303; critical values, page 302.

Table 13 Critical Values of U in the Mann-Whitney Test

A. The entries are the critical values of U for a one-tailed test at 0.025 or for a two-tailed test at 0.05.

n_2 \ n_1	1	2	3	4	5	6	7	8	9	10	11	12	13	14	15	16	17	18	19	20
1																				
2								0	0	0	0	1	1	1	1	1	2	2	2	2
3					0	1	1	2	2	3	3	4	4	5	5	6	6	7	7	8
4				0	1	2	3	4	4	5	6	7	8	9	10	11	11	12	13	14
5			0	1	2	3	5	6	7	8	9	11	12	13	14	15	17	18	19	20
6			1	2	3	5	6	8	10	11	13	14	16	17	19	21	22	24	25	27
7			1	3	5	6	8	10	12	14	16	18	20	22	24	26	28	30	32	34
8		0	2	4	6	8	10	13	15	17	19	22	24	26	29	31	34	36	38	41
9		0	2	4	7	10	12	15	17	20	23	26	28	31	34	37	39	42	45	48
10		0	3	5	8	11	14	17	20	23	26	29	33	36	39	42	45	48	52	55
11		0	3	6	9	13	16	19	23	26	30	33	37	40	44	47	51	55	58	62
12		1	4	7	11	14	18	22	26	29	33	37	41	45	49	53	57	61	65	69
13		1	4	8	12	16	20	24	28	33	37	41	45	50	54	59	63	67	72	76
14		1	5	9	13	17	22	26	31	36	40	45	50	55	59	64	67	74	78	83
15		1	5	10	14	19	24	29	34	39	44	49	54	59	64	70	75	80	85	90
16		1	6	11	15	21	26	31	37	42	47	53	59	64	70	75	81	86	92	98
17		2	6	11	17	22	28	34	39	45	51	57	63	67	75	81	87	93	99	105
18		2	7	12	18	24	30	36	42	48	55	61	67	74	80	86	93	99	106	112
19		2	7	13	19	25	32	38	45	52	58	65	72	78	85	92	99	106	113	119
20		2	8	13	20	27	34	41	48	55	62	69	76	83	90	98	105	112	119	127

B. The entries are the critical values of U for a one-tailed test at 0.05 or for a two-tailed test at 0.10.

n_2 \ n_1	1	2	3	4	5	6	7	8	9	10	11	12	13	14	15	16	17	18	19	20
1																			0	0
2					0	0	0	1	1	1	1	2	2	2	3	3	3	4	4	4
3			0	0	1	2	2	3	3	4	5	5	6	7	7	8	9	9	10	11
4			0	1	2	3	4	5	6	7	8	9	10	11	12	14	15	16	17	18
5		0	1	2	4	5	6	8	9	11	12	13	15	16	18	19	20	22	23	25
6		0	2	3	5	7	8	10	12	14	16	17	19	21	23	25	26	28	30	32
7		0	2	4	6	8	11	13	15	17	19	21	24	26	28	30	33	35	37	39
8		1	3	5	8	10	13	15	18	20	23	26	28	31	33	36	39	41	44	47
9		1	3	6	9	12	15	18	21	24	27	30	33	36	39	42	45	48	51	54
10		1	4	7	11	14	17	20	24	27	31	34	37	41	44	48	51	55	58	62
11		1	5	8	12	16	19	23	27	31	34	38	42	46	50	54	57	61	65	69
12		2	5	9	13	17	21	26	30	34	38	42	47	51	55	60	64	68	72	77
13		2	6	10	15	19	24	28	33	37	42	47	51	56	61	65	70	75	80	84
14		2	7	11	16	21	26	31	36	41	46	51	56	61	66	71	77	82	87	92
15		3	7	12	18	23	28	33	39	44	50	55	61	66	72	77	83	88	94	100
16		3	8	14	19	25	30	36	42	48	54	60	65	71	77	83	89	95	101	107
17		3	9	15	20	26	33	39	45	51	57	64	70	77	83	89	96	102	109	115
18		4	9	16	22	28	35	41	48	55	61	68	75	82	88	95	102	109	116	123
19	0	4	10	17	23	30	37	44	51	58	65	72	80	87	94	101	109	116	123	130
20	0	4	11	18	25	32	39	47	54	62	69	77	84	92	100	107	115	123	130	138

Reproduced from the *Bulletin of the Institute of Educational Research at Indiana University,* vol. 1, no. 2; with the permission of the author and the publisher.
For specific details about using this table to find p-values, see pages 309–310; critical values, page 309.

Table 14 Critical Values for Total Number of Runs (V)

The entries in this table are the critical values for a two-tailed test using $\alpha = 0.05$. For a one-tailed test, at $\alpha = 0.025$ use only one of the critical values: the smaller critical value for a left-hand critical region, the larger for a right-hand critical region.

The larger of n_1 and n_2

The smaller of n_1 and n_2	5	6	7	8	9	10	11	12	13	14	15	16	17	18	19	20
2								2	2	2	2	2	2	2	2	2
								6	6	6	6	6	6	6	6	6
3		2	2	2	2	2	2	2	2	2	3	3	3	3	3	3
		8	8	8	8	8	8	8	8	8	8	8	8	8	8	8
4	2	2	2	3	3	3	3	3	3	3	4	4	4	4	4	4
	9	9	10	10	10	10	10	10	10	10	10	10	10	10	10	10
5	2	3	3	3	3	3	4	4	4	4	4	4	4	5	5	5
	10	10	11	11	12	12	12	12	12	12	12	12	12	12	12	12
6		3	3	3	4	4	4	4	5	5	5	5	5	5	6	6
		11	12	12	13	13	13	13	14	14	14	14	14	14	14	14
7			3	4	4	5	5	5	5	5	6	6	6	6	6	6
			13	13	14	14	14	14	15	15	15	16	16	16	16	16
8				4	5	5	5	6	6	6	6	6	7	7	7	7
				14	14	15	15	16	16	16	16	17	17	17	17	17
9					5	5	6	6	6	7	7	7	7	8	8	8
					15	16	16	16	17	17	18	18	18	18	18	18
10						6	6	6	7	7	7	7	8	8	8	9
						16	17	17	18	18	18	19	19	19	20	20
11							7	7	7	8	8	8	9	9	9	9
							17	18	19	19	19	20	20	20	21	21
12								7	8	8	8	9	9	9	10	10
								19	19	20	20	21	21	21	22	22
13									8	9	9	9	10	10	10	10
									20	20	21	21	22	22	23	23
14										9	9	10	10	10	11	11
										21	22	22	23	23	23	24
15											10	10	11	11	11	12
											22	23	23	24	24	25
16												11	11	11	12	12
												23	24	25	25	25
17													11	12	12	13
													25	25	26	26
18														12	13	13
														26	26	27
19															13	13
															27	27
20																14
																28

From C. Eisenhart and F. Swed, "Tables for testing randomness of grouping in a sequence of alternatives,i *Annals of Statistics,* vol. 14 (1943): 66–87. Reprinted by permission.
For specific details about using this table to find *p*-values, see pages 313–314; critical values, page 313.

Table 15 Critical Values of Spearman's Rank Correlation Coefficient

The entries in this table are the critical values of r_s for a two-tailed test at α. For a one-tailed test, the value of α shown at the top of the table is double the value of α being used in the hypothesis test.

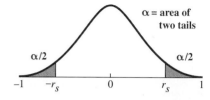

n	$\alpha = 0.10$	$\alpha = 0.05$	$\alpha = 0.02$	$\alpha = 0.01$
5	0.900	—	—	—
6	0.829	0.886	0.943	—
7	0.714	0.786	0.893	0.929
8	0.643	0.738	0.833	0.881
9	0.600	0.700	0.783	0.833
10	0.564	0.648	0.745	0.794
11	0.536	0.618	0.709	0.755
12	0.503	0.587	0.678	0.727
13	0.484	0.560	0.648	0.703
14	0.464	0.538	0.626	0.679
15	0.446	0.521	0.604	0.654
16	0.429	0.503	0.582	0.635
17	0.414	0.485	0.566	0.615
18	0.401	0.472	0.550	0.600
19	0.391	0.460	0.535	0.584
20	0.380	0.447	0.520	0.570
21	0.370	0.435	0.508	0.556
22	0.361	0.425	0.496	0.544
23	0.353	0.415	0.486	0.532
24	0.344	0.406	0.476	0.521
25	0.337	0.398	0.466	0.511
26	0.331	0.390	0.457	0.501
27	0.324	0.382	0.448	0.491
28	0.317	0.375	0.440	0.483
29	0.312	0.368	0.433	0.475
30	0.306	0.362	0.425	0.467

Adapted in part from J. H. Zar, Significance testing of the Spearman rank correlation coeffient, *Journal of the American Statistical Association* **67** (1972): 578–580. Reprinted with permission from the *Journal of the American Statistical Association*. Copyright, © (1972) by the American Statistical Association. All rights reserved; and, in part, from A. Otten, Note on the Spearman rank correlation coefficient, *Journal of the American Statistical Association* **68** (1973): 585. Reprinted with permission from the *Journal of the American Statistical Association*. Copyright, © (1973) by the American Statistical Association. All rights reserved.

For specific details about using this table to find *p*-values, see page 319; critical values, page 319.

LEARN YOUR WAY

With **STAT2**, students have a multitude of study aids at their fingertips.

The **Students Solutions Manual** contains the worked-out solutions to every odd-numbered exercise, further reinforcing students understanding of statistical concepts in **STAT2**.

In addition to the Student Solutions Manual, **CourseMate** offers exercises and questions that correspond to every section and chapter in the text for students to practice the concepts they learned.

Students can sign in at **www.cengagebrain.com**.

grab (convenience) samples, 11
graphic display
  box-and-whiskers display, 45
  deception with, 49
  dotplots and stem-and-leaf displays
    for quantitative data, 25–29
  frequency distributions and
    histograms, 29–33
  ogives, 34–35
  pie charts and bar graphs for
    qualitative data, 24–25
  purpose of, 23–24
  tree diagrams, 77–78
grouped frequency distributions,
  29–31

H

histograms, 31–33
  of binomial distributions, 111–112
  probability, 103–104
  of sampling distribution of sample
    means, 142–143
homogeneity, test of, 247–249
hypothesis, 162
hypothesis testing
  for μ with σ unknown, 187–191
  binomial parameter p, 195–198
  chi-square, 199–203
  classical approach, 173–177
  difference between proportions,
    223–225
  difference between two means,
    217–221
  equality of variances, 226–229
  homogeneity test, 248–249
  independence test, 244–247
  linear correlation, 277–279
  Mann–Whitney U test, 307–312
  mean of paired differences,
    212–214
  multinomial experiments, 239–244
  nature of, 161–166
  probability-value approach, 166–173
  rank correlation, 315–319
  regression, 286–287
  runs test, 312–315
  for several means, 256–260,
    264–266
  sign test, 301–306

I

independence, test of, 244–247
independent events, 90–93, 93–94
independent (input) variables, 58. See
  also linear correlation analysis

independent means, 215
independent samples, 208–210
inferences in correlation and regression
  linear correlation coefficient,
    276–279
  slope of regression line, 284–287
inferences with one population
  binomial parameter, 191–198
  population mean, 182–191
  variance and standard deviation,
    198–203
inferences with two populations
  dependent and independent
    samples, 208–210
  difference between means, 215–221
  difference between proportions,
    221–225
  mean of paired differences,
    210–214
  ratio of variances, 225–229
inferential statistics
  descriptive statistics vs., 4
  estimation, concept of, 153–155
  objective of, 152
infinite vs. finite populations, 7
input (independent) variables, 58. See
  also linear correlation analysis
interquartile range, 44
interval estimates, 155, 156–161

J

J-shaped distribution, 33
judgment samples, 13

L

large numbers, law of, 79–80
law of large numbers, 79–80
least squares criterion, 65
level of confidence $(1 − α)$, 155
level of significance. See alpha (α) or
  level of significance
lies, statistical, 49
line of best fit, 65–68, 279–281
linear correlation analysis
  bivariate data and covariance of x
    and y, 273–275
  causation and lurking variables,
    63–64
  coefficient of linear correlation, r,
    61–63, 275–276
  inferences about linear correlation
    coefficient, 276–279
  purpose of, 60
  regression, relationship with, 292
  types of relationships, 60–61

linear regression. See regression
  analysis
long-term average, 80
lurking variables, 63–64

M

Mann–Whitney U test, 307–312, 347
marginal totals (marginals), 56
mathematical models, 263
maximum error of estimate (E)
  in binomial parameter p inference,
    193–194
  in estimation of population mean,
    157, 160–161
mean, 35. See also population mean;
  sampling distribution of sample
  means (SDSM)
  as "average," 38
  of binomial distributions,
    110–112
  deviation from, 40
  difference between two, 215–221,
    307–312
  of discrete random variables,
    104–106
  hypothesis testing for several,
    256–260, 264–266
  of paired differences, 210–214
  sampling distribution of, 138
  t in relation to, 185–186
  t-statistic for, 187
mean square (MS), 259, 260–262
measurability, 10–11
measures of central tendency, 35–39.
  See also mean
measures of dispersion, 39–41. See
  also standard deviation; variance
median, 35–36, 38, 301
method of least squares, 65
midquartiles, 44
midrange, 37, 38
misrepresentation, superimposed, 49
mode, 37, 38
multinomial experiments, 239–244
multiple regression, 281
multiplication rule, general, 85–86
multiplication rule, special, 92
multistage random sampling, 15–16
mutually exclusive events, 87–90,
  93–94

N

nominal variables, 8
noncritical region (acceptance region),
  174

quantitative data, 25–29
quantitative variables, 8, 57–60
quartiles, 42

R

$r \times c$ contingency table, 247
random numbers table, 327–328
random samples. *See also* samples
  multistage random sampling, 15–16
  sampling distribution of sample means and, 139, 141
  simple, 13–14
  systematic samples, 14–15
random variables
  binomial, 106–107, 191–192
  discrete vs. continuous, 100–102
randomness, 14, 312
range, 39
rank correlation, 315–319, 349
regression analysis
  confidence intervals for, 287–292
  correlation, relationship with, 292
  curvilinear and multiple, 281
  line of best fit and linear model, 65–68, 279–284
  making predictions with, 68–69
  precautions, 291
  purpose of, 64
  slope inferences, 284–287
relative frequency, 32, 74
relative frequency histograms, 32
replicates, 256
representative samples, 13
research (alternative) hypothesis, 162. *See also* hypothesis testing
response (observed) variable, 260
round-off rule, 37
runs test, 312–315, 348

S

sample means, 154–155. *See also* sampling distribution of sample means (SDSM)
sample size
  for binomial parameter p, 193–195
  confidence intervals and, 159–161
  efficiency and, 300
sample space, 75, 76–78, 89
sample variance, 40
samples, 7
  between-sample and within-sample variation, 262
  cluster, 16
  convenience (grab), 11

judgment, 13
probability, 13–15
stratified random, 15
systematic, 14–15
volunteer, 11
sampling design, 13–16
sampling distribution of means, 138
sampling distribution of a sample statistic, 137
sampling distribution of sample means (SDSM), 138
  application of, 145–147
  central limit theorem and, 141–142
  comparison by sample size, 144
  constructing, 142–145
  creating, 138–140
  estimation of population mean with, 156
  hypothesis testing and, 166
  point estimates and, 155
  standard error of the mean, 141
sampling distribution of the slope, 284
sampling distributions, 136–140, 144
sampling frames, 12–13
sampling methods, 11
scatter diagrams, 58
  constructing, 58–60
  correlation and, 60–61
  line of best fit and, 66, 281
SDSM. *See* sampling distribution of sample means (SDSM)
side-by-side dotplots, 29
sign test, 300–306, 346
simple random samples, 13–14
single-stage sampling, 13–15
skewed distribution, 33
slope
  inferences, 284–287
  line of best fit and, 65–66, 279
sources, 5, 208
Spearman, Charles, 315
Spearman's rank correlation coefficient, r_s, 315–319, 349
special addition rule, 90
special multiplication rule, 92
standard deviation (σ), 46
  of binomial distributions, 110–112
  Chebyshev's theorem and, 47–48
  of discrete random variables, 105–106
  empirical rule and, 46–47
  inferences with chi-square, 198–200, 201–203
  known vs. unknown, and z or t, 183–184
  maximum error and, 161
  of a sample, 40–41
standard error of the estimate, 282–284
standard error of the mean, 141, 145–146, 157

standard error of regression (slope), 285
standard normal distribution, 120–123. *See also* normal distribution; normal probability distributions
standard scores (z-scores), 45–46
  for bounded areas, 128–129
  converting sample means into, 145
  corresponding z values for $z(\alpha)$, 127–128
  cut-off point, locating, 125–126
  normal distribution and, 120–123
  Table 4 and commonly used z values, 128
  visual interpretation of $z(\alpha)$, 127
star ($\star$), use of, 169
statistic, 8
Statistical Abstract of the United States, 2
statistical hypothesis tests, 162. *See also* hypothesis testing
statistics, 3–4
  descriptive vs. inferential, 4
  everyday examples, 4–7
  probability compared to, 81–82
  technology and, 16
  terminology of, 7–8, 9
  uses of, 7
stem-and-leaf displays, 26–30, 32
stratified random samples, 15
Student's t-statistic, 183–184, 215, 335, 336. *See also* t-statistic
subjective probability, 78
sum of squares, 41
  in ANOVA, 257–258
  in Pearson's product moment formula, 62
surveys, 12
symmetrical distribution, 33
systematic samples, 14–15
systematic sampling method, 14–15

T

t-distribution
  confidence interval, 186–187
  degrees of freedom and critical values, 185
  difference between two means, 216
  hypothesis testing, 187–191
  mean of paired differences and, 211
  properties of, 184–185
t-distribution table, 185–186
t-statistic
  critical values, 345
  hypothesis testing, 187–191
  Student's, 183–184, 215, 335, 336
  z vs., 183–184
tallies, 30–31

Learning Objectives and Outcomes

What's a review card?

To help you refresh key concepts from each chapter, we've developed a review card. Each card can be torn out and carried easily for quick review before quizzes and tests.

(This column contains the chapter objectives with related learning outcomes and brief reviews.)

(18–120)

... d the normal ... curve ...

The normal probability distribution is considered the single most importan... distribution. An unlimited number of continuous random variables have ei... ... normal distribution. Several other probability distribut... ...random variables are also approximately normal under certain ...roportion, and probability are basically the same concepts. ...entation of all three. The empirical rule is a fairly crude ...t we are able to find probabilities associated only with whole number multiples of the standard deviation.

6.2 The Standard Normal Distribution (pp. 120–123)

Understand that the normal curve is symmetrical about the mean with an area of 0.5000 on each side of the mean

1. The total area under the standard normal curve is equal to 1.
2. The distribution is mounded and symmetrical; it extends indefinitely in both directions, approaching but never touching the horizontal axis.
3. The distribution has a mean of 0 and a standard deviation of 1.
4. The mean divides the area in half—0.5000 on each side.
5. Nearly all the area is between $z = -3.00$ and $z = 3.00$.

(Key pieces of art from the chapter support the summaries when relevant.)

...ons of Normal Distributions (pp. 124–126)

...es for intervals defined on the st... ...pute, describe, and interpret a z... ...al distribution * Compute z-scores... for applications of the normal distribution

We can convert information about the standard normal variable z into probability, so we can also convert probability information about the standard normal distribution into z-scores. That means we can apply this methodology to all normal distributions using the standard score, z.

6.4 Notation (pp. 127–129)

z will be used with great frequency, and the convention that we will use as an "algebraic name" for a specific z-score is $z(\alpha)$, where α represents the "area to the right" of the z being named.

6.5 Normal Approximation of the Bi...

Compute z-scores and probabilities for normal a... binomial

The binomial distribution is a probability distribution of the discrete random variable x, the number of successes observed in n repeated independent trials. Binomial probabilities can be reasonably approximated by using the normal probability distribution. The binomial random variable is discrete, whereas the normal random variable is continuous. The continuity correction factor allows a discrete variable to be converted into a continuous variable.

Key Concepts

Here, you'll find the key terms in the order they appear in the chapter. These are the boldface terms in the chapter.

Vocabulary

normal probability distribution (p. 118)

...uous random variables (...8)

...l distribution (p. 118)

...te random variables (p. 118)

percentage (p. 120)

proportion (p. 120)

probability (p. 120)

standard normal distribution (p. 120)

standard score, or z-score (p. 120)

cumulative area (p. 120)

normal curve (p. 121)

binomial distribution (p. 129)

binomial probability (p. 129)

discrete (p. 129)

continuous (p. 129)

continuity correction factor (p. 130)

Key Formulas

(Formulas from the chapter appear next.)

(6.1) Normal probability distribution function

$$y = f(x) = \frac{e^{-\frac{1}{2}\left(\frac{x-\mu}{\sigma}\right)^2}}{\sigma\sqrt{2\pi}} \quad \text{for all real } x$$

(6.2) Probability associated with interval from $x = a$ to $x = b$

$$P(a \leq x \leq b) = \int_a^b f(x)\,dx$$

(6.3) Standard score

$$z = \frac{x - \mu}{\sigma}$$

Rule

(Finally, this column ends with rules and assumptions described in the chapter.)

The normal distribution provides a reasonable approximation to a binomial probability distribution whenever the values of np and $n(1-p)$ both equal or exceed 5.

Practice Test

PART I–Knowing the Definitions

Answer "True" if the statement is always true. If the statement is not always true, replace the words printed in bold with words that make the statement always true.

_____ 6.1 The normal probability distribution is symmetric about **zero.**

_____ 6.2 The total area under the curve of any normal distribution is **1.0.**

_____ 6.3 The theoretical probability that a p⁻ of a **continuous** random variable exactly zero.

_____ 6.4 The unit of measure for the stan... **same as the unit of measure of t.**

_____ 6.5 All **normal** distributions have the same general probability function and distribution.

_____ 6.6 In the notation z(0.05), the number in parentheses is the measure of the area to the **left** of the z-score.

_____ 6.7 Standard normal scores have a mean of **one** and a standard deviation of **zero.**

_____ 6.8 Probability distributions of **all** continuous random variables are normally distributed.

_____ 6.9 We are able to add and subtract the areas under the curve of a continuous distribution because these areas represent probabilities of **independent** events.

_____ 6.10 The most common distribution of a continuous random variable is the **binomial** probability.

PART II–Applying the Concepts

6.11 Find the following probabilities for z, the standard normal score:

a. $P(0 < z < 2.42)$ b. $P(z < 1.38)$

c. $P(z < -1.27)$ d. $P(-1.35 < z < 2.72)$

6.12 Find the value of each z-score:

a. $P(z > ?) = 0.2643$ b. $P(z < ?) = 0.17$ c. $z(0.04)$

6.13 Use the symbolic notation $z(\alpha)$ to give the symbolic name for each z-score shown in the figure.

a.

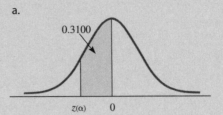

b.

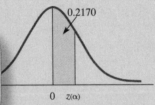

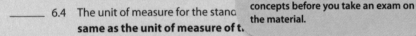

...es of flashlight batteries are normally distributed about a mean of 35.6 hours with a standard deviation of 5.4 hours. Kevin selected one of these batteries at random and tested it. What is the probability that this one battery will last less than 40.0 hours?

6.15 The lengths of time, x, spent by students commuting one way to college each day are believed to have a mean of 22 minutes with a standard deviation of 9 minutes. If the lengths of time spent commuting are approximately normally distributed, find the time, x, that separates the 25% who spend the most time commuting from the rest of the commuters.

6.16 Thousands of high school students take the SAT each year. The scores attained by the students in a certain city are approximately normally distributed with a mean of 490 and a standard deviation of 70. Find:

a. the percentage of students who score between 600 and 700

b. the percentage of students who score less than 650

c. the third quartile

d. the 15th percentile, P_{15}

e. the 95th percentile, P_{95}

PART III–Understanding the Concepts

6.17 In 50 words, describe the standard normal distribution.

6.18 Describe the meaning of the symbol $z(\alpha)$.

6.19 Explain why the standard normal distribution, as computed in Table 3 in Appendix B, can be used to find probabilities for all normal distributions.

Learning Objectives and Outcomes

1.1 What Is Statistics? (pp. 3–10)

Understand and be able to describe the difference between descriptive and inferential statistics * Understand and be able to identify and interpret the relationships between sample and population, and statistic and parameter * Know and be able to identify and describe the different types of variables

Descriptive statistics includes the collection, presentation, and description of sample data. *Inferential statistics* refers to the technique of interpreting the values resulting from the descriptive techniques and making decisions and drawing conclusions about the population. Because large populations are difficult to study, statisticians study the data from a subset of the population, which is called a sample. Statisticians are interested in particular variables of that sample. Variables can be either qualitative or quantitative.

1.2 Measurability and Variability (pp. 10–11)

Understand that variability is inherent in everything, including the sample process

One of the primary objectives of statistical analysis is to measure variability. That's because within a set of data, there is always variability. Limited or no variability would indicate that the measuring device is not calibrated to a small enough unit of measure.

1.3 Data Collection (pp. 11–16)

Understand how convenience and volunteer samples result in biased samples * Understand the differences among and be able to identify experiments, observational studies, and judgment samples * Understand and be able to describe the single-stage sampling methods of "simple random sample" and "systematic sampling" * Understand and be able to describe the multistage sampling methods of "stratified sampling" and "cluster sampling"

Sampling methods should produce data that are representative of the population and are unbiased. The five steps of the data-collection process include: (1) defining objectives, (2) variables and population of interest, (3) data-collection and measurement schemes, (4) collecting the data, and (5) reviewing the sampling process to ensure techniques were appropriate and produced good data. Sample designs can either be judgment or probability samples, and sampling methods can be either single-stage or multistage.

1.4 Statistics and Technology (p. 16)

Understand the role of technology in responsible statistical methodology

Technology makes easier the long and sometimes tedious calculations required in statistics, but it's important to remember that your results are only as accurate as the data you put in.

Key Concepts

Vocabulary

statistics (p. 4)

population (p. 7)

finite population (p. 7)

infinite population (p. 7)

sample (p. 7)

variable (p. 7)

data value (p. 7)

data (p. 7)

experiment (p. 7)

parameter (p. 7)

statistic (p. 8)

qualitative (or attribute or categorical) variable (p. 8)

quantitative (or numerical) variable (p. 8)

nominal variable (p. 8)

ordinal variable (p. 8)

discrete variable (p. 8)

continuous variable (p. 8)

variability (p. 11)

sampling method (p. 11)

biased sampling method (p. 11)

unbiased sampling method (p. 11)

experiments (p. 12)

observational studies (p. 12)

sampling frame (p. 12)

representative (p. 13)

judgment samples (p. 13)

probability samples (p. 13)

single-stage sampling (p. 13)

simple random sample (p. 13)

systematic sampling method (p. 14)

multistage random sampling (p. 15)

stratified random sample (p. 15)

proportional stratified sampling (p. 15)

cluster sample (p. 16)

PART I–Knowing the Definitions

Answer "True" if the statement is always true. If the statement is not always true, replace the words printed in bold with words that make the statement always true.

_____ 1.1 **Inferential** statistics is the study and description of data that result from an experiment.

_____

_____ 1.2 **Descriptive statistics** is the study of a sample that enables us to make projections or estimates about the population from which the sample is drawn.

_____

_____ 1.3 A **population** is typically a very large collection of individuals or objects about which we desire information.

_____

_____ 1.4 A statistic is the calculated measure of some characteristic of a **population.**

_____

_____ 1.5 A parameter is the measure of some characteristic of a **sample.**

_____

_____ 1.6 As a result of surveying 50 freshmen, it was found that 16 had participated in interscholastic sports, 23 had served as officers of classes and clubs, and 18 had been in school plays during their high school years. This is an example of **numerical data.**

_____

_____ 1.7 The "number of rotten apples per shipping crate" is an example of a **qualitative** variable.

_____

_____ 1.8 The "thickness of a sheet of sheet metal" used in a manufacturing process is an example of a **quantitative** variable.

_____

_____ 1.9 A **representative** sample is a sample obtained in such a way that all individuals had an equal chance of being selected.

_____

_____ 1.10 The basic objectives of **statistics** are obtaining a sample, inspecting this sample, and then making inferences about the unknown characteristics of the population from which the sample was drawn.

_____

PART II–Applying the Concepts

The owners of Corner Convenience Store are concerned about the quality of service their customers receive. In order to study the service, they collected samples for each of several variables.

1.11 Classify each of the following variables as nominal, ordinal, discrete, or continuous:

   a. Method of payment for purchases (cash, credit card, check)

   b. Customer satisfaction (very satisfied, satisfied, not satisfied)

   c. Amount of sales tax on purchase

   d. Number of items purchased

   e. Customer's driver's license number

1.12 The mean checkout time for all customers at Corner Convenience Store is to be estimated by using the mean checkout time for 75 randomly selected customers. Match the items with the statistical terms in the columns below.

_____ data value   (a) the 75 customers

_____ data   (b) the mean time for all customers

_____ experiment   (c) 2 minutes, one customer's checkout time

_____ parameter

_____ population   (d) the mean time for the 75 customers

_____ sample   (e) all customers at Corner Convenience Store

_____ statistic

_____ variable   (f) the checkout time for each customer

   (g) the 75 checkout times

   (h) the process used to select 75 customers and measure their times

PART III–Understanding the Concepts

Write a brief paragraph in response to each problem.

1.13 The population and the sample are both sets of objects. Describe the relationship between them and give an example.

1.14 The variable and the data for a specific situation are closely related. Explain this relationship and give an example.

1.15 The data, the statistic, and the parameter are all values used to describe a statistical situation. How does one distinguish among these three terms? Give an example.

1.16 What conditions are required for a sample to be a random sample? Explain and include an example of a sample that is random and one that is not random.

_____

Practice problems can be found at the end of Chapter 1, and solutions for the practice test can be found on the CourseMate for STAT2 site—login at underlinecengagebrain.com.

Learning Objectives and Outcomes

2.1 Graphs, Pareto Diagrams, and Stem-and-Leaf Displays (pp. 23–29)

Create and interpret graphical displays, including circle graphs, bar graphs, Pareto diagrams, dotplots, and stem-and-leaf diagrams

Both qualitative and quantitative data can be summarized visually in graphical depictions. There are several graphic ways to describe data, but regardless of the type of data being displayed, graphic representations should be completely self-explanatory.

2.2 Frequency Distributions and Histograms (pp. 29–35)

Create and interpret frequency histograms and relative frequency histograms * Identify the shapes of distributions

Data sets are often large. Frequency distributions are tabular depictions that make volumes of data more manageable. A histogram can depict a frequency distribution or a relative frequency distribution. Cumulative frequency distributions pair cumulative frequencies with the values of the variables and can be displayed graphically using an ogive.

2.3 Measures of Central Tendency (pp. 35–39)

Compute, describe, and compare the four measures of central tendency: mean, median, mode, and midrange

Measures of central tendency are numerical values that locate, in some sense, the center of the data. Common measures are the mean, median, mode, and midrange.

2.4 Measures of Dispersion (pp. 39–41)

Compute, describe, compare, and interpret the two measures of dispersion: range and standard deviation (variance)

Measures of dispersion describe the amount of spread or variability that is found among the data. Such measures include the range, variance, and standard deviation. There is no limit to how spread out the data can be, so measures of dispersion can be very large.

2.5 Measures of Position (pp. 41–46)

Compute, describe, and interpret the measures of position: quartiles, percentiles, and z-scores

Measures of position describe the position of a specific data value in relation to the rest of the data. Quartiles and percentiles are two of the most popular measures of position. Other measures of position include midquartiles, 5-number summaries, and z-scores and are related to quartiles and percentiles.

2.6 Interpreting and Understanding Standard Deviation (pp. 46–48)

Understand the empirical rule and Chebyshev's theorem and be able to assess a set of data's compliance to these rules

Standard deviation allows the comparison of one set of data with another. According to the empirical rule, if a variable is normally distributed, then 68% of the data will fall within one standard deviation, 95% will fall within two standard deviations, and 99.7% of the data will fall within three. For all data, whether normally distributed or not, Chebyshev's theorem states that at least 75% of the data will fall within two standard deviations.

Key Concepts

Vocabulary

pie charts (circle graphs) (p. 24)
bar graphs (p. 24)
Pareto diagram (p. 24)
distribution (p. 25)
dotplot display (p. 26)
stem-and-leaf display (p. 26)
frequency distribution (p. 29)
frequency (p. 29)
ungrouped frequency distribution (p. 29)
grouped frequency distribution (p. 29)
class width (p. 30)
class midpoint (class mark) (p. 31)
histogram (p. 31)
cumulative frequency distribution (p. 34)
ogive (p. 34)
mean (arithmetic mean) (p. 35)
median (p. 35)
mode (p. 37)
midrange (p. 37)
range (p. 39)
deviation from the mean (p. 40)
sample variance (p. 40)
standard deviation of a sample (p. 40)
quartiles (p. 42)
percentiles (p. 42)
midquartile (p. 44)
5-number summary (p. 44)
interquartile range (p. 44)
box-and-whiskers display (p. 45)
standard score or z-score (p. 45)
empirical rule (p. 46)
Chebyshev's theorem (p. 47)

Key Formulas

(2.1) Mean (arithmetic mean)

$$x\text{-}bar = \frac{sum\ of\ all\ x}{number\ of\ x}$$

$$\bar{x} = \frac{\Sigma x}{n}$$

(2.2) Depth of median

$$depth\ of\ median = \frac{sample\ size + 1}{2}$$

$$d(\tilde{x}) = \frac{n+1}{2}$$

(2.3) Midrange

$$midrange = \frac{low\ value + high\ value}{2}$$

$$midrange = \frac{L + H}{2}$$

(2.4) Range

$$range = high\ value - low\ value$$

$$range = H - L$$

(2.5) Sample variance

$$s\text{-}squared = \frac{sum\ of\ (deviations\ squared)}{number - 1}$$

$$s^2 = \frac{\Sigma(x - \bar{x})^2}{n - 1}$$

(2.6) Sample standard deviation

$$s = square\ root\ of\ sample\ variance$$

$$s = \sqrt{s^2}$$

(2.7) Sample variance

$$s^2 = \frac{SS(x)}{n - 1}$$

(2.8) Sum of squares for x

$$SS(x) = \Sigma x^2 - \frac{(\Sigma x)^2}{n}$$

(2.9) Sample variance, "short-cut formula"

$$s\text{-}squared = \frac{(sum\ of\ x^2) - \left[\frac{(sum\ of\ x)^2}{number}\right]}{number - 1}$$

$$s^2 = \frac{\Sigma x^2 - \frac{(\Sigma x)^2}{n}}{n - 1}$$

(2.10) Midquartile

$$midquartile = \frac{Q_1 + Q_3}{2}$$

(2.11) Standard score, or z-score

$$z = \frac{value - mean}{std.\ dev.} = \frac{x - \bar{x}}{s}$$

2.7 The Art of Statistical Deception (p. 49)

Graphical displays of statistics can be tricky and misleading when they are designed to show only a portion of and not the whole picture. Inadequate or inaccurate labeling, uneven frequency scales, superimposed information, and truncated scales lead to misleading or deceptive visual representations.

Practice Test

PART I–Knowing the Definitions

Answer "True" if the statement is always true. If the statement is not always true, replace the words printed in bold with words that make the statement always true.

_____ 2.1 The **mean** of a sample always divides the data into two halves (half larger and half smaller in value than itself).

_____

_____ 2.2 A measure of **central tendency** is a quantitative value that describes how widely the data are dispersed about a central value.

_____

_____ 2.3 The sum of the squares of the deviations from the mean, $\Sigma(x - \bar{x})^2$, will **sometimes** be negative.

_____

_____ 2.4 For any distribution, the sum of the deviations from the mean equals **zero.**

_____

_____ 2.5 The standard deviation for the set of values 2, 2, 2, 2, and 2 is **2.**

_____

_____ 2.6 On a test John scored at the 50th percentile and Jorge scored at the 25th percentile; therefore, John's test score was **twice** Jorge's test score.

_____

PART II–Applying the Concepts

2.7 A sample of the purchases of several Corner Convenience Store customers resulted in the following sample data (x = number of items purchased per customer):

x	1	2	3	4	5
f	6	10	9	8	7

a. What does the 2 represent?

b. What does the 9 represent?

c. How many customers were used to form this sample?

d. How many items were purchased by the customers in this sample?

e. What is the largest number of items purchased by one customer?

Find each of the following (show formulae and work):

f. mode   g. median   h. midrange

i. mean   j. variance   k. standard deviation

PART III–Understanding the Concepts

Answer all questions.

2.8 The Corner Convenience Store kept track of the number of paying customers it had during the noon hour each day for 100 days. The resulting statistics are rounded to the nearest integer:

mean = 95   third quartile = 107

median = 97   midrange = 93

mode = 98   range = 56

first quartile = 85   standard deviation = 12

a. The Corner Convenience Store served what number of paying customers during the noon hour more often than any other number? Explain how you determined your answer.

b. On how many days were there between 85 and 107 paying customers during the noon hour? Explain how you determined your answer.

c. What was the greatest number of paying customers during any one noon hour? Explain how you determined your answer.

d. For how many of the 100 days was the number of paying customers within three standard deviations of the mean ($\bar{x} \pm 3s$)? Explain how you determined your answer.

Practice problems can be found at the end of Chapter 2, and solutions for the practice test can be found on the CourseMate for STAT2 site—login at <u>cengagebrain.com</u>.

Learning Objectives and Outcomes

3.1 Bivariate Data (pp. 54–60)

Understand and be able to present and describe the relationship between two quantitative variables using a scatter diagram

Bivariate data are the values of two different variables that are obtained from the population. Bivariate data can be both qualitative, both quantitative, or one of each type.

3.2 Linear Correlation (pp. 60–64)

Define and understand the difference between correlation and causation * Determine and explain possible lurking variables and their effects on a linear relationship

Linear correlation analysis measures the strength of the linear relationship between two variables. Correlation is positive when y tends to increase as x increases, and is negative when y tends to decrease as x increases. A strong correlation does not necessarily imply causation.

3.3 Linear Regression (pp. 64–69)

Compute, describe, and interpret a line of best fit * Create a scatter diagram with the line of best fit drawn on it * Compute prediction values based on the line of best fit

Regression analysis finds the equation of the line that best describes the relationship between the two variables under examination. That is, regression analysis describes the mathematical relationship between the two variables. One of the main reasons for finding a regression equation is to make predictions.

Key Concepts

Vocabulary

bivariate data (p. 54)

cross-tabulation (p. 54)

contingency table (p. 54)

ordered pairs (p. 58)

input (independent) variable (p. 58)

output (dependent) variable (p. 58)

scatter diagram (p. 58)

linear correlation analysis (p. 60)

no correlation (p. 60)

positive correlation (p. 60)

negative correlation (p. 60)

linear correlation (p. 60)

coefficient of linear correlation, *r* (p. 61)

Pearson's product moment formula (p. 62)

cause-and-effect relationship (p. 63)

lurking variable (p. 63)

regression analysis (p. 64)

method of least squares (p. 65)

predicted value of y (p. 65)

least squares criterion (p. 65)

predicted value of $\hat{y}$ (p. 65)

line of best fit (p. 65)

slope (b_1) (p. 65)

y-intercept (b_0) (p. 65)

Key Formulas

(3.1) Linear correlation coefficient (definition formula)

$$r = \frac{\Sigma(x - \bar{x})(y - \bar{y})}{(n - 1)s_x s_y}$$

(3.2) Linear correlation coefficient (computational formula)

$$r = \frac{SS(xy)}{\sqrt{SS(x)SS(y)}}$$

(2.8) Sum of squares for x

$$SS(x) = \Sigma x^2 - \frac{(\Sigma x)^2}{n}$$

(3.3) Sum of squares for y

$$SS(y) = \Sigma y^2 - \frac{(\Sigma y)^2}{n}$$

(3.4) Sum of squares for xy

$$SS(xy) = \Sigma xy - \frac{\Sigma x \Sigma y}{n}$$

(3.5) Slope, b_1 (definition formula)

$$b_1 = \frac{\Sigma(x - \bar{x})(y - \bar{y})}{\Sigma(x - \bar{x})^2}$$

(3.6) Slope, b_1 (computational formula)

$$b_1 = \frac{SS(xy)}{SS(x)}$$

(3.7) y-intercept (computational formula)

$$b_0 = \frac{\Sigma y - (b_1 \cdot \Sigma x)}{n}$$

(3.7a) y-intercept (alternative computational formula)

$$b_0 = \bar{y} - (b_1 \cdot \bar{x})$$

Practice Test

PART I–Knowing the Definitions

Answer "True" if the statement is always true. If the statement is not always true, replace the words printed in bold with words that make the statement always true.

_____ 3.1 **Correlation** analysis is a method of obtaining the equation that represents the relationship between two variables.

_____

_____ 3.2 The linear correlation coefficient is used to determine the **equation that represents** the relationship between two variables.

_____

_____ 3.3 A correlation coefficient of **zero** means that the two variables are perfectly correlated.

_____

_____ 3.4 Whenever the slope of the regression line is zero, the **correlation coefficient** will also be zero.

_____

_____ 3.5 When r is positive, b_1 will always be **negative.**

_____

_____ 3.6 The **slope** of the regression line represents the amount of change expected to take place in y when x increases by one unit.

_____

_____ 3.7 When the calculated value of r is positive, the calculated value of b_1 will be **negative.**

_____

_____ 3.8 Correlation coefficients range between **0 and −1.**

_____

_____ 3.9 The value being predicted is called the **input variable.**

_____

_____ 3.10 The line of best fit is used to predict the **average value of y** that can be expected to occur at a given value of x.

_____

PART II–Applying the Concepts

3.11 Refer to the scatter diagram that follows.

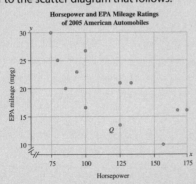

Horsepower and EPA Mileage Ratings of 2005 American Automobiles

a. Match the descriptions in the column on the right with the terms in the column on the left.

___ population   (a) the horsepower rating for an automobile

___ sample   (b) all 2005 American-made automobiles

___ input variable   (c) the EPA mileage rating for an automobile

___ output variable   (d) the 2005 automobiles with ratings shown on the scatter diagram

b. Find the sample size.

c. What is the smallest value reported for the output variable?

d. What is the largest value reported for the input variable?

e. Does the scatter diagram suggest a positive, negative, or zero linear correlation coefficient?

f. What are the coordinates of point Q?

g. Will the slope for the line of best fit be positive, negative, or zero?

h. Will the intercept for the line of best fit be positive, negative, or zero?

3.12 For the bivariate data, the extensions, and the totals shown on the table, find the following:

a. $SS(x)$

b. $SS(y)$

c. $SS(xy)$

d. The linear correlation coefficient, r

e. The slope, b_1

f. The y-intercept, b_0

g. The equation of the line of best fit

x	y	x^2	xy	y^2
2	6	4	12	36
3	5	9	15	25
3	7	9	21	49
4	7	16	28	49
5	7	25	35	49
5	9	25	45	81
6	8	36	48	64
28	49	124	204	353

PART III–Understanding the Concepts

3.13 A test was administered to measure the mathematics ability of the people in a certain town. Some of the towns-people were surprised to find out that their test results and their shoe sizes correlated strongly. Explain why a strong positive correlation should not have been a surprise.

Practice problems can be found at the end of Chapter 3, and solutions for the practice test can be found on the CourseMate for STAT2 site—login at <u>cengagebrain.com</u>.

Learning Objectives and Outcomes

4.1 Probability of Events (pp. 74–82)

Understand and be able to describe the differences between empirical, theoretical, and subjective probabilities * Understand the properties of probability numbers * Understand and be able to explain the difference between probability and statistics

Empirical probability is the *observed relative frequency* with which an event occurs. Theoretical probability is the proportion of a sample space that represents the events occurring. (Equally likely sample spaces are the most convenient sample spaces to use.) Subjective probability results from a personal judgment (a gut feeling or hunch). Whether empirical, theoretical, or subjective, for a probability experiment, the probability of each outcome is always a numerical value between zero and one, and the sum of all probabilities for all outcomes is equal to exactly one. Odds are an alternative way to express probabilities. Odds express the number of ways an event *can* happen compared with the number of ways it *cannot* happen. Probability and statistics are related but separate fields of mathematics. Probability is the chance that something specific will occur when the possibilities are known. Statistics requires drawing a sample, describing it, then making inferences about the population based on the information found.

4.2 Conditional Probability of Events (pp. 82–83)

Construct, describe, compute, and interpret a conditional probability

Probabilities are affected by conditions existing at the time. Because conditional probabilities are subject to certain conditions, some outcomes from the list of possible outcomes will be eliminated as possibilities as soon as the condition is known.

4.3 Rules of Probability (pp. 84–87)

Understand and be able to utilize the complement rule * Compute probabilities of compound events using the addition rule

Compound events are combinations of more than one simple event. Complementary events are one way to examine compound events. Examples of complements:

Event	Complement
Success	Failure
Yes	No
Zero outcomes of heads on set of coin tosses	At least one outcome of heads on set of coin tosses

The general addition rule is useful in finding the probability of "A or B"; the general multiplication rule is useful in finding the probability of "A and B."

Key Concepts

Vocabulary

probability of an event (p. 74)

experimental (empirical) probability (p. 74)

observed relative frequency (p. 74)

sample space (p. 75)

equally likely (p. 76)

outcome (p. 76)

event (p. 76)

ordered pair (p. 76)

tree diagram (p. 77)

subjective probability (p. 78)

all-inclusive (p. 79)

long-term average (p. 80)

law of large numbers (p. 80)

odds (p. 81)

probability (p. 81)

statistics (p. 81)

conditional probability an event will occur (p. 82)

compound event (p. 84)

complementary event (p. 84)

addition rule (p. 84)

multiplication rule (p. 85)

mutually exclusive events (p. 87)

intersection (p. 88)

independent events (p. 90)

dependent events (p. 91)

special multiplication rule (p. 93)

Key Formulas

(4.1) Empirical (observed) probability $P'(A)$

$$P'(A) = \frac{n(A)}{n}$$

(4.2) Theoretical (expected) probability $P(A)$

$$P(A) = \frac{n(A)}{n(S)}$$

(4.3) Complement rule

$$P(\bar{A}) = 1 - P(A)$$

(4.4) General addition rule

$$P(A \text{ or } B) = P(A) + P(B) - P(A \text{ and } B)$$

(4.5) General multiplication rule

$$P(A \text{ and } B) = P(A) \cdot P(B \mid A)$$

(4.6) Special addition rule (if A and B are mutually exclusive)

$$P(A \text{ or } B) = P(A) + P(B)$$

(4.7) Special multiplication rule (if A and B are independent events)

$$P(A \text{ and } B) = P(A) \cdot P(B)$$

Properties and Rules

Property 1

In words: A probability is always a numerical value between zero and one.

In algebra: $0 \leq$ each $P(A) \leq 1$ or
$\qquad 0 \leq$ each $P'(A) \leq 1$

Property 2

In words: The sum of the probabilities for all outcomes of an experiment is equal to exactly one.

In algebra: $\displaystyle\sum_{\text{all outcomes}} P(A) = 1$ or

$\displaystyle\sum_{\text{all outcomes}} P'(A) = 1$

Complement Rule

probability of the complement of A = one − probability of A

General Addition Rule

Let A and B be two events defined in a sample space S.

probability of A or B = probability of A + probability of B − probability of A and B

General Multiplication Rule

Let A and B be two events defined in sample space S.

probability of A and B = probability of A × probability of B, knowing A

Special Addition Rule

Let A and B be two mutually exclusive events defined in a sample space S.

probability of A or B = probability of A + probability of B

This formula can be expanded to consider more than two mutually exclusive events:

$$P(A \text{ or } B \text{ or } C \text{ or } \ldots \text{ or } E) = P(A) + P(B) + P(C) + \ldots + P(E)$$

Special Multiplication Rule

Let A and B be two independent events defined in a sample space S.

probability of A and B = probability of A × probability of B

This formula can be expanded to consider more than two independent events:

$$P(A \text{ and } B \text{ and } C \text{ and } \ldots \text{ and } E) = P(A) \cdot P(B) \cdot P(C) \cdot \ldots \cdot P(E)$$

4.4 Mutually Exclusive Events (pp. 87–90)

Compute probabilities of compound events using the addition rule for mutually exclusive events

Mutually exclusive events share no common elements. For example, if either one of the events has occurred, then by definition the other cannot have or is excluded. The special addition rule helps calculate probabilities involving mutually exclusive events.

4.5 Independent Events (pp. 90–93)

Compute probabilities of compound events using the multiplication rule for independent events

When events are independent, the occurrence of one gives us no information about the occurrence of the other. In other words, changes to one have no impact on the other. When events are not independent, they are called dependent. For dependent events, a change to one affects the other. The special multiplication rule is helpful in calculating probabilities involving independent events.

4.6 Are Mutual Exclusiveness and Independence Related? (pp. 93–94)

Recognize and compare the differences between mutually exclusive events and independent events

Mutually exclusive events are two nonempty events defined on the same sample space that share no common elements. Mutually exclusive is not a probability concept by definition—it just happens to be easy to express the concept using a probability statement. Independent events are two nonempty events defined on the same sample space that are related in such a way that the occurrence of either event does not affect the probability of the other event.

Practice Test

PART I–Knowing the Definitions

Answer "True" if the statement is always true. If the statement is not always true, replace the words printed in bold with words that make the statement always true.

_____ 4.1 The probability of an event is a **whole number.**

_____ 4.2 The concepts of probability and relative frequency as related to an event are **very similar.**

_____ 4.3 The **sample space** is the theoretical population for probability problems.

_____ 4.4 The sample points of a sample space are **equally likely** events.

_____ 4.5 The value found for experimental probability will **always be** exactly equal to the theoretical probability assigned to the same event.

_____ 4.6 The probabilities of complementary events always **are equal.**

_____ 4.7 If two events are mutually exclusive, they are also **independent.**

_____ 4.8 If events A and B are **mutually exclusive,** the sum of their probabilities must be exactly 1.

_____ 4.9 If the sets of sample points that belong to two different events do not intersect, the events are **independent.**

_____ 4.10 A compound event formed with the word "and" requires the use of the **addition rule.**

PART II–Applying the Concepts

4.11 A computer is programmed to generate the eight single-digit integers 1, 2, 3, 4, 5, 6, 7, and 8 with equal frequency. Consider the experiment "the next integer generated" and these events:

A: odd number, {1, 3, 5, 7}
B: number greater than 4, {5, 6, 7, 8}
C: 1 or 2, {1, 2}

  a. Find $P(A)$.

  b. Find $P(B)$.

  c. Find $P(C)$.

  d. Find $P(\overline{C})$.

  e. Find $P(A$ and $B)$.

  f. Find $P(A$ or $B)$.

  g. Find $P(B$ and $C)$.

  h. Find $P(B$ or $C)$.

  i. Find $P(A$ and $C)$.

  j. Find $P(A$ or $C)$.

  k. Find $P(A \mid B)$.

  l. Find $P(B \mid C)$.

  m. Find $P(A \mid C)$.

  n. Are events A and B mutually exclusive? Explain.

  o. Are events B and C mutually exclusive? Explain.

  p. Are events A and C mutually exclusive? Explain.

  q. Are events A and B independent? Explain.

  r. Are events B and C independent? Explain.

  s. Are events A and C independent? Explain.

4.12 Events A and B are mutually exclusive, and $P(A) = 0.4$ and $P(B) = 0.3$.

  a. Find $P(A$ and $B)$.

  b. Find $P(A$ or $B)$.

  c. Find $P(A \mid B)$.

  d. Are events A and B independent? Explain.

4.13 Events E and F have probabilities $P(E) = 0.5$, $P(F) = 0.4$, and $P(E$ and $F) = 0.2$.

  a. Find $P(E$ or $F)$.

  b. Find $P(E \mid F)$.

  c. Are E and F mutually exclusive? Explain.

  d. Are E and F independent? Explain.

4.14 Janice wants to become a police officer. She must pass a physical exam and then a written exam. Records show that the probability of passing the physical exam is 0.85 and that once the physical is passed, the probability of passing the written exam is 0.60. What is the probability that Janice will pass both exams?

PART III–Understanding the Concepts

4.15 Student A says that independence and mutual exclusiveness are basically the same thing; namely, both mean neither event has anything to do with the other one. Student B argues that although Student A's statement has some truth in it, Student A has missed the point of these two properties. Student B is correct. Carefully explain why.

4.16 Using complete sentences, describe the following in your own words:

  a. Mutually exclusive events

  b. Independent events

  c. The probability of an event

  d. A conditional probability

> Practice problems can be found at the end of Chapter 4, and solutions for the practice test can be found on the CourseMate for STAT2 site—login at cengagebrain.com.

Putting Chapter 4 to Work:
Sweet Statistics

The chapter project takes us back to the M&M example on page 74, as a way to assess what we have learned in this chapter. And what a better way to do that than with some candy! We can explore the differences between theoretical and experimental probabilities as well as see the law of large numbers in action— all with M&M's. Now that is "sweet statistics." Let's begin.

Let's take a theoretical look at the expected. Mars, Inc., currently uses the following percentages to mix the colors for M&M's Milk Chocolate Candies: 13% brown, 13% red, 14% yellow, 16% green, 20% orange, 24% blue.

a. Construct a bar graph showing the expected (theoretical) proportion of M&M's for each color.
b. Theoretically, what percentage of red M&M's should you expect in a bag of M&M's?
c. If you opened a bag of M&M's right now, would you be surprised to find color percentages different from those given by Mars? Explain.

Let's take an empirical (experimental) look at what happened.

d. Obtain a pack of M&M's (at least a 1.69 oz. size— approximately $0.50 in cost).
e. Record the number of each color in a frequency distribution with the headings "Color" and "Frequency."
f. Verify the total number of M&M's with the sum of the Frequency column.
g. Now you may snack! ☺
h. Present the frequency distribution as a relative frequency distribution, using the heading "Empirical Probability."
i. Verify that the sum of the Empirical Probability column is equal to 1. Explain the meaning of this sum.
j. Construct a bar graph showing the relative frequency for each color. Use the same color order as in part (a).
k. Empirically, what percentage of red M&M's should you expect in a bag of M&M's?
l. What other statistical displays could you use to present the data from the bag of M&M's? Present them.
m. Compare your empirical (experimental) findings to the expectations (theoretical) expressed in part (a).

Learning Objectives and Outcomes

5.1 Random Variables (pp. 100–102)

Understand the difference between a discrete and a continuous random variable

Random variables denote the outcomes of a probability experiment. The events in a probability experiment are both mutually exclusive and all inclusive. Discrete random variables assume a countable number of events, and continuous random variables assume an uncountable number of events.

5.2 Probability Distributions of a Discrete Random Variable (pp. 102–106)

Be able to construct a discrete probability distribution based on an experiment or given function * Understand and be able to utilize the two main properties of probability distributions to verify compliance * Compute, describe, and interpret the mean and standard deviation of a probability distribution

A probability distribution organizes probability events in a table format. Every probability function must display the two basic properties of a probability: the probability assigned to each value is between zero and one, and the sum of the probabilities must equal one. A common way to represent a probability function graphically is by using a histogram. In much the same way that sample statistics describe samples, population parameters like mean, variance, and standard deviation can be used to describe probability distributions.

5.3 Binomial Probability Distribution (pp. 106–112)

Know and be able to calculate binomial probabilities using the binomial probability function * Understand and be able to use Table 2 in Appendix B, Binomial Probabilities, to determine binomial probabilities * Compute, describe, and interpret the mean and standard deviation of a binomial probability distribution

Experiments made up of multiple trials are binomial experiments if there are n repeated identical trials, each trial has one of two possible outcomes, the sum of probability of success and the probability of failure equals one, and the number of successful trials x is an integer from zero to n. All binomial experiments have the same properties and the binomial probability function can be used to represent them all. Using formulas (5.7) and (5.8), it is possible to find the mean and standard deviation of a binomial distribution. These formulas are much easier to use when x is a binomial random variable.

Key Concepts

Vocabulary
experiment (p. 100)
random variable (p. 100)
mutually exclusive events (p. 100)
all-inclusive events (p. 100)
discrete random variable (p. 101)
continuous random variable (p. 101)
probability distribution (p. 102)
probability function (p. 102)
constant function (p. 103)
probability histogram (p. 103)
population parameters (p. 104)
sample statistics (p. 104)
mean of a discrete random variable (expected value) (p. 104)
variance of a discrete random variable (p. 105)
standard deviation of a discrete random variable (p. 105)
success (p. 106)
failure (p. 106)
binomial probability experiment (p. 106)
binomial random variable (p. 106)
binomial coefficient (p. 107)
factorial notation (p. 107)
independent trials (p. 107)

Key Formulas
(5.1) Mean of x
$$\mu = \Sigma[xP(x)]$$
(5.2) Variance
$$\sigma^2 = \Sigma[(x - \mu)^2 P(x)]$$
(5.3a) Variance
$$\sigma^2 = \Sigma[x^2 P(x)] - \{\Sigma[xP(x)]\}^2$$
(5.3b) Variance
$$\sigma^2 = \Sigma[x^2 P(x)] - \mu^2$$
(5.4) Standard deviation
$$\sigma = \sqrt{\sigma^2}$$
(5.5) Binomial probability function
$$P(x) = \binom{n}{x}(p^x)(q^{n-x})$$
for $x = 0, 1, 2, \ldots, n$
(5.6) Binomial coefficient
$$\binom{n}{x} = \frac{n!}{x!(n-x)!}$$
(5.7) Mean of binomial distribution
$$\mu = np$$
(5.8) Standard deviation of binomial distribution
$$\sigma = \sqrt{npq}$$

Practice Test

PART I–Knowing the Definitions

Answer "True" if the statement is always true. If the statement is not always true, replace the words printed in bold with words that make the statement always true.

_____ 5.1 The number of hours you waited in line to register this semester is an example of a **discrete** random variable.

_____

_____ 5.2 The number of automobile accidents you were involved in as a driver last year is an example of a **discrete** random variable.

_____

_____ 5.3 The sum of all the probabilities in any probability distribution is always exactly **two.**

_____

_____ 5.4 The various values of a random variable form a list of **mutually exclusive events.**

_____

_____ 5.5 A binomial experiment always has **three or more** possible outcomes to each trial.

_____

_____ 5.6 The formula $\mu = np$ may be used to compute the mean of a **discrete** population.

_____

_____ 5.7 The binomial parameter p is the probability of **one success occurring in n trials** when a binomial experiment is performed.

_____

_____ 5.8 A parameter is a statistical measure of some aspect of a **sample.**

_____

_____ 5.9 **Sample statistics** are represented by letters from the Greek alphabet.

_____

_____ 5.10 The probability of event A or B is equal to the sum of the probability of event A and the probability of event B when A and B are **mutually exclusive events.**

_____

PART II–Applying the Concepts

5.11 a. Show that the following is a probability distribution:

x	1	3	4	5
$P(x)$	0.2	0.3	0.4	0.1

  b. Find $P(x = 1)$.

  c. Find $P(x = 2)$.

  d. Find $P(x > 2)$.

  e. Find the mean of x.

  f. Find the standard deviation of x.

5.12 A T-shirt manufacturing company advertises that the probability of an individual T-shirt being irregular is 0.1. A box of 12 T-shirts is randomly selected and inspected.

  a. What is the probability that exactly 2 of these 12 T-shirts are irregular?

  b. What is the probability that exactly 9 of these 12 T-shirts are not irregular?

Let x be the number of T-shirts that are irregular in all boxes of 12 T-shirts.

  c. Find the mean of x.

  d. Find the standard deviation of x.

PART III–Understanding the Concepts

5.13 What properties must an experiment possess in order for it to be a binomial probability experiment?

5.14 Student A uses a relative frequency distribution for a set of sample data and calculates the mean and standard deviation using formulas from Chapter 5. Student A justifies her choice of formulas by saying that since relative frequencies are empirical probabilities, her sample is represented by a probability distribution and therefore her choice of formulas was correct. Student B argues that since the distribution represented a sample, the mean and standard deviation involved are known as $\bar{x}$ and s and must be calculated using formulas from Chapter 2. Who is correct, A or B? Justify your choice.

5.15 Student A and Student B were discussing one entry in a probability distribution chart:

x	$P(x)$
-2	0.1

Student B thought this entry was okay because $P(x)$ was a value between 0.0 and 1.0. Student A argued that this entry was impossible for a probability distribution because x was -2 and negatives are not possible. Who is correct, A or B? Justify your choice.

Practice problems can be found at the end of Chapter 5, and solutions for the practice test can be found on the CourseMate for STAT2 site—login at cengagebrain.com.

Learning Objectives and Outcomes

6.1 Normal Probability Distributions (pp. 118–120)

Understand the relationship between the empirical rule and the normal curve * Understand that a normal curve is a bell-shaped curve, with total area under the curve equal to 1

The normal probability distribution is considered the single most important probability distribution. An unlimited number of continuous random variables have either a normal or an approximately normal distribution. Several other probability distributions of both discrete and continuous random variables are also approximately normal under certain conditions. Percentage, proportion, and probability are basically the same concepts. Area is the graphic representation of all three. The empirical rule is a fairly crude measuring device; with it we are able to find probabilities associated only with whole number multiples of the standard deviation.

6.2 The Standard Normal Distribution (pp. 120–123)

Understand that the normal curve is symmetrical about the mean with an area of 0.5000 on each side of the mean

> 1. The total area under the standard normal curve is equal to 1.
> 2. The distribution is mounded and symmetrical; it extends indefinitely in both directions, approaching but never touching the horizontal axis.
> 3. The distribution has a mean of 0 and a standard deviation of 1.
> 4. The mean divides the area in half—0.5000 on each side.
> 5. Nearly all the area is between $z = -3.00$ and $z = 3.00$.

6.3 Applications of Normal Distributions (pp. 124–126)

Calculate probabilities for intervals defined on the standard normal distribution * Compute, describe, and interpret a z value for a data value from a normal distribution * Compute z-scores and probabilities for applications of the normal distribution

We can convert information about the standard normal variable z into probability, so we can also convert probability information about the standard normal distribution into z-scores. That means we can apply this methodology to all normal distributions using the standard score, z.

6.4 Notation (pp. 127–129)

z will be used with great frequency, and the convention that we will use as an "algebraic name" for a specific z-score is $z(\alpha)$, where α represents the "area to the right" of the z being named.

6.5 Normal Approximation of the Binomial (pp. 129–131)

Compute z-scores and probabilities for normal approximations to the binomial

The binomial distribution is a probability distribution of the discrete random variable x, the number of successes observed in n repeated independent trials. Binomial probabilities can be reasonably approximated by using the normal probability distribution. The binomial random variable is discrete, whereas the normal random variable is continuous. The continuity correction factor allows a discrete variable to be converted into a continuous variable.

Key Concepts

Vocabulary

normal probability distribution (p. 118)

continuous random variables (p. 118)

normal distribution (p. 118)

discrete random variables (p. 118)

percentage (p. 120)

proportion (p. 120)

probability (p. 120)

standard normal distribution (p. 120)

standard score, or z-score (p. 120)

cumulative area (p. 120)

normal curve (p. 121)

binomial distribution (p. 129)

binomial probability (p. 129)

discrete (p. 129)

continuous (p. 129)

continuity correction factor (p. 130)

Key Formulas

(6.1) Normal probability distribution function

$$y = f(x) = \frac{e^{-\frac{1}{2}\left(\frac{x - \mu}{\sigma}\right)^2}}{\sigma\sqrt{2\pi}} \text{ for all real } x$$

(6.2) Probability associated with interval from $x = a$ to $x = b$

$$P(a \le x \le b) = \int_a^b f(x)\, dx$$

(6.3) Standard score

$$z = \frac{x - \mu}{\sigma}$$

Rule

The normal distribution provides a reasonable approximation to a binomial probability distribution whenever the values of np and $n(1 - p)$ both equal or exceed 5.

PART I–Knowing the Definitions

Answer "True" if the statement is always true. If the statement is not always true, replace the words printed in bold with words that make the statement always true.

_____ 6.1 The normal probability distribution is symmetric about **zero.**

_____

_____ 6.2 The total area under the curve of any normal distribution is **1.0.**

_____

_____ 6.3 The theoretical probability that a particular value of a **continuous** random variable will occur is exactly zero.

_____

_____ 6.4 The unit of measure for the standard score is the **same as the unit of measure of the data.**

_____

_____ 6.5 All **normal** distributions have the same general probability function and distribution.

_____

_____ 6.6 In the notation z(0.05), the number in parentheses is the measure of the area to the **left** of the z-score.

_____

_____ 6.7 Standard normal scores have a mean of **one** and a standard deviation of **zero.**

_____

_____ 6.8 Probability distributions of **all** continuous random variables are normally distributed.

_____

_____ 6.9 We are able to add and subtract the areas under the curve of a continuous distribution because these areas represent probabilities of **independent** events.

_____

_____ 6.10 The most common distribution of a continuous random variable is the **binomial** probability.

_____

PART II–Applying the Concepts

6.11 Find the following probabilities for z, the standard normal score:

　　a. $P(0 < z < 2.42)$　　　　b. $P(z < 1.38)$

　　c. $P(z < -1.27)$　　　　d. $P(-1.35 < z < 2.72)$

6.12 Find the value of each z-score:

　　a. $P(z > ?) = 0.2643$　b. $P(z < ?) = 0.17$　c. $z(0.04)$

6.13 Use the symbolic notation z(α) to give the symbolic name for each z-score shown in the figure.

　a.

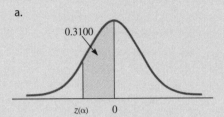

　b.

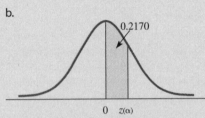

6.14 The lifetimes of flashlight batteries are normally distributed about a mean of 35.6 hours with a standard deviation of 5.4 hours. Kevin selected one of these batteries at random and tested it. What is the probability that this one battery will last less than 40.0 hours?

6.15 The lengths of time, x, spent by students commuting one way to college each day are believed to have a mean of 22 minutes with a standard deviation of 9 minutes. If the lengths of time spent commuting are approximately normally distributed, find the time, x, that separates the 25% who spend the most time commuting from the rest of the commuters.

6.16 Thousands of high school students take the SAT each year. The scores attained by the students in a certain city are approximately normally distributed with a mean of 490 and a standard deviation of 70. Find:

　a. the percentage of students who score between 600 and 700

　b. the percentage of students who score less than 650

　c. the third quartile

　d. the 15th percentile, P_{15}

　e. the 95th percentile, P_{95}

PART III–Understanding the Concepts

6.17 In 50 words, describe the standard normal distribution.

6.18 Describe the meaning of the symbol z(α).

6.19 Explain why the standard normal distribution, as computed in Table 3 in Appendix B, can be used to find probabilities for all normal distributions.

Practice problems can be found at the end of Chapter 6, and solutions for the practice test can be found on the CourseMate for STAT2 site—login at cengagebrain.com.

Learning Objectives and Outcomes

7.1 Sampling Distributions (pp. 136–140)

Understand what a sampling distribution of a sample statistic is and that the distribution is obtained from repeated samples, all of the same size

The basic purpose for considering what happens when a population is repeatedly sampled is to form sampling distributions. The sampling distribution is then used to describe the variability that occurs from one sample to the next.

Repeated samples are commonly used in the field of production control, in which samples are taken to determine whether a product is of the proper size or quantity. When the sample statistic does not fit the standards, a mechanical adjustment of the machinery is necessary. The adjustment is then followed by another sampling to be sure the production process is in control.

7.2 The Sampling Distribution of Sample Means (pp. 141–145)

Understand and be able to explain the relationship between the sampling distribution of sample means and the central limit theorem * Determine and be able to explain the effect of sample size on the standard error of the mean

The basic purpose for considering what happens when a population is repeatedly sampled is to form sampling distributions. The sampling distribution is then used to describe the variability that occurs from one sample to the next. Once this pattern of variability is known and understood for a specific sample statistic, we are able to make predictions about the corresponding population parameter with a measure of how accurate the prediction is. The SDSM and the central limit theorem help describe the distribution for sample means.

The "standard error of the _____" is the name used for the standard deviation of the sampling distribution for whatever statistic is named in the blank. In this chapter we have been concerned with the standard error of the mean. However, we could also work with the standard error of the proportion, median, or any other statistic.

7.3 Application of the Sampling Distribution of Sample Means (pp. 145–147)

Understand when and how the normal distribution can be used to find probabilities corresponding to sample means * Compute z-scores and probabilities for applications of the sampling distribution of sample means

Calculating probabilities is one way we are able to make predictions about the corresponding population parameter we are looking for (recall the example about the height of kindergarteners). When the population is normally distributed, the sampling distribution of $\bar{x}$'s is normally distributed. To determine probabilities, you need to format a probability statement involving a z-score.

You must be careful to distinguish between the two formulas for calculating a z-score. The first gives the standard score when we have individual values from a normal distribution (x values). The second formula deals with a sample mean ($\bar{x}$ value). The key to distinguishing between the formulas is to decide whether the problem deals with an individual x or a sample mean $\bar{x}$. If it deals with the individual values of x, we use the first formula, as presented in Chapter 6. If the problem deals with a sample mean, $\bar{x}$, we use the second formula and proceed as illustrated in this chapter.

Key Concepts

Vocabulary

repeated sample statistics (p. 137)

sampling distribution of a sample statistic (p. 137)

sampling distribution of sample means (SDSM) (p. 138)

probability distribution (p. 138)

frequency distribution (p. 138)

random sample (p. 139)

standard error of the mean ($\sigma_{\bar{x}}$) (p. 141)

central limit theorem (CLT) (p. 141)

sampling distribution (p. 144)

z-score (p. 145)

Key Formulas

(7.1) $z = \dfrac{\bar{x} - \mu_{\bar{x}}}{\sigma_{\bar{x}}}$

(7.2) $z = \dfrac{\bar{x} - \mu}{\sigma/\sqrt{n}}$

Practice Test

PART I–Knowing the Definitions

Answer "True" if the statement is always true. If the statement is not always true, replace the words printed in bold with words that make the statement always true.

_____ 7.1   A sampling distribution **is** a distribution listing all the sample statistics that describe a particular sample.

_____

_____ 7.2   The histograms of **all** sampling distributions are symmetrical.

_____

_____ 7.3   The mean of the sampling distribution of $\bar{x}$'s is equal to the mean of the **sample.**

_____

_____ 7.4   The standard error of the mean is the standard deviation of the population **from which the samples have been taken.**

_____

_____ 7.5   The standard error of the mean **increases** as the sample size increases.

_____

_____ 7.6   The shape of the distribution of sample means is always that of a **normal** distribution.

_____

_____ 7.7   A **probability** distribution of a sample statistic is a distribution of all the values of that statistic that were obtained from all possible samples.

_____

_____ 7.8   The sampling distribution of sample means provides us with a description of the three characteristics of a sampling distribution of sample **medians.**

_____

_____ 7.9   A **frequency** sample is obtained in such a way that all possible samples of a given size have an equal chance of being selected.

_____

_____ 7.10   We **do not need** to take repeated samples in order to use the concept of the sampling distribution.

_____

PART II–Applying the Concepts

7.11   The lengths of the lake trout in Conesus Lake are believed to have a normal distribution with a mean of 15.6 inches and a standard deviation of 3.8 inches.

a. Kevin is going fishing at Conesus Lake tomorrow. If he catches one lake trout, what is the probability that it is less than 15.0 inches long?

b. If Captain Brian's fishing boat takes 10 people fishing on Conesus Lake tomorrow and they catch a random sample of 16 lake trout, what is the probability that the mean length of their total catch is less than 15 inches?

7.12   Cigarette lighters manufactured by EasyVice Company are claimed to have a mean lifetime of 20 months with a standard deviation of 6 months. The money-back guarantee allows you to return the lighter if it does not last at least 12 months from the date of purchase.

a. If the lifetimes of these lighters are normally distributed, what percentage of the lighters will be returned to the company?

b. If a random sample of 25 lighters is tested, what is the probability that the sample mean lifetime will be more than 18 months?

7.13   Aluminum rivets produced by Rivets Forever, Inc., are believed to have shearing strengths that are distributed about a mean of 13.75 with a standard deviation of 2.4. If this information is true and a sample of 64 such rivets is tested for shear strength, what is the probability that the mean strength will be between 13.6 and 14.2?

PART III–Understanding the Concepts

7.14   "Two heads are better than one." If that's true, then how good would several heads be? To find out, a statistics instructor drew a line across the chalkboard and asked her class to estimate its length to the nearest inch. She collected their estimates, which ranged from 33 to 61 inches, and calculated the mean value. She reported that the mean was 42.25 inches. She then measured the line and found it to be 41.75 inches long.

Does this show that "several heads are better than one"? What statistical theory supports this occurrence? Explain how.

7.15   The sampling distribution of sample means is more than just a distribution of the mean values that occur from many repeated samples taken from the same population. Describe what other specific condition must be met in order to have a sampling distribution of sample means.

7.16   Student A states, "A sampling distribution of the standard deviations tells you how the standard deviation varies from sample to sample." Student B argues, "A population distribution tells you that." Who is right? Justify your answer.

7.17   Student A says it is the "size of each sample used" and Student B says it is the "number of samples used" that determines the spread of an empirical sampling distribution. Who is right? Justify your choice.

Practice problems can be found at the end of Chapter 7, and solutions for the practice test can be found on the CourseMate for STAT2 site—login at cengagebrain.com.

Learning Objectives and Outcomes

8.1 The Nature of Estimation (pp. 152–155)

Understand that a confidence interval is an interval estimate of a population parameter, with a degree of certainty, used when the population parameter is unknown

Estimating the value of a population parameter is a type of inference. The basic concepts of estimation are point estimate, interval estimate, level of confidence, and confidence interval. The quality of an estimation procedure (or method) is greatly enhanced if the sample statistic is both *less variable* and *unbiased*.

8.2 Estimation of Mean μ (σ Known) (pp. 156–161)

The assumption for estimating mean μ with a known σ:
The sampling distribution of $\bar{x}$ has a normal distribution.

Understand and be able to describe the key components for a confidence interval: point estimate, level of confidence, confidence coefficient, maximum error of estimate, lower confidence limit, and upper confidence limit * Compute, describe, and interpret a confidence interval for the population mean, μ

The estimation procedure is organized into a five-step process that takes into account confidence coefficient, standard error of the mean, the maximum error of estimate, E, and the lower and upper confidence limits and produces both the point estimate and the confidence interval.

8.3 The Nature of Hypothesis Testing (pp. 161–166)

Understand that a hypothesis test is used to make a decision about the value of a population parameter * Understand and be able to describe the relationship between the four possible outcomes of a hypothesis test—the two types of errors and the two types of correct decisions * Determine and know the proper format for stating a decision in a hypothesis test

The decision-making process starts by identifying something of concern and formulating a null hypothesis and the alternative hypothesis. The hypothesis test does not *prove* or *disprove* anything. The decision reached in a hypothesis test has probabilities associated with the four various situations. If "fail to reject H_o" is the decision, it is possible that an error has occurred. Furthermore, if "reject H_o" is the decision reached, it is possible for this to be an error. Both errors have probabilities greater than zero.

8.4 Hypothesis Test of Mean μ (σ Known): A Probability-Value Approach (pp. 166–173)

The assumption for hypothesis tests about mean μ using a known σ:
The sampling distribution of $\bar{x}$ has a normal distribution.

Key Concepts

Vocabulary

estimation question (p. 152)

hypothesis testing question (p. 152)

estimation (p. 153)

point estimate for a parameter (p. 153)

unbiased statistic (p. 154)

population mean, μ (p. 155)

interval estimate (p. 155)

level of confidence (1 − α) (p. 155)

confidence interval (p. 155)

confidence coefficient [z(α/2)] (p. 156)

standard error of the mean (p. 157)

maximum error of estimate, E (p. 157)

hypothesis (p. 162)

statistical hypothesis test (p. 162)

null hypothesis (p. 162)

alternative hypothesis (p. 162)

type A correct decision (p. 162)

type B correct decision (p. 162)

type I error (p. 163)

type II error (p. 163)

alpha (α) (p. 163)

beta (β) (p. 163)

level of significance (p. 165)

test statistic (p. 165)

decision rule (p. 165)

probability-value approach (p. 166)

calculated value of the test statistic, z★ (p. 170)

probability (p-) value (p. 170)

critical region (p. 174)

noncritical region (acceptance region) (p. 174)

critical value(s) (p. 175)

Demonstrate and understand the three possible combinations for the null and alternative hypotheses * Compute and understand the value of the test statistic. Compute the p-value for the test statistic * Determine and know the proper format for stating a decision in a hypothesis test

The probability-value approach, or simply p-value approach, is the hypothesis test process that has gained popularity in recent years, largely as a result of the convenience and the "number crunching" ability of the computer. This approach is organized as a five-step procedure:

Step 1–The Set-Up:

a. Describe the population parameter of interest.
b. State the null hypothesis (H_o) and the alternative hypothesis (H_a).

Step 2–The Hypothesis Test Criteria:

a. Check the assumptions.
b. Identify the probability distribution and the test statistic to be used.
c. Determine the level of significance, α.

Step 3–The Sample Evidence:

a. Collect the sample information.
b. Calculate the value of the test statistic.

Step 4–The Probability Distribution:

a. Calculate the p-value for the test statistic.
b. Determine whether or not the p-value is smaller than α.

Step 5–The Results:

a. State the decision about H_o.
b. State the conclusion about H_a.

8.5 Hypothesis Test of Mean μ (σ Known): A Classical Approach (pp. 173–177)

Demonstrate and understand the three possible combinations for the null and alternative hypotheses * Compute and understand the value of the test statistic * Determine the critical region and critical values * Determine and know the proper format for stating a decision in a hypothesis test

The *classical approach* is the hypothesis test process that has enjoyed popularity for many years. It is also organized as a five-step procedure and shares the same first three steps. For the classical approach, however:

Step 4–The Probability Distribution:

a. Determine the critical region and critical value(s).
b. Determine whether or not the calculated test statistic is in the critical region.

Step 5–The Results:

a. State the decision about H_o.
b. State the conclusion about H_a.

Key Formulas

(8.1) Confidence interval for mean

$$\bar{x} - z_{(\alpha/2)}\left(\frac{\sigma}{\sqrt{n}}\right) \quad \text{to} \quad \bar{x} + z_{(\alpha/2)}\left(\frac{\sigma}{\sqrt{n}}\right)$$

(8.2) Maximum error of estimate

$$E = z_{(\alpha/2)}\left(\frac{\sigma}{\sqrt{n}}\right)$$

(8.3) Sample size

$$n = \left[\frac{z_{(\alpha/2)} \cdot \sigma}{E}\right]^2$$

(8.4) Test statistic for mean

$$z\star = \frac{\bar{x} - \mu}{\sigma/\sqrt{n}}$$

Decision Rule

a. If the p-value is *less than or equal to* the level of significance α, then the decision must be to **reject H_o**.

b. If the p-value is *greater than* the level of significance α, then the decision must be to **fail to reject H_o**.

Practice problems can be found at the end of Chapter 8, and solutions for the practice test can be found on the CourseMate for STAT2 site—login at cengagebrain.com.

Practice Test

PART I–Knowing the Definitions

Answer "True" if the statement is always true. If the statement is not always true, replace the words printed in bold with words that make the statement always true.

_____ **8.1** **Beta** is the probability of a type I error.

_____ **8.2** $1 - \alpha$ is known as the level of significance of a hypothesis test.

_____ **8.3** The standard error of the mean is the standard deviation of the **sample selected.**

_____ **8.4** The maximum error of estimate is controlled by three factors: **level of confidence, sample size,** and **standard deviation.**

_____ **8.5** Alpha is the measure of the area under the curve of the standard score that lies in the **rejection region** for H_o.

_____ **8.6** The risk of making a **type I error** is directly controlled in a hypothesis test by establishing a level for α.

_____ **8.7** Failing to reject the null hypothesis when it is false is a **correct decision.**

_____ **8.8** If the noncritical region in a hypothesis test is made wider (assuming σ and n remain fixed), α becomes **larger.**

_____ **8.9** Rejection of a null hypothesis that is false is a **type II error.**

_____ **8.10** To conclude that the mean is greater (or less) than a claimed value, the value of the test statistic must fall in the **acceptance region.**

PART II–Applying the Concepts

Answer all questions, showing all formulas, substitutions, and work.

8.11 An unhappy post office customer is frustrated with the waiting time to buy stamps. Upon registering his complaint, he was told, "The average waiting time in the past has been about 4 minutes with a standard deviation of 2 minutes." The customer collected a sample of $n = 45$ customers and found the mean wait was 5.3 minutes. Find the 95% confidence interval for the mean waiting time.

8.12 State the null (H_o) and the alternative (H_a) hypotheses that would be used to test each of these claims:

 a. The mean weight of professional football players is more than 245 pounds.

 b. The mean monthly amount of rainfall in Monroe County is less than 4.5 inches.

 c. The mean weight of the baseball bats used by major league players is not equal to 35 ounces.

8.13 Determine the level of significance, test statistic, critical region, and critical value(s) that would be used in completing each hypothesis test using $\alpha = 0.05$:

 a. $H_o: \mu = 43$ b. $H_o: \mu = 0.80$ c. $H_o: \mu = 95$
 $H_a: \mu < 43$ $H_a: \mu > 0.80$ $H_a: \mu \neq 95$
 (given $\sigma = 6$) (given $\sigma = 0.13$) (given $\sigma = 12$)

8.14 Find each value:

 a. $z(0.05)$ b. $z(0.01)$ c. $z(0.12)$

8.15 In the past, the grapefruits grown in a particular orchard have had a mean diameter of 5.50 inches and a standard deviation of 0.6 inch. The owner believes this year's crop is larger than those in the past. He collected a random sample of 100 grapefruits and found a sample mean diameter of 5.65 inches.

 a. Find the value of the test statistic, $z\bigstar$, that corresponds to $\bar{x} = 5.65$.

 b. Calculate the p-value for the owner's hypothesis.

8.16 A manufacturer claims that its light bulbs have a mean lifetime of 1,520 hours with a standard deviation of 85 hours. A random sample of 40 such bulbs is selected for testing. If the sample produces a mean value of 1,498.3 hours, is there sufficient evidence to claim that the mean lifetime is less than the manufacturer claimed? Use $\alpha = 0.01$.

PART III–Understanding the Concepts

8.17 Sugar Creek Convenience Stores has commissioned a statistics firm to survey its customers in order to estimate the mean amount spent per customer. According to previous records, the standard deviation is believed to be $\sigma = \$5$. In its proposal to Sugar Creek, the statistics firm states that it plans to base the estimate for the mean amount spent on a sample of size 100 and use the 95% confidence level. Sugar Creek's president has suggested that the sample size be increased to 400. If nothing else changes, what effect will this increase in the sample size have on the following:

a. The point estimate for the mean

b. The maximum error of estimation

c. The confidence interval

The CEO wants the level of confidence increased to 99%. If nothing else changes, what effect will this increase in level of confidence have on the following:

d. The point estimate for the mean

e. The maximum error of estimation

f. The confidence interval

8.18 Determine the critical region and the critical values used to test the following null hypotheses:

a. $H_o: \mu = 55 \ (\geq)$, $H_a: \mu < 55$, $\alpha = 0.02$

b. $H_o: \mu = -86 \ (\geq)$, $H_a: \mu < -86$, $\alpha = 0.01$

c. $H_o: \mu = 107$, $H_a: \mu \neq 107$, $\alpha = 0.05$

d. $H_o: \mu = 17.4 \ (\leq)$, $H_a: \mu > 17.4$, $\alpha = 0.10$

8.19 The length of major league baseball games are approximately normally distributed and average 2 hours and 50.1 minutes, with a standard deviation of 21.0 minutes. It has been claimed that New York Yankee baseball games last, on the average, longer than the games of the other major league teams. To test the truth of this statement, a sample of eight Yankee games was randomly identified and the "time of game" (in minutes) for each obtained:

| 199 | 196 | 202 | 213 | 187 | 169 | 169 | 188 |

SOURCE: MLB.com

At the 0.05 level of significance, does this data show sufficient evidence to conclude that the mean time of Yankee baseball games is longer than that of other major league baseball teams?

Learning Objectives and Outcomes

9.1 Inferences about the Mean μ (σ Unknown) (pp. 182–191)

Compute, describe, and interpret a confidence interval for the population mean, μ, using the *t*-distribution * Perform, describe, and interpret a hypothesis test for the population mean, μ, using the *t*-distribution with the *p*-value approach and classical approach

Inferences about the population mean μ are based on the sample mean $\bar{x}$ and information obtained from the sampling distribution of sample means. When a known σ is being used to make an inference about the mean μ, a sample provides one value for use in the formulas; that one value is $\bar{x}$. When the sample standard deviation *s* is also used, the sample provides two values: the sample mean $\bar{x}$ and the estimated standard error $s/\sqrt{n}$. As a result, the *z*-statistic will be replaced with the *t*-statistic.

9.2 Inferences about the Binomial Probability of Success (pp. 191–198)

Compute, describe, and interpret a confidence interval for the population proportion, *p*, using the *z*-distribution * Perform, describe, and interpret a hypothesis test for the population proportion, *p*, using the *z*-distribution with the *p*-value approach and classical approach

The binomial parameter *p* is called the "probability of success." Binomial experiments are probability experiments in which there are many (*n*) repeated independent trials that each have two possible outcomes called "success" and "failure." In Chapter 5, the emphasis was on the variable *x* and its probability distribution; in this section, the emphasis is on the sample statistic *p'* and its use in inferences about *p*.

9.3 Inferences about the Variance and Standard Deviation (pp. 198–203)

Perform, describe, and interpret a hypothesis test for the population variance, σ^2, or standard deviation, σ, using the χ^2-distribution with the *p*-value approach and classical approach

To make inferences about variability of a normally distributed population, use the chi-square distributions.

Most inferences about a single population parameter are concerned with mean μ, proportion *p*, or standard deviation σ.

Key Concepts

Vocabulary

standard error (p. 182)
sample size (p. 182)
σ unknown (p. 183)
σ known (p. 183)
Student's *t*-statistic (p. 183)
properties of the *t*-distribution (p. 184)
standard normal distribution (p. 185)
degrees of freedom, df (p. 185)
critical values of *t* (p. 185)
test statistic *t⋆* (p. 187)
calculated *t* (p. 187)
binomial parameter, *p* (p. 191)
binomial experiment (p. 191)
random variable *x* (p. 191)
proportions (p. 192)
unbiased estimator for *p* (p. 192)
population proportion *p* (p. 192)
point estimate, *p'* (p. 192)
level of confidence (p. 192)
maximum error of the estimate for a proportion (p. 193)
chi-square, χ^2 (p. 198)
critical values for chi-square (p. 199)

Key Formulas

(9.1) Confidence interval for mean

$$\bar{x} - t_{(df, \alpha/2)}\left(\frac{s}{\sqrt{n}}\right) \text{ to }$$

$$\bar{x} + t_{(df, \alpha/2)}\left(\frac{s}{\sqrt{n}}\right)$$
with df = n − 1

(9.2) Test statistic for mean

$$t⋆ = \frac{\bar{x} - \mu}{s/\sqrt{n}} \text{ with df} = n - 1$$

(9.3) Sample binomial probability

$$p' = \frac{x}{n}$$

(9.4) Mean of binomial probability

$$\mu_{p'} = p$$

(9.5) Standard deviation of binomial probability

$$\sigma_{p'} = \sqrt{\frac{pq}{n}}$$

(9.6) Confidence interval for a proportion

$$p' \pm z_{(\alpha/2)}\left(\sqrt{\frac{p'q'}{n}}\right)$$

(9.7) Maximum error of estimate for a proportion

$$E = z_{(\alpha/2)}\left(\sqrt{\frac{pq}{n}}\right)$$

(9.8) Sample size for 1 − α confidence interval for *p*

$$n = \frac{[z_{(\alpha/2)}]^2 \cdot p^* \cdot q^*}{E^2}$$

(9.9) Test statistic for a proportion

$$z⋆ = \frac{p' - p}{\sqrt{\frac{pq}{n}}} \text{ with } p' = \frac{x}{n}$$

(9.10) Test statistic for variance and standard deviation

$$\chi^2⋆ = \frac{(n-1)s^2}{\sigma^2} \text{ with df} = n - 1$$

Rule

The distribution of *p'* is considered to be approximately normal if *n* is greater than 20 and if *np* and *nq* are both greater than 5.

The assumption for inferences about the mean μ when σ is unknown

The sampled population is normally distributed.

The assumption for inferences about the binomial parameter *p*

The *n* random observations that form the sample are selected independently from a population that is not changing during the sampling.

The assumption for inferences about the variance σ^2 or standard deviation σ

The sampled population is normally distributed.

Practice Test

PART I–Knowing the Definitions

Answer "True" if the statement is always true. If the statement is not always true, replace the words printed in bold with words that make the statement always true.

_____ 9.1 Student's t-distributions have an approximately normal distribution but are **more** dispersed than the standard normal distribution.

_____ 9.2 The **chi-square** distribution is used for inferences about the mean when σ is unknown.

_____ 9.3 **Student's t-distribution** is used for all inferences about a population's variance.

_____ 9.4 If the test statistic falls in the critical region, the null hypothesis has **been proved true.**

_____ 9.5 When the test statistic is t and the number of degrees of freedom gets very large, the critical value of t is very close to that of the **standard normal z.**

_____ 9.6 When making inferences about one mean when the value of σ is not known, the **z-score** is the test statistic.

_____ 9.7 The chi-square distribution is a skewed distribution whose mean value is **2** for df > 2.

_____ 9.8 Often, the concern with testing the variance (or standard deviation) is to keep its size under control or relatively small. Therefore, many of the hypothesis tests with chi-square are **one-tailed.**

_____ 9.9 $\sqrt{npq}$ is the standard error of proportion.

_____ 9.10 The sampling distribution of p' is distributed approximately as a **Student's t-distribution.**

PART II–Applying the Concepts

Answer all questions, showing all formulas, substitutions, and work.

9.11 Find each value:

a. $z_{(0.02)}$ b. $t_{(18, 0.95)}$ c. $\chi^2_{(25, 0.95)}$

9.12 A random sample of 25 data values was selected from a normally distributed population for the purpose of estimating the population mean, μ. The sample statistics are $n = 25, \bar{x} = 28.6$, and $s = 3.50$.

a. Find the point estimate for μ.

b. Find the maximum error of estimate for the 0.95 confidence interval.

c. Find the lower confidence limit (LCL) and the upper confidence limit (UCL) for the 0.95 confidence interval for μ.

9.13 Thousands of area elementary school students were recently given a nationwide standardized exam to test their composition skills. If 64 of a random sample of 100 students passed this exam, construct the 0.98 confidence interval for the true proportion of all area students who passed the exam.

9.14 The automobile manufacturer of the Alero claims that the typical Alero will average 32 mpg of gasoline. An independent consumer group is somewhat skeptical of this claim and thinks the mean gas mileage is less than the 32 claimed. A sample of 24 randomly selected Aleros produced these sample statistics: mean 30.15 and standard deviation 4.87. At the 0.05 level of significance, does the consumer group have sufficient evidence to refute the manufacturer's claim?

9.15 A coffee machine is supposed to dispense 6 fluid ounces of coffee into a paper cup. In reality, the amount dispensed varies from cup to cup. However, if the machine is operating properly, the standard deviation of the amounts dispensed should be 0.1 ounce or less. A random sample of 15 cups produced a standard deviation of 0.13 ounce. Does this represent sufficient evidence at the 0.10 level of significance to conclude that the machine is not operating properly?

PART III–Understanding the Concepts

9.16 Student B says the range of a set of data may be used to obtain a crude estimate for the standard deviation of a population. Student A is not sure. How will student B correctly explain how and under what circumstances his statement is true?

9.17 When you reject a null hypothesis, student A says that you are expressing disbelief in the value of the parameter as claimed in the null hypothesis. Student B says that instead you are expressing the belief that the sample statistic came from a population other than the one related to the parameter claimed in the null hypothesis. Who is correct? Explain.

9.18 Student A says that the best way to improve a confidence interval estimate is to increase the level of confidence. Student B argues that using a high confidence level does not really improve the resulting interval estimate. Who is right? Explain.

Practice problems can be found at the end of Chapter 9, and solutions for the practice test can be found on the CourseMate for STAT2 site—login at cengagebrain.com.

Learning Objectives and Outcomes

10.1 Dependent and Independent Samples (pp. 208–210)

Discuss the terminology that would be used to indicate independent samples

Comparing populations requires samples from the populations under study. Samples can be either dependent or independent. Dependence or independence is determined by the relationship between sources of the data.

10.2 Inferences Concerning the Mean Difference Using Two Dependent Samples (pp. 210–214)

Compute, describe, and interpret a confidence interval for the population mean difference * Perform, describe, and interpret a hypothesis test for the population mean difference, μ_d, using the p-value approach and classical approach

10.3 Inferences Concerning the Difference between Means Using Two Independent Samples (pp. 215–221)

Compute, describe, and interpret a confidence interval for the difference between two means using independent samples * Perform, describe, and interpret a hypothesis test for the difference between two population means, $\mu_1 - \mu_2$, using the p-value approach and classical approach

10.4 Inferences Concerning the Difference between Proportions Using Two Independent Samples (pp. 221–225)

Compute, describe, and interpret a confidence interval for the difference between two population proportions, $p_1 - p_2$ * Perform, describe, and interpret a hypothesis test for the difference between two population proportions, $p_1 - p_2$, using the p-value approach and classical approach

10.5 Inferences Concerning the Ratio of Variances Using Two Independent Samples (pp. 225–229)

Perform, describe, and interpret a hypothesis test for the ratio of two population variances, $\dfrac{\sigma_1^2}{\sigma_2^2}$, using the F-distribution with the p-value approach and the classical approach

Formulas to Use for Inferences Involving Two Populations

Situations	Test Statistic	Formula to Be Used — Confidence Interval	Formula to Be Used — Hypothesis Test
Difference between two means			
Dependent samples	t	Formula (10.2) (p. 211)	Formula (10.5) (p. 212)
Independent samples	t	Formula (10.8) (p. 216)	Formula (10.9) (p. 217)
Difference between two proportions	z	Formula (10.11) (p. 222)	Formula (10.12) (p. 223)
Difference between two variances	F		Formula (10.16) (p. 227)

Key Concepts

Vocabulary

source (p. 208)
dependent samples (p. 208)
independent samples (p. 208)
paired difference (p. 210)
dependent means (p. 211)
mean of the paired differences (p. 211)
t-distribution (p. 211)
confidence interval for estimating the mean difference (p. 211)
hypothesis about the mean difference (p. 212)
standard error (p. 212)
test statistic $t\star$ (p. 212)
independent means (p. 215)
$1 - \alpha$ confidence interval (p. 216)
proportion (p. 221)
percentage (p. 221)
probability (p. 221)
binomial experiment (p. 221)
F-distribution (p. 226)

Key Formulas

(10.1) Paired difference

$$d = x_1 - x_2$$

(10.2) Confidence interval for mean difference (dependent samples)

$$\bar{d} - t_{(df, \alpha/2)} \cdot \frac{s_d}{\sqrt{n}} \quad \text{to}$$
$$\bar{d} + t_{(df, \alpha/2)} \cdot \frac{s_d}{\sqrt{n}}$$

where $df = n - 1$

(10.3) Mean of sample differences

$$\bar{d} = \frac{\Sigma d}{n}$$

(10.4) Standard deviation of sample differences

$$s_d = \sqrt{\frac{\Sigma d^2 - \frac{(\Sigma d)^2}{n}}{n - 1}}$$

(10.5) Test statistic for mean difference (dependent samples)

$$t\star = \frac{\bar{d} - \mu_d}{s_d / \sqrt{n}} \quad \text{where } df = n - 1$$

(10.6) Standard error of independent means

$$\sigma_{\bar{x}_1 - \bar{x}_2} = \sqrt{\left(\frac{\sigma_1^2}{n_1}\right) + \left(\frac{\sigma_2^2}{n_2}\right)}$$

(10.7) Estimated standard error

$$\sqrt{\left(\frac{s_1^2}{n_1}\right) + \left(\frac{s_2^2}{n_2}\right)}$$

(10.8) Confidence interval for the difference between two means (independent samples)

$$(\bar{x}_1 - \bar{x}_2) - t(df, \alpha/2) \cdot \sqrt{\left(\frac{s_1^2}{n_1}\right) + \left(\frac{s_2^2}{n_2}\right)}$$

to

$$(\bar{x}_1 - \bar{x}_2) + t(df, \alpha/2) \cdot \sqrt{\left(\frac{s_1^2}{n_1}\right) + \left(\frac{s_2^2}{n_2}\right)}$$

(10.9) Test statistic for the difference between two means (independent samples)

$$t\star = \frac{(\bar{x}_1 - \bar{x}_2) - (\mu_1 - \mu_2)}{\sqrt{\left(\frac{s_1^2}{n_1}\right) + \left(\frac{s_2^2}{n_2}\right)}}$$

(10.10) Standard error of the difference between two proportions

$$\sigma_{p'_1 - p'_2} = \sqrt{\left(\frac{p_1 q_1}{n_1}\right) + \left(\frac{p_2 q_2}{n_2}\right)}$$

(10.11) Confidence interval for the difference between two proportions

$$(p'_1 - p'_2) - z(\alpha/2) \cdot \sqrt{\left(\frac{p'_1 q'_1}{n_1}\right) + \left(\frac{p'_2 q'_2}{n_2}\right)}$$

to

$$(p'_1 - p'_2) + z(\alpha/2) \cdot \sqrt{\left(\frac{p'_1 q'_1}{n_1}\right) + \left(\frac{p'_2 q'_2}{n_2}\right)}$$

(10.12) Test statistic for the difference between two proportions—population proportion known

$$z\star = \frac{p'_1 - p'_2}{\sqrt{pq\left[\left(\frac{1}{n_1}\right) + \left(\frac{1}{n_2}\right)\right]}}$$

(10.13) Pooled probability

$$p'_p = \frac{x_1 + x_2}{n_1 + n_2}$$

(10.14) Complement to pooled probability

$$q'_p = 1 - p'_p$$

(10.15) Test statistic for the difference between two

proportions—population proportion unknown

$$z\star = \frac{p'_1 - p'_2}{\sqrt{(p'_p)(q'_p)\left[\left(\frac{1}{n_1}\right) + \left(\frac{1}{n_2}\right)\right]}}$$

(10.16) Test statistic for equality of variances

$$F\star = \frac{s_n^2}{s_d^2}$$

with $df_n = n_n - 1$ and $df_d = n_d - 1$

Rule

Sampling Distribution of $\bar{d}$

When paired observations are randomly selected from normal populations, the paired difference, $d = x_1 - x_2$, will be approximately normally distributed about a mean μ_d with a standard deviation of σ_d.

Assumption for inferences about the mean of paired differences, μ_d

The paired data are randomly selected from normally distributed populations.

Assumptions for inferences about the difference between two means, $\mu_1 - \mu_2$

The samples are randomly selected from normally distributed populations, and the samples are selected in an independent manner. No assumptions are made about the population variances.

Assumption for inferences about the difference between two proportions, $p_1 - p_2$

The n_1 random observations and the n_2 random observations that form the two samples are selected independently from two populations that are not changing during the sampling.

Assumptions for inferences about the ratio of two variances

The samples are randomly selected from normally distributed populations, and the two samples are selected in an independent manner.

PART I–Knowing the Definitions

Answer "True" if the statement is always true. If the statement is not always true, replace the words printed in bold with words that make the statement always true.

____ 10.1 When the means of two unrelated samples are used to compare two populations, we are dealing with **two dependent means.**

____ 10.2 The use of **paired data (dependent means)** often allows for the control of unmeasurable or confounding variables because each pair is subjected to these confounding effects equally.

____ 10.3 The **chi-square distribution** is used for making inferences about the ratio of the variances of two populations.

____ 10.4 The **z-distribution** is used when two dependent means are to be compared.

PART II–Applying the Concepts

Answer all questions, showing all formulas, substitutions, and work.

10.5 State the null (H_o) and the alternative (H_a) hypotheses that would be used to test each of these claims:

a. There is no significant difference in the mean batting averages for the baseball players of the two major leagues.

b. There is a significant difference between the percentages of male and female college students who own their own car.

10.6 In a nationwide sample of 600 school-age boys and 500 school-age girls, 288 boys and 175 girls admitted to having committed a destruction-of-property offense. Use these sample data to construct a 95% confidence interval for the difference between the proportions of boys and girls who have committed this type of offense.

PART III–Understanding the Concepts

10.7 To compare the accuracy of two short-range missiles, 8 of the first kind and 10 of the second kind are fired at a target. Let x be the distance by which the missile missed the target. Do these two sets of data (8 distances and 10 distances) represent dependent or independent samples? Explain.

10.8 Let's assume that 400 students in our college are taking elementary statistics this semester. Describe how you could obtain two dependent samples of size 20 from these students to test some precourse skill against the same skill after completing the course. Be very specific.

Practice problems can be found at the end of Chapter 10, and solutions for the practice test can be found on the CourseMate for STAT2 site—login at <u>cengagebrain.com</u>.

Learning Objectives and Outcomes

11.1 Chi-Square Statistic (pp. 236–239)

Understand that enumerative data are data that can be counted and placed into categories * Understand that the chi-square distribution will be used to test hypotheses involving enumerative data

Assumptions for using chi-square to make inferences based on enumerative data:

The sample information is obtained using a random sample drawn from a population in which each individual is classified according to the categorical variable(s) involved in the test.

The chi-square statistic is useful in comparing observed frequencies to the expected frequencies, or two sets of frequencies in general. Small values of chi-square indicate agreement between the two sets; large values indicate disagreement. Chi-square test statistics are customarily one-tailed with the critical region on the right. Chi-square distributions are a family of probability distributions, each one being identified by the parameter number of degrees of freedom.

11.2 Inferences Concerning Multinomial Experiments (pp. 239–244)

Perform, describe, and interpret a hypothesis test for a multinomial experiment, using the chi-square distribution with the p-value approach and classical approach

A multinomial experiment has n independent trials, the outcomes of which each fit into exactly one of k possible cells. The probability associated with each cell remains constant, and the sum of probabilities equals exactly one. Multinomial experiments will always use a one-tailed critical region, and it will be the right-hand tail of the χ^2 distribution because larger deviations (positive or negative) from the expected values lead to an increase in the calculated $\chi^2\star$.

11.3 Inferences Concerning Contingency Tables (pp. 244–249)

Perform, describe, and interpret a hypothesis test for a test of independence or homogeneity, using the chi-square distribution with the p-value approach and classical approach * Understand the differences and similarities between tests of independence and tests of homogeneity

Contingency tables are a cross-tabulation of data resulting in an enumerative summary of sample data. The contingency table is a convenient organization not only to display the sample results, but to use when testing for independence and homogeneity. A test of independence is about the independence, or lack of, between the two factors (or variables) used to form the contingency table. A test of homogeneity is a side-by-side comparison of several multinomial experiments. For homogeneity, the experimenter fixes one of the sets of marginal totals before the data is collected.

The test for homogeneity and the test for independence look very similar and, in fact, are carried out in exactly the same way. The concepts being tested, however—same distributions and independence—are quite different.

Key Concepts

Vocabulary

enumerative data (p. 236)

cells (p. 237)

observed frequencies (p. 237)

expected frequencies (p. 238)

hypothesis test (p. 238)

chi-square (p. 238)

degrees of freedom (p. 238)

multinomial experiment (p. 239)

contingency table (p. 244)

test of independence (p. 244)

marginal totals (p. 245)

$r \times c$ contingency table (p. 247)

rows (p. 247)

columns (p. 247)

test of homogeneity (p. 247)

Key Formulas

(11.1) Test statistic for chi-square

$$\chi^2\star = \sum_{\text{all cells}} \frac{(O - E)^2}{E}$$

(11.2) Degrees of freedom for multinomial experiments

$$df = k - 1$$

(11.3) Expected value for multinomial experiments

$$E_i = n \cdot p_i$$

(11.4) Degrees of freedom for contingency tables

$$df = (r - 1) \cdot (c - 1)$$

(11.5) Expected frequencies for contingency tables

$$E_{i,j} = \frac{\text{row total} \times \text{column total}}{\text{grand total}} = \frac{R_i \times C_j}{n}$$

Practice Test

Part I–Knowing the Definitions

Answer "True" if the statement is always true. If the statement is not always true, replace the words printed in bold with words that make the statement always true.

_____ 11.1 The number of degrees of freedom for a test of a multinomial experiment is **equal to** the number of cells in the experimental data.

_____ 11.2 The **expected frequency** in a chi-square test is found by multiplying the hypothesized probability of a cell by the total number of observations in the sample.

_____ 11.3 The **observed** frequency of a cell should not be allowed to be smaller than 5 when a chi-square test is being conducted.

_____ 11.4 In a **multinomial experiment** we have $(r - 1)(c - 1)$ degrees of freedom (r is the number of rows, and c is the number of columns).

_____ 11.5 A multinomial experiment consists of n **identical independent trials.**

_____ 11.6 A **multinomial experiment** arranges the data in a two-way classification such that the totals in one direction are predetermined.

_____ 11.7 The charts for both the multinomial experiment and the contingency table **must** be set in such a way that each piece of data will fall into exactly one of the categories.

_____ 11.8 The test statistic $\frac{\Sigma(O - E)^2}{E}$ has a distribution that is **approximately normal.**

_____ 11.9 The data used in a chi-square multinomial test are always **enumerative.**

_____ 11.10 The null hypothesis being tested by a test of **homogeneity** is that the distribution of proportions is the same for each of the subpopulations.

Part II–Applying the Concepts

Answer all questions. Show formulas, substitutions, and work.

11.11 State the null and alternative hypotheses that would be used to test each of these claims:

a. The single-digit numerals generated by a certain random-number generator were not equally likely.

b. The results of the last election in our city suggest that the votes cast were not independent of the voter's registered party.

c. The distributions of types of crimes committed against society are the same in the four largest U.S. cities.

11.12 Find each value:

a. $\chi^2(12, 0.025)$ b. $\chi^2(17, 0.005)$

11.13 Three hundred consumers were asked to identify which one of three different items they found to be the most appealing. The table shows the number that preferred each item.

Item	1	2	3
Number	85	103	112

Do these data present sufficient evidence at the 0.05 level of significance to indicate that the three items are not equally preferred?

11.14 To study the effect of the type of soil on the amount of growth attained by a new hybrid plant, saplings were planted in three different types of soil and their subsequent amounts of growth classified into three categories:

Growth	Soil Type		
	Clay	Sand	Loam
Poor	16	8	14
Average	31	16	21
Good	18	36	25
Total	65	60	60

Does the quality of growth appear to be distributed differently for the tested soil types at the 0.05 level?

a. State the null and alternative hypotheses.

b. Find the expected value for the cell containing 36.

c. Calculate the value of chi-square for these data.

d. Find the p-value.

e. Find the test criteria [level of significance, test statistic, its distribution, critical region, and critical value(s)].

f. State the decision and the conclusion for this hypothesis test.

Part III–Understanding the Concepts

11.15 Explain how a multinomial experiment and a binomial experiment are similar and also how they are different.

11.16 Explain the distinction between a test for independence and a test for homogeneity.

11.17 Student A says that tests for independence and homogeneity are the same, and student B says that they are not at all alike because they are tests of different concepts. Both students are partially right and partially wrong. Explain.

> **Practice problems can be found at the end of Chapter 11, and solutions for the practice test can be found on the CourseMate for STAT2 site—login at <u>cengagebrain.com</u>.**

Learning Objectives and Outcomes

12.1 Analysis of Variance Technique—An Introduction (pp. 254–260)

Understand that analysis of variance techniques (ANOVA) are used to test differences among more than two means

When testing a hypothesis about several means, analysis of variance (ANOVA) is useful. ANOVA techniques allow you to test the null hypothesis against an alternative hypothesis. This chapter addresses only single-factor ANOVA. The test of multiple means is done by partitioning the sum of squares into two segments: (1) the sum of squares due to variation between the levels of the factor being tested and (2) the sum of squares due to variation between the replicates within each level.

The ANOVA technique separates the variance among the sample data into two measures of variance: (1) MS(factor), the measure of variance between the levels being tested, and (2) MS(error), the measure of variance within the levels being tested. Then these measures of variance can be compared.

12.2 The Logic behind ANOVA (pp. 260–263)

Understand that if the variation between the means is significantly more than the variation within the samples, then the means are considered unequal

The design for the single-factor ANOVA is to obtain independent random samples at each of the several levels of the factor being tested. Using ANOVA, if MS(factor) is significantly larger than MS(error), you can conclude that the means for the factor levels being tested are not all the same. That is, the factor being tested does have a significant effect on the response variable. If, however, MS(factor) is not significantly larger than MS(error), you can't reject the null hypothesis that all means are equal.

12.3 Applications of Single-Factor ANOVA (pp. 263–266)

Compute, describe, and interpret a hypothesis test for the differences among several means, using the F-distribution with the p-value approach and classical approach

In the mathematical model for ANOVA: $x_{c,k} = \mu + F_c + \epsilon_{k(c)}$

- $x_{c,k}$ is the value of the variable at the kth replicate of level c.
- μ is the mean value for all the data without respect to the test factor.
- F_c is the effect that the factor being tested has on the response variable at each different level c.
- $\epsilon_{k(c)}$ (ϵ is the lowercase Greek letter epsilon) is the *experimental error* that occurs among the k replicates in each of the c columns.

Remember that one-factor techniques can be developed further and applied to more complex experiments.

Key Concepts

Vocabulary

ANOVA (p. 256)
replicates (p. 256)
sample variance (p. 257)
sum of squares, SS(total) (p. 257)
partitioning (p. 258)
variation between levels, SS(factor) (p. 258)
variation within the rows, SS(error) (p. 258)
degrees of freedom, df (p. 258)
mean square for a factor, MS(factor) (p. 259)
mean square for an error, MS(error) (p. 259)
response (observed) variable (p. 260)
between-sample variation (p. 262)
within-sample variation (p. 262)
assumptions (p. 262)
randomized (p. 263)
mathematical model (p. 263)
experimental error (p. 264)

Key Formulas

(12.1) Total sum of squares

$$\text{sum of squares} = \Sigma(x - \bar{x})^2$$

(12.2) Shortcut for total sum of squares

$$SS(\text{total}) = \Sigma(x^2) - \frac{(\Sigma x)^2}{n}$$

(12.3) Sum of squares due to factor

$$SS(\text{factor}) = \left(\frac{C_1^2}{k_1} + \frac{C_2^2}{k_2} + \frac{C_3^2}{k_3} + \ldots \right) - \frac{(\Sigma x)^2}{n}$$

(12.4) Sum of squares due to error

$$SS(\text{error}) = \Sigma(x^2) - \left(\frac{C_1^2}{k_1} + \frac{C_2^2}{k_2} + \frac{C_3^2}{k_3} + \ldots \right)$$

(12.5) Degrees of freedom for factor

$$df(\text{factor}) = c - 1$$

(12.6) Degrees of freedom for total

$$df(\text{total}) = n - 1$$

(12.7) Degrees of freedom for error

$$df(\text{error}) = n - c$$

(12.8) Total sum of squares

$$SS(\text{factor}) + SS(\text{error}) = SS(\text{total})$$

(12.9) Total sum of degrees of freedom

$$df(\text{factor}) + df(\text{error}) = df(\text{total})$$

(12.10) Mean square for factor

$$MS(factor) = \frac{SS(factor)}{df(factor)}$$

(12.11) Mean square for error

$$MS(error) = \frac{SS(error)}{df(error)}$$

(12.12) Test statistic for ANOVA

$$F\star = \frac{MS(factor)}{MS(error)}$$

(12.13) Mathematical model for single-factor ANOVA

$$x_{c,k} = \mu + F_c + \epsilon_{k(c)}$$

Assumptions when using ANOVA:

1. Our goal is to investigate the effect that various levels of the factor being tested have on the response variable. Typically, we want to find the level that yields the most advantageous values of the response variable. This, of course, means that we probably will want to reject the null hypothesis in favor of the alternative hypothesis. Then a follow-up study could determine the "best" level of the factor.

2. We must assume that the effects due to chance and due to untested factors are normally distributed and that the variance caused by these effects is constant throughout the experiment.

3. We must assume independence among all observations of the experiment. We will usually conduct the tests in a randomized order to ensure independence. This technique also helps avoid data contamination.

Practice Test

Part I—Knowing The Definitions

Answer "True" if the statement is always true. If the statement is not always true, replace the words printed in bold with words that make the statement always true.

_____ 12.1 To partition the sum of squares for the total is to separate the numerical value of SS(total) into two values such that the **sum** of these two values is equal to SS(total).

_____ 12.2 A **sum of squares** is actually a measure of variance.

_____ 12.3 **Experimental error** is the name given to the variability that takes place between the levels of the test factor.

_____ 12.4 **Experimental error** is the name given to the variability that takes place among the replicates of an experiment as it is repeated under constant conditions.

_____ 12.5 **Fail to reject** H_o is the desired decision when the means for the levels of the factor being tested are all different.

_____ 12.6 The **mathematical model** for a particular problem is an equational statement showing the anticipated makeup of an individual piece of data.

_____ 12.7 The degrees of freedom for the factor are equal to the **number of factors tested.**

_____ 12.8 The measure of a specific level of a factor being tested in an ANOVA is the **variance** of that factor level.

_____ 12.9 We **need not** assume that the observations are independent to do analysis of variance.

_____ 12.10 The rejection of H_o **indicates** that you have identified the level(s) of the factor that is (are) different from the others.

Part II—Applying the Concepts

12.11 Consider this table:

	SS	df	MS	F★
Factor	A	4	18	E
Error	B	18	D	
Total	144	C		

Find the values:

a. A   b. B   c. C   d. D   e. E

Part III—Understanding the Concepts

12.12 A state environmental agency tested three different scrubbers used to reduce the resulting air pollution in the generation of electricity. The primary concern was the emission of particulate matter. Several trials were run with each scrubber. The amount of particulate emission was recorded for each trial.

	Amounts of Emission					
Scrubber I	11	10	12	9	13	12
Scrubber II	12	10	12	8	9	
Scrubber III	9	11	10	7	8	

a. State the mathematical model for this experiment.

b. State the null and alternative hypotheses.

c. Calculate and form the ANOVA table.

d. Complete the testing of H_o using a 0.05 level of significance. State the decision and conclusion clearly.

e. Construct a graph representing the data that is helpful in picturing the results of the hypothesis test.

Practice problems can be found at the end of Chapter 12, and solutions for the practice test can be found on the CourseMate for STAT2 site—login at cengagebrain.com.

Learning Objectives and Outcomes

13.1 Linear Correlation Analysis (pp. 273–276)

Understand that the correlation coefficient, *r*, standardizes covariance so that relative strengths can be compared

One measure of linear dependency is covariance, the sum of the products of the distances of all values of *x* and *y* from the centroid divided by *n* − 1. The biggest disadvantage of covariance is that it does not have a standardized unit of measure. The coefficient of linear correlation standardizes the measure of dependency so you can compare the relative strengths of dependency for different sets of data.

13.2 Inferences about the Linear Correlation Coefficient (pp. 276–279)

Compute, describe, and interpret a confidence interval for the population correlation coefficient, ρ, using Table 10 in Appendix B * Perform, describe, and interpret a hypothesis test of the population correlation coefficient, ρ, using the calculated *r*

13.3 Linear Regression Analysis (pp. 279–284)

Regression analysis produces the mathematical equation for the line of best fit. When one input and one output variable produce a straight line of best fit, it is a simple linear regression. If the scatter diagram suggests something other than a straight line, it is curvilinear regression. When two or three input variables are used to increase the usefulness of the regression equation, it is multiple regression.

To assess the accuracy of a regression line, you need to estimate the experimental error and determine its variance.

13.4 Inferences Concerning the Slope of the Regression Line (pp. 284–287)

Compute, describe, and interpret a confidence interval for the population slope of the regression line, β_1, using the *t*-distribution * Perform, describe, and interpret a hypothesis test for the population slope of the regression line, β_1, using the *t*-distribution with the *p*-value approach and the classical approach

13.5 Confidence Intervals for Regression (pp. 287–292)

Compute, describe, and interpret a confidence interval for the mean value of *y* for a particular *x*, ($\mu_{y|x_o}$), using the *t*-distribution * Compute, describe, and interpret a prediction interval for an individual value of *y* for a particular *x*, (y_{x_o}), using the *t*-distribution * Understand the difference between a confidence interval and a prediction interval for a *y* value at a particular *x* value

Confidence and prediction intervals for the mean at a given value of *x* are constructed similarly to those of mean μ. The regression equation is meaningful only in the domain of the *x* variable studied, so the results of one sample should not be used to make inferences about a population other than the one from which the sample was drawn. Regression only measures movement between *x* and *y*; it does not prove *x* causes *y* to change.

13.6 Understanding the Relationship between Correlation and Regression (p. 292)

The linear correlation coefficient is used to determine *if* two variables are linearly related. Linear regression analysis is used to answer questions related to *how* the two variables are related.

Key Concepts

Vocabulary
linear correlation (p. 272)
bivariate data (p. 274)
centroid (p. 274)
covariance of x and y (p. 274)
coefficient of linear correlation (p. 275)
Pearson's product moment, r (p. 276)
confidence interval (p. 277)
confidence belts (p. 277)
hypothesis test (p. 277)
line of best fit (p. 279)
intercept (β_0 or b_0) (p. 279)
slope (β_1 or b_1) (p. 279)
linear regression (p. 279)
scatter diagram (p. 281)
curvilinear regression (p. 281)
multiple regression (p. 281)
experimental error (*e* or ε) (p. 281)
regression line (p. 282)
predicted value of y ($\hat{y}$) (p. 281)
variance of the experimental error (σ_ϵ^2) (p. 281)
sum of squares for error (SSE) (p. 282)
sampling distribution (p. 284)
standard error (p. 285)
confidence interval for $\mu_{y|x_o}$ (p. 287)
prediction interval for y_{x_o} (p. 287)
sampling distribution of $\hat{y}$ (p. 287)

Key Formulas

(13.1) Covariance of *x* and *y*
$$\text{covar}(x, y) = \frac{\sum_{i=1}^{n}(x_i - \bar{x})(y_i - \bar{y})}{n - 1}$$

(13.2) Coefficient of linear correlation
$$r = \text{covar}(x', y') = \frac{\text{covar}(x, y)}{s_x \cdot s_y}$$

(13.3) Shortcut for coefficient of linear correlation
$$r = \frac{\text{covar}(x, y)}{s_x \cdot s_y} = \frac{\frac{\sum[(x - \bar{x})(y - \bar{y})]}{n - 1}}{s_x \cdot s_y}$$
$$= \frac{SS(xy)}{\sqrt{SS(x) \cdot SS(y)}}$$

(13.4) Linear model
$$y = \beta_0 + \beta_1 x + \varepsilon$$

(13.5) Estimate of the experimental error
$$e = y - \hat{y}$$

Key Concepts

(13.6) Variance of the estimated error, e

$$s_e^2 = \frac{\Sigma(y - \hat{y})^2}{n - 2}$$

(13.7)

$$s_e^2 = \frac{\Sigma(y - b_0 - b_1 x)^2}{n - 2}$$

(13.8) Variance of the estimated error, e

$$s_e^2 = \frac{(\Sigma y^2) - (b_0)(\Sigma y) - (b_1)(\Sigma xy)}{n - 2}$$

(13.9) Standard deviation of the estimated error, e (standard error of the estimate)

$$s_e = \sqrt{s_e^2}$$

(13.10)

$$\sigma_{b_1}^2 = \frac{\sigma_\epsilon^2}{\Sigma(x - \bar{x})^2}$$

(13.11)

$$s_{b_1}^2 = \frac{s_e^2}{\Sigma(x - \bar{x})^2}$$

(13.12) Estimate for variance of slope

$$s_{b_1}^2 = \frac{s_e^2}{\Sigma x^2 - \frac{(\Sigma x)^2}{n}}$$

(13.13) Estimate for standard error of regression (slope)

$$s_{b_1} = \sqrt{s_{b_1}^2}$$

(13.14) Confidence interval for slope

$$b_1 \pm t(n - 2, \alpha/2) \cdot s_{b_1}$$

(13.15) Test statistic for slope

$$t\star = \frac{b_1 - \beta_1}{s_{b_1}}$$

(13.16)

$$\hat{y} \pm t(n - 2, \alpha/2) \cdot s_e \cdot \sqrt{\frac{1}{n} + \frac{(x_0 - \bar{x})^2}{\Sigma(x - \bar{x})^2}}$$

(13.17) Confidence interval for $\mu_{y|x_0}$

$$\hat{y} \pm t(n - 2, \alpha/2) \cdot s_e \cdot \sqrt{\frac{1}{n} + \frac{(x_0 - \bar{x})^2}{SS(x)}}$$

(13.18) Prediction interval for $y_{x = x_0}$

$$\hat{y} \pm t(n - 2, \alpha/2) \cdot s_e \cdot \sqrt{1 + \frac{1}{n} + \frac{(x_0 - \bar{x})^2}{SS(x)}}$$

Assumptions for inferences about the linear correlation coefficient:

The set of (x, y) ordered pairs forms a random sample, and the y values at each x have a normal distribution. Inferences use the t-distribution with $n - 2$ degrees of freedom.

Caution: *Significance of the linear correlation coefficient does not mean that you have established a cause-and-effect relationship.* Cause and effect is a separate issue.

Assumptions for inferences about the linear regression:

The set of (x, y) ordered pairs forms a random sample, and the y values at each x have a normal distribution. Since the population standard deviation is unknown and replaced with the sample standard deviation, the t-distribution will be used with $n - 2$ degrees of freedom.

Practice Test

Part I–Knowing the Definitions

Answer "True" if the statement is always true. If the statement is not always true, replace the words printed in bold with words that make the statement always true.

_____ 13.1 **Covariance** measures the strength of the linear relationship and is a standardized measure.

_____ 13.2 The **sum of squares for error** is the name given to the numerator of the formula used to calculate the variance of y about the line of regression.

_____ 13.3 **Correlation** analysis attempts to find the equation of the line of best fit for two variables.

_____ 13.4 There are $n - 3$ degrees of freedom involved with the inferences about the regression line.

_____ 13.5 $\hat{y}$ serves as the **point estimate** for both $\mu_{y|x_0}$ and y_{x_0}.

Part II—Applying the Concepts

How does nitrogen fertilizer affect wheat yield per acre? The following data show the amount of nitrogen fertilizer used per test plot and the amount of wheat harvested per test plot. All test plots were the same size.

x, Pounds of Fertilizer	y, 100 Pounds of Wheat	x, Pounds of Fertilizer	y, 100 Pounds of Wheat
30	9	70	19
30	11	70	22
30	14	70	31
50	12	90	29
50	14	90	33
50	23	90	35

13.6 Draw a scatter diagram of the data. Be sure to label completely.

13.7 Complete an extensions table.

13.8 Calculate SS(x), SS(xy), and SS(y).

13.9 Calculate the linear correlation coefficient, r.

13.10 Determine the 95% confidence interval estimate for the population linear correlation coefficient.

13.11 Calculate the equation of the line of best fit.

13.12 Draw the line of best fit on the scatter diagram.

13.13 Calculate the standard deviation of the y values about the line of best fit.

13.14 Does the value of b_1 show strength significant enough to conclude that the slope is greater than zero at the 0.05 level?

13.15 Determine the 0.95 confidence interval for the mean yield when 85 pounds of fertilizer are used per plot.

13.16 Draw a line on the scatter diagram representing the 95% confidence interval found in question 13.10.

Part III–Understanding the Concepts

13.17 "There is a high correlation between how frequently skiers have their bindings tested and the incidence of lower-leg injuries, according to researchers at the Rochester Institute of Technology. To make sure your bindings release properly when you begin to fall, you should have them serviced by a ski mechanic every 15 to 30 ski days or at least at the start of each ski season" (University of California, Berkeley, "Wellness Letter"). Explain what two variables are discussed in this statement, and interpret the "high correlation" mentioned.

13.18 If a "moment" is defined as the distance from the mean, describe why the method used to define the correlation coefficient is referred to as "a product moment."

13.19 If you know that the value of r is very close to zero, what value would you anticipate for b_1? Explain why.

Practice problems can be found at the end of Chapter 13, and solutions for the practice test can be found on the CourseMate for STAT2 site—login at cengagebrain.com.

Learning Objectives and Outcomes

14.1 Nonparametric Statistics (pp. 298–300)

Understand that parametric methods are statistical methods that assume that the parent population is approximately normal or that the central limit theorem gives (at least approximately) a normal distribution of a test statistic * Understand that nonparametric methods (distribution-free methods) do not depend on the distribution of the population being sampled

14.2 The Sign Test (pp. 300–306)

Understand that the sign test is the nonparametric alternative to the *t*-test for one mean and the difference between paired data that result from two dependent samples

Assumptions for inferences about the population single-sample median using the sign test
The *n* random observations that form the sample are selected independently, and the population is continuous in the vicinity of the median *M*.

Assumptions for inferences about the median of paired differences using the sign test
The paired data are selected independently, and the variables are ordinal or numerical.

14.3 The Mann–Whitney *U* Test (pp. 307–312)

Understand that the Mann–Whitney *U* test is the nonparametric alternative to the *t*-test for the difference between two independent means

Assumptions for inferences about two populations using the Mann–Whitney *U* test
The two independent random samples are independent within each sample as well as between samples, and the random variables are ordinal or numerical.

14.4 The Runs Test (pp. 312–315)

Perform, describe, and interpret a hypothesis test for the randomness of data using the runs test with the *p*-value approach and classical approach

Assumption for inferences about randomness using the runs test
Each piece of sample data can be classified into one of two categories.

14.5 Rank Correlation (pp. 315–319)

Perform, describe, and interpret a hypothesis test for the significance of correlation between two variables using the Spearman rank correlation coefficient with the *p*-value approach and classical approach

Assumptions for inferences about rank correlation
The *n* ordered pairs of data form a random sample, and the variables are ordinal or numerical.

Key Concepts

Vocabulary
parametric methods (p. 298)
nonparametric methods (distribution-free methods) (p. 298)
power of a statistical test (p. 300)
efficiency (p. 300)
sign test (p. 300)
population median, *M* (p. 301)
paired data (p. 303)
two dependent samples (p. 303)
binomial random variable (p. 304)
continuity correction (p. 305)
Mann–Whitney *U* test (p. 307)
independent random samples (p. 307)
rank (p. 307)
test statistic *U* (p. 308)
test statistic *z*★ (p. 310)
runs test (p. 312)
randomness (p. 312)
run (p. 312)
test statistic *V* (p. 312)
Spearman rank correlation coefficient, r_s (p. 315)
paired rankings (p. 315)

Key Formulas

(14.1) $1 - \alpha$ confidence interval for *M*

$$\frac{1}{2}(n) \pm \left[\frac{1}{2} + \frac{1}{2} \cdot z(\alpha/2) \cdot \sqrt{n}\right]$$

(14.2) Sign test *z* test statistic

$$z\star = \frac{x' - \frac{n}{2}}{\frac{1}{2} \cdot \sqrt{n}}$$

(14.3) Mann–Whitney *U* test statistic

$$U_a = n_a \cdot n_b + \frac{(n_b)(n_b + 1)}{2} - R_b$$

(14.4) Mann–Whitney *U* test statistic

$$U_b = n_a \cdot n_b + \frac{(n_a)(n_a + 1)}{2} - R_a$$

(14.5) Mean of *U*

$$\mu_U = \frac{n_a \cdot n_b}{2}$$

(14.6) Standard deviation of *U*

$$\sigma_U = \sqrt{\frac{n_a \cdot n_b \cdot (n_a + n_b + 1)}{12}}$$

Key Concepts

(14.7) Mann–Whitney z test statistic

$$z\star = \frac{U\star - \mu_U}{\sigma_U}$$

(14.8) Mean of V

$$\mu_V = \frac{2n_1 \cdot n_2}{n_1 + n_2} + 1$$

(14.9) Standard deviation of V

$$\sigma_V = \sqrt{\frac{(2n_1 \cdot n_2) \cdot (2n_1 \cdot n_2 - n_1 - n_2)}{(n_1 + n_2)^2(n_1 + n_2 - 1)}}$$

(14.10) Runs test z test statistic

$$z\star = \frac{V\star - \mu_V}{\sigma_V}$$

(14.11) Spearman rank correlation coefficient

$$r_s = 1 - \frac{6\Sigma(d_i)^2}{n(n^2 - 1)}$$

Practice Test

PART I–Knowing the Definitions

Answer "True" if the statement is always true. If the statement is not always true, replace the words printed in bold with words that make the statement always true.

_____ 14.1 One of the advantages of the nonparametric tests is the necessity for **less restrictive** assumptions.

_____ 14.2 The sign test is a possible replacement for the **F-test.**

_____ 14.3 The **sign test** can be used to test the randomness of a set of data.

_____ 14.4 Two dependent **means** can be compared nonparametrically by using the sign test.

_____ 14.5 The sign test is a possible alternative to Student's t-test for **one mean value.**

_____ 14.6 If a tie occurs in a set of ranked data, the data that form the tie are **removed from the set.**

_____ 14.7 The **runs test** is a nonparametric alternative to the difference between two independent means.

_____ 14.8 The **confidence level** of a statistical hypothesis test is measured by $1 - \beta$.

_____ 14.9 Spearman's rank correlation coefficient is an alternative to using the **linear correlation coefficient.**

_____ 14.10 The **efficiency** of a nonparametric test is the probability that a false null hypothesis is rejected.

PART II–Applying the Concepts

14.11 The weights (in pounds) of nine people before they stopped smoking and five weeks after they stopped smoking are listed here:

Person	1	2	3	4	5	6	7	8	9
Before	148	176	153	116	128	129	120	132	154
After	155	178	151	120	130	136	126	128	158

Find the 95% confidence interval for the average weight change.

14.12 The following data show the weight gains (in ounces) for 20 laboratory mice, half of which were fed one diet and half a different diet. Test to determine whether the difference in weight gain is significant at $\alpha = 0.05$.

Diet A	41	40	36	43	36	43	39	36	24	41
Diet B	35	34	27	39	31	41	37	34	42	38

14.13 A large textbook publishing company hired nine new sales representatives three years ago. At the time of hire, the nine were ranked according to their potential. Now three years later, the company president wants to know how well their potential ranks correlate with their sales totals for the three years.

Sales Representative	a	b	c	d	e	f	g	h	i
Potential	2	5	6	1	4	3	9	8	7
Sales Total	450	410	350	345	330	400	250	310	270

Is there significant correlation at the 0.05 level?

14.14 The new school principal thought that there might be a pattern to the order in which discipline problems arrived at his office. He had his secretary record the grade levels of the students as they arrived.

9	10	11	9	12	11	9	10	10	11
10	11	10	10	11	12	12	9	9	11
12	10	9	12	10	11	12	11	10	10

At the 0.05 level, is there significant evidence of randomness?

PART III–Understanding the Concepts

14.15 What advantages do nonparametric statistics have over parametric statistics?

14.16 Explain how the sign test is based on the binomial distribution and is often approximated by the normal distribution.

14.17 Why does the sign test use a null hypothesis about the median instead of the mean like a t-test uses?

14.18 Explain why a nonparametric test is not as sensitive to extreme data as a parametric test might be.

14.19 A restaurant has collected data on which of two seating arrangements its customers prefer. In a sign test to determine whether one seating arrangement is significantly preferred, which null hypothesis would be used?

a. $M = 0$ b. $M = 0.5$ c. $p = 0$ d. $p = 0.5$

Explain your choice.

Practice problems can be found at the end of Chapter 14, and solutions for the practice test can be found on the CourseMate for STAT2 site—login at cengagebrain.com.

techcard CHAPTER 1
STATISTICS

Minitab Instructions For:

BASIC CONVENTIONS

Choose: tells you to make a menu selection by a mouse "point and click" entry.

For example:

Choose: **Stat > Quality Tools > Pareto Chart** *instructs you to, in sequence, "point and click on"* **Stat** *on the menu bar, "followed by"* **Quality Tools** *on the pull-down, and then "followed by"* **Pareto Chart** *on the second pull-down.*

Select: indicates that you should click on the small box or circle to the left of a specified item.

Enter: instructs you to type or select information needed for a specific item.

FYI For information about obtaining MINITAB, check the Internet at http://www.minitab.com.

TI-83/84 Plus Instructions For:

BASIC CONVENTIONS

Choose: tells you which keys to press or menu selections to make.

For example:

Choose: **Zoom > 9:ZoomStat > Trace > > >** *instructs you to press the* **Zoom** *key, followed by selecting* **9:ZoomStat** *from the menu, followed by pressing the* **Trace** *key; > > > indicates that you should press arrow keys repeatedly to move along a graph to obtain important points.*

Enter: instructs you to type or select information needed for a specific item.

Screen

Capture: gives pictures of what your calculator screen should look like with chosen specifications highlighted.

FYI For information about obtaining TI-83/84 Plus, check the Internet at http://www.ti.com/calc.

Excel Instructions For:

BASIC CONVENTIONS

Excel 2007 is laid out in a tabbed format, with pull-down menus and buttons in each tab and pop-out boxes along the bottom of some tabs. You access different selections of tabs through the Standard Menu bar along the top of the tabs. We will indicate tabs and pop-out boxes in this style:

Choose: **Home > Number > Number**

So, you would click the Home menu, find the Number tab, click the bottom of the small arrow to the right of the word Number, and then click number on the pop-out box that appears.

Select: indicates that you should click on the button or box next to a specified item. This will often be followed by a **> OK, > Close,** or **> Finish.**

Enter: instructs you to type or select information.

FYI Excel is part of Microsoft Office and can be found on many personal computers.

ADDITIONAL TOOLS

You will need to download the Analysis ToolPak for many of the problems in this textbook and in order to follow the Excel instructions on your Tech Cards.

Choose: **Office Button > Excel Options > Add-Ins**

Next to Manage, Select: **Excel Add-Ins > Go > Analysis ToolPak > OK**

NOTE: Not all chapters have technology instructions. For example, Chapters 4 and 7 utilize technology instructions from the previous chapters, so your cards will jump right to the next chapter.

1.1 *Use a random-number table or a computer to simulate rolling a pair of dice 100 times.*

a. *List the results of each roll as an ordered pair and the sum.*
b. *Prepare an ungrouped frequency distribution and a histogram of the sums.*
c. *Describe how these results compare with what you expect to occur when two dice are rolled.*

Technology Instructions: Simulate Dice

MINITAB

Choose:	**Calc > Random Data > Integer**
Enter:	*Number of rows to generate:* **100**
	Store in column(s): **C1 C2**
	Minimum value: **1** *Maximum value:* **6 > OK**
Choose:	**Calc > Calculator**
Enter:	*Store result in variable:* **C3**
	Expression: **C1 + C2 > OK**
Choose:	**Stat > Tables > Tally Individual Variables**
Enter:	*Variable:* **C3**
Select:	**Counts > OK**

Use the MINITAB commands from the Chapter 2 Tech Card to construct a frequency histogram of the data in C3. (Use Binning > midpoint and midpoint positions 2:12/1 if necessary.)

Excel

Enter 1, 2, 3, 4, 5, 6 into column A; Label Column J **Sums** *and input the numbers 2, 3, 4...12 into it; label C1:* **Die1***; D1:* **Die2***; E1:* **Dice***, and activate B1.*

Choose:	**Home > Number > Number > Category: Number**
Enter:	*Decimal places:* **8 > OK**
Enter:	**1/6** *in B1*
Drag:	**Bottom right corner of B1 down for 6 entries**
Choose:	**Data > Data Analysis > Random Number Generation > OK**
Enter:	*Number of Variables:* **2**
	Number of Random Numbers: **100**
	Distribution: **Discrete**
	Value and Probability Input Range: **(A1:B6 or select cells)**
Select:	**Output Range**
Enter:	**(C2 or select cells) > OK**
Activate the **E2** *cell.*	
Enter:	**=C2+D2 > Enter**
Drag:	**Bottom right corner of E2 down for 100 entries**
Choose:	**Data > Data Analysis > Histogram > OK**
Enter:	*Input Range:* **(E2:E101 or select cells)**
	Bin Range: **J2:J12**
Select:	**Output Range**
Enter:	**F1 > OK**

To generate a Histogram, follow all the previous instructions, then:

Choose:	**Data > Data Analysis > Histogram > OK**
Select:	**Chart Output > OK**

Select OK for pop-up warnings about overwriting data.

TI-83/84 Plus

Choose:	**MATH > PRB > 5:randInt(**
Enter:	**1,6,100)**
Choose:	**STO→ > 2nd L1**
Repeat preceding for L2.	
Choose:	**STAT > EDIT > 1:Edit**
Highlight:	**L3**
Enter:	*L3* **= L1 + L2**
Choose:	**2nd > STAT PLOT**
	> 1:Plot1
Choose:	**WINDOW**
Enter:	**−0.5, 12.5, 1,**
	−10, 40, 10, 1
Choose:	**TRACE > > >**

Multiple formulas: *Statisticians have multiple formulas for convenience— that is, convenience relative to the situation. The following statements will help you decide which formula to use:*

1. *When you are working on a computer and using statistical software, you will generally store all the data values first. The computer handles repeated operations easily and can "revisit" the stored data as often as necessary to complete a procedure. The computations for sample variance will be explained in Chapter 2 using formula (2.5), following the process shown in Table 2.12.*

2. *When you are working on a calculator with built-in statistical functions, the calculator must perform all necessary operations on each data value as the values are entered (most handheld nongraphing calculators do not have the ability to store data). Then after all data have been entered, the computations will be completed using the appropriate summations. The computations for sample variance will be done using formula (2.9), following the procedure shown in Table 2.13.*

3. *If you are doing the computations either by hand or with the aid of a calculator without statistical functions, the most convenient formula to use will depend on how many data there are and how convenient the numerical values are to work with.*

Minitab Instructions For:

PIE CHART

Input the categories into C1 and the corresponding frequencies into C2; then continue with:

Choose:	**Graph > Pie Chart . . .**
Select:	**Chart values from a table**
Enter:	Categorical variable: **C1** Summary variables: **C2**
Select:	**Labels > Titles/Footnotes** Enter: Title: **your title**
Select:	**Slice Labels >** Select desired labels **> OK > OK**

PARETO DIAGRAM

Input the categories into C1 and the corresponding frequencies into C2; then continue with:

Choose:	**Stat > Quality Tools > Pareto Chart**
Select:	**Chart defects table**
Enter:	Labels in: **C1** Frequencies in: **C2**
Select:	**Options**
Enter:	Title: **your title > OK > OK**

DOTPLOT

Input the data into C1; then continue with:

Choose:	**Graph > Dotplot . . . > One Y > Simple > OK**
Enter:	Graph Variables: **C1 > OK**

STEM-AND-LEAF DISPLAY

Input the data into C1; then continue with:

Choose:	**Graph > Stem-and-Leaf . . .**
Enter:	Graph variables: **C1**
	Increment: **stem width** (optional) **> OK**

MULTIPLE DOTPLOTS

Input the data into C1 and the corresponding numerical categories into C2; then continue with:

Choose:	**Graph > Dotplot . . .**
Select:	**One Y, With Groups > OK**
Enter:	Graph variable: **C1**
	Categorical variables for grouping: **C2 > OK**

If the various categories are in separate columns, select Multiple Y's Simple and enter all of the columns under Graph variables.

HISTOGRAM

Input the data into C1; then continue with:

Choose:	**Graph > Histogram > Simple > OK**
Enter:	Graph variables: **C1**
Choose:	**Labels > Titles/Footnote**
Enter:	**Your title and/or footnote > OK**
Choose:	**Scale > Y-Scale Type**
Select:	Y scale Type: **Frequency or Percent or Density > OK > OK**

To adjust histogram: Double click anywhere on bars of histogram.

Select:	**Binning**
Select:	Interval Type: **Midpoint or Cutpoint**
	Interval Definitions: **Automatic or Number of intervals**
	Enter: **N** or Midpt/cutpt positions
	Enter: **A:B/C > OK**

Notes:

1. Midpoints are the class midpoints, and cutpoints are the class boundaries.
2. Percent is relative frequency.

3. Automatic means MINITAB will make all the choices; N = number of intervals, that is, the number of classes you want used.
4. A = smallest class midpoint or boundary, B = largest class midpoint or boundary, C = class width you want to specify.

The following commands will draw the histogram of a frequency distribution. The end classes can be made full width by adding an extra class with frequency zero to each end of the frequency distribution. Input the class midpoints into C1 and the corresponding frequencies into C2.

Choose:	**Graph > Scatterplot > With Connect Line > OK**
Enter:	Y variables: **C2** X variables: **C1**
Select:	Data View: Data Display: **Symbols Connect > OK > OK**

Double click on a connect line.

Select:	**Options**
	Connection Function: **Step > OK**

OGIVE

Input the class boundaries into C1 and the cumulative percentages into C2 (enter 0 [zero] for the percentage paired with the lower boundary of the first class and pair each cumulative percentage with the class upper boundary). Use percentages; that is, use 25% in place of 0.25.

Choose:	**Graph > Scatterplot > With Connect Line > OK**
Enter:	Y variables: **C2** X variables: **C1**
Select:	Data View: Data Display: **Symbols Connect > OK**
Select:	**Labels > Titles/Footnotes**
Enter:	**your title or footnotes > OK > OK**

MEAN

Input the data into C1; then continue with:

Choose:	**Calc > Column Statistics**
Select:	**Mean**
Enter:	Input variable: **C1 > OK**

MEDIAN

Input the data into C1; then continue with:

Choose:	**Calc > Column Statistics**
Select:	**Median**
Enter:	Input variable: **C1 > OK**

STANDARD DEVIATION

Input the data into C1; then continue with:

Choose:	**Calc > Column Statistics**
Select:	**Standard deviation**
Enter:	Input variable: **C1 > OK**

ADDITIONAL STATISTICS

Input the data into C1, then continue with:

Choose:	**Calc > Column Statistics**
	Then one at a time select the desired statistics

Select:	**N total**	Number of data in column
	Sum	Sum of the data in column
	Minimum	Smallest value in column
	Maximum	Largest value in column
	Range	Range of values in column
	Sum of squares	Sum of squared x-values, Σx^2
Enter:	Input variable: **C1 > OK**	

5-NUMBER SUMMARY

Input the data into C1; then continue with:

Choose: **Stat > Basic Statistics > Display Descriptive Statistics . . .**
Enter: **Variables: C1 > OK**

BOX-AND-WHISKERS DISPLAY

Input the data into C1; then continue with:

Choose: **Graph > Boxplot . . . > One Y, Simple > OK**
Enter: Graph variables: **C1**
Optional:
Select: **Labels > Titles/Footnoes**
Enter: **your title, footnotes > OK**
Select: **Scale > Axes and Ticks**
Select: **Transpose value and category scales > OK > OK**
For multiple boxplots, enter additional set of data into C2; then do as just described plus:
Choose: **Graph > Boxplot . . . > Multiple Y's, Simple > OK**
Enter: Graph variables: **C1 C2 > OK**
Optional: See above.

ADDITIONAL COMMANDS

Input the data into C1; then:
To sort the data into ascending order and store them in C2, continue with:
Choose: **Data > Sort . . .**
Enter: Sort column(s):   **C1**   By column: **C1**
Select: Store sorted data in: **Column(s) of current worksheet**
Enter: **C2 > OK**
To form an ungrouped frequency distribution of integer data, continue with:
Choose: **Stat > Tables > Tally Individual Variables**
Enter: Variables: **C1**
Select: **Counts > OK**
To print data on the session window, continue with:
Choose: **Data > Display Data**
Enter: Columns to display: **C1 or C1 C2 or C1–C2 > OK**

GENERATE RANDOM SAMPLES

The data will be put into C1:
Choose: **Calc > Random Data > {Normal, Uniform, Integer, etc.}**
Enter: Number of rows of data to generate: **K**
          Store in column(s):   **C1**
          Population parameters needed: **(μ, σ, L, H, A, or B) > OK**
          (Required parameters will vary depending on the distribution.)

SELECT RANDOM SAMPLES

The existing data to be selected from should be in C1; then continue with:
Choose: **Calc > Random Data > Sample from Columns**
Enter: Number of row to sample: **K**
          From columns: **C1**
          Store samples in: **C2**
Select: **Sample with replacement** (optional) **> OK**

TESTING FOR NORMALITY

Input the data into C1; then continue with:
Choose: **Stat > Basic Statistics > Normality Test**
Enter: Variable: **C1**
          Title: **your title > OK**

2.1 When a study of how people use the Internet was performed, it appeared that the variable x, the number of Internet activities in a week, had an approximately normal distribution. That distribution is approximated by this relative frequency distribution:

Internet Activities/ Week, x	Relative Frequency	Internet Activities/ Week, x	Relative Frequency
1	0.01	8	0.14
2	0.03	9	0.11
3	0.05	10	0.08
4	0.09	11	0.05
5	0.10	12	0.04
6	0.14	13	0.03
7	0.13		

a. Select a random sample of size 40 from this relative frequency representation of the population of all Internet users.
b. Construct a histogram of the sample obtained in part (a). Do not group the data. (See instructions that follow.)

MINITAB

Input the x values into C1 and the corresponding relative frequencies into C2; then continue with:

Choose: **Calc > Random Data > Discrete**
Enter: Number of rows of data to generate: **40**
          Store in column(s):   **C3**
          Values (of x) in: **C1**
          Probabilities in: **C2 > OK**

Excel

Input the x values into column A and the corresponding relative frequencies into column B; then continue with:

Choose: **Data > Data Analysis > Random Number Generation > OK**
Enter: Number of Variables: **1**
          Number of Random Numbers: **40**
          Distribution: **Discrete**
          Value & Prob. Input Range:
          (A1:B13 or select data cells not labels)
Select: **Output Range**
Enter: **(C1 or select cell) > OK**

c. Find the mean, median, and standard deviation of the sample obtained in part (a).
d. Repeat parts (a)–(c) three more times, being sure to keep the answers for each set of data together.
e. Describe the similarities and differences between the distributions shown on the four histograms.
f. Make a chart displaying the numerical statistics for each of the four samples and describe the variability from sample to sample of each statistic.
g. The four samples were all drawn randomly from the same distribution. Write a statement describing the overall variability between these four random samples.

2.2 Use a random-number table or a computer to simulate rolling a pair of dice 100 times.
a. List the results of each roll as an ordered pair and the sum.
b. Prepare an ungrouped frequency distribution and a histogram of the sums.
c. Describe how these results compare with what you expect to occur when two dice are rolled.

Excel Instructions For:

PIE CHART

Input the categories into column A and the corresponding frequencies into column B; activate both columns of data by highlighting and selecting the column names and data cells, then continue with:

Choose:	**Insert > Charts > Pie > 1st picture** *(usually)*
Choose:	**Chart Layouts – Layout 1**
Enter:	*Chart title:* **Your title**

To edit the pie chart:

Click on:	*Anywhere clear on the chart—use handles to size*
	Any cell in the category or frequency column
	and type in different name or amount > **ENTER**

PARETO DIAGRAM

Input the categories into column A and the corresponding frequencies into column B (column headings are optional); then continue with:

First, sorting the table:

Activate both columns of the distribution

Choose:	**Data > AZ|ZA Sort**
Select:	*Sort by:* **frequency column**
	Order: **Largest to Smallest** > **OK**
Choose:	**Insert > Charts > Column > 1st picture** *(usually)*
Choose:	*Chart Layouts –* **Layout 9**
Enter:	*Chart title:* **your title**
	Category (x) axis title: **title for x-axis**
	Value (y) axis title: **title for y-axis**

To edit the Pareto diagram:

Click on:	*Anywhere clear on the chart—use handles to size*
	Any title name to change
	Any cell in the category column and type in a name > **ENTER**

Excel does not include the line graph.

DOTPLOT

The dotplot display is not available, but the initial step of ranking the data can be done. Input the data into column A and activate the column of data; then continue with:

Choose:	**Data > AZ↓**

Use the sorted data to finish constructing the dotplot display.

STEM-AND-LEAF DISPLAY

Input the data into column A; then continue with:

Choose:	**Add-Ins > Data Analysis Plus* > Stem and Leaf Display > OK**
Enter:	*Input Range:* **(A2:A6 or select cells)**
	Increment: **Stem Increment**

*Data Analysis Plus is a collection of statistical macros for Excel. They can be downloaded onto your computer from cengagebrain.com.

MULTIPLE DOTPLOTS

Multiple dotplots are not available, but the initial step of ranking the data can be done. Use the commands shown under Dotplot in the previous column, then finish constructing the dotplots by hand.

HISTOGRAM

Input the data into column A and the upper class limits* into column B (optional) (column headings are optional); then continue with:

Choose:	**Data > Data Analysis** > Histogram > OK**
Enter:	*Input Range:* **Data (A1:A6 or select cells)**
	Bin Range: **upper class limits (B1:B6 or select cells)**
	[leave blank if Excel determines the intervals]
Select:	**Labels** *(if column headings are used)*
	Output Range
Enter:	*area for freq. distr. & graph:* **(C1 or select cell)**
Select:	**Chart Output > OK**

To remove gaps between bars:

Click on:	**Any bar on graph**
Click on:	**Right mouse button**
Choose:	**Format Data Series**
Enter:	*Gap Width:* **0% > Close**

To edit the histogram:

Click on:	*Anywhere clear on the chart—use handles to size*
	Any title or axis name to change
	Any upper class limit[†] or frequency in the frequency distribution to change value > **ENTER**
	Delete 'Frequency' box on right

*If boundary = 50, then limit = 49.9 (depending on the number of decimal places in the data).

**If Data Analysis does not show on the Data menu:

Choose:	**Office Button > Excel Options (bottom) > Add-Ins**
Select:	**Analysis ToolPak**
	Analysis ToolPak-VBA

[†]Note that the upper class limits appear in the center of the bars. Replace with class midpoints. The "More" cell in the frequency distribution may also be deleted.

For tabled data, input the classes into column A (ex. 30–40) and the frequencies into column B; activate both columns then continue with:

Choose:	**Insert > Column > 1st picture** *(usually)*
Choose:	**Chart Layouts > Layout 8**
Enter:	*Chart title:* **your title**
	Category (x) axis: **title for x-axis**
	Value (y) axis: **title for y-axis**

Do as just described to remove gaps and adjust.

Note: Practice Problems for this chapter are found on the back of the Minitab Instructions Tech Card.

OGIVE

Input the data into column A and the upper class limits* into column B (include an additional class at the beginning).

Choose: Data > Data Analysis > Histogram > OK
Enter: Input Range: data (A1:A6 or select cells)
           Bin Range: upper class limits (B1:B6 or select cells)
Select: Labels (if column headings were used)
           Output Range
Enter: area for freq. distr. & graph: (C1 or select cell)
           Cumulative Percentage
           Chart Output > OK

To close gaps and edit, see the histogram commands in the previous section. For tabled data, input the upper class boundaries into column A and the cumulative relative frequencies into column B (include an additional class boundary at the beginning with a cumulative relative frequency equal to 0 [zero]); activate column B then continue with:

Choose: Insert > Line > 1st picture (usually)
Right click on chart area
Choose: Select Data > Horizontal (Category) Axis Labels Edit
Enter: (A2:A8 or select cells)
Choose: Chart Tools > Layout > Labels
Enter: Chart title: your title
           Axis titles: title for x-axis; title for y-axis
      For editing, see the histogram commands.

*If the boundary = 50, then the limit = 49.9 (depending on the number of decimal places in the data).

MEAN

Input the data into column A and activate a cell for the answer; then continue with:

Choose: Formulas > Insert Function, f_x > Statistical > Average > OK
Enter: Number 1:   (A2:A6 or select cells) > OK
           [Start at A1 if no header row (column title) is used.]

MEDIAN

Input the data into column A and activate a cell for the answer; then continue with:

Choose: Formulas > Insert Function, f_x > Statistical > Median > OK
Enter: Number 1:   (A2:A6 or select cells) > OK

STANDARD DEVIATION

Input the data into column A and activate a cell for the answer; then continue with:

Choose: Formulas > Insert Function, f_x > Statistical > STDEV > OK
Enter: Number 1:   (A2:A6 or select cells) > OK

ADDITIONAL STATISTICS

Input the data into column A and activate a cell for the answer; then continue with:

Choose: Formulas > Insert Function, f_x > Statistical > COUNT > OK
                                                           > MIN > OK
                                                           > MAX > OK

           OR

                                                           > ALL > SUM
                                                           > SUMSQ

Enter: Number 1:   (A2:A6 or select cells) > OK
For range, write a formula: Max( )−Min( )

5-NUMBER SUMMARY

Input the data into column A; then continue with:

Choose: Data > Data Analysis* > Descriptive Statistics > OK
Enter: Input Range: (A2:A6 or select cells)
Select: Labels in First Row (if necessary)
           Output Range
Enter: (B1 or select cell)
Select: Summary Statistics > OK
To make output readable:
Choose: Home > Cells > Format > AutoFit Column Width
*If Data Analysis does not show on the Data menu, see histogram section on previous page.

BOX-AND-WHISKERS DISPLAY

Input the data into column A; then continue with:

Choose: Add-Ins > Data Analysis Plus* > BoxPlot > OK
Enter: (A2:A6 or select cells) > OK
To edit the boxplot, review options shown with editing histograms.
*If Data Analysis Plus does not show on the Data menu, see stem-and-leaf display section on previous page

ADDITIONAL COMMANDS

Input the data into column A, activate the data; then continue with the following to sort the data:

Choose: Data > AZ↓

GENERATE RANDOM SAMPLES

Choose: Data > Data Analysis > Random Number Generation > OK
Enter: Number of Variables: 1
           Number of Random Numbers: (desired quantity)
Select: Distribution: Normal, Discrete, or others
Enter: Parameters: (μ, σ, L, H, A, or B)
           (Required parameters will vary depending on the distribution.)
Select: Output Range
Enter: (A1 or select cell) > OK

SELECT RANDOM SAMPLES

The existing data to be selected from should be in column A; then continue with:

Choose: Data > Data Analysis > Sampling > OK
Enter: Input range: (A2:A10 or select cells)
Select: Labels (optional)
           Random
Enter: Number of Samples: K
Select: Output range:
Enter: (B1 or select cell) > OK

TESTING FOR NORMALITY

Excel uses a test for normality, not the probability plot.
Input the data into column A; then continue with:

Choose: Add-Ins > Data Analysis Plus > Chi-Squared Test of
           Normality > OK
Enter: Input Range: data (A1:A6 or select cells)
Select: Labels (if column headings were used) > OK
Expected values for a normal distribution are given versus the given distribution. If the p-value is greater than 0.05, then the given distribution is approximately normal.

TI-83/84 Plus Instructions For:

PIE CHART

Input the frequencies for the various categories into L1; then continue with:

Choose:	PRGM > EXEC > CIRCLE*
Enter:	LIST: L1 > ENTER
	DATA DISPLAYED?: 1:PERCENTAGES
	OR
	2:DATA

*The TI-83/84 Plus program "CIRCLE" and others can be downloaded from cengagebrain.com. The TI-83/84 Plus programs and data files are jkprogs.zip and jklists.zip. Save the files to your computer and uncompress them using a zip utility. Download the programs to your calculator using TI-Graph Link Software.

PARETO DIAGRAM

Input the numbered categories into L1 and the corresponding frequencies into L2; then continue with:

Choose:	PRGM > EXEC > PARETO*
Enter:	LIST: L2 > ENTER
	Ymax: at least the sum of the frequencies > ENTER
	Yscl: increment for y-axis > ENTER

*Program "PARETO" is one of many programs that are available for downloading from cengagebrain.com. See above.

DOTPLOT

Input the data into L1; then continue with:

Choose:	PRGM > EXEC > DOTPLOT*
Enter:	LIST: L1 > ENTER
	Xmin: at most the lowest x value
	Xmax: at least the highest x value
	Xscl: 0 or increment
	Ymax: at least the highest frequency

*Program "DOTPLOT" is one of many programs that are available for downloading from cengagebrain.com. See above.

STEM-AND-LEAF DISPLAY

Input the data into L1; then continue with:

Choose:	STAT > EDIT > 2:SortA(
Enter:	L1

Use sorted data to finish constructing the stem-and-leaf diagram by hand.

MULTIPLE DOTPLOTS

Input the data for the first dotplot into L1 and the data for the second dotplot into L3; then continue with:

Choose:	STAT > EDIT > 2:SortA(	
Enter:	L1 > ENTER	
	In L2, enter counting numbers for each category.	
	Ex.	L1	L2
		15	1
		16	1
		16	2
		17	1
Choose:	STAT > EDIT > 2:SortA(	
Enter:	L3 > ENTER	
	In L4, enter counting numbers for each category. You should use a higher set, such as 10,10,11,10,10,11,11,12... This offsets the dotplots.	
Choose:	2nd > FORMAT > AxesOff (Optional— must return to AxesOn)	
Choose:	2nd > STAT PLOT > 1:PLOT1	

HISTOGRAM

Input the data into L1; then continue with:

Choose:	2nd > STAT PLOT > 2:PLOT2
Choose:	Window
Enter:	at most lowest value for both, at least highest value for both, 0 or increment, −2, at least highest counting number, 1, 1
Choose:	Graph > Trace > > > > *(gives data values)*

HISTOGRAM

Input the data into L1; then continue with:

Choose:	2nd > STAT PLOT > 1:Plot1

Calculator selects classes:

Choose:	Zoom > 9:ZoomStat > Trace > > >

Individual selects classes:

Choose:	Window
Enter:	at most lowest value, at least highest value, class width, −1, at least highest frequency, 1 (depends on frequency numbers), 1
Choose:	Graph > Trace *(use values to construct frequency distribution)*

For tabled data, input the class midpoints into L1 and the frequencies into L2; then continue with:

Choose:	2nd > STAT PLOT > 1:Plot1
Choose:	Window
Enter:	smallest lower class boundary, largest upper class boundary, class width, −ymax/4, highest frequency, 0 (for no tick marks), 1
Choose:	Graph > Trace > > >

To obtain a relative frequency histogram of tabled data instead:

Choose:	STAT > EDIT > 1:EDIT . . .
Highlight:	L3
Enter:	L3 = L2/SUM(L2) [SUM - 2nd LIST > MATH > 5:sum]
Choose:	2nd > STAT PLOT > 1:Plot1
Choose:	Window
Enter:	smallest lower class boundary, largest upper class boundary, class width, −ymax/4, highest rel. frequency, 0 (for no tick marks), 1
Choose:	Graph > Trace > > >

OGIVE

Input the class boundaries into L1 and the frequencies into L2 (include an extra class boundary at the beginning with a frequency of zero); then continue with:

Choose:	STAT > EDIT > 1:EDIT . . .
Highlight:	L3
Enter:	L3 = 2nd > LIST > OPS > 6:cum sum(L2)
Highlight:	L4
Enter:	L4 = L3 / 2nd > LIST > Math > 5:sum (L2)
Choose:	2nd > STAT PLOT > 1:Plot
Choose:	Zoom > 9:ZoomStat > Trace > > >

Adjust window if needed for better readability.

MEAN

Input the data into L1; then continue with:

Choose: **2nd > LIST > Math > 3:mean(**
Enter: **L1**

MEDIAN

Input the data into L1; then continue with:

Choose: **2nd > LIST > Math > 4:median(**
Enter: **L1**

STANDARD DEVIATION

Input the data into L1; then continue with:

Choose: **2nd > LIST > Math > 7:StdDev(**
Enter: **L1**

Standard deviation on your calculator: *Most calculators have two formulas for finding the standard deviation and mindlessly calculate both, fully expecting the user to decide which one is correct for the given data. How do you decide?*

The sample standard deviation is denoted by s and uses the "divide by $n - 1$" formula.

The population standard deviation is denoted by σ and uses the "divide by n" formula.

      When you have sample data, always use the s or "divide by $n - 1$" formula. Having the population data is a situation that will probably never occur, other than in a textbook exercise. If you don't know whether you have sample data or population data, it is a "safe bet" that they are sample data—use the s or "divide by $n - 1$" formula!

ADDITIONAL STATISTICS

Input the data into L1; then continue with:

Choose: **2nd > LIST > Math > 5:sum(**
                            > 1:min(
                            > 2:max(
Enter: **L1**

5-NUMBER SUMMARY

Input the data into L1; then continue with:

Choose: **STAT > CALC > 1:1-VAR STATS**
Enter: **L1**

BOX-AND-WHISKERS DISPLAY

Input the data into L1; then continue with:

Choose: **2nd > STAT PLOT >**
               1:Plot1 . . .
Choose: **ZOOM > 9:ZoomStat > TRACE > > >**
If class midpoints are in L1 and frequencies are in L2, do as just described except for:
Enter: **Freq: L2**
For multiple boxplots, enter additional set of data into L2 or L3; do as just described plus:
Choose: **2nd > STAT PLOT > 2:Plot2 . . .**

ADDITIONAL COMMANDS

Input the data into L1; then continue with the following to sort the data:

Choose: **2nd > STAT > OPS > 1:SortA(**
Enter: **L1**

To form a frequency distribution of the data in L1, continue with:

Choose: **PRGM > EXEC > FREQDIST***
Enter: **L1 > ENTER**
              LW BOUND = **first lower class boundary**
              UP BOUND = **last upper class boundary**
              WIDTH = **class width** *(use 1 for ungrouped distribution)*

*The program "FREQDIST" is one of many programs available for downloading from <u>cengagebrain.com</u>. See side one of this card.

GENERATE RANDOM SAMPLES

Choose: **STAT > 1:EDIT**
Highlight: **L1**
Choose: **MATH > PRB > 6:randNorm(**   *or*   **5:randInt(**
Enter: μ, σ, **# of trials**   *or*   **L, H, # of trials**

TESTING FOR NORMALITY

Input the data into L1; then continue with:

Choose: **Window**
Enter: **at most the smallest data value, at least the largest data value, x scale, −5, 5, 1, 1**
Choose: **2nd > STAT PLOT > 1:Plot**

> **Note: Practice Problems for this chapter are found on the back of the Minitab Instructions Tech Card.**

Minitab Instructions For:

CROSS-TABULATION TABLES

Input the row-variable single categorical values into C1 and the corresponding single column-variable categorical values into C2; then continue with:

Choose: **Stat > Tables > Cross Tabulation and Chi-Square**
Enter: *Categorical variables: For rows:* **C1** *For columns:* **C2**
Select: **Counts**
        Row Percents
        Column Percents
        Total Percents > OK

Suggestion: The four subcommands that are available for "Display" can be used together; however, the resulting table will be much easier to read if one subcommand at a time is used.

SIDE-BY-SIDE BOXPLOTS AND DOTPLOTS

Input the numerical values into C1 and the corresponding categories into C2; then continue with:

Choose: **Graph > Boxplot . . . > One Y, With Groups > OK**
Enter: *Graph variables:* **C1** *Categorical variables:* **C2 > OK**

MINITAB commands to construct side-by-side dotplots for data in this form are located on your Chapter 2 Tech Card.

If the data for the various categories are in separate columns, use the MINITAB commands for multiple boxplots on your Chapter 2 Tech Card. If side-by-side dotplots are needed for data in this form, continue with:

Choose: **Graph > Dotplots**
Select: **Multiple Y's, Simple > OK**
Enter: *Graph variables:* **C1 C2 > OK**

SCATTER DIAGRAM

Input the x-variable values into C1 and the corresponding y-variable values into C2; then continue with:

Choose: **Graph > ScatterPlot. . . > Simple > OK**
Enter: *Y variables:* **C2** *X variables:* **C1**
Select: **Labels > Titles/Footnotes**
Enter: *Title:* **your title > OK > OK**

CORRELATION COEFFICIENT

Input the x-variable data into C1 and the corresponding y-variable data into C2; then continue with:

Choose: **Stat > Basic Statistics > Correlation. . .**
Enter: *Variables:* **C1 C2 > OK**

LINE OF BEST FIT

Input the x values into C1 and the corresponding y values into C2; then to obtain the equation for the line of best fit, continue with:

Method 1—
Choose: **Stat > Regression > Regression . . .**
Enter: *Response (y):* **C2**
        Predictors (x): **C1 > OK**

To draw the scatter diagram with the line of best fit superimposed on the data points, continue with:

Choose: **Graph > Scatterplot**
Select: **With Regression > OK**
Enter: *Y variable:* **C2** *X variable:* **C1**
Select: **Labels > Titles/Footnotes**
Enter: *Title:* **your title > OK > OK**
OR

Method 2—
Choose: **Stat > Regression > Fitted Line Plot**
Enter: *Response (Y):* **C2**
        Response (X): **C1**
Select: **Linear**
Select: **Options**
Enter: *Title:* **your title > OK > OK**

Excel Instructions For:

CROSS-TABULATION TABLES

Using column headings or titles, input the row-variable categorical values into column A and the corresponding column-variable categorical values into column B; then continue with:

Choose: **Insert > Tables > Pivot Table pulldown > Pivot Chart**
Select: **Select a table or range**
Enter: *Table/Range:* **(A1:B5 or select cells)**
Select: **Existing Worksheet**
Enter: **(C1 or select cell) > OK**
Drag: **Headings to row or column (depends on preference) into chart box formed**
        One heading into data area*

*For other summations, double-click "Count of" in data area box; then continue with:
Choose: *Summarize by:* **Count**
        Show values as: **% of row or % of column or % of total > OK**

SIDE-BY-SIDE BOXPLOTS AND DOTPLOTS

Excel commands to construct a single boxplot are on your Chapter 2 Tech Card.

SCATTER DIAGRAM

Input the x-variable values into column A and the corresponding y-variable values into column B; activate columns of data then continue with:

Choose: **Insert > Charts > Scatter > 1st picture** (usually)
Choose: **Chart Layouts > Layout 1**
Enter: *Chart title:* **your title**
        (x) axis title: **title for x axis**
        (y) axis title: **title for y axis***

*To remove gridlines:
Choose: **Chart Tools > Layout > Axes > Gridlines > Primary Horizontal Gridlines > None**

To edit the scatter diagram, follow the basic editing commands shown for a histogram on your Chapter 2 Tech Card.

To change the scale, and/or show tick marks, double click on the axis; then continue with:

Choose: **Chart Tools > Layout > Current Selection > Horizontal/Vertical Axes > Format Selection**
Select: *Major tick mark type:* **Cross > OK**

CORRELATION COEFFICIENT

Input the x-variable data into column A and the corresponding y-variable data into column B, activate a cell for the answer; then continue with:

Choose: **Insert Function, f_x > Statistical > CORREL > OK**
Enter: *Array 1:* **x data range**
        Array 2: **y data range > OK**

LINE OF BEST FIT

Input the x-variable data into column A and the corresponding y-variable data into column B; then continue with:

Choose: **Data > Data Analysis* > Regression > OK**
Enter: Input Y Range: **(B1:B10 or select cells)**
Input X Range: **(A1:A10 or select cells)**
Select: **Labels** (if necessary)
Output Range
Enter: **(C1 or select cell)**
Select: **Line Fit Plots > OK**

To make the output readable; continue with:

Choose: **Home > Cells > Format > AutoFit Column Width**
* See Chapter 2 Excel histogram command.

To form the regression equation, the y-intercept is located at the intersection of the intercept and coefficients columns, whereas the slope is located at the intersection of the x variable and the coefficients columns.

To draw the line of best fit on the scatter diagram, activate the chart; then continue with:

Choose: **Chart Tools > Layout > Analysis > Trendline > Linear Trendline > OK**

OR

Choose: **Chart Tools > Design > Chart Layouts > Layout 9**
(This command also works with the scatter diagram Excel commands above.)

TI-83/84 Plus Instructions For:

CROSS-TABULATION TABLES

The categorical data must be numerically coded first; use 1, 2, 3, . . . for the various row variables and 1, 2, 3, . . . for the various column variables. Input the numeric row-variable values into L1 and the corresponding numeric column-variable values into L2; then continue with:

Choose: **PRGM > EXEC > CROSSTAB***
Enter: **ROWS: L1 > ENTER**
COLS: L2 > ENTER

The cross-tabulation table showing frequencies is stored in matrix [A], the cross-tabulation table showing row percentages is in matrix [B], column percentages in matrix [C], and percentages based on the grand total in matrix [D]. All matrices contain marginal totals. To view the matrices, continue with:

Choose: **MATRX > NAMES**
Enter: **1:[A] or 2:[B] or 3:[C] or 4:[D] > ENTER**
*Program 'CROSSTAB' is one of many programs that are available for downloading from cengagebrain.com. See your Chapter 2 Tech Card for specific instructions.

SIDE-BY-SIDE BOXPLOTS AND DOTPLOTS

TI-83/84 commands to construct multiple boxplots are on your Chapter 2 Tech Card.
TI-83/84 commands to construct multiple dotplots are on your Chapter 2 Tech Card.

SCATTER DIAGRAM

Input the x-variable values into L1 and the corresponding y-variable values into L2; then continue with:

Choose: **2nd > STATPLOT > 1:Plot1**
Choose: **ZOOM > 9:ZoomStat**
> TRACE > > >
OR
WINDOW
Enter: at most lowest x value, at least highest x value, x-scale, − y-scale, at least highest y value, y-scale,1
TRACE > > >

CORRELATION COEFFICIENT

Input the x-variable data into L1 and the corresponding y-variable data into L2; then continue with:

Choose: **2nd > CATALOG > DiagnosticOn* > ENTER > ENTER**
Choose: **STAT > CALC > 8:LinReg(a + bx)**
Enter: **L1, L2**
*Diagnostic On must be selected for r and r^2 to show. Once set, omit this step.

LINE OF BEST FIT

Input the x-variable data into L1 and the corresponding y-variable data into L2; then continue with:

If just the equation is desired:

Choose: **STAT > CALC > 8:LinReg(a + bx)**
Enter: **L1, L2***
*If the equation and graph on the scatter diagram are desired, use:
Enter: **L1, L2, Y1†**
then continue with the same commands for a scatter diagram as shown above.
†To enter Y1, use:
Choose: **VARS > Y-VARS > 1:Function > 1:Y1 > ENTER**

Practice Problem

Movie production companies spend millions of dollars to produce movies with the great hope of attracting millions of people to the theater. The success of a movie can be measured in many ways, two of which are the box office receipts and the number of Oscar nominations received. Below is a list of ten 2008 movies with their "report cards." Each movie is measured by its budget cost (in millions of dollars), its box office receipts (in millions of dollars), and the number of Oscar nominations it received.

Movie	Budget	Box Office	Nominations
The Curious Case of Benjamin Button	150	127.5	13
Slumdog Millionaire	15	141.3	10
Milk	20	31.8	8
The Dark Knight	185	533.3	8
WALL-E	180	223.8	6
Frost/Nixon	25	18.6	5
The Reader	32	34.2	5
Doubt	20	33.4	5
Changeling	55	35.7	3
The Wrestler	6	26.2	2

SOURCE: http://www.boxofficemojo.com/

a. Draw a scatter diagram using x = budget and y = box office.
b. Does there appear to be a linear relationship?
c. Calculate the linear correlation coefficient, r.
d. What does this value of correlation seem to be telling us? Explain.
e. Repeat parts (a) through (d) using x = box office and y = nominations.

*remember,
Chapter 5 is your next Tech Card.

Minitab Instructions For:

GENERATE RANDOM DATA

Input the possible values of the random variable into C1 and the corresponding probabilities into C2; then continue with:

Choose:	**Calc > Random Data > Discrete**
Enter:	*Number of rows of data to generate:* **25** *(number wanted)*
	Store in column(s): **C3**
	Values (of x) in: **C1**
	Probabilities in: **C2 > OK**

BINOMIAL AND CUMULATIVE BINOMIAL PROBABILITIES

For binomial probabilities, input x values into C1; then continue with:

Choose:	**Calc > Probability Distributions > Binomial**
Select:	**Probability ***
Enter:	*Number of trials:* **n**
	Event probability: **p**
Select:	**Input column**
Enter:	**C1**
	Optional Storage: **C2** *(not necessary)* **> OK**
Or	
Select:	**Input constant**
Enter:	**One single x value > OK**

*For cumulative binomial probabilities, repeat the preceding commands but replace the probability selection with:

Select:	**Cumulative Probability**

Excel Instructions For:

GENERATE RANDOM DATA

Input the possible values of the random variable into column A and the corresponding probabilities into column B; then continue with:

Choose:	**Data > Data Analysis > Random Number Generation > OK**
Enter:	*Number of Variables:* **1**
	Number of Random Numbers: **25** *(# wanted)*
	Distribution: **Discrete**
	Value & Prob. Input Range: **(A2:B5 select data cells, not labels)**
Select:	**Output Range**
Enter:	**(C1 or select cell) > OK**

BINOMIAL AND CUMULATIVE BINOMIAL PROBABILITIES

For binomial probabilities, input x values into column A and activate the column B cell across from the first x value; then continue with:

Choose:	**Insert Function, f_x > Statistical> BINOMDIST > OK**
Enter:	*Number_s:* **(A1:A4 or select 'x value' cells)**
	Trials: **n**
	Probability_s: **p**
	Cumulative: **false*** *(gives individual probabilities)* **> OK**
Drag:	**Bottom right corner of probability value cell in column B down to give other probabilities**

*For cumulative binomial probabilities, repeat the preceding commands but replace the false cumulative with:

Cumulative:	**true (gives cumulative probabilities) > OK**

TI-83/84 Plus Instructions For:

BINOMIAL AND CUMULATIVE BINOMIAL PROBABILITIES

To obtain a complete list of probabilities for a particular n and p, continue with:

Choose:	**2nd > DISTR > 0:binompdf(**
Enter:	**n, p)**

Use the right arrow key to scroll through the probabilities.
To scroll through a vertical list in L1:

Choose:	**STO→ > L1 > ENTER**
	STAT > EDIT > 1:Edit

To obtain individual probabilities for a particular n, p, and x, continue with:

Choose:	**2nd > DISTR > 0:binompdf(**
Enter:	**n, p, x)**

To obtain cumulative probabilities for x = 0 to x = n for a particular n and p, continue with:

Choose:	**2nd > DISTR > A:binomcdf(**
Enter:	**n, p)*** *(repeat above commands to scroll through probabilities)*

*To obtain individual cumulative probabilities for a particular n, p, and x, repeat the preceding commands but replace the enter with:

Enter:	**n, p, x)**

Practice Problems

5.1 *Use a computer to find the probabilities for all possible x values for a binomial experiment where n = 30 and p = 0.35.*

MINITAB

Choose:	**Calc > Make Patterned Data > Simple Set of Numbers**
Enter:	*Store patterned data in:* **C1**
	From first value: **0**
	To last value: **30**
	In steps of: **1 > OK**

Continue with the binomial probability MINITAB commands explained above, using n = 30, p = 0.35, and C2 for optional storage.

Excel

Enter:	**0, 1, 2, ... , 30 into column A**

Continue with the binomial probability Excel commands explained at left, using n = 30 and p = 0.35.

TI-83/84 Plus

Use the binomial probability TI-83/84 commands explained above, using n = 30 and p = 0.35.

5.2 *Use a computer to find the cumulative probabilities for all possible x values for a binomial experiment where n = 45 and p = 0.125.*

a. *Explain why there are so many 1.000s listed.*

b. *Explain what is represented by each number listed.*

MINITAB

Choose:	**Calc > Make Patterned Data > Simple Set of Numbers ...**
Enter:	*Store patterned data in:* **C1**
	From first value: **0**
	To last value: **45**
	In steps of: **1 > OK**

Continue with the cumulative binomial probability MINITAB commands explained above, using n = 45, p = 0.125, and C2 as optional storage.

Excel

Enter:	**0, 1, 2, ... , 45 into column A**

Continue with the cumulative binomial probability Excel commands explained at left, using n = 45 and p = 0.125.

Use the cumulative binomial probability TI-83/84 commands explained above, using n = 45 and p = 0.125.

5.3 Imprints Galore buys T-shirts (to be imprinted with an item of the customer's choice) from a manufacturer who guarantees that the shirts have been inspected and that no more than 1% are imperfect in any way. The shirts arrive in boxes of 12. Let x be the number of imperfect shirts found in any one box.

a. List the probability distribution and draw the histogram of x.

b. What is the probability that any one box has no imperfect shirts?

c. What is the probability that any one box has no more than one imperfect shirt?

d. Find the mean and standard deviation of x.

e. What proportion of the distribution is between $\mu - \sigma$ and $\mu + \sigma$?

f. What proportion of the distribution is between $\mu - 2\sigma$ and $\mu + 2\sigma$?

g. How does this information relate to the empirical rule and Chebyshev's theorem? Explain.

h. Use a computer to simulate Imprints Galore's buying 200 boxes of shirts and observing x, the number of imperfect shirts per box of 12. Describe how the information from the simulation compares with what was expected [answers to parts (a)–(g) describe the expected results].

i. Repeat part (h) several times. Describe how these results compare with those of parts (a)–(g) and with part (h).

MINITAB

a.

Choose:	**Calc > Make Patterned Data > Simple Set of Numbers ...**
Enter:	Store patterned data in: **C1**
	From first value: **−1 (see note)**
	To last value: **12**
	In steps of: **1 > OK**

c. Continue with the binomial probability MINITAB commands on the other side of this card, using n = 12, p = 0.01, and C2 for optional storage.

Choose:	**Graph > Scatterplot > Simple > OK**
Enter:	Y variables: **C2** X variables: **C1**
Select:	Data view: Data Display: **Area > OK**

The graph is not a histogram, but can be converted to a histogram by double clicking on "area" of graph.

Select:	**Options** Select: **Step > OK > OK**

h. Continue with the cumulative binomial probability MINITAB commands on the other side of this card, using n = 12, p = 0.01, and C3 for optional storage.

Choose:	**Calc > Random Data > Binomial**
Enter:	Number of rows to generate: **200**
	Store in column **C4**
	Number of trials: **12**
	Probability: **.01 > OK**
Choose:	**Stat > Tables > Cross Tabulation**
Enter:	Categorical variables: For rows: **C4**
Select:	Display: **Total percents > OK**
Choose:	**Calc > Column Statistics**
Select:	Statistic: **Mean**
Enter:	Input variable: **C4 > OK**
Choose:	**Calc > Column Statistics**
Select:	Statistic: **Standard deviation**
Enter:	Input variable: **C4 > OK**

Continue with the histogram MINITAB commands on your Chapter 2 Tech Card, using the data in C4 and selecting the options: percent and midpoint with intervals 0:12/1.

Note: The binomial variable x cannot take on the value −1. The use of −1 (the next would-be class midpoint to left of 0) allows MINITAB to draw the histogram of a probability distribution. Without −1, PLOT will draw only half of the bar representing x = 0.

Excel

a.

Enter:	**0, 1, 2, ... ,12 into column A**

Continue with the binomial probability Excel commands on side one, using n = 12 and p = 0.01. Activate columns A and B; then continue with:

Choose:	**Insert > Charts > Column > 1st picture** (usually)
Choose:	**Data > Select Data > Series 1 > Remove > OK**

If necessary:

Click on:	**Anywhere clear on the chart**
	—use handles to size so x values fall under corresponding bars

Continue with the cumulative binomial probability Excel commands on side one, using n = 12, p = 0.01, and column C1 for the activated cell.

h.

Choose:	**Data > Data Analysis > Random Number Generation > OK**
Enter:	Number of Variables: **1**
	Number of Random Numbers: **200**
	Distribution: **Binomial**
	p Value = **0.01**
	Number of Trials = **12**
Select:	Output Options: **Output Range**
Enter:	**(D1 or select cell) > OK**

Activate the E1 cell, then:

Choose:	**Insert Function, f_x > Statistical > AVERAGE > OK**
Enter:	Number 1: **D1:D200 > OK**

Activate the E2 cell, then:

Choose:	**Insert Function, f_x > Statistical > STDEV > OK**
Enter:	Number 1: **D1:D200 > OK**

Continue with the histogram Excel commands on your Chapter 2 Tech Card, using the data in column D and the bin range in column A.

TI-83/84 Plus

a.

Choose:	**STAT > EDIT > 1:Edit**
Enter:	L1: **0, 1, 2, 3, 4, 5, 6, 7, 8, 9, 10, 11, 12**
Choose:	**2nd QUIT > 2nd DISTR > 0:binompdf(**
Enter:	**12, 0.01) > ENTER**
Choose:	**STO→ > L2 > ENTER**
Choose:	**2nd > STAT PLOT > 1:Plot1**
Choose:	**WINDOW**
Enter:	**0, 13, 1, −.1, .9, .1, 1**
Choose:	**TRACE > > >**

c.

Choose:	**2nd > DISTR > A:binomcdf(**
Enter:	**12, 0.01)**
Choose:	**STO→ > L3 > ENTER**
	STAT > EDIT > 1:Edit

h.

Choose:	**MATH > PRB > 7:randBin(**
Enter:	**12, .01, 200)** (takes a while to process)
Choose:	**STO→ > L4 > ENTER**
Choose:	**2nd LIST > Math > 3:mean(**
Enter:	**L4**
Choose:	**2nd LIST > Math > 7:StdDev(**
Enter:	**L4**

Continue with the histogram TI-83/84 commands on your Chapter 2 Tech Card, using the data in column L4 and adjusting the window after the initial look using ZoomStat.

Minitab Instructions For:

Excel Instructions For:

GENERATING RANDOM DATA FROM A NORMAL DISTRIBUTION

Choose: **Calc > Random Data > Normal**
Enter: *Number of rows of data to generate:* **n**
    Store in column(s): **C1**
    Mean: **μ**
    Stand. dev.: **σ > OK**

If multiple samples (say, 12), all of the same size, are wanted, modify the preceding commands: Store in column(s): C1–C12.

Note: To find descriptive statistics for each of these samples, use the commands: Stat > Basic Statistics > Display Descriptive Statistics for C1–C12.

CALCULATING ORDINATE (y) VALUES FOR A NORMAL DISTRIBUTION CURVE

Input the desired abscissas (x values) into C1; then continue with:

Choose: **Calc > Probability Distributions > Normal**
Select: **Probability Density**
Enter: *Mean:* **μ**
    Stand. dev.: **σ**
    Input column: **C1**
    Optional Storage: **C2 > OK**

To draw the graph of a normal probability curve with the x values in C1 and the y values in C2, continue with:

Choose: **Graph > Scatterplot**
Select: **With Connect Line > OK**
Enter: **Y variables: C2**
    X variables: C1 > OK

CUMULATIVE PROBABILITY FOR NORMAL DISTRIBUTIONS

Input the desired abscissas (x values) into C1; then continue with:

Choose: **Calc > Probability Distributions > Normal**
Select: **Cumulative probability**
Enter: *Mean:* **μ**
    Stand. dev.: **σ**
    Input column: **C1**
    Optional Storage: **C3 > OK**

Notes:

1. To find the probability between two x values, enter the two values into C1, use the preceding commands, and subtract using the numbers in C3.
2. To draw a graph of the cumulative probability distribution (ogive), use the Scatterplot commands with C3 as the y variable.

GENERATING RANDOM DATA FROM A NORMAL DISTRIBUTION

Choose: **Data > Data Analysis > Random Number Generation > OK**
Enter: *Number of Variables:* **1**
    Number of Random Numbers: **n**
    Distribution: **Normal**
    Mean: **μ**
    Standard Deviation: **σ**
Select: *Output Options:* **Output Range**
Enter: **(A1 or select cell) > OK**

If multiple samples (say, 12) all of the same size are wanted, modify the preceding commands: Number of variables: 12.

Note: To find descriptive statistics for each of these samples, use the commands: Data > Data Analysis > Descriptive Statistics for columns A through L.

CALCULATING ORDINATE (y) VALUES FOR A NORMAL DISTRIBUTION CURVE

Input the desired abscissas (x values) into column A and activate B1; then continue with:

Choose: **Insert Function, f_x > Statistical > NORMDIST > OK**
Enter: *X:* **(A1:A100 or select 'x value' cells)**
    Mean: **μ**
    Standard_dev: **σ**
    Cumulative: **False > OK**
Drag: **Bottom right corner of the ordinate value box down to give other ordinates**

To draw the graph of a normal probability curve with the x values in column A and the y values in column B, activate both columns and continue with:

Choose: **Insert > Charts > Scatter > 1st picture**

CUMULATIVE PROBABILITY FOR NORMAL DISTRIBUTIONS

Input the desired abscissas (x values) into column A and activate C1; then continue with:

Choose: **Insert Function, f_x > Statistical > NORMDIST > OK**
Enter: *X:* **(A1:A100 or select 'x value' cells)**
    Mean: **μ**
    Standard_dev: **σ**
    Cumulative: **True > OK**
Drag: **Bottom right corner of the cumulative probability box down to give other cumulative probabilities**

Notes:

1. To find the probability between two x values, enter the two values into column A, use the preceding commands, and subtract using the numbers in column C.
2. To draw a graph of the cumulative probability distribution (ogive), activate the x values (col A) and the y values (col C), then use the Insert > Charts > Scatter > 1st picture commands above.

GENERATING RANDOM DATA FROM A NORMAL DISTRIBUTION

Choose: **MATH > PRB > 6:randNorm(**
Enter: **μ, σ, # of trials)**
Choose: **STO→ > L1 > ENTER**

If multiple samples (say, six) all of the same size are wanted, repeat the preceding commands six times and store in L1–L6.

Note: To find descriptive statistics for each of these samples, use the commands: STAT > CALC > 1:1-Var Stats for L1–L6.

CALCULATING ORDINATE (y) VALUES FOR A NORMAL DISTRIBUTION CURVE

The ordinate values can be calculated for individual abscissa values, x:

Choose: **2nd > DISTR > 1:normalpdf(**
Enter: **x, μ, σ)**

To draw the graph of the normal probability curve for a particular μ and σ, continue with:

Choose: **WINDOW**
Enter: **μ – 3σ, μ + 3σ, σ, – .05, 1, .1, 0)**
Choose: **Y = > 2nd > DISTR > 1:normalpdf(**
Enter: **x, μ, σ)**

After an initial graph, adjust with 0:ZoomFit from the ZOOM menu.

CUMULATIVE PROBABILITY FOR NORMAL DISTRIBUTIONS

The cumulative probabilities can be calculated for individual abscissa values, x:

Choose: **2nd > DISTR > 2:normalcdf(**
Enter: **–1 EE 99, x, μ, σ)**

Notes:

1. *To find the probability between two x values, enter the two values in place of –1 EE 99 and the x.*
2. *To draw a graph of the cumulative probability distribution (ogive), use either the Scatter command under STATPLOTS, with the x values and their cumulative probabilities in a pair of lists, or normalcdf(–1EE99, x, μ, σ) in the Y = editor.*

*remember,

Chapter 8 is your next Tech Card.

6.1 *Use a computer or calculator to find the probability that one randomly selected value of x from a normal distribution (mean of 584.2 and standard deviation of 37.3) will have a value that corresponds to the following:*

a. *Less than 525*
b. *Between 525 and 590*
c. *At least 590*
d. *Verify the results of parts (a)–(c) using Table 3.*
e. *Explain any differences you may find between answers in part (d) and those in parts (a)–(c).*

MINITAB

Input 525 and 590 into C1; then continue with the cumulative probability commands on side one of this card, using 584.2 as μ, 37.3 as σ, and C2 as optional storage.

Excel

Input 525 and 590 into column A and activate the B1 cell; then continue with the cumulative probability commands on side one of this card, using 584.2 as μ and 37.3 as σ.

TI-83/84 Plus

Input 525 and 590 into L1; then continue with the cumulative probability commands in the left column in L2, using 584.2 as μ and 37.3 as σ.

6.2 *A soft-drink vending machine can be regulated to ensure that it dispenses an average of μ oz. of soft drink per glass.*

a. *If the ounces dispensed per glass are normally distributed with a standard deviation of 0.2 oz., find the setting for μ that will allow a 6-oz. glass to hold (without overflowing) the amount dispensed 99% of the time.*
b. *Use a computer or calculator to simulate drawing a sample of 40 glasses of soft drink from the machine [set using your answer to part (a)].*

MINITAB

Use the GENERATING RANDOM DATA FROM A NORMAL DISTRIBUTION commands on side one of this card, replacing n with 40, store in with C1, mean with the value calculated in part (a), and standard deviation with 0.2.

Use the histogram commands on the Chapter 2 Tech Card for the data in C1. To adjust the histogram, select Binning with cutpoint and cutpoint positions 5:6.2/0.05.

Excel

Use the GENERATING RANDOM DATA FROM A NORMAL DISTRIBUTION commands on side one of this card, replacing n with 40, the mean with the value calculated in part (a), the standard deviation with 0.2, and the output range with A1.

Use the GENERATING RANDOM DATA FROM A NORMAL DISTRIBUTION commands on side one of this card, replacing the distribution with Patterned, the first value with 5, the last value with 6.2, the steps with 0.05, and the output range with B1.

Use the histogram commands on the Chapter 2 Tech Card with column A as the input range and column B as the bin range.

TI-83/84 Plus

Use the 6:randNorm commands to the left, replacing the mean with the value calculated in part (a), the standard deviation with 0.2, and the number of trials with 40. Store in with L1.

Use the histogram commands on the Chapter 2 Tech Card for the data in L1, entering the following WINDOW VALUES: 5, 6.2, 0.05, −1, 10, 1, 1.

c. *What percentage of your sample would have overflowed the cup?*
d. *Does your sample seem to indicate the setting for μ is going to work? Explain.*

FYI *Repeat part (b) a few times. Try a different value for the mean amount dispensed and repeat part (b). Observe how many would overflow in each set of 40.*

Minitab Instructions For:

CONFIDENCE INTERVAL FOR MEAN μ WITH A GIVEN σ

Input the data into C1; then continue with:

Choose:	**Stat > Basic Statistics > 1-Sample Z**
Enter:	*Samples in columns:* **C1**
	Standard deviation: σ
Select:	**Options**
Enter:	*Confidence Level:* **1** − α *(ex.: 0.95 or 95.0)*
Select:	*Alternative:* **not equal > OK > OK**

HYPOTHESIS TEST FOR MEAN μ WITH A GIVEN σ

Input the data into C1; then continue with:

Choose:	**Stat > Basic Statistics > 1-Sample Z**
Enter:	*Samples in columns:* **C1**
	Standard deviation: σ
Select:	**Perform hypothesis test**
Enter:	*Hypothesized mean:* μ
Select:	**Options**
Select:	*Alternative:* **less than** *or* **not equal to** *or* **greater than** **> OK > OK**

Excel Instructions For:

CONFIDENCE INTERVAL FOR MEAN μ WITH A GIVEN σ

Input the data into column A; then continue with:

Choose:	**Add-Ins > Data Analysis Plus* > Z-Test: Mean > OK**
Enter:	*Input Range:* **(A1:A20 or select cells)**
	Standard Deviation (SIGMA): σ
	Alpha: α *(ex.: 0.05)* **> OK**

HYPOTHESIS TEST FOR MEAN μ WITH A GIVEN σ

Input the data into column A; then continue with:

Choose:	**Add-Ins > Data Analysis Plus* > Z-Test: Mean > OK**
Enter:	*Input Range:* **(A1:A20 or select cells)**
	Hypothesized Mean: μ
	Standard Deviation (SIGMA): σ **> OK**

Gives p-values for both one-tailed and two-tailed tests.

*Data Analysis Plus is a collection of statistical macros for Excel. They can be downloaded onto your computer after logging in to cengagebrain.com.

CONFIDENCE INTERVAL FOR MEAN μ WITH A GIVEN σ

Input the data into L1; then continue with the following, entering the appropriate values and highlighting Calculate:

Choose: **STAT > TESTS > 7:ZInterval**

HYPOTHESIS TEST FOR MEAN μ WITH A GIVEN σ

Input the data into L1; then continue with the following, entering the appropriate values and highlighting Calculate:

Choose: **STAT > TESTS > 1:Z-Test**

8.1 *Using the commands for generating integer data from Chapter 2, randomly select a sample of 40 single-digit numbers. Then, continue with the confidence interval commands and find the 90% confidence interval for μ. Repeat several times, observing whether or not 4.5 is in the interval each time. Describe your results.*

8.2 *Using the commands for generating integer data from Chapter 2, select 40 random single-digit numbers, find the sample mean, z★, and p-value for testing $H_o: \mu = 4.5$ against a two-tailed alternative. Repeat several times. Describe your findings.*

8.3 *Using the commands for generating data from Chapter 2, select 36 random numbers from a normal distribution with mean 100 and standard deviation 15. Then continue with the hypothesis test commands and find the sample mean, z★, and p-value for testing a two-tailed hypothesis test of $\mu = 100$. Repeat several times. Describe your findings.*

8.4 *Using the commands for generating integer data from Chapter 2, select 40 random single-digit numbers. Then continue with the hypothesis test commands and find the sample mean and z★. Using $\alpha = 0.05$, state the decision for testing $H_o: \mu = 4.5$ against a two-tailed alternative. Repeat it several times. Describe your findings after several tries.*

8.5 *Using the commands for generating integer data from Chapter 2, select 36 random numbers from a normal distribution with mean 100 and standard deviation 15. Then continue with the hypothesis test commands and find the sample mean and z★ for testing a two-tailed hypothesis test of $\mu = 100$. Using $\alpha = 0.05$, state the decision. Repeat several times. Describe your findings.*

Minitab Instructions For:

PROBABILITY ASSOCIATED WITH A SPECIFIED VALUE OF t

Cumulative probability for a specified value of t:

Choose: **Calc > Probability Distribution > t**
Select: **Cumulative Probability**
     Noncentrality parameter: **0.0**
Enter: Degrees of freedom: **df**
Select: **Input constant***
Enter: **t-value** (ex. 1.74) **> OK**

*Select Input column if several t-values are stored in C1. Use C2 for optional storage. If the area in the right tail is needed, subtract the calculated probability from 1.

$1 - \alpha$ CONFIDENCE INTERVAL FOR MEAN μ WITH σ UNKNOWN

Input the data into C1; then continue with:

Choose: **Stat > Basic Statistics > 1-Sample t**
Enter: Samples in columns: **C1**
Select: **Options**
Enter: Confidence level: **$1 - \alpha$** (ex. 95.0)
Select: Alternative: **not equal > OK > OK**

HYPOTHESIS TEST FOR MEAN μ WITH σ UNKNOWN

Input the data into C1; then continue with:

Choose: **Stat > Basic Statistics > 1-Sample t**
Enter: Samples in columns: **C1**
Select: **Perform Hypothesis test**
Enter: Hypothesized mean: **μ**
Select: **Options**
Select: Alternative: **less than** or **not equal** or **greater than > OK > OK**

$1 - \alpha$ CONFIDENCE INTERVAL FOR A PROPORTION p

Choose: **Stat > Basic Statistics > 1 Proportion**
Select: **Summarized Data**
Enter: Number of events: **x**
     Number of trials: **n**
Select: **Options**
Enter: Confidence level: **$1 - \alpha$** (ex. 95.0)
Select: Alternative: **not equal**
     Use test and interval based on normal distribution. > OK > OK

HYPOTHESIS TEST FOR PROPORTION p

Choose: **Stat > Basic Statistics > 1 Proportion**
Select: **Summarized Data**
Enter: Number of events: **x**
     Number of trials: **n**
Select: **Perform hypothesis test**
Enter: Hypothesized proportion: **p**
Select: **Options**
Select: Alternative: **less than** or **not equal** or **greater than**
     Use test and interval based on normal distribution. > OK > OK

CUMULATIVE PROBABILITIES FOR χ^2

Input the data into C1; then continue with:

Choose: **Calc > Probability Distributions > Chi-Square**
Select: **Cumulative Probability**
     Noncentrality Parameter: **0.0**
Enter: Degrees of freedom: **df**
Select: **Input constant***
Enter: **χ^2-value** (ex. 47.25) **> OK**

*Select Input column if several χ^2-values are stored in C1. Use C2 for optional storage.
If the area in the right tail is needed, subtract the calculated probability from 1.

HYPOTHESIS TEST FOR STANDARD DEVIATION σ

Input the data into C1; then continue with:

Choose: **Stat > Basic Statistics > 1 Variance**
Enter: Samples in columns: **C1**
Select: **Perform Hypothesis test**
Enter: Hypothesized standard deviation: **σ**
Select: **Options**
Select: Alternative: **less than** or **not equal** or **greater than > OK > OK**

Excel Instructions For:

PROBABILITY ASSOCIATED WITH A SPECIFIED VALUE OF t

Probability in one or two tails for a given t-value:

     If several t-values (nonnegative) are to be used, input the values into column A and activate B1; continue with:

Choose: **Insert Function, f_x > Statistical > TDIST > OK**
Enter: X: **individual t-value** or **(A1:A5** or select "t-value" cells)*
     Deg_freedom: **df**
     Tails: **1** or **2** (one or two-tailed distribution) **> OK**

*For more than one t-value, drag bottom right corner of the B1 cell down to give other probabilities.
To find the probability within the two tails or the cumulative probability for one tail, subtract the calculated probability from 1.

$1 - \alpha$ CONFIDENCE INTERVAL FOR MEAN μ WITH σ UNKNOWN

Input the data into column A; then continue with:

Choose: **Add-Ins > Data Analysis Plus** > t-Estimate : Mean > OK**
Enter: Input Range: **(A1:A20** or select cells)
Enter: Alpha: **α** (ex. 0.05) **> OK**

HYPOTHESIS TEST FOR MEAN μ WITH σ UNKNOWN

Input the data into column A; then continue with:

Choose: **Add-Ins > Data Analysis Plus** > t-Test: Mean > OK**
Enter: Input Range: **(A1:A20** or select cells)
     Hypothesized Mean: **μ**
     Alpha: **α** (ex. 0.05) **> OK**

Gives p-values and critical values for both one-tailed and two-tailed tests.

$1 - \alpha$ CONFIDENCE INTERVAL FOR A PROPORTION p

Input the data into column A using 0's for failures (or no's) and 1's for successes (or yes's); then continue with:

Choose: **Add-Ins > Data Analysis Plus** > Z Estimate : Proportion > OK**
Enter: Input Range: **(A2:A20** or select cells)
     Code for success: **1**
     Alpha: **α** (ex. 0.05) **> OK**

HYPOTHESIS TEST FOR PROPORTION p

Input the data into column A using 0's for failures (or no's) and 1's for successes (or yes's); then continue with:

Choose: **Add-Ins > Data Analysis Plus** > Z-Test: Proportion > OK**
Enter: Input Range: **(A2:A20** or select cells)
     Code for success: **1**
     Hypothesized Proportion: **p**
     Alpha: **α** (ex. 0.05) **> OK**

Gives p-values and critical values for both one-tailed and two-tailed tests.

CUMULATIVE PROBABILITIES FOR χ^2

If several χ^2 values are to be used, input the values into column A and activate B1; then continue with:

Choose: **Insert Function, f_x > Statistical > CHIDIST > OK**

Enter: X: individual χ^2-value *or* (A1:A5 or select "χ^2-value" cells)*

          Deg_freedom: df > OK

Drag bottom right corner of the B1 cell down to give other probabilities.

HYPOTHESIS TEST FOR STANDARD DEVIATION σ

Input the data into column A; then continue with:

Choose: **Add-Ins > Data Analysis Plus* > Chi-Squared Test: Variance > OK**

Enter: Input Range: (A1:A20 or select cells)

          Hypothesized Variance: σ^2

          Alpha: α *(ex. 0.05)* > OK

Gives p-values and critical values for both one-tailed and two-tailed tests.

Data Analysis Plus is a collection of statistical macros for Excel. They can be downloaded onto your computer after logging in to cengagebrain.com.

TI-83/84 Plus Instructions For:

CUMULATIVE PROBABILITY FOR A SPECIFIED VALUE OF t:

Choose: **2nd > DISTR > 5:tcdf(**

Enter: −1EE99, t-value, df)

* To find the probability between two *t*-values, enter the two values in place of −1EE99 and *t*-value.

If the area in the right tail is needed, subtract the calculated probability from 1.

1 − α CONFIDENCE INTERVAL FOR MEAN μ WITH σ UNKNOWN

Input the data into L1; then continue with the following, entering the appropriate values and highlighting Calculate:

Choose: **STAT > TESTS > 8:TInterval**

```
TInterval
Inpt:Data Stats
List:L1
Freq:1
C-Level:.95
Calculate
```

HYPOTHESIS TEST FOR MEAN μ WITH σ UNKNOWN

Input the data into L1; then continue with the following, entering the appropriate values and highlighting Calculate:

Choose: **STAT > TESTS > 2:T-Test**

```
T-Test
Inpt:Data Stats
μ0:0
List:L1
Freq:1
μ:≠μ0 <μ0 >μ0
Calculate Draw
```

1 − α CONFIDENCE INTERVAL FOR A PROPORTION p

Choose: **STAT > TESTS > A:1-PropZInt**
          Enter the appropriate values and highlight Calculate.

```
1-PropZInt
x:0
n:0
C-Level:.95
Calculate
```

HYPOTHESIS TEST FOR PROPORTION p

Choose: **STAT > TESTS > 5:1-PropZTest**
          Enter the appropriate values and highlight Calculate.

```
1-PropZTest
p0:0
x:0
n:0
prop≠p0 <p0 >p0
Calculate Draw
```

CUMULATIVE PROBABILITIES FOR χ^2

Choose: **2nd > DISTR > 7: χ^2cdf(**

Enter: 0, χ^2-value, df)

If the area in the right tail is needed, subtract the calculated probability from 1.

9.1 *Karl Pearson once tossed a coin 24,000 times and recorded 12,012 heads.*

a. *Calculate the point estimate for p = P(head) based on Pearson's results.*

b. *Determine the standard error of proportion.*

c. *Determine the 95% confidence interval estimate for p = P(head).*

d. *It must have taken Mr. Pearson many hours to toss a coin 24,000 times. You can simulate 24,000 coin tosses using the computer and calculator commands that follow. (Note: A Bernoulli experiment is like a "single" trial binomial experiment. That is, one toss of a coin is one Bernoulli experiment with p = 0.5; and 24,000 tosses of a coin either is a binomial experiment with n = 24,000 or is 24,000 Bernoulli experiments. Code: 0 = tail, 1 = head. The sum of the 1s will be the number of heads in the 24,000 tosses.)*

MINITAB

Choose Calc > Random Data > Bernoulli, entering 24000 for generate, C1 for Store in column(s) and 0.5 for Probability of success. Sum the data and divide by 24,000.

Excel

Choose Data > Data Analysis > Random Number Generation > OK. Then choose Bernoulli for the Distribution and enter 1 for Number of Variables, 24000 for Number of Random Numbers and 0.5 for p Value. Sum the data and divide by 24,000.

TI-83/84 PLUS

Choose MATH > PRB > 5:randInt, then enter 0, 1, number of trials. The maximum number of elements (trials) in a list is 999 (slow process for large n). Sum the data and divide by n.

e. *How do your simulated results compare with Pearson's?*

f. *Use the commands [part (d)] and generate another set of 24,000 coin tosses. Compare these results to those obtained by Pearson. Also, compare the two simulated samples to each other. Explain what you can conclude from these results.*

9.2 *The "rule of thumb," stated on page 192, indicated that we would expect the sampling distribution of p' to be approximately normal when "n > 20 and both np and nq are greater than 5." What happens when these guidelines are not followed?*

a. *Use the following set of computer or calculator commands to see what happens. Try n = 15 and p = 0.1 (K1 = n and K2 = p). Do the distributions look normal? Explain what causes the "gaps." Why do the histograms look alike? Try some different combinations of n (K1) and p (K2):*

MINITAB

Choose Calc > Random Data > Binomial to simulate 1,000 trials for an n of 15 and a p of 0.5. Divide each generated value by n, forming a column of sample p's. Calculate a z-value for each sample p by using $z = (p' - p)/\sqrt{p(1 - p)/n}$. Construct a histogram for the sample p's and another histogram for the z's.

Excel

Choose Data > Data Analysis > Random Number Generation > OK. Select Binomial for the distribution to simulate 1,000 trials for an n of 15 and a p of 0.5. Divide each generated value by n, forming a column of sample p's. Calculate a z-value for each sample p by using $z = (p' - p)/\sqrt{p(1 - p)/n}$. Construct a histogram for the sample p's and another histogram for the z's.

TI-83/84 Plus

Choose MATH > PRB > 7:randBin, then enter n, p, number of trials. The maximum number of elements (trials) in a list is 999 (slow process for large n). Divide each generated value by n, forming a list of sample p's. Calculate a z-value for each sample p by using $z = (p' - p)/\sqrt{p(1 - p)/n}$. Construct a histogram for the sample p's and another histogram for the z's.

b. *Try n = 15 and p = 0.01.*

c. *Try n = 50 and p = 0.03.*

d. *Try n = 20 and p = 0.2.*

e. *Try n = 20 and p = 0.8.*

f. *What happens when the rule of thumb is not followed?*

Minitab Instructions For:

1 − α CONFIDENCE INTERVAL FOR MEAN μ_d WITH UNKNOWN STANDARD DEVIATION FOR TWO DEPENDENT SETS OF SAMPLE DATA

Input the paired data into C1 and C2; then continue with:

Choose: **Stat > Basic Statistics > Paired t**
Select: **Samples in columns**
Enter: First sample: **C1***
            Second sample: **C2**
Select: **Options**
Enter: Confidence level: **1 − α** *(ex. 0.95 or 95.0)*
Select: Alternative: **not equal > OK > OK**

Paired t evaluates the first sample minus the second sample.

The solution to the example on page 212 looks like this when solved in MINITAB:

Paired T for Brand B − Brand A

	N	Mean	StDev	SE Mean
Brand B	6	92.5	35.2	14.4
Brand A	6	86.2	30.9	12.6
Difference	6	6.33	5.13	2.09

95% CI for mean difference: (0.95, 11.71)

HYPOTHESIS TEST FOR THE MEAN μ_d, WITH UNKNOWN STANDARD DEVIATION FOR TWO DEPENDENT SETS OF SAMPLE DATA

Input the paired data into C1 and C2; then continue with:

Choose: **Stat > Basic Statistics > Paired t**
Select: **Samples in columns**
Enter: First sample: **C1***
            Second sample: **C2**
Select: **Options**
Enter: Test mean: **0.0 or μ_d**
Select: Alternative: **less than** *or* **not equal** *or* **greater than > OK > OK**

Paired t evaluates the first sample minus the second sample.

The solution to the example on page 213 looks like this when solved in MINITAB:

Paired T for Before − After

	N	Mean	StDev	SE Mean
Difference	26	1.07	1.74	0.34

T-Test of mean difference = 0 (vs > 0): T-Value = 3.14
P-Value = 0.002

HYPOTHESIS TEST FOR THE DIFFERENCE BETWEEN TWO POPULATION MEANS WITH UNKNOWN STANDARD DEVIATION GIVEN TWO INDEPENDENT SETS OF SAMPLE DATA

MINITAB's 2-Sample t (Test and Confidence Interval) command performs both the confidence interval and the hypothesis test at the same time.
Input the two independent sets of data into C1 and C2; then continue with:

Choose: **Stat > Basic Statistics > 2-Sample t**
Select: **Samples in different columns***
Enter: First: **C1** Second: **C2**
Select: **Assume equal variances** *(if known)*
Select: **Options**
Enter: Confidence level: **1 − α** *(ex. 0.95 or 95.0)*
            Test mean: **0.0**
Choose: Alternative: **less than** *or* **not equal** *or* **greater than > OK > OK**

Note the other possible data formats.

The example on pages 217–218 was solved using MINITAB. With 40 cumulative GPAs for nonmembers in C1 and 40 averages for fraternity members in C2, the preceding commands resulted in the output shown here. Compare these results to the solution of for the example on pages 217–218. Notice the difference in P and df values. Explain.

Two-Sample T-Test and CI

Sample	N	Mean	StDev	SE Mean
1	40	2.210	0.590	0.093
2	40	2.030	0.680	0.11

Difference = mu (1) − mu (2) Est. diff.: 0.180
95% CI for difference: (−0.10, 0.46)
T-Test diff. = 0 (vs >): T = 1.26 P = 0.105 DF = 76

CONFIDENCE INTERVALS FOR THE DIFFERENCE BETWEEN TWO PROPORTIONS GIVEN TWO INDEPENDENT SETS OF SAMPLE DATA

Choose: **Stat > Basic Statistics > 2 Proportions**
Select: **Summarized data:**
Enter: First: **x** *(events)* **n** *(trials)*
            Second: **x** *(events)* **n** *(trials)*
Select: **Options**
Enter: Confidence level: **1 − α** *(ex. 0.95 or 95.0)*
Select: Alternative: **not equal > OK > OK**

HYPOTHESIS TEST FOR THE DIFFERENCE BETWEEN TWO PROPORTIONS, $p_1 − p_2$, FOR TWO INDEPENDENT SETS OF SAMPLE DATA

Choose: **Stat > Basic Statistics > 2 Proportions**
Select: **Summarized data:**
Enter: First: **x** *(events)* **n** *(trials)*
            Second: **x** *(events)* **n** *(trials)*
Select: **Options**
Enter: Test difference: **0.0**
Select: Alternative: **less than** *or* **not equal** *or* **greater than**
Select: **Use pooled estimate of p for test > OK > OK**

CUMULATIVE PROBABILITY ASSOCIATED WITH A SPECIFIED VALUE OF F

Choose: **Calc > Probability Distributions > F**
Select: **Cumulative Probability**
            Noncentrality parameter: **0.0**
Enter: Numerator degrees of freedom: **df$_n$**
            Denominator degrees of freedom: **df$_d$**
Select: **Input constant***
Enter: **F-value** *(ex. 1.74)* **> OK**

Select the Input column if several F-values are stored in C1. Use C2 for optional storage. If the area in the right tail is needed, subtract the calculated probability from 1.

HYPOTHESIS TEST FOR THE RATIO BETWEEN TWO POPULATION VARIANCES, σ_1^2/σ_2^2, FOR TWO INDEPENDENT SETS OF SAMPLE DATA

Choose: **Stat > Basic Statistics > 2 Variances***
Select: **Samples in one column:**
Enter: Samples: **C1**   Subscripts: **C2**
Or
Select: **Samples in different columns:**
Enter: First: **C1** Second: **C2**
Or
Select: **Summarized data**
Enter: Sample size and Variance for each sample:
Select: Storage: **Standard Deviations > OK > OK**

The 2 Variances procedure evaluates the first sample divided by the second sample.

1 — α CONFIDENCE INTERVAL FOR MEAN μ_d WITH UNKNOWN STANDARD DEVIATION FOR TWO DEPENDENT SETS OF SAMPLE DATA

Input the paired data into columns A and B; activate C1 or C2 (depending on whether column headings are used or not); then continue with:

Enter: **= A2–B2*** *(if column headings are used)*

Drag: **Bottom right corner of C2 down to give other differences**

Choose: **Add-Ins > Data Analysis Plus > t-Estimate : Mean > OK**

Enter: *Input range:* **(C2:C20 or select cells)**

Select: **Labels** *(if necessary)*

Enter: *Alpha:* **α** *(ex. 0.05)* **> OK**

*Enter the expression in the order that is needed: A2—B2 or B2—A2.

HYPOTHESIS TEST FOR THE MEAN μ_d, WITH UNKNOWN STANDARD DEVIATION FOR TWO DEPENDENT SETS OF SAMPLE DATA

Input the paired data into columns A and B; then continue with:

Choose: **Data > Data Analysis > t-Test: Paired Two Sample for Means > OK**

Enter: *Variable 1 Range:* **(A1:A20 or select cells)**
            Variable 2 Range: **(B1:B20 or select cells)**
            (subtracts: Var1 — Var2)
            Hypothesized Mean Difference: **μ_d** *(usually 0)*

Select: **Labels** *(if necessary)*

Enter: **α** *(ex. 0.05)*

Select: **Output Range**

Enter: **(C1 or select cell) > OK**

Use Home > Cells > Format > AutoFit Column Width to make the output more readable. The output shows p-values and critical values for one- and two-tailed tests. The hypothesis test may also be done by first subtracting the two columns and then using the inference about a mean (sigma unknown) commands on the Chapter 9 Tech Card on the differences.

HYPOTHESIS TEST FOR THE DIFFERENCE BETWEEN TWO POPULATION MEANS WITH UNKNOWN STANDARD DEVIATION GIVEN TWO INDEPENDENT SETS OF SAMPLE DATA

Input the two independent sets of data into columns A and B; then continue with:

Choose: **Data > Data Analysis > t-Test: Two-Sample Assuming Unequal Variances > OK**

Enter: *Variable 1 Range:* **(A1:A20 or select cells)**
            Variable 2 Range: **(B1:B20 or select cells)**
            Hypothesized Mean Difference: **μ_B − μ_A** *(usually 0)*

Select: **Labels** *(if necessary)*

Enter: **α** *(ex. 0.05)*

Select: **Output Range**

Enter: **(C1 or select cell) > OK**

Use Home > Cells > Format > AutoFit Column Width to make the output more readable. The output shows p-values and critical values for one- and two-tailed tests.

CONFIDENCE INTERVALS FOR THE DIFFERENCE BETWEEN TWO PROPORTIONS GIVEN TWO INDEPENDENT SETS OF SAMPLE DATA

Input the data for the first sample into column A using 0s for failures (or no's) and 1s for successes (or yes's), then repeat the same procedure for the second sample in column B; then continue with:

Choose: **Add-Ins > Data Analysis Plus > Z-Estimate: Two Proportions > OK**

Enter: *Variable 1 Range:* **(A2:A20 or select cells)**
            Variable 2 Range: **(B1:B20 or select cells)**
            Code for success: **1**

Select: **Labels** *(if necessary)*

Enter: *Alpha:* **α** *(ex. 0.05)* **> OK**

HYPOTHESIS TEST FOR THE DIFFERENCE BETWEEN TWO PROPORTIONS, p_1 − p_2, FOR TWO INDEPENDENT SETS OF SAMPLE DATA

Input the data for the first sample into column A using 0s for failures (or no's) and 1s for successes (or yes's), then repeat the same procedure for the second sample in column B; then continue with:

Choose: **Add-Ins > Data Analysis Plus > Z-Test: Two Proportions > OK**

Enter: *Variable 1 Range:* **(A1:A20 or select cells)**
            Variable 2 Range: **(B1:B20 or select cells)**
            Code for success: **1**
            Hypothesized difference: **0**

Select: **Labels** *(if necessary)*

Enter: *Alpha:* **α** *(ex. 0.05)* **> OK**

CUMULATIVE PROBABILITY ASSOCIATED WITH A SPECIFIED VALUE OF F

If several F-values are to be used, input the values into column A and activate B1; then continue with:

Choose: **Insert Function, f_x > Statistical > FDIST > OK**

Enter: *X:* **individual F-value** *or* **(A1:A5 or select "F-value" cells)***
            Deg_freedom 1: **df_n**
            Deg_freedom 2: **df_d > OK**

*Drag bottom right corner of the B1 cell down to give other probabilities.

To find the probability for the left tail (the cumulative probability up to the F-value), subtract the calculated probability from 1.

HYPOTHESIS TEST FOR THE RATIO BETWEEN TWO POPULATION VARIANCES, σ_1^2/σ_2^2, FOR TWO INDEPENDENT SETS OF SAMPLE DATA

Input the data for the numerator (larger spread) into column A and the data for the denominator (smaller spread) into column B; then continue with:

Choose: **Data > Data Analysis > F-Test Two-Sample for Variances > OK**

Enter: *Variable 1 Range:* **(A1:A20 or select cells)**
            Variable 2 Range: **(B1:B20 or select cells)**

Select: **Labels** *(if necessary)*

Enter: **α** *(ex. 0.05)*

Select: **Output Range**

Enter: **(C1 or select cell) > OK**

Use Home > Cells > Format > AutoFit Column Width to make the output more readable. The output shows the p-value and the critical value for a one-tailed test.

Note: Practice Problems for this chapter are found on the back of the TI-83/84 Plus Instructions Tech Card.

1 − α CONFIDENCE INTERVAL FOR MEAN μ_d WITH UNKNOWN STANDARD DEVIATION FOR TWO DEPENDENT SETS OF SAMPLE DATA

Input the paired data into L1 and L2; then continue with the following, entering the appropriate values and highlighting Calculate:

Highlight: **L3**
Enter: **L3 = L1 − L2***
Choose: **STAT > TESTS > 8:TInterval**

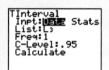

*Enter the expression in the order that is needed: L1 − L2 or L2 − L1.

HYPOTHESIS TEST FOR THE MEAN μ_d, WITH UNKNOWN STANDARD DEVIATION FOR TWO DEPENDENT SETS OF SAMPLE DATA

Input the paired data into L1 and L2; then continue with the following, entering the appropriate values and highlighting Calculate:

Highlight: **L3**
Enter: **L3 = L1 − L2***
Choose: **STAT > TESTS > 2:T-Test > . . .**

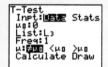

*Enter the expression in the order that is needed: L1 − L2 or L2 − L1.

HYPOTHESIS TEST FOR THE DIFFERENCE BETWEEN TWO POPULATION MEANS WITH UNKNOWN STANDARD DEVIATION GIVEN TWO INDEPENDENT SETS OF SAMPLE DATA

*Input the two independent sets of data into L1 and L2.**

To construct a 1 − α confidence interval for the difference between two means, continue with the following, entering the appropriate values and highlighting Calculate:

Choose: **STAT > TESTS > 0:2-SampTInt . . .**

 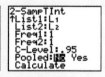

To complete a hypothesis test for the difference between two means, continue with the following, entering the appropriate values and highlighting Calculate:

Choose: **STAT > TESTS > 4:2-SampTTest . . .**

 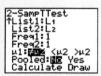

*Enter the data in the order that is needed; the program subtracts as L1 − L2.

Highlight No for Pooled if there are no assumptions about the equality of variances.

CONFIDENCE INTERVALS FOR THE DIFFERENCE BETWEEN TWO PROPORTIONS GIVEN TWO INDEPENDENT SETS OF SAMPLE DATA

Choose: **STAT > TESTS > B:2-PropZInt**
Enter the appropriate values and highlight Calculate.

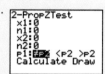

HYPOTHESIS TEST FOR THE DIFFERENCE BETWEEN TWO PROPORTIONS, $p_1 − p_2$, FOR TWO INDEPENDENT SETS OF SAMPLE DATA

Choose: **STAT > TESTS > 6:2-PropZTest > . . .**
Enter the appropriate values and highlight Calculate.

CUMULATIVE PROBABILITY ASSOCIATED WITH A SPECIFIED VALUE OF F

Choose: **2nd > DISTR > 9:Fcdf(**
Enter: **0, F-value, df$_n$, df$_d$)**

Note: To find the probability between two F-values, enter the two values in place of 0 and the F-value.

If the area in the right tail is needed, subtract the calculated probability from one.

HYPOTHESIS TEST FOR THE RATIO BETWEEN TWO POPULATION VARIANCES, σ_1^2/σ_2^2, FOR TWO INDEPENDENT SETS OF SAMPLE DATA

Input the data for the numerator (larger spread) into L1 and the data for the denominator (smaller spread) into L2; then continue with the following, entering the appropriate values and highlighting Calculate:

Choose: **STAT > TESTS > D:2-SampFTest > . . .**

10.1 Use a computer to demonstrate the truth of the theory presented in Objective 10.5.

a. The underlying assumptions are "the populations are normally distributed," and while conducting a hypothesis test for the equality of two standard deviations, it is assumed that the standard deviations are equal. Generate very large samples of two theoretical populations: N(100, 20) and N(120, 20). Find graphic and numerical evidence that the populations satisfy the assumptions.

b. Randomly select 100 samples, each of size eight, from both populations and find the standard deviation of each sample.

c. Using the first sample drawn from each population as a pair, calculate the F★-statistic. Repeat for all samples. Describe the sampling distribution of the 100 F★-values using both graphic and numerical statistics.

d. Generate the probability distribution for F (7, 7), and compare it with the observed distribution of F★. Do the two graphs agree? Explain.

10.2 Use a computer or calculator to complete the hypothesis test with alternative hypothesis $\mu_d < 0$ based on the paired data that follow and d = M − N. Use $\alpha = 0.02$. Assume normality.

M	58	78	45	38	49	62
N	62	86	42	39	47	68

*10.3 Ten subjects with borderline-high cholesterol levels were recruited for a study. The study involved taking a nutrition education class. Cholesterol readings were taken before the class and 3 months after the class.

Subject	1	2	3	4	5	6	7	8	9	10
Preclass	295	279	250	235	255	290	310	260	275	240
Postclass	265	266	245	240	230	230	235	250	250	215

Let d = preclass cholesterol - postclass cholesterol. Use the following Excel output to test the null hypothesis that the population mean difference equals zero versus the alternative hypothesis that the population mean difference is positive at $\alpha = 0.05$. Rejection of the null hypothesis would indicate that the (population) average cholesterol level after the class is lower than the average level before the class. Assume normality.

t-Test: Paired Two-Sample for Means		
	Pretest	Posttest
Mean	268.9	242.6
Variance	618.7666667	256.4888889
Observations	10	10
Hypothesized mean difference	0	
df	9	
t Stat	3.394655392	
$P(T \leq t)$ one-tail	0.003970146	
t Critical one-tail	1.833113856	

*10.4 A study was designed to estimate the difference in diastolic blood pressure readings between men and women. MINITAB was used to construct a 99% confidence interval for the difference between the means based on the following sample data.

Males	76	76	74	70	80	68	90	70
	90	72	76	80	68	72	96	80

Females	76	70	82	90	68	60	62	68
	80	74	60	62	72			

Two-Sample T for Males vs Females				
	N	Mean	StDev	SE Mean
Males	16	77.37	8.35	2.1
Females	13	71.08	9.22	2.6
99% C.I. for mu males — mu females: (−2.9, 15.5)				

Verify the results (the two sample means and standard deviations, and the confidence interval bounds) by calculating the values yourself. Assume normality of blood pressure readings.

*10.5 MINITAB was used to complete a t-test of the difference between the two means using the following two independent samples.

Sample 1	33.7	21.6	32.1	38.2	33.2	35.9	34.1	39.8
	23.5	21.2	23.3	18.9	30.3			

Sample 2	28.0	59.9	22.3	43.3	43.6	24.1	6.9	14.1
	30.2	3.1	13.9	19.7	16.6	13.8	62.1	28.1

Two-Sample T for Sample 1 vs Sample 2				
	N	Mean	StDev	SE Mean
Sample 1	13	29.68	7.07	2.0
Sample 2	16	26.9	17.4	4.4
T-Test mu sample1 = mu sample2 (vs not =): T=0.59 P=0.56 DF=20				

a. Assuming normality, verify the results (two sample means and standard deviations, and the calculated t★) by calculating the values yourself.

b. Use Table 7 in Appendix B to verify the p-value based on the calculated df.

c. Find the p-value using the smaller number of degrees of freedom. Compare the two p-values.

*10.6 According to the College Board (http://www.collegeboard.com/), the average 2008–2009 cost (tuition, fees, and room and board) for a public college is $14,333 versus $34,132 for a private college. Is there also a difference in the average cost of required textbooks between public and private colleges? The following samples of size 10 were taken.

Public	Private	Public	Private	Public	Private
64.69	71.00	103.59	98.56	110.69	112.58
89.60	96.19	106.38	98.94	118.94	114.00
101.49	96.47	106.77	107.79	135.94	116.55
101.75	97.14				

Using the Excel output that follows and $\alpha = 0.05$, determine whether the average cost of required textbooks per class is different between public and private colleges. Assume normality.

a. Solve using the p-value approach.

b. Solve using the classical approach.

t-Test: Two-Sample Assuming Unequal Variances		
	Public	Private
Mean	103.984	100.922
Variance	340.6249822	173.2995511
Observations	10	10
Hypothesized mean difference	0	
df	16	
t Stat	0.427125511	
$P(T \leq t)$ two-tail	0.674980208	
t Critical two-tail	2.119904821	

*remember

Problems marked with an asterisk (*) have data sets available on the CourseMate for STAT2 site. Login at cengagebrain.com.

Minitab Instructions For:

Excel Instructions For:

GOODNESS OF FIT TEST

Input the observed frequencies into C1. If performing a test with unequal expected frequencies, input the specific proportions into C2. Then continue with:

Choose: **Stat > Tables > Chi-Square Goodness-of-Fit Test (One Variable)…**
Enter: *Observed counts:* **C1**
Select: **Equal Proportions > OK**
           or
           Specific Proportions
Enter: **C2 > OK**

HYPOTHESIS TEST OF INDEPENDENCE OR HOMOGENEITY

Input each column of observed frequencies from the contingency table into C1, C2, …; then continue with:

Choose: **Stat > Tables > Chi-Square Test (Two-Way Table in Worksheet)**
Enter: *Columns containing the table:* **C1 C2 > OK**

COMPUTER SOLUTION MINITAB Printout for the voting example on pages 247–249:

Chi-Square Test: C1, C2
Expected counts are printed below observed counts.
Chi-square contributions are printed below expected counts.

	C1	C2	Total
1	143	57	200
	101.60	98.40	
	16.870	17.418	
2	98	102	200
	101.60	98.40	
	0.128	0.132	
3	13	87	100
	50.80	49.20	
	28.127	29.041	
Total	254	246	500

Chi-Sq = 91.715, DF = 2, P-Value = 0.000

GOODNESS OF FIT TEST

Input the observed frequencies into column A and the corresponding expected frequencies into column B. (You can use Excel to convert probabilities into expected frequencies.) Then continue with:

Choose: **Insert Function, f_x > Statistical > CHITEST > OK**
Enter: *Actual Range:*   **(A1:A6 or select cells)**
           Expected Range   **(B1:B6 or select cells) > OK**
Excel output only provides the p-value for the test.

HYPOTHESIS TEST OF INDEPENDENCE OR HOMOGENEITY

Input each column of observed frequencies from the contingency table into columns A, B, …; then continue with:

Choose: **Add-Ins > Data Analysis Plus > Contingency Table > OK**
Enter: *Input range:*   **(A1:B4 or select cells)**
Select: **Labels** *(if necessary)*
Enter: *Alpha:*   **α (ex. 0.05)**

GOODNESS OF FIT TEST*

Input the observed frequencies into L and the expected frequencies into L2; then continue with:

Choose: **Stat > Tables > D:χ² GOF-Test...**

Enter: Observed: **L1**

Expected: **L2**

df: **k-1**

*Goodness of Fit Test is only available on the TI-84 Plus.

```
χ²GOF-Test
 Observed:L₁
 Expected:L₂
 df:5
 Calculate Draw
```

HYPOTHESIS TEST OF INDEPENDENCE OR HOMOGENEITY

Input the observed frequencies from the r × c contingency table into an r × c matrix A. Set up matrix B as an empty r × c matrix for the expected frequencies.

Choose: **MATRIX > EDIT > 1:[A]**

Enter: **r > ENTER > c > ENTER**

Each observed frequency with an ENTER afterward

Then continue with:

Choose: **MATRIX > EDIT > 2[B]**

Enter: **r > ENTER > c > ENTER**

Choose: **STAT > TESTS > C:χ² – Test. . .**

Enter: Observed: **[A]** or wherever the contingency table is located

Expected: **[B]** place for expected frequencies

Highlight: **Calculate > ENTER**

11.1 MINITAB was used to complete a chi-square test of independence between the number of boat-related manatee deaths and two Florida counties.

County	Boat-Related Deaths	Non–Boat-Related Deaths	Total Deaths
Lee County	23	25	48
Collier County	8	23	31

Chi-Square Test: Boat-Related Deaths, Non–Boat-Related Deaths
Expected counts are printed below observed counts
Chi-Square contributions are printed below expected counts

	Boat-Related Deaths	Non–Boat-Related Deaths	Total
1	23	25	48
	18.84	29.16	
	0.921	0.595	
2	8	23	31
	12.16	18.84	
	1.426	0.921	
Total	31	48	79

Chi-Sq = 3.862, DF = 1, P-Value = 0.049

a. Verify the results (expected values and the calculated $\chi^2\star$) by calculating the values yourself.
b. Use Table 8 to verify the p-value based on the calculated df.
c. Is the proportion of boat-related deaths independent of the county? Use $\alpha = 0.05$.

11.2 To demonstrate/explore the effect increased sample size has on the calculated chi-square value, let's consider the Skittles candies in exercise 11.8 and sample some larger bags of the candy.

a. Suppose we purchase a 16-oz bag of Skittles, count the colors, and observe exactly the same proportion of colors as found in problem 11.8 at the end of the chapter.

Red	Orange	Yellow	Green	Purple
72	84	92	68	108

Calculate the value of chi-square for these data. How is the new chi-square value related to the one found in problem 11.8? What effect does this new value have on the test results? Explain.

b. To continue this demonstration/exploration, suppose we purchase a 48-oz bag, count the colors, and observe exactly the same proportion of colors as found in problem 11.8 and part (a) of this problem:

Red	Orange	Yellow	Green	Purple
216	252	276	204	324

Calculate the value of chi-square for these data. How is the new chi-square value related to the one found in problem 11.8? Explain.

c. What effect does the size of the sample have on the calculated chi-square value when the proportion of observed frequencies stays the same as the sample size increases?

d. Explain in what way this indicates that if a large enough sample is taken, the hypothesis test will eventually result in a rejection.

Minitab Instructions For:

ONE-WAY ANALYSIS OF VARIANCE

Input the data for each level into columns C1, C2, . . . ; then continue with:

Choose: **Stat > ANOVA > One-Way (Unstacked)**
Enter: Responses: **C1 C2 . . .** * **> OK**
OR
Input all of the data into C1 with the corresponding levels of factors into C2; then continue with:

Choose: **Stat > ANOVA > One-Way**
Enter: Response: **C1**
            Factor: **C2** * **> OK**
*Optional for either method:
Choose   **Graphs . . .**
Select: **Individual value plot** and/or **Boxplots of data > OK > OK**

Computer Solution MINITAB Printout for Target Shooting Example on pages 264–266:

Information given to computer →	Row	Right eye	Left eye	Both eyes
	1	12	10	16
	2	10	17	14
	3	18	16	16
	4	12	13	11
	5	14		20
	6			21

ANALYSIS OF VARIANCE

	SOURCE	DF	SS	MS	F	P
The ANOVA table → compare with Table 12.10	FACTOR	2	29.2	14.6	1.29	0.312
	ERROR	12	136.1	11.3		
	TOTAL	14	165.3			

The calculated value of F, $F_{\ast}$

	LEVEL	N	MEAN	ST. DEV.
Sample statistics for each factor level →	1	5	13.200	3.033
	2	4	14.000	3.162
The calculated p-value	3	6	16.333	3.724

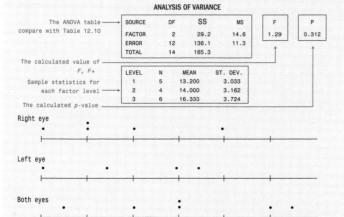

Note: *Side-by-side dotplots are very useful for visualizing the within-sample variation, the between-sample variation, and the relationship between them. Commands for side-by-side dotplots can be found on the Chapter 2 Tech Card.*

Excel Instructions For:

ONE-WAY ANALYSIS OF VARIANCE

Input the data for each level into columns A, B, . . . ; then continue with:

Choose: **Data > Data Analysis > Anova: Single Factor > OK**
Enter: Input Range: **(A1:C4 or select cells)**
Select: Grouped By: **Columns**
            Labels in First Row *(if necessary)*
Enter: Alpha: **α**
Select: **Output Range:**
Enter: **(D1 or select cell) > OK**
To make the output more readable, continue with: Home > Cells > Format > AutoFit Column Width.

ONE-WAY ANALYSIS OF VARIANCE

Input the data for each level into lists L1, L2, . . . ; then continue with:

Choose: **STAT > TESTS > F: ANOVA(**

Enter: **L1, L2, . . .)**

Note for all three technologies: Side-by-side dotplots are very useful for visualizing the within-sample variation, the between-sample variation, and the relationship between them. Commands for side-by-side dotplots can be found on the Chapter 2 Tech Card.

12.1 *An analysis of variance experiment with level A containing 10 data values; level B, 12 data values; level C, 10 values; level D, 12 values; level E, 9 values; and level F, 10 values was analyzed using MINITAB.*

One-way ANOVA: Level A, Level B, Level C, Level D, Level E, Level F				
Source	DF	SS	MS	F	P
Factor	5	6355	1271	3.15	0.014
Error	57	22964	403		
Total	62	29319			

a. Verify the three values for df shown on the printout. Also verify the relationship between the three numbers.
b. Verify the two MS values reported on the printout.
c. Verify the F-value.
d. Verify the p-value.

Minitab Instructions For:

REGRESSION ANALYSIS

Output includes the equation for the regression line, information for a t-test concerning the slope of the regression line, the standard deviation of error, r and/or r², and a scatter diagram showing the regression line.
MINITAB output also includes the predicted y values for given x values and residuals.
Input the x-variable data into C1 and the corresponding y-variable data into C2; then continue with:

Choose:	**Stat > Regression > Regression . . .**
Enter:	*Response* (y): **C2**
	Predictors (x): **C1**
Select:	**Results**
	Regression equation, table of coefficients, s, R-squared, . . .
	OR
	In addition, the full table of fits and residuals > OK
Select:	**Storage**
	Residuals and **Fits > OK > OK**
Choose:	**Graph > Scatterplot**
Select:	**With Regression > OK**
Enter:	*Y variables:* **C2** *X variables:* **C1**
Select:	**Labels > Title/Footnotes**
Enter:	**your title > OK > OK**

Here is the MINITAB printout with explanations for parts of the one-way commute example in Section 13.3.

Equation of line of best fit
$\hat{y} = 3.64 + 1.89x$;
see p. 284

Calculated values of b_0 *and* b_1

Calculated value of s_{b_1}
$s_{b_1} = 0.285$; *compare to*
$s_{b_1}^2 = 0.0813$ *see p. 285*
$(\sqrt{0.0813} = 0.285)$

Calculated t★ *and p-value for* H_o: $\beta_1 = 0$ *as found in steps 3 and 4 on p. 286*

Calculated value of s_e,
$s_e = 5.401$: *compare to*
$s_e^2 = 29.1723$ *as found on p. 284* $(\sqrt{29.1723} = 5.401)$

Given data

Values of $\hat{y}$ *for each given x-value using*
$\hat{y} = 3.643 + 1.8932x$

Regression Analysis: y, minutes versus x, miles

The regression equation is
y, minutes = 3.64 + 1.89 x, miles

Predictor	Coef	SECoeff	T	P
Constant	3.643	3.765	0.97	0.351
x, miles	1.8932	0.2851	6.64	0.000

s = 5.401   R − Sq = 77.2%   R − Sq (adj) = 75.5%

Obs	x, miles	y, minute	Fit	Residual
1	3.0	7.00	9.32	−2.32
2	5.0	20.00	13.11	6.89
3	7.0	20.00	16.90	3.10
4	8.0	15.00	18.79	−3.79
5	10.0	25.00	22.58	2.42
6	11.0	17.00	24.47	−7.47
7	12.0	20.00	26.36	−6.36
8	12.0	35.00	26.36	8.64
9	13.0	26.00	28.26	−2.26
10	15.0	25.00	32.04	−7.04
11	15.0	35.00	32.04	2.96
12	16.0	32.00	33.93	−1.93
13	18.0	44.00	37.72	6.28
14	19.0	37.00	39.61	−2.61
15	20.0	45.00	41.51	3.49

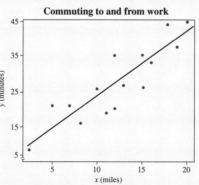

Commuting to and from work

CONFIDENCE AND PREDICTION INTERVALS CALCULATION AND GRAPH

Input the x-variable data into C1 and the corresponding y-variable data into C2; then continue with:

Choose:	**Stat > Regression > Regression . . .**
Enter:	*Response* (y): **C2**
	Predictors (x): **C1**
Select:	**Options**
Enter:	*Prediction intervals for new observations:*
	x-value or C1 *(C1–list of x values)*
	Confidence level: **1 − α** *(ex. 95.0)*
Select:	**Confidence limits**
	Prediction limits > OK > OK
Choose:	**Stat > Regression > Fitted Line Plot**
Enter:	*Response* (y): **C2**
	Predictor (x): **C1**
Select:	*Type of Regression Model:* **Linear**
Select:	**Options**
	Display options: **Confidence limits**
	Prediction limits
Enter:	*Confidence level:* **1 − α** *(ex. 95.0) > OK*
Select:	**Storage**
	Residuals
	Fits > OK > OK

Here is the MINITAB printout for parts of the one-way travel example used throughout the chapter.

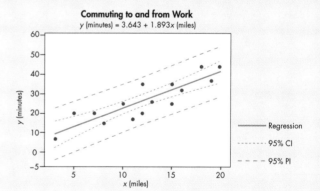

Commuting to and from Work
y (minutes) = 3.643 + 1.893x (miles)

REGRESSION ANALYSIS

Output includes the equation for the regression line, information for a t-test concerning the slope of the regression line, the standard deviation of error, r and/or r², and a scatter diagram showing the regression line.
Excel output also includes predicted y values for given x values, residuals, and a 1 – α confidence interval for the slope.
Input the x-variable data into column A and the corresponding y-variable data into column B; then continue with:

Choose:	**Data > Data Analysis > Regression > OK**
Enter:	*Input Y Range:* **(B1:B10 or select cells)**
	Input X Range: **(A1:A10 or select cells)**
Select:	**Labels** *(if necessary)*
	Confidence Level:
Enter:	**95%** *(desired level)*
Select:	**Output Range:**
Enter:	**(C1 or select cell)**
Select:	**Line Fit Plots > OK**

To make the output more readable, continue with:
Home > Cells > Format > AutoFit Column Width

TI-83/84 Plus Instructions For:

REGRESSION ANALYSIS

Output includes the equation for the regression line, information for a t-test concerning the slope of the regression line, the standard deviation of error, r and/or r², and a scatter diagram showing the regression line.

Input the x-variable data into L1 and the corresponding y-variable data into L2; then continue with the following, entering the apppropriate values and highlighting Calculate:

Choose: **STAT > TESTS > E:LinRegTTest**
(To enter Y1, use: **VARS > YVARS > 1:Function . . . .1:Y1.)**

```
LinRegTTest
Xlist:L1
Ylist:L2
Freq:1
B & ρ:≠0 <0 >0
RegEQ:Y1
Calculate
```

Enter the following to obtain a scatter diagram with regression line:

Choose: **2nd > STATPLOT > 1:Plot1 . . . On**
Choose: **ZOOM > 9:ZoomStat > Trace**

```
Plot1 Plot2 Plot3
On Off
Type: ⬚ ⬚ ⬚
         ⬚ ⬚ ⬚
Xlist:L1
Ylist:L2
Mark: ▫ + ■
```

13.1 *Mr. B, store manager in problem 13.32 at the end of the chapter, found the data from the months of November and December to be different from the data for the other months. Since the data that are separate from the rest in the scatter diagram in problem 13.32 are from November and December, let's remove the November and December values and investigate the relationship between the number of customers per day and the number of items purchased per day for the first 10 months of the year.*

January to October				
Day J-O	Month J-O	Customers J-O	Items J-O	Sales J-O
2	1	425	1,311	12,707.00

••• Remainder of data online at cengagebrain.com.

Day Code: 1 = M, 2 = Tu, 3 = W, 4 = Th, 5 = F, 6 = Sa
Month Code: 1 = Jan, 2 = Feb, 3 = Mar, . . . , 10 = Oct

a. *Use your calculator or computer to construct the scatter diagram for the data for January to October.*
b. *Describe the graphical evidence found and discuss the linearity. Are there any ordered pairs that appear to be different from the others?*
c. *What is the relationship between the number of customers per day and the number of items purchased per day for the first 10 months of the year?*
d. *Is the slope of the regression line significant at α = 0.05?*
e. *Give the 95% prediction interval for the number of items that one would expect to be purchased if the number of customers were 600.*

13.2 *An issue of Popular Mechanics gives specifications and dimensions for various jet boats. The following table summarizes some of this information.*

Model	Base Price	Engine Horsepower
Baja Blast	$8,395	120
Bayliner Jazz	$8,495	90
Boston Whaler Rage 15	$11,495	115
Dynasty Jet Storm	$8,495	90
Four Winds Fling	$9,568	115
Regal Rush	$9,995	90
Sea-Doo Speedster	$11,499	160
Sea Ray Sea Rayder	$8,495	90
Seaswirl Squirt	$8,495	115
Suga Sand Mirage	$8,395	120

Using the Excel output at the bottom of the page:
a. *Determine the equation for the line of best fit.*
b. *Verify the calculation of t★ (t Stat) for engine horsepower.*
c. *Determine whether horsepower is an effective predictor of base price.*
d. *Verify the 95% confidence interval for β₁.*

Table for Problem 13.2 Excel Summary Output

	Coefficients	Standard Error	t Stat	p-Value	Lower 95%	Upper 95%
Intercept	5936.793025	1929.63032	3.076647876	0.01519394	1487.05465	10386.5314
Engine horsepower	30.73218982	17.15820176	1.791107847	0.111051486	−8.834719985	70.29909963

Minitab Instructions For:

Excel Instructions For:

SIGN TEST FOR A SINGLE-SAMPLE HYPOTHESIS TEST OF THE MEDIAN

Input the set of data into C1; then continue with:

Choose: **Stat > Nonparametrics > 1-Sample Sign**
Enter: *Variables:*   **C1**
Select: **Test median:***
Enter: **M** *(hypothesized median value)*
Select: *Alternative:*   **less than** or **not equal** or **greater than > OK**

*A confidence interval may also be selected.

(If original data are not given, just the number of plus and minus signs, then input data values above and below the median that will compute into the correct number of each sign.)

SIGN TEST FOR THE MEDIAN OF PAIRED DIFFERENCES

Input the paired set of data into C1 and C2; then continue with:

Choose: **Calc > Calculator**
Enter: *Store result in variable:*   **C3**
          Expression: **C1–C2** *(whichever order is needed, based on H$_a$)* > **OK**
Choose: **Stat > Nonparametrics > 1-Sample Sign . . .**
Enter: *Variables:*   **C3**
Select: **Test median:***
Enter: **0** *(hypothesized median value)*
Select: *Alternative:*   **less than** or **not equal** or **greater than > OK**

*As before, the confidence interval may be selected.

MANN–WHITNEY *U* TEST FOR THE DIFFERENCE BETWEEN TWO INDEPENDENT DISTRIBUTIONS

Input the two independent sets of data into C1 and C2; then continue with:

Choose: **Stat > Nonparametrics > Mann–Whitney**
Enter: *First Sample:* **C1**   *Second Sample:* **C2**
          Confidence level: **1 – α**
Select: *Alternative:*   **less than** or **not equal** or **greater than > OK**

With respect to the p-value approach, the p-value is given. With respect to the classical approach, just the sum of the ranks for one of the samples, W, is given. Use this to find U for that one sample. The U for the other sample is found by subtracting U from the product of n$_1$ and n$_2$.

RUNS TEST FOR TESTING RANDOMNESS ABOVE AND BELOW THE MEDIAN

Input the set of data into C1; then continue with:

Choose: **Stat > Nonparametrics > Runs Test**
Enter: *Variable:* **C1**
Select: **Above and below mean > OK**
          or
          Above and below:
Enter: **Median value > OK**

SPEARMAN'S RANK CORRELATION COEFFICIENT

Input the set of data for the first variable into C1 and the corresponding data values for the second variable into C2; then continue with:

Choose: **Data > Rank . . .**
Enter: *Rank data in:*   **C1**
          Store ranks in: **C3 > OK**
Repeat the preceding commands for the data in C2 and store in C4.
Choose: **Stat > Basic Statistics > Correlation**
Enter: *Variables:*   **C3 C4 > OK**

SIGN TEST FOR A SINGLE-SAMPLE HYPOTHESIS TEST OF THE MEDIAN

The following Excel commands will compute the differences between the data values and the hypothesized median. The data will then be sorted so that the number of + and – signs can be easily counted.

Input the data into column A and select cell B1; then continue with:

Choose: **Insert Function, f_x > All > SIGN > OK**
Enter: *Number:*   **A1–hypothesized median value > OK**
Drag: **Bottom right corner of the B1 cell down to give other differences**
Select the data in columns A and B; then continue with:
Choose: **Data > Sort**
Select: *Sort by:*   **Column B**
          Order: **Smallest to Largest > OK**

SIGN TEST FOR THE MEDIAN OF PAIRED DIFFERENCES

Input the paired data into columns A and B; then continue with:

Choose: **Add-Ins > Data Analysis Plus > Sign Test > OK**
Enter: *Variable 1 Range:*   **(A1:A20 or select cells)**
          Variable 2 Range: **(B1:B20 or select cells)**
Select: **Labels** *(if necessary)*
Enter: *Alpha:*   **α** *(ex. 0.05) > OK**

MANN–WHITNEY *U* TEST FOR THE DIFFERENCE BETWEEN TWO INDEPENDENT DISTRIBUTIONS

Input the two independent sets of data into column A and column B; then continue with:

Choose: **Add-Ins > Data Analysis Plus > Wilcoxon Rank Sum Test***
          > OK
Enter: *Variable 1 Range:*   **(A1:A20 or select cells)**
          Variable 2 Range: **(B1:B20 or select cells)**
Select: **Labels** *(if necessary)*
Enter: *Alpha:*   **α** *(ex. 0.05) > OK**

The sum of the ranks is given for both samples and also the p-value.

*The Wilcoxon rank sum test is equivalent to the Mann–Whitney test.

RUNS TEST FOR TESTING RANDOMNESS ABOVE AND BELOW THE MEDIAN

The following commands compute differences between the data values and median. Count the number of runs created by the sequence of + and – signs to complete the runs test.

Input the data into column A; select cell B1 and continue with:

Enter: **= median(A1:A20 or select cells) > Enter**
Select cell C1, then continue with:
Enter: **= A1 – [actual B1 median value]**   *(ex.A1 –5.5) > Enter**
Drag: **Bottom right corner of C1 cell down to give other differences**

SPEARMAN'S RANK CORRELATION COEFFICIENT

Input the set of data for the first variable into column A and the corresponding data values for the second variable into column B; then continue with:

Choose: **Add-Ins > Data Analysis Plus > Correlation (Spearman) > OK**
Enter: *Variable 1 range:*   **(A1:A10 or select cells)**
          Variable 2 range: **(B1:B10 or select cells)**
Select: **Labels** *(if necessary)*
Enter: *Alpha:*   **α** *(ex. 0.05) > OK**

SIGN TEST FOR A SINGLE-SAMPLE HYPOTHESIS TEST OF THE MEDIAN

Input the data into L1; then continue with:

Choose: PRGM > EXEC > SIGNTEST*
Select: PROCEDURE: 3: HYP TEST   INPUT? 2:DATA: 1 LIST
Enter: DATA: **L1**
          MED0: **hypothesized median value**
Select: ALT HYP? **1:** > or **2:** < or **3:** ≠

*Program SIGNTEST is one of many programs that are available for downloading from cengagebrain.com. See your Chapter 2 Tech Card for specific directions.

SIGN TEST FOR THE MEDIAN OF PAIRED DIFFERENCES

Input the paired data into L1 and L2; then continue with:

Highlight: **L3**
Enter: **L1–L2** (whichever order is needed, based on H_a)
Choose: PRGM > EXEC > SIGNTEST*
Select: PROCEDURE: 3: HYP TEST
          INPUT? 2:DATA: 1 LIST
Enter: DATA: **L3**
          MED0: **hypothesized median value**
Select: ALT HYP? **1:** > or **2:** < or **3:** ≠

*Program SIGNTEST is one of many programs that are available for downloading from cengagebrain.com. See your Chapter 2 Tech Card for specific directions.

MANN–WHITNEY *U* TEST FOR THE DIFFERENCE BETWEEN TWO INDEPENDENT DISTRIBUTIONS

Input the two independent sets of data into L1 and L2; then continue with:

Choose: PRGM > EXEC > MANNWHIT
Enter: XLIST: **L1**
          YLIST: **L2**
          NULL HYPOTHESIS D0 = **difference amount** (ex. 0)
Select: ALT HYP? **1:U1-U2 > D0** or **2:U1-U2 < D0** or **3:U1-U2 ≠ D0**

*Program MANNWHIT is one of many programs that are available for downloading from cengagebrain.com. See your Chapter 2 Tech Card for specific directions.

RUNS TEST FOR TESTING RANDOMNESS ABOVE AND BELOW THE MEDIAN

Input the data into L1; then continue with:

Highlight: **L2**
Enter: **L1 – median*(L1)**   (*2nd LIST > MATH > 4:median( )
Choose: PRGM > EXEC > RUNSTEST*
Enter: n1 = **# of observations with particular characteristic**
          (ex. below median)
          n2 = **# of observations with other characteristic**
          (ex. above median)
          V = **# of runs**

*Program RUNSTEST is one of many programs that are available for downloading from cengagebrain.com. See your Chapter 2 Tech Card for specific directions.

SPEARMAN'S RANK CORRELATION COEFFICIENT

Input the set of data for the first variable into L1 and the corresponding data values for the second variable into L2; then continue with:

Choose: PRGM > EXEC > SPEARMAN*
Enter: XLIST: **L1**
          YLIST: **L2**
Select: DATA?: **1:UNRANKED**
          ALT HYP? **1:RHO > 0** or **2:RHO < 0** or **3:RHO ≠ 0**

*Program SPEARMAN is one of many programs that are available for downloading from cengagebrain.com. See your Chapter 2 Tech Card for specific directions.

14.1 *The following are 24 consecutive downtimes (in minutes) of a particular machine.*

| 20 | 33 | 33 | 35 | 36 | 36 | 22 | 22 | 25 | 27 | 30 | 30 |
| 30 | 31 | 31 | 32 | 32 | 36 | 40 | 40 | 50 | 45 | 45 | 40 |

The null hypothesis of randomness is to be tested against the alternative that there is a trend. A MINITAB analysis of the number of runs above and below the median follows.

Runs Test: Downtime
Runs test for Downtime
Runs above and below K = 32.5
The observed number of runs = 4
The expected number of runs = 13.0000
12 Observations above K 12 below
The test is significant at 0.0002

a. *Confirm the values reported for the median and the number of runs by calculating them yourself.*
b. *Compute the value of z★ and the p-value.*
c. *Would you reject the hypothesis of randomness? Explain.*
d. *Construct a graph that displays the sample data and visually supports your answer to part c.*

Both Excel and TI83-84 Plus use the normal approximation to complete the Spearman rank correlation test.